Introduction to Quality by Design (QbD)

N. K. Jain • Neha Bajwa
Editors

Introduction to Quality by Design (QbD)

From Theory to Practice

Springer

Editors
N. K. Jain
Department of Pharmaceutical Sciences
Dr. H. S. Gour University
Sagar, Madhya Pradesh, India

Neha Bajwa
University Centre of Excellence in Research
Baba Farid University of Health Sciences
Faridkot, Punjab, India

ISBN 978-981-99-8033-8 ISBN 978-981-99-8034-5 (eBook)
https://doi.org/10.1007/978-981-99-8034-5

This Springer imprint is published by the registered company Springer Nature Singapore Pte Ltd.
The registered company address is: 152 Beach Road, #21-01/04 Gateway East, Singapore 189721, Singapore

Paper in this product is recyclable.

Preface

Welcome to *Introduction to Quality by Design (QbD): From Theory to Practice*. This book aims to provide a comprehensive understanding of Quality by Design (QbD) principles and practices in the pharmaceutical industry. QbD is a systematic approach that emphasizes the importance of designing and controlling the manufacturing process to ensure product quality.

In this book, we will explore various aspects of QbD, starting with an introduction to the concept and its regulatory aspects. We will delve into the guidelines provided by regulatory bodies such as the US Food and Drug Administration (USFDA), the Pharmaceuticals and Medical Devices Agency (PMDA), the International Council for Harmonisation (ICH), and the European Medicines Agency (EMEA).

Furthermore, we will discuss the criteria for selecting the design elements, including Process Analytical Technology (PAT), Critical Material Attributes (CMA), Critical Process Parameters (CPA), Critical Quality Attributes (CQA), and Risk Management and Experimental Approaches (RMEA). These considerations are crucial in ensuring the development of robust and efficient manufacturing processes.

The book will also cover specific areas of QbD implementation, such as Analytical Quality by Design, API Quality by Design, and Quality by Design for various formulation types, including oral formulations like tablets and spheroids, topical formulations, parenteral formulations, and novel drug delivery systems. We will also explore the application of QbD principles in the development of herbal products and formulations.

Additionally, we will examine the role of QbD in Quality Assurance, Pharmaceutical Packaging, and Clinical Trials. We will explore how QbD can be integrated into these areas to enhance product quality and safety.

Throughout the book, real-world examples and case studies will be provided to illustrate the practical application of QbD principles. We will also discuss the impact of QbD on marketing strategies and its potential benefits for the pharmaceutical industry.

In conclusion, *Introduction to Quality by Design (QbD): From Theory to Practice* offers a comprehensive overview of QbD principles and their practical implementation in the pharmaceutical industry. It is our hope that this book will serve as a valuable resource for researchers, practitioners, and students seeking to enhance their understanding and application of QbD concepts.

We would like to express our gratitude to all the contributors who have shared their expertise and insights, making this book possible. We hope that readers find this book informative, engaging, and inspiring as they embark on their journey towards incorporating QbD principles into their own work.

Best regards

Sagar, Madhya Pradesh, India N. K. Jain
Faridkot, Punjab, India Neha Bajwa

Contents

About the Editors

N. K. Jain is an internationally renowned academician and researcher and a senior professor in India who superannuated in June 2014 after rendering more than 40 years of dedicated and distinguished service as a teacher and 25 years as full professor. Starting his academic career in 1973, he worked as Lecturer in Pharmacy, Holkar College, Indore; Reader in Pharmaceutical Technology, M.S. University, Baroda; Professor and Principal, College of Pharmacy, Nasik; Professor and Head, Department of Pharmaceutical Sciences, Dr. H. S. Gour University, Sagar; Principal, College of Pharmacy, New Delhi; Visiting Professor, Dubai Pharmacy College, Dubai (UAE); and Head, Pharmacy Department, Delhi University. Subsequent to his superannuation, Prof. Jain rendered his services as Emeritus Fellow (U.G.C.) at Rajiv Gandhi Technological University, Bhopal, M.P., India. He is the author of two dozen celebrated books in pharmaceutical sciences and has contributed over 60 chapters in national and international books. He has to his credit over 500 publications in reputed pharmaceutical journals. According to a ranking announced by the Stanford University scientists, he is included in top 2% scientists in the world in the area of Pharmacology and Pharmacy, consecutively for 4 years.

Neha Bajwa currently serves as a Research Officer at the University Centre of Excellence in Research, Baba Farid University of Health Sciences, Faridkot. With over 30 publications in both international and national journals, her cumulative impact factor stands at 156, featuring articles in renowned publications such as *Nature Materials* and *Advanced Drug Delivery Reviews*. Dr. Bajwa is currently engaged in editing two upcoming books with prestigious publishers, namely Springer Nature and Taylor & Francis.

Her prolific writing extends beyond academic journals, as she contributes regularly to various magazines and newspapers. Dr. Bajwa recently secured the Young Investigator Training Program, enabling her to work at the University of Turin for a 6-month tenure. Further, she received a travel grant from FIP to participate in the FIP-PSWC 2020 conference in Montreal, Canada. Actively involved in professional associations, she holds Fellow and Life Member statuses with the Indian Pharmacy Graduate Associations and the Association of Pharmaceutical Teachers of India.

Dr. Bajwa has presented her research at over 20 national and international conferences, earning accolades such as the Best Poster Research Award at the 6th World Congress on Nanomedical Sciences-ISNSCON-2018 held at Vigyan Bhawan, New Delhi. During her Ph.D., she focused on the ICMR-sponsored project titled "Intensifying Solubility and Bioavailability of Arteether through Molecular Encapsulation based Oral Drug Delivery System." Her research portfolio extends to collaborations with government agencies like AICTE, UGC, DST, ICMR, CYCLONET, and FIP, showcasing her commitment to impactful projects in the pharmaceutical domain.

Chapter 1
Introduction to Quality by Design

Rakhee Kapadia, Ganesh Shevalkar, Ushasi Das, Vikas Singhai, Dipak Bari, and Chandrakantsing V. Pardeshi

Abstract The pharmaceutical sector is concerned with product efficacy, safety, and quality. FDA decided to implement various Designs of Experiments (DoE) in the pharmaceutical industries. Quality by Design helps in the development of high-quality products and guides to manage product quality across its entire life cycle. This method quickly gains popularity worldwide due to its many benefits and use of numerous high-quality statistical tools. QbD is a meticulous, proactive, risk-based strategy for pharmaceutical development that commences with predetermined goals and emphasizes product and process understanding and process control based on reliable research. The QTPP, knowledge of product and process design, scale-up, control strategy, and constant improvement are considered as crucial components of pharmaceutical QbD. After approval, during the product lifecycle management process, the competency of the product and the process is evaluated and continuously enhanced.

R. Kapadia
SIRT-Pharmacy, Sanjeev Agarwal Global Educational University, Bhopal, Madhya Pradesh, India

G. Shevalkar
Department of Pharmaceutics, Sandip Institute of Pharmaceutical Sciences, Nashik, Maharashtra, India

U. Das
Department of Pharmaceutical Technology, Jadavpur University, Kolkata, West Bengal, India

V. Singhai
Alembic Research Center, Vadodara, Gujarat, India

D. Bari · C. V. Pardeshi (✉)
Department of Pharmaceutics, R. C. Patel Institute of Pharmaceutical Education and Research, Shirpur, Maharashtra, India
http://www.rcpatelpharmacy.co.in

N. K. Jain, N. Bajwa (eds.), *Introduction to Quality by Design (QbD)*,
https://doi.org/10.1007/978-981-99-8034-5_1

Keywords Quality by design · ICH Q8 (R2) · Quality target product profile · Critical quality attributes · Risk assessment · Pharmaceutical products

1.1 Introduction

One of the fundamental criteria is "Quality", along with the safety and efficacy for any entity to be considered and approved as a drug. The pharmaceutical sector has always struggled to provide high-quality drug products. Major problems with the poor quality of pharmaceutical products can be ascribed to several factors, including variable starting materials, a lack of manufacturing process automation and control, and erroneous knowledge regarding the product and process characteristics. The principal sources of variability associated with the development of pharmaceutical goods are shown in Fig. 1.1, and they play a significant role in product recall [1].

In the pharmaceutical industry, it is crucial to consistently produce high-quality products because only these may deliver the desired therapeutic effects. The regulatory authorities have been further motivated to opt for this product development method by the rising number of research studies based on the quality by design (QbD) concept.

The United States Food and Drug Administration (USFDA) first endeavored to incorporate quality paradigms into pharmaceutical development and regulation practice due to poor pharmaceutical product quality. According to the European Medicines Agency (EMA), QbD is an approach that aims to ensure the quality of medicines by employing statistical, analytical, and risk-management methodology in the design, development, and manufacturing of medicines [2]. The International conference on harmonization's (ICH) quality guidelines, ICH Q8 Pharmaceutical Development, ICH Q9 Quality Risk Management, and ICH Q10 Pharmaceutical

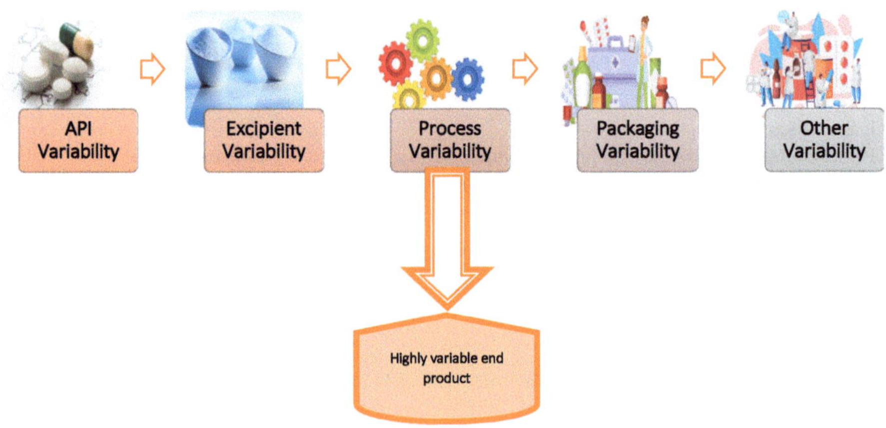

Fig. 1.1 Sources that contribute to quality variation among pharmaceutical products

Quality System, explain the fundamental QbD principles of risk- and science-based product development, risk assessment, lifecycle approach, and method design.

1.2 History of QbD: Indian Scenario

Crosby first proposed the idea of quality in the 1950s with the idea of "zero defects," which was acknowledged as a significant advancement in industrial quality control [3]. Dr. Joseph M. Juran, an American Engineer, was the first to coin the phrase "quality by design" (QbD) [4].

The concept of QbD was mentioned in the ICH Q8 guidance, which states that "quality cannot be tested into products, i.e., quality should be built in by design". In the (ICH) Q8 content, the QbD approach is defined as: "A systematic approach to drug development that emphasizes the need to start with predefined goals and understand the product and process based on sound science and quality risk management" [5–7]. The progression of QbD in the pharmaceutical industry is depicted in Fig. 1.2.

QbD entered the pharmaceutical sector very late, in 2004, when the USFDA took measures to raise the bar for pharmaceutical manufacturing. The FDA released Pharmaceutical cGMP for the twenty-first Century—A Risk-Based Approach in 2004 to outline industry expectations about target product quality with the ultimate robustness and efficacy for the benefit of patients [8]. To develop a comprehensive quality system for drug development, the FDA also released a process analytical tool (PAT)—A Framework for Innovative Pharmaceutical Development, Manufacturing, and Quality Assurance in the same year [9]. This was done in accordance with the FDA's Quality Systems Approach to Pharmaceutical cGMP Regulations.

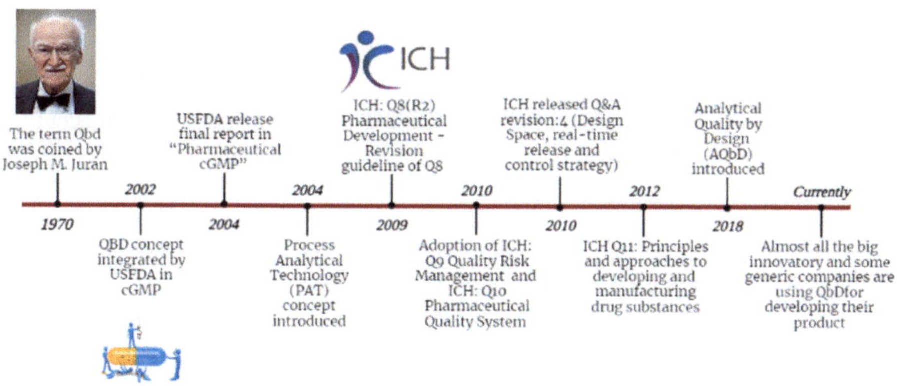

Fig. 1.2 Historical perspective of QbD concept

1.3 Fundamental Aspects Relevant to QbD in Pharmaceuticals

1.3.1 Pharmaceutical QbD: Desirable Attributes

Pharmaceutical QbD aims to identify the crucial quality characteristics or variables, modifying both the final product and the pharmaceutical product development cycle. Subcritical quality characteristics are referred to as critical quality attributes (CQAs). The interaction between CQAs and formulation/manufacturing factors is crucial because it helps to ensure that the drug products are of the right quality [10, 11].

Objectives of pharmaceutical QbD:

- To achieve clinical performance-based meaningful product quality standards.
- To improve product and process design, understanding, and control to boost the capability and decrease product variability and errors.
- To improve production and product development efficiency.
- To enhance post-approval change management and root cause analysis.

These objectives can frequently be met under QbD by connecting product quality to the required clinical performance and then developing a reliable formulation and manufacturing procedure to consistently supply the desired product quality [12].

The following principles are therefore crucial to the QbD approach Fig. 1.3 [13]:

1. A quality target product profile (QTPP) that includes (CQAs) that have been defined.
2. Determining critical material attributes (CMAs) to ensure the CQA of the product.
3. Designing the process to ensure the detection of critical process parameters (CPPs) while also taking scale-up into mind to connect CMAs and CPPs to CQAs.
4. The QbD approach specifies the parameters at every stage of the product manufacturing process and includes specifications for the drug substance(s), excipients, and drug products.
5. Process competence and recurrent development for ongoing product and process improvement.

Compared to the conventional approach, QbD's advancements in pharmaceutical development and manufacturing can be stated as follows (Table 1.1).

1.3.2 Advantages of (QbD) [14–16]

- Delivers cost savings and efficiency for the pharmaceutical sector.
- Offers a higher level of assurance of medication product quality.

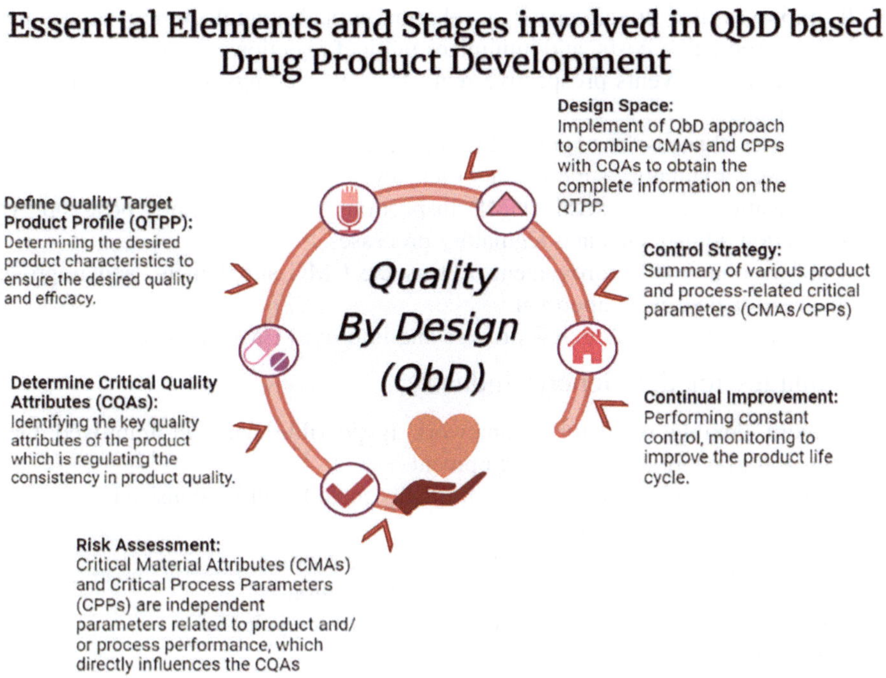

Fig. 1.3 Essential elements and stages involved in QbD

Table 1.1 Conventional *vs.* QbD approach of pharmaceutical product development

Aspects	Conventional approach	QbD approach
Pharmaceutical development	Empirical	Systematic, multivariate experiments.
Manufacturing process	Fixed	Adjustable within design space
Process control	Offline analysis wide and slow response	PAT is utilized for feedback and provided in realtime
Product specification	Based on batch data	Based on the desired product performance
Control strategy	Mainly by intermediate product and end product testing	Risk-based, controlled shifted upstream, real-time release
Lifecycle management	Post-approval changes needed	Continual improvement enables design space

- It makes the sponsor's understanding of the control approach for getting the drug product approved and eventually commercialized more transparent.
- Promotes transparency, rationality, and predictability in the scale-up, validation, and commercialization processes.
- It promotes medical innovation for unmet needs.

- It improves the effectiveness of pharmaceutical manufacturing procedures, lowers production costs, and minimizes product rejections.
- It reduces or prevents prospective regulatory proceedings, expensive fines, and drug recalls.
- It provides chances for ongoing development.
- It increases regulatory supervision efficiency.
- It intensifies post-approval cGMP inspections and streamlines post-approval production adjustments and regulatory processes.
- It allows continual improvement, lowers the CMC supplement, and improves chances for first-cycle approval.
- It minimizes the CMC review process and improves CMC quality.

Advantages for the Generic Industry:

- A more flexible regulatory framework is possible with an improved QbD approach to pharmaceutical development.
- Minimizes deviations to ensure high-quality manufacturing consistency.
- Lowest possible batch failures.
- Better comprehension of the procedure and the product.
- Improved comprehension of the risks and mitigation.
- Continual development over the entire product life cycle.
- More efficient technology transfers.

1.3.3 Demerits of QbD

Although QbD has several benefits over the traditional design technique, there are some restrictions and difficulties that act as roadblocks to its application. The problems include:

- Corporate internal resistance.
- Lack of confidence in a business case.
- Inadequate use of the latest technologies to ensure proper implementation.
- The participation of third parties, such as suppliers and manufacturers.
- The FDA's inconsistent QbD channelization.
- The lack of specific industry guidelines from the FDA.
- The regulators' unprepared approach to handling QbD applications.
- The lack of regulatory benefits presented to encourage further QbD.
- The misalignment of international regulatory bodies.
- The lack of flexibility in regulatory approaches for all regions.

1.3.4 Opportunities of QbD

Beyond all doubt, QbD has been an unrivalled quality-targeted approach to achieving excellence while developing effective, cost-effective, safe, and robust drug products. In addition, it facilitates macroscopic and microscopic knowledge of products or processes and helps to accomplish regulatory compliance with extraordinary ease and economy, whether for generics or innovators.

Additionally, QbD offers a variety of opportunities, such as a way to simplify post-approval change requirements, establishing clinically relevant specifications, outlining the requirements for the quality system's implementation and change management, aligning the previous version of the ICH guidelines with the new quality system, achieving global harmonization, etc. [17].

Today, a pharmaceutical scientist on industrial fronts has been benefiting from its outstanding benefits not only during the entire product development lifecycle but also afterward. Developing and implementing analytical methods is a crucial area in which QbD concepts are regularly applied. Furthermore, QbD has been progressively gaining interest in several other interdisciplinary fields, including active pharmaceutical ingredients (API) formulation, dissolution testing, manufacturing, bioequivalence research, and stability testing [18]. Figure 1.4 represents how the strategic QbD principles are applied at various stages of the pharmaceutical product development cycle.

1.4 Principles/Elements of QbD in Pharmaceutical Product Development

QbD is a scientific, systematic, proactive, holistic, and risk-based approach to product development that begins with predefined objectives and quality risk management throughout the life cycle of a product. QbD is a novel concept adopted recently in the pharmaceutical industry. International Council took several initiatives for Harmonization (ICH Q8, Q9 and Q10), FDAs cGMP, PAT, and new regulatory documents, which played a dominant role in its implementation. The paramount aspiration of QbD is to have a paradigm shift from quality by testing to quality by design, upgrading process, and product understanding consequently resulting in enhanced product quality, process efficacy and better regulatory flexibility [19, 20] Concept of QbD emphasizes on:

- Product understanding
- Process understanding
- Process control

ICH Q8 (R2) guideline describes six basic principles or elements of QbD Fig. 1.5

Fig. 1.4 Application of QbD at various stages of the pharmaceutical product development cycle

1.4.1 Quality Target Product Profile (QTPP)

This is the first component of QbD. QTPP commences with the customer "in mind". As per FDA, QTPP is related to quality, safety, and efficacy of dosage form, administration route, strength, bioavailability, and stability. QTPP forms the crux of design for product development which is defined as "A prospective summary of the quality characteristics of a drug product that ideally will be achieved to ensure the desired quality, taking into account safety and efficacy of the drug product" by ICH Q8. QTPP is a crucial component and forms the basis for product development. The QTPP includes all product attributes that will be needed to ensure the safety and efficacy of a drug product. QTPP could include but not limited to dosage form, delivery system/dosage design, route of administration, intended use in a clinical setting, dosage strength, pharmacokinetics, drug product quality criteria (including sterility, stability, purity, drug release), drug product quality attributes (dissolution, identification, microbial limits), container and closure system, administration/concurrence with labeling etc. QTPP defines the expectations in the final product hence it will vary for different pharmaceutical dosage forms [21, 22]. QTPP for a modified-release tablet dosage form is tabulated below (Table 1.2).

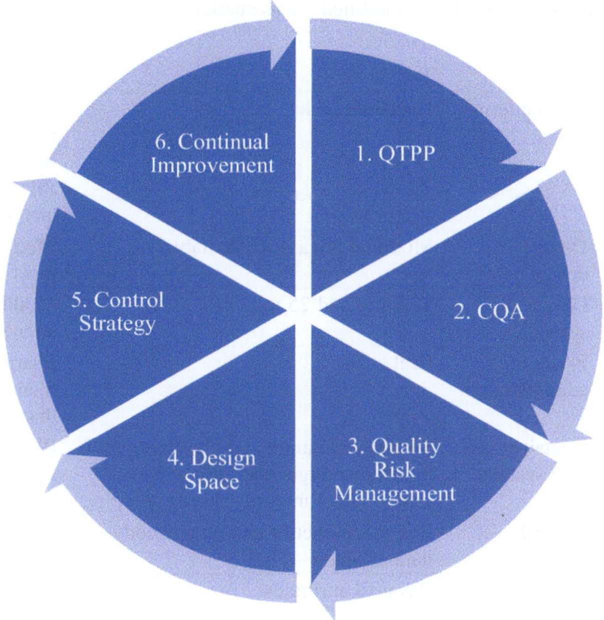

Fig. 1.5 Elements of QbD

1.4.2 Critical Quality Attributes (CQA)

Critical Quality Attributes are the second building block of QbD. They are defined by ICH Q8 (R2) as "a physical, chemical, biological or microbiological property or characteristics that should be within appropriate range, limit or distribution, ensuring desired product quality". Identification of the Quality Targets Product Profile is followed by recognition of appropriate CQAs, which are associated with active pharmaceutical ingredients, pharmaceutical excipients, and pharmaceutical drug products. For pharmaceutical dosage forms, the CQAs typically comprise potency, purity, bioavailability, stability, etc. CQAs are subsets of the QTPP, which has the potential to be altered by the change in formulation or process variables. ICH Q9 provides the risk assessment for the identification of CQAs. Recognition of CQAs can be accomplished based on preceding knowledge attained by literature review, manufacturing experience, technology transfer, stability data, raw material testing, adverse events and product recalls. Identification can also be achieved by applying Quality Risk Management which is discussed below. CQAs diverge subject to

Table 1.2 QTPP of modified release tablet formulation

QTPP element	Target	Justification
Dosage form	Uncoated tablet	Pharmaceutical equivalence requirement: same dosage form as that of immediate release. Literature: FOI, PIL, chemistry review, research article, review article
Dosage design	Modified release matrix tablet	Modified release design needed to meet label claims
Dosage strength	12.5 mg and 25 mg	Modified release tablet of lower strength to be developed
Route of administration	Oral	Pharmaceutical equivalence requirement: same route of administration as that of immediate release.
Container closure system	Suitable container and closure system to achieve target shelf life and to ensure tablet integrity during shipping	PVC/PVdC, Alu-Alu blister and HDPE bottles
Animal studies	Pharmacokinetics and pharmacodynamics study	To confirm the findings of theoretical kinetic calculations.
Drug product quality attributes	Physical attributes	Pharmaceutical equivalence requirement: meeting the same or compendial or other applicable (quality) standards (identity, assay, purity, and quality)
	Identification	
	Water	
	Assay	
	Content uniformity	
	Residual solvent	
	Dissolution	
	Microbial limits	
Stability	Minimum 3 month shelf-life at room temperature.3 month shelf-life at accelerated conditions.	Equivalent to or better than immediate release shelf life and for commercialization

dosage form, manufacturing process and analytical method selection [23, 24]. CQA for a modified-release tablet dosage form is tabulated below (Table 1.3). These CQAs are identified using Quality Risk Assessment.

1.4.3 Quality Risk Management

Before understanding Quality Risk Management, we must understand a few other terminologies, such as CPP, CMA, Risk, Risk Management, Risk Assessment and Risk Analysis.

Table 1.3 CQA of modified release tablet

Drug product quality attributes		Target	Is this critical	Justification of criticality
Physical attributes	Appearance	Color acceptable to the patient & visual tablet defects should be observed	No	Appearance is not directly associated with efficacy and safety and efficacy. Hence it is not crucial. The objective is to ensure acceptability by patients
	Odor	No unpleasant odor	No	Odor is not directly associated with efficacy and safety and efficacy. Odor can impact patient acceptability. Here neither the raw material nor the finished product has any unpleasant odor
	Size and shape	Acceptable to patient. Easily swallowed	No	Preliminary trial batches were planned, considering the dimensions of the reference product (immediate release formulation). Tablet size and shape are decided considering average weight, bulk density and tooling available so patients can easily swallow it
Identification		Meet identification requirements as defined in the IH method	No	QMS will control the identification
Water		To comply IH limit	Yes	The presence of water may accelerate the degradation rate during storage affecting impurities
Assay		95–105% of label claims	Yes	Variations in the assay of the finished product will impact safety and efficacy. Process variables might have an impact on drug product assay. Therefore this CQA will be evaluated during formulation development
Content uniformity		Conforms to USP <905>	Yes	The assay will be directly affected by CU
Dissolution	Whole tablet	Similar drug release profile as calculated	Yes	The bioavailability of solid oral dosage form is directly linked to its dissolution profile. Any major change in this CQA can impact in vivo properties. Dissolution behavior is impacted by formulation and process variables. Hence this CQA will be evaluated during formulation development

(continued)

Table 1.3 (continued)

Drug product quality attributes	Target	Is this critical	Justification of criticality
Residual solvent	To comply with ICH guideline	No	The solvent will not be used during the manufacturing process. Water will be used as a solvent during granulation
Microbial limits	To comply with regulatory guidelines	No	Patient safety and to comply regulatory requirements. Microbial limit is essential for products that avoid first-pass metabolism such as injectable and ophthalmic preparations

1.4.3.1 Critical Material Attributes (CMAs)

CMAs of raw material (active pharmaceutical ingredient and excipients) and in-process material is a physical, chemical, biological, and micro-biological attributes of an input material that should be consistent within acceptance criteria to ensure the predetermined quality of raw material and in-process material. CMAs will likely impact drug product quality attributes [25].

1.4.3.2 Critical Process Parameters (CPPs)

CPPs are defined as "parameters whose variability has an impact on CQAs, and thus it should be monitored/controlled, ensuring that process produces predetermined quality" [26, 27].

1.4.3.3 Risk

Risk is the combination of the probability of occurrence of harm and the severity of that harm as defined by ICH Q9.

1.4.3.4 Risk Management

Risk management is the macro-level technique of assessing, analyzing, prioritizing, and making a strategy for mitigating threats and managing risk to an organization's assets and earnings. ICH Q9 defined risk management as "The systematic application of quality management policies, procedures, and practices to the tasks of assessing, controlling, communicating, and reviewing risk" [28].

1.4.3.5 Risk Assessment

Risk assessment is a meso-level technique within risk management. Risk assessment is a science–based process used in quality risk management, aiding in identifying which process parameter or material attribute will potentially impact product CQAs. There are many attributes of API and excipients used in the formulation that can potentially impact the CQAs of finished drug products. It is impractical to investigate all the identified process parameters and material attributes during formulation optimization studies. Thus risk assessment is precious in identifying process parameters and material attributes that are critical. Generally, risk assessment is performed at an earlier stage in product development and is repeated later when more information is available, along with more excellent knowledge. Based on prior knowledge, a literature survey and initial experimental data risk assessment tool can be used to identify and rank process parameters, equipment, and input materials potentially impacting product quality. A process and material attributes are considered to be critical when some change in their attributes has a significant impact on product quality (CQAs) [29]. The initial risk assessment is considerably exhaustive but can be amended and emphasized by additional studies.

1.4.3.6 Risk Analysis

Risk analysis is the micro-level process of measuring risks and their associated impact. ICH Q9 defined risk analysis as "Estimating the risk associated with the identified hazards."

Quality risk management is defined as "A systematic process for the assessment, control, communication, and review of risks to the quality of the drug (medicinal) product across the product lifecycle" by ICH Q9. Quality Risk Management is an indispensable component of QbD as it assists in recognizing the magnitude of the impact of the (CMAs) and (CPPs) on (CQAs). This can eventually support categorizing the (CQAs) as per their priority. There are two principles of quality risk management defined in ICH Q9.

- The evaluation of the risk to quality should be based on scientific knowledge and ultimately link to the protection of the patient; and
- The level of effort, formality, and documentation of the quality risk management process should be commensurate with the level of risk.

In the pharmaceutical industry, quality risk management is a collective responsibility of different departments like formulation development, analytical development, packaging development, production, supply chain management, technology transfer, quality control, quality assurance, engineering, intellectual property, regulatory affairs, clinical research, legal departments, etc. The pharmaceutical industry can access and manage risk using recognized risk management tools or their in-house techniques. A non-exhaustive list of some tools defined in ICH Q9 for risk assessment is given below.

- Basic risk management facilitation methods (flowcharts, and check sheets)
- Failure Mode Effects Analysis (FMEA)
- Failure Mode, Effects, and Criticality Analysis (FMECA)
- Fault Tree Analysis (FTA)
- Hazard Analysis and Critical Control Points (HACCP)
- Hazard Operability Analysis (HAZOP)
- Preliminary Hazard Analysis (PHA)
- Risk ranking and filtering
- Supporting statistical tools

An overview of a typical quality risk management process is given below Fig. 1.6:

1.4.4 Design Space

ICH Q8 guideline defines design space as "Multidimensional interaction and combination of different input variables such as process parameters and material attributes which have been demonstrated to provide quality assurance." In the Design space process, understanding is represented that has been established. Design space describes the linkage between process input and CQAs. Design space can be established for the entire process or even single unit operation [30]. One factor at a time (OFAT) is a traditional approach to product development in which only one factor or variable is altered while keeping all other factors constant. The outcomes of OFAT are highly dependent on the starting point; hence it rarely uncovers the optimal conditions in large part. OFAT can also not separate the "noise" unless many trials are repeated using the same conditions. OFAT cannot intrinsically detect interactions among factors. Design of experiment (DoE) is a branch of applied statistics. DoE is the crucial element of the statistical toolbox and the most extensively used model for the implementation of QbD in pharmaceutical research and industrial setting among the several mathematical models available. In DoE, controlled input factors are purposely and systematically changed to establish their effects on responses. Here Critical Process Parameters (CPPs) x1, x2, x3 are connected with Critical quality attributes (CQAs) y1 through a mathematical function $y = f(x)$ [31]. DoE extends the following key over OFAT.

- Comprehensive data by performing the minimum number of experimental trials.
- Analyzing the effects independently by simultaneously altering all parameters.
- The variability of input materials, process parameters, and operators can be studied simultaneously.
- Cause and effect relationship is established between CPP, CMA, and CQA
- Interactions between the variables can be studied.

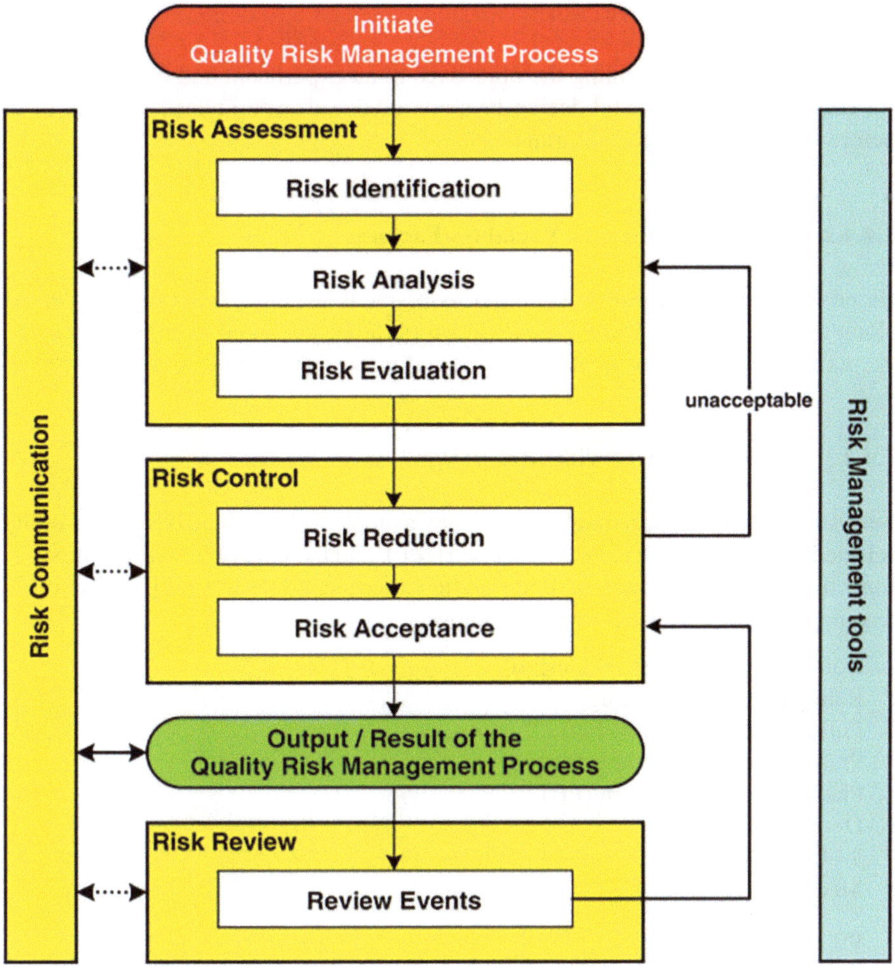

Fig. 1.6 Overview of a typical quality risk management process

- Proven Acceptable Range (PAR) can be established through optimization of CQA and appropriate selection of CPP and CMA.
- There are several CQAs for pharmaceutical products. DoE allows multiple response optimizations.
- Robust processes and products can be developed which will be insensitive to external and non-controllable influences.

The fundamental steps involved in the DoE approach are as follows.

1.4.4.1 Define Objective

The DoE begins with defining the objectives of an experiment and selecting factors for study. The experimental design lays out advanced experimental plans in detail before initiating experimental trials.

1.4.4.2 Selection of Process Variables/Factors

Based on preceding knowledge and risk assessment, the input factors can be defined along with their range. These variables/factors can be classified into three categories *viz:* experimental factors, control factors, and noise factors (Fig. 1.7)

1.4.4.3 Selection of Experimental Design

Several types of experimental designs are available. The objective of the study, selected factors and their range (CMA and CPP), and the kind of interaction involved form the basis for selecting an appropriate experimental design [32]. A few of the experimental designs used are given below.

1. Completely Randomized Design
2. Randomized Block Design
3. Full Factorial Design (FFD)
4. Fractional Factorial Design
5. Plackett-Burman Design (PB)
6. D—Optimal design
7. I—Optimal Design
8. Mixture Design
   (a) Axial Design
   (b) Simplex Lattice Design
   (c) Simplex Centroid Design

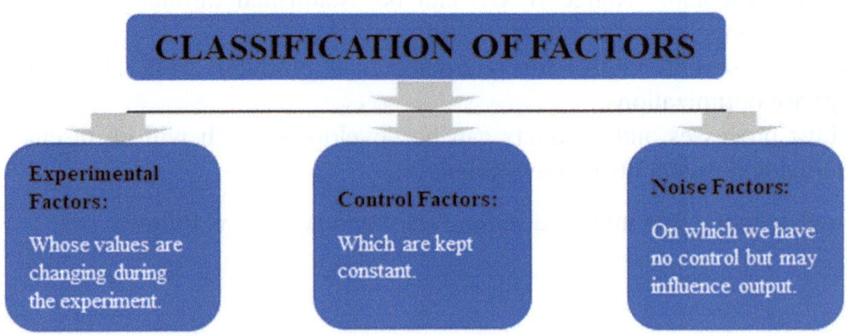

Fig. 1.7 Classification of factors in QbD

9. Response Surface (Second order) design (RSM)
   (a) Central Composite Design
      - Circumscribed Central Composite Design (CCD)
      - Inscribed Central Composite Design (CCI)
      - Face Centered Central Composite Design (CCF)
   (b) Box-Behnken Design (BBD)

These designs can be applied at the screening stage or optimization stage.

DoE at the screening stage: These designs are applied to screen a higher number of factors in fewer experiments to identify the significant factors that may impact the quality of the finished product. The main objective of these designs is to identify the main effects and not the interactions. Designs such as Plackett-Burman and Fractional Factorial are used for screening purposes [33].

DoE at Optimization stage: These designs are only applied once screening is done, and significant factors are identified which will be contributing to the process or formulation. These design generally studies the main and interaction effects, which may be quadratic or cubic, to obtain curvature. Experimental designs such as Full factorial, Surface Response methodology (Central composite design, Box-Behnken) and Mixture designs are considered to carry out optimization [34].

The selection of an appropriate design is based on the number of design variables, experimental trials, and files of use. Table 1.4 describes the rationale for using available designs.

Further details of the selected experimental design have been covered in the experimental work.

1.4.4.4 Analysis of Response Via Computational Design Expert Tool

The analysis of results is performed with graphical and statistical interpretation. The graphical interpretation is done by plotting different graphs like half normal plot, normal plot, and Pareto chart. The fit of the half-normal and normal plot is used to determine the significance of the effect. After selecting critical factors, statistical evaluation is performed using Analysis of Variance (ANOVA) for the selected factorial model. Different statistical tools such as the p-value of the model, lack of fit value, adjacent R^2 value, predicted R^2 value, adequate precision etc. are studied.

1.4.4.5 Illustration of Design Space

The design space can be tabulated or graphically displayed using various methods. Graphically the design space can be illustrated by Contour plots, Three-dimensional plots, and Overlay plots.

Table 1.4 Various types of design and their use

Design	Screening	Factor influence	Optimization	Area of application	Number of design variables
FFD		✓		Investigate the consequence of a low number of design variables that are independent of each other. This is the only design that accommodates categorical variables with a level of 3 or more	2–9
Fractional factorial design	✓	✓		Choose to study independent effects of lower order depending on the number of variables. It creates a screening design to discover the most critical main effect among plenty	3–13
Plackett-Burman	✓			It is an economical substitute for fractional factorial design, which investigates only the main effects	8–35
CCD (central composite design)			✓	Discovers foremost levels of design variable with the inclusion of certain additional experimental trials to FFD	2–6
D—optimal design	✓	✓	✓	Some of the designs are not orthogonal, having multilinear constraints. Partial least square regression must be used to evaluate them	2–9
BBD (Box Behnken design)			✓	It is a substitute for CCD when the optimum response is not located at the extremes of the experimental region and results from an FFD are not in hand	3–6
Axial	✓			The axial design consists of only a mixture of variables. The region of design is simplex. This design studies only linear (first order) effects	3–30
Simplex lattice	✓	✓	✓	The simplex design consists of only a mixture of variables. Region of design in simplex with a tunable lattice degree (order)	3–7
Simplex centroid			✓	The design consists of only a mixture of variables. Region of design in simplex	3–6

1.4.5 Control Strategy

As defined in ICH Q10, the control strategy is "a planned set of controls derived from process and product understanding that ensures process performance and product quality. The controls include attributes and parameters related to API, excipients, solvents, equipment, facility, operating conditions, in-process controls, finished product specification, and the associated methods and frequency of monitoring and control." Control strategy ensures that the process is maintained within the boundaries described by the design space [21, 35]. Control strategy may include.

- Check input material attributes (e.g., active pharmaceutical ingredient, excipients, primary and secondary packaging materials) based on an understanding of their impact on process-ability or product quality.
- Finished product release and shelf-life specifications
- Finished product standard test procedure.
- Facility controls, such as utilities, environmental systems, and operating conditions (temperature and relative humidity).
- Controls for unit operations that have an impact on downstream processing or end-product quality (the impact of drying on degradation, particle size distribution of the granulate on dissolution).
- Standard operating procedures (SOPs) or guidelines to monitor Out of specification (OOS), Out of trend (OOT), deviations, and so on.

1.4.6 Continual Improvement

The quality of the product can be enhanced throughout the lifecycle of product. Pharmaceutical industries have opportunities to choose innovative routes to improve quality. The performance of a manufacturing process can be monitored to make sure that there is consistency in the quality of the product. Additional knowledge and experience are acquired during the routine manufacturing of the product, which contributes to method/process development. Preventive and periodic maintenance can be done within a company's internal quality system, but design space should be unaffected. The QbD approach benefits the continuous improvement throughout the product's life cycle. This is a distinguishing point from the conventional method which is a much-frozen process.

1.5 Pharmaceutical Applications of QbD

Currently, QbD has numerous applications, and it is widely used in all phases of developing a pharmaceutical product. It is used to develop analytical methods, drug substances, formulation, dissolution testing, bioequivalence study, clinical trials, and manufacturing (Table 1.5).

Table 1.5 Applications of QbD

Application of formulation by design (FbD)				
DDS	**Drug**	**Factors**	**Design**	**Ref**
Self-nanoemulsifying	Bosentan	Oil phase percentage, surfactant percentage, and co-surfactant percentage	Box-Behnken	[36]
Floating Gastro retentive	Famotidine	Concentrations of Guar gum and the concentrations of Rice Bran Wax	3^2 factorial	[37]
Self-microemulsifying (SMEDDS)	Teriflunomide	Concentration of oil (Sefsol 218), surfactant (Acrysol EL-135), and cosurfactant (PEG 400)	Mixture design	[38]
Nano encapsulation	Crocin component of saffron (*Crocus sativus* L.)	pH and the concentrations of CS, ALG, and calcium chloride ($CaCl_2$)	Taguchi	[39]
Gastro-retentive (GRDDS)	Itopride hydrochloride	Concentrations of Eudragit L 100, PEG, and sodium bicarbonate	Box-Behnken	[40]
Magnetite nanoparticles	Ciprofloxacin	Concentrations of CIP (35–80 mg/L), adsorbent doses (20–60 mg), and pH values (4–10) at reaction time (80 min)	Box-Behnken	[41]
Nano invasomal gel	Glibenclamide (GLB) and Atenolol (ATN)	Amount of phospholipid (mg), ethanol (%), and terpene (%)	Box-Behnken	[42]
Gastro retentive	Ranitidine hydrochloride	Concentration of HPMC/NaCMC and concentration of $NaHCO_3$	Central Composite design	[43]
Emulsomes	Bifonazole	Phospholipid to bifonazole ratio; phospholipid to tristearin ratio and phospholipid to stearylamine ratio	Box-Behnken	[44]
	Triamcinolone acetonide	DPPC:GMS ratio, lipid:drug ratio, sonication time	Box-Behnken	[45]
Self-nanoemulsifying (SNEDDS)	Candesartan	Oil percentage (Capmul PG-8), surfactant percentage (Kolliphor EL), and a co-surfactant percentage (Transcutol P)	D-optimal mixture	[46]
Self-nanoemulsifying (SNEDDS)	Andrographolide	Amount of Capryol-90 as the oil phase (20–50%), Kolliphor RH 40 as the surfactant (40–70%), and propylene glycol	Simplex lattice	[47]
Solid lipid nanoparticle	Pioglitazone	Concentration of lipid (Compritol® 888 ATO), surfactant (tween80) and homogenization speed	Box-Behnken	[48]
Nanoparticles	Benzylisothiocyanate	Amount of polymer Concentration of surfactant	Central Composite design	[49]

Fast disintegrating oral film	Zolmitriptan	Amount of polymer and plasticizer		2^2 factorial design	[50]
Neuronanoemulsions	Ropinirole hydrochloridedextran sulfate nanoplex	Concentration of flaxseed oil; Concentration of surfactant (TPGS)		Central Composite design	[51]
Nanostructured lipid carriers	Ropinirole hydrochloride-dextran sulfate nanoplex	Concentration of flaxseed oil; Concentration of solid lipid (tristearin) and concentration of surfactant (TPGS)		Box-Behnken design	[52]
Nano-emulsomes	Ganciclovir	DPPC:GMS ratio, lipid:drug ratio, sonication time		Box-Behnken	[53]
Application of analytical quality by design (AQbD)					
Drug	**Quality target method profile**	**Critical method variables**	**Critical analytical attributes**	**Design**	**Ref**
Tamoxifen	Estimation of tamoxifen	Mobile phase ratio, buffer pH, and oven temperature	Peak area, retention time, theoretical plates, peak tailing	Taguchi	[54]
Ceftriaxone sodium	HPLC method development and validation	Mobile phase and pH	Retention time, theoretical plates, and peak asymmetry	Central composite	[55]
Eberconazole nitrate	Stability indicating HPLC method	Mobile phase and pH	Retention time, theoretical plates, and peak asymmetry	Response surface methodology	[56]
β-artemether and lumefantrine	Stability indicating HPLC method	Mobile phase ratio, flow rate	Retention factor, peak symmetry	Plackett-Burman	[57]
Ebastine	Degradation product characterization	Buffer strength, pH	Peak resolution	Factorial design	[58]
Darifenacin hydrochloride	Stability indicating UPLC method	Mobile phase ratio, pH, column oven temperature	Peak resolution and retention time of degradation products	Central composite	[59]
Rosuvastatin, Telmisartan, Ezetimibe, Atorvastatin	Simultaneous estimation using HPLC method	Mobile phase ratio, buffer strength, flow rate	Peak resolution, peak asymmetry	Rotatable central composite	[60]
Protamine sulfate	HPLC method development	Flow rate, temperature, pH	Peak resolution, tailing factor	Central composite	[61]

1.5.1 Bulk Drug Development

The ICH Q11 guidance explicitly outlines the implementation of QbD concepts for bulk drug manufacture. Such an approach primarily gives a rational knowledge of drug substance CQAs such as assay, particle size, purity, flow property, polymorphs, related substances, etc. These variables are primarily regulated by various CMAs involved in drug substance synthesis, such as reactant, solvent, initiator, and stabilizer concentrations, as well as process parameters such as solvent type, solvent flow, solvent addition time, reaction time, filtration process, crystallization temperature cooling patterns, and many more [62].

1.5.2 Formulation Development

Pharmaceutical goods are required to deliver medications with the appropriate purity, potency, and delivery rate. Hence, the current Q8 recommendations for pharmaceutical research constitute a significant step toward producing high-quality goods. The QbD has been found as a flexible approach to producing both pharmaceutically identical generic pharmaceuticals that match the performance of the innovator and safe and reliable new therapeutic products [63]. The formulation scientist may maximize the benefits of QbD to save time, effort, and resources. Aside from this, QbD may be used to existing items on the market to obtain product and process understanding and troubleshooting at the commercial level. Regulatory agencies now urge "continuous improvement" beyond validation with three batches because it may not be feasible to have complete information to implement all potential QbD-related issues at submission effectively.

Similar to generic products, the development of novel nano-drug delivery systems has made extensive use of QbD. To date, QbD has been used to develop various nanoformulations like liposomes, nanoparticles, nanocrystals, nanoemulsion, nanogels, carbon nanotubes, etc. QbD aids in optimizing the therapeutic performance of the pharmaceuticals from these nanosize systems for unmet patient demands, coupled with the product and process performance. The vast amount of literature published in this field attests to the increased acceptability of QbD for producing nano-pharmaceuticals [63, 64].

1.5.3 Biosimilar Development

Biosimilars are generic versions of biological products that are now being studied for disease therapy. The production of biosimilars, on the other hand, is thought to be a complex process with a significant degree of variability in end-product quality. Several QbD studies in recent years have proven the importance of organized approaches to biopharmaceutical development. Only a few studies, however, identified biosimilars as a unique example of product development. The innovator's product characteristic firmly defines the targeted quality profile of biosimilars [65].

1.5.4 Analytical Method Development

To date QbD approach has been employed for developing various analytical methods like chromatographic techniques (determination of impurity, screening of column, capillary electrophoresis, and stability studies), bioanalytical method development, spectroscopic measurements (mass spectroscopy and IR spectroscopy).

Analytical QbD (AQbD) is a subset of QbD that draws on QbD concepts to provide enhanced analytical knowledge. AQbD enables systematic analytical development based on established analytical objectives [66, 67].

The AQbD approach's key processes are creating the quality target method profile (QTMP) and identifying critical method variables (CMVs) influencing critical analytical attributes (CAAs) for improved method performance. The composition of mobile phase, flow rate, gradient time, the temperature of column oven, and pH are a few examples of CMVs, whereas peak area, retention time, theoretical plates, asymmetry factor, and capacity factor are examples of CAAs that are considered during AQbD optimization. Through analytical design space, AQbD improves the resilience, ruggedness, and flexibility of the analytical technique, allowing for continual improvement in the post-development stages. AQbD offers a wide range of applications, including the quantification of drugs, process contaminants, and degradation products in bulk samples and pharmaceutical formulations, as well as stability and clinical samples [68, 69].

1.5.5 Application in Stability Studies

Compliance with stringent regulatory standards is essential in the testing of pharmaceuticals, particularly for generics. In this context, applying QbD in stability testing can aid in optimizing product performance. The QbD techniques carefully monitored the critical product quality attributes, such as assay, related compounds, and dissolution, that determine the stability of therapeutic products during accelerated, intermediate, and long-term testing [70, 71].

In their research, Karmarkar et al. used QbD concepts to create a stability-indicating HPLC method for a complex pain medication that contained a drug component, two preservatives, and their degradants. The first approach lacked any resolution in the peaks for preservative and another drug degradant, as well as drug degradant and preservative oxidative degradation peaks. Fusion AETM software, which employs a DOE methodology, was used to optimize the process. The development of a design space, an operating space, and details of all method performance characteristics, constraints, and robustness within the operating space was made possible by the QbD-based method development [72].

1.5.6 Application in Dissolution Studies

QbD applications are currently being used to develop drug dissolution procedures that aid in investigating generic formulations under discriminating dissolution settings. The QbD aids in the optimization of CMAs such as concentration of release rate-controlling polymers, disintegrants, diluents, and type of dissolution medium that impact the amount of drug released at the specified time, mean dissolution time, dissolution efficiency, release exponent, and so on [73].

Miroslav et al. designed a HPLC technique for quantifying digoxin in dissolution samples. The experimental design is utilized in this case to demonstrate robustness. The influence of minor modifications in acetonitrile percentage, mobile phase flow rate, column temperature, and column length on the properties of the digoxin peak is investigated using a full factorial design. The presented HPLC technique was used to assess the quality and stability of digoxin [74]. In another study, Jun et al. used the QbD approach to investigate the changes in the dissolution profile of tablets during an accelerated stability study employing a multivariate method [75].

1.5.7 Application in Bioequivalence Studies

Optimizing the pharmacokinetic characteristics of the generic medications and creating a meaningful match with the innovator products can be accomplished by using QbD concepts in bioequivalence studies. Applying QbD minimizes the variability caused by formulation parameters and human factors, significantly reducing the time and cost associated with bioequivalence investigations.

1.5.8 Application in Clinical trials

Clinical trials should be designed to be of high quality; audits, inspections, or post-hoc rework cannot be relied upon to do so. The FDA encourages and promotes the implementation of organized strategies meant to boost the effectiveness and quality of clinical trials. The FDA has recently advised the use of the QRM method in clinical studies. The use of the QbD technique in clinical studies aids in detecting serious errors that impede the capacity to safeguard patients during the study and produce accurate findings and useful information. Furthermore, with QbD approaches, quality monitoring of a clinical trial is feasible, which eventually benefits meeting trial outcomes with consumer opinion and preventing crucial shortcomings. In addition, to assure the quality of a clinical trial, it is critical to do quality planning and process monitoring before trial execution.

Additionally, systematic methods can enable better flexibility in quality planning and regulatory control of trial outcomes, resulting in continual improvement [1]. Steven et al. in their study on theophylline extended-release tablets, explained the relationship between the critical control parameters of processes and their impact on clinical performance [76]. It is also possible to establish relationships between CQAs and pharmacokinetic parameters in healthy volunteer trials and also to establish relationships between pharmacokinetics and safety and efficacy [77].

1.6 ICH Q8, Q9, Q10 Guidelines: The Foundation of QbD

In the field of QbD, which uses regulatory intelligence to modernize the understanding and management of pharmaceutical manufacturing processes, the FDA is at the epicenter of effort. The pharmaceutical sector started to modernize with the introduction of cGMP applications in 2002 and the publishing of the PAT Guide by the FDA in 2004. Because of the basic differences between the EMA and the FDA in the United States have different approaches to design and quality management. For the first time, EMA has designated PAT as its focal area, and it addresses design by quality through participation in ICH (The International Conference on Harmonisation of Technical Requirements for Registration of Pharmaceuticals for Human Use) [78–81].

1.6.1 ICH Q8 and Q8 (R2) Pharmaceutical Development

Pharmaceutical development aims to provide “quality products” and “production systems that enable these products to be produced in a repeatable manner.” The design space (DS) is defined in the Q8 Guide as “the multidimensional combination and relationship of input variables and process parameters introduced to provide quality assurance.” After clearance, the operation that occurs outside of the DS will modify the situation. The establishment of a manufacturing process is a step in the pharmaceutical development process that guarantees reliable access to products of superior quality. Development studies and experience-based knowledge enable the development of specifications, controls, and risk management. Design is a means of achieving quality.

ICH Q8 (R2) is an annex to ICH Q8 Pharmaceutical Development and explains the important ideas in the core guideline. This annex also discusses the QbD concepts. This handbook is an annex to the ICH Q8 Pharmaceutical Development guide and defines certain important ideas. In addition, this annex defines the QbD principles. There is a chance for a science- and risk-based regulatory approach when a business implements QbD and quality risk management in conjunction with an adequate pharmaceutical quality system. Concerning the PAT that came up to the fore with the QbD approach, the FDA has published a guide named "PAT-A Framework for Innovative Drug Development, Manufacturing, and Quality Assurance." This guideline covers "scientific principles and tools that support innovation" and "adaptation of regulatory strategies to bring about innovation" [6].

1.6.2 ICH Q9 Quality Risk Management

The two fundamental principles of the risk management model are described in this document, together with the terminology and tools used in risk assessment. Additionally, there is a brief list of resources [such as Failure Mode Effects Analysis (FMEA)] where more specific details on risk management techniques can be discovered. Such tools are helpful in prioritizing the PAT application [82].

1.6.3 ICH Q10 Pharmaceutical Quality System

This manual is an addition to constructing a pharmaceutical quality system based on the principles outlined in ISO 9001 and 9004. The guideline is also anticipated to provide a unified description of the crucial topics to PAT. These include ongoing improvement initiatives, data collection techniques, and system validation measurement strategies [83].

1.7 Software Used in QbD

For implementing QbD, assessing risks, and designing experiments, there is a wide variety of tools on the market. The functionality also changes depending on the software being used. Some software is utilized for the DoE, while other software is used for initial risk assessment. The software can also be used for all QbD phases, such as QbD experts, Design-Expert®, MODDE®, Unscrambler®, JMP®, Statistica®, Minitab®, and other pertinent software applications are at the fingertips for DoE optimization, and they often provide interface guidance at each stage throughout the whole product development cycle. MODDE®, Unscrambler®, SIMCA®, and CODDESA® are examples of software that supports chemometric analysis through multivariate techniques like MNLRA, PCA, PLS, etc. Software like Minitab®, Risk®, Statgraphics, FMEA-Pro, iGrafx, etc. can be used for QRM execution employing Fish-bone diagrams, REM and FMEA matrices during risk assessment studies, etc. Figure 1.8 lists the software applications that are commercially accessible for conducting QbD research in an industrial setting.

Fig. 1.8 Software used in QbD

1.8 Conclusion and Future Prospective

QbD is a cost- and time-effective breakthrough in modeling and formulating with DoE, risk assessment, and PAT because its tools enable a better understanding of the resources and procedures that make QbD accessible and practical for the pharmaceutical industry. When QbD ideology is applied correctly, it results in successful product development, rapid regulatory approval, reduced validation difficulties, and much fewer post-approval adjustments. The philosophy of QbD in the novel ICH guidance, ICHQ8 [R2], will aid the pharmaceutical industry's manufacturing of solid-dosage products in successfully advancing their processes and speeding up regulatory approval. The QbD mindset can change the industrialized process, and chemistry, and regulate the regulatory transition to a scientific and risk-dependent patient-oriented approach.

The aging population and the rising expense of drugs for society provide the healthcare industry with a number of significant obstacles that must be overcome. A well-defined process development endeavor aims to create a dependable procedure that can be regularly produced at satisfying stated requirements when operated with ambiguous margins, as demonstrated with an acceptable level of assurance. Additionally, the creation and evaluation of formulation methods can be done using QbD.

According to ICH Q8 and Q9 guidance, the essential analytical factors are particularly predictable in an accession; therefore parallel explanations are intended for efficient process control and improvement. This chapter introduces the key methods used in the QbD paradigm, somewhat obscures the underlying science, and gives an overview of upcoming difficulties.

References

1. Beg, S., Hasnain, M. S., Rahman, M., & Swain, S. (2019). Introduction to quality by design (QbD): Fundamentals, principles, and applications. In *Pharmaceutical quality by design*. Academic. https://doi.org/10.1016/C2017-0-03874-0
2. EMA. (2018). *Quality by design*. European Medicines Agency. Retrieved May 2023.
3. Saxonaug, W. (2001). Philip Crosby, 75, developer of the zero-defects concept. New York Times XX Month.
4. Juran, J. M. (1992). *Quality by design the new steps for planning quality into goods and services*. Free Press.
5. Das, U., Kumar Panda, D., & Mandal, S. (2023). Formulation by design: An overview. In *Drug formulation design*. IntechOpen. https://doi.org/10.5772/intechopen.109702
6. ICH. (2009). Pharmaceutical development Q8(R2). In *International Conference on Harmonisation, IFPMA*, Geneva (Switzerland). Retrieved May 2023
7. Kalyane, D., Raval, N., Polaka, S., & Tekade, R. K. (2020). Quality by design as an emerging concept in the development of pharmaceuticals. In *The future of pharmaceutical product development and research* (pp. 1–25). Elsevier. https://doi.org/10.1016/b978-0-12-814455-8.00001-3
8. FDA. (2004). Pharmaceutical CGMPs for the 21st century: A risk-based approach.
9. PAT. (2004). A framework for innovative pharmaceutical development, manufacturing, and quality assurance, FD.
10. Pallagi, E., Ambrus, R., Szabo-Revesz, P., & Csoka, I. (2015). Adaptation of the quality by design concept in early pharmaceutical development of an intranasal nanosized formulation. *International Journal of Pharmaceutics, 491*, 384–392. https://doi.org/10.1016/j.ijpharm.2015.06.018
11. Zhang, L., & Mao, S. (2017). Application of quality by design in the current drug development. *Asian Journal of Pharmaceutical Sciences., 12*(1), 1–8. https://doi.org/10.1016/j.ajps.2016.07.006
12. Yu, L. X., Amidon, G., Khan, M. A., Hoag, S. W., Polli, J., Raju, G. K., et al. (2014). Understanding pharmaceutical quality by design. *The AAPS Journal, 16*(4), 771–783. https://doi.org/10.1208/s12248-014-9598-3
13. Sitre, D., & Kamble, R. (2020). Articulation of quality by design elements for product development and its unique applications. *Journal of Drug Delivery and Therapeutics, 10*(3), 253–226. https://doi.org/10.22270/jddt.v10i3.4080
14. Bhise, J. J., Bhusnure, O. G., Mule, S. T., Mujewar, I. N., Gholve, S. B., & Jadhav, P. U. (2019). A review on quality by design approach (QBD) for pharmaceuticals. *Journal of Drug Delivery and Therapeutics, 9*(3-s), 1137–1146. https://doi.org/10.22270/jddt.v9i3-s.3014
15. Nwoko, V. E. (2013). Quality by design (Qbd): Manufacturing and product quality of generics drugs perspective. *Journal of Global Trends in Pharmaceutical Sciences, 4*, 1257–1262.
16. Rojhe, K., & Kapoor, M. (2020). Quality by design approach in Indian pharmaceutical industry: A brief status appraisal. *International Journal of Pharmacy Research, 12*(3). https://doi.org/10.31838/ijpr/2020.12.03.014

17. Varu, R. K., & Khanna, A. (2010). Opportunities and challenges to implementing quality by design approach in generic drug development. *Journal of Generic Medicines, 7*(1), 60–73. https://doi.org/10.1057/jgm.2009.37
18. Singh, B., Beg, S., Sharma, G., Jain, A., & Negi, P. (2018). Holistic application of quality by design (QbD) for pharma product development excellence and regulatory compliance. *Nirma University Journal of Pharmaceutical Sciences, 1*, 19–35.
19. Singhai, V. D., Maheshwari, R., Sharma, S., & Paliwal, S. (2021). Employment of quality by design approach via response surface methodology to optimize and develop modified-release formulation of hydrochlorothiazide. *Current Computer-Aided Drug Design, 17*, 266–280. https://doi.org/10.2174/1573409916666200226114517
20. Suresh, S., Roy, S., & Ahuja, B. K. (2015). Quality by design: An overview. *Indian Drugs, 52*, 5–11. https://doi.org/10.53879/id.52.02.p0005
21. Yan, B., Li, Y., Guo, Z., & Qu, H. (2014). Quality by design for herbal drugs: A feedforward control strategy and an approach to define the acceptable ranges of critical quality attributes: Methods promoting quality-by-design for herbal drugs. *Phytochemical Analysis, 25*(1), 59–65. https://doi.org/10.1002/pca.2463
22. Zalai, D., Dietzsch, C., & Herwig, C. (2013). Risk-based process development of biosimilars as part of the quality by design paradigm. *PDA Journal of Pharmaceutical Science and Technology, 67*(6), 569–580. https://doi.org/10.5731/pdajpst.2013.00943
23. Kelley, B., Cromwell, M., & Jerkins, J. (2016). Integration of QbD risk assessment tools and overall risk management. *Biologicals, 44*, 341–351. https://doi.org/10.1016/j.biologicals.2016.06.001
24. Matji, A., Donato, N., Gagol, A., Morales, E., Carvajal, L., Serrano, D. R., et al. (2019). Predicting the critical quality attributes of ibuprofen tablets via modelling of process parameters for roller compaction and tabletting. *International Journal of Pharmaceutics, 565*, 209–218. https://doi.org/10.1016/j.ijpharm.2019.05.011
25. Lionberger, R. A., Lee, S. L., Lee, L., Raw, A., & Yu, L. X. (2008). Quality by design: Concepts for ANDAs. *The AAPS Journal, 10*, 268–276. https://doi.org/10.1208/s12248-008-9026-7
26. Evans, R. C., Bochmann, E. S., Kyeremateng, S. O., Gryczke, A., & Wagner, K. G. (2019). Holistic QbD approach for hot-melt extrusion process design space evaluation: Linking materials science, experimentation and process modeling. *European Journal of Pharmaceutics and Biopharmaceutics, 141*, 149–160. https://doi.org/10.1016/j.ejpb.2019.05.021
27. Rivnay, B., Wakim, J., Avery, K., Petrochenko, P., Myung, J. H., Kozak, D., et al. (2019). Critical process parameters in manufacturing of liposomal formulations of amphotericin B. *International Journal of Pharmaceutics, 565*, 447–457. https://doi.org/10.1016/j.ijpharm.2019.04.052
28. Mishra, V., Thakur, S., Patil, A., & Shukla, A. (2018). Quality by design (QbD) approaches in current pharmaceutical set-up. *Expert Opinion on Drug Delivery, 15*, 737–758. https://doi.org/10.1080/17425247.2018.1504768
29. Singhai, V. D., Sharma, S., Paliwal, S., Tekade, R., & Maheshwari, R. (2022). Scalable design and development of modified release hydrochlorothiazide formulation employing quality by design approach. *Indian Journal of Pharmaceutical Sciences, 84*, 84. https://doi.org/10.36468/pharmaceutical-sciences.986
30. Hakemeyer, C., McKnight, N., St. John, R., Meier, S., Trexler-Schmidt, M., Kelley, B., et al. (2016). Process characterization and design space definition. *Biologicals, 44*, 306–318. https://doi.org/10.1016/j.biologicals.2016.06.004
31. Csóka, I., Pallagi, E., & Paál, T. L. (2018). Extension of quality-by-design concept to the early development phase of pharmaceutical R&D processes. *Drug Discovery Today, 23*, 1340–1343. https://doi.org/10.1016/j.drudis.2018.03.012
32. Singhai, V. D., Sharma, S., & Paliwal, S. (2019). Implementation of fraction of design space to assess prediction capability of response surface design to optimize modified release tablets of hydrochlorothiazide. *International Journal of Pharmaceutical Science and Research, 10*(9), 4207–4214.

33. Politis, N. S., Colombo, P., Colombo, G., & Rekkas, D. M. (2017). Design of experiments (DoE) in pharmaceutical development. *Drug Development and Industrial Pharmacy, 43*, 889–901. https://doi.org/10.1080/03639045.2017.1291672
34. Singh, B., Kumar, R., & Ahuja, N. (2005). Optimizing drug delivery systems using systematic "design of experiments." Part I: Fundamental aspects. *Critical Reviews in Therapeutic Drug Carrier Systems, 22*, 27–105. https://doi.org/10.1615/critrevtherdrugcarriersyst.v22.i1.20
35. Coughlin, K., & Posencheg, M. A. (2019). Quality improvement methods—Part II. *Journal of Perinatology, 39*, 1000–1007. https://doi.org/10.1038/s41372-019-0382-1
36. Usta, D. Y., Timur, B., & Teksin, Z. S. (2022). Formulation development, optimization by Box-Behnken design, characterization, in vitro, ex-vivo, and in vivo evaluation of bosentan-loaded self-nanoemulsifying drug delivery system: A novel alternative dosage form for pulmonary arterial hypertension treatment. *European Journal of Pharmaceutical Sciences, 174*, 106159. https://doi.org/10.1016/j.ejps.2022.106159
37. Sabale, V. P., & Gadge, G. G. (2022). Factorial design approach to fabricate and optimize floating tablets based on combination of natural polymer and rice bran wax. *Beni-Suef University Journal of Basic and Applied Sciences, 11*(1), 12. https://doi.org/10.1186/s43088-021-00186-9
38. Patel, A., Shah, S., Patel, M., & Vyas, G. (2022). Quality by design approach to the development of self-microemulsifying systems for oral delivery of teriflunomide: Design, optimization, and in vitro and in vivo evaluation. *Egyptian Pharmaceutical Journal, 21*(2), 167. https://doi.org/10.4103/epj.epj_84_21
39. Nasrpour, S., Yousefi, G., Niakosari, M., & Aminlari, M. (2022). Nanoencapsulation of saffron crocin into chitosan/alginate interpolyelectrolyte complexes for oral delivery: A Taguchi approach to design optimization. *Journal of Food Science, 87*(3), 1148–1160. https://doi.org/10.1111/1750-3841.16052
40. Alaithan, S., Naveen, N. R., Goudanavar, P. S., Bhavani, P. D., Ramesh, B., Koppuravuri, N. P., et al. (2022). Development of novel unfolding film system of itopride hydrochloride using Box-Behnken design—A gastro retentive approach. *Pharmaceuticals (Basel), 15*(8), 981. https://doi.org/10.3390/ph15080981
41. Salari, M. (2022). Optimization by Box–Behnken design and synthesis of magnetite nanoparticles for removal of the antibiotic from an aqueous phase. *Adsorption Science & Technology, 2022*, 1–13. https://doi.org/10.1155/2022/1267460
42. Anitha, P., & Satyanarayana, S. V. (2021). Design and optimization of nano invasomal gel of glibenclamide and atenolol combination: In vitro and in vivo evaluation. *Future Journal of Pharmaceutical Sciences, 7*(1), 1–18. https://doi.org/10.1186/s43094-021-00240-4
43. Nigusse, B., Gebre-Mariam, T., & Belete, A. (2021). Design, development and optimization of sustained release floating, bioadhesive and swellable matrix tablet of ranitidine hydrochloride. *PLoS One, 16*(6), e0253391. https://doi.org/10.1371/journal.pone.0253391
44. Malviya, V. (2021). Preparation and evaluation of emulsomes as a drug delivery system for bifonazole. *Indian Journal of Pharmaceutical Education and Research, 55*(1), 86–94. https://doi.org/10.5530/ijper.55.1.12
45. Kapadia, R., Parikh, K., Jain, M., & Sawant, K. (2021). Topical instillation of triamcinolone acetonide-loaded emulsomes for posterior ocular delivery: Statistical optimization and in vitro-in vivo studies. *Drug Delivery and Translational Research, 11*(3), 984–999. https://doi.org/10.1007/s13346-020-00810-8
46. Verma, R., & Kaushik, D. (2020). Design and optimization of candesartan loaded self-nanoemulsifying drug delivery system for improving its dissolution rate and pharmacodynamic potential. *Drug Delivery, 27*(1), 756–771. https://doi.org/10.1080/10717544.2020.1760961
47. Indrati, O., Martien, R., Rohman, A., & Nugroho, A. K. (2020). Application of simplex lattice design on the optimization of andrographolide self nanoemulsifying drug delivery system (SNEDDS). *Indonesian Journal of Pharmacy, 31*(2), 124. https://doi.org/10.14499/indonesianjpharm31iss2pp124

48. Shaveta, S., Singh, J., Afzal, M., Kaur, R., Imam, S. S., Alruwaili, N. K., et al. (2020). Development of solid lipid nanoparticle as carrier of pioglitazone for amplification of oral efficacy: Formulation design optimization, in-vitro characterization and in-vivo biological evaluation. *Journal of Drug Delivery Science and Technology, 57*, 101674. https://doi.org/10.1016/j.jddst.2020.101674
49. Parmar, A., Kaur, G., Kapil, S., Sharma, V., & Sharma, S. (2020). Central composite design-based optimization and fabrication of benzylisothiocynate-loaded PLGA nanoparticles for enhanced antimicrobial attributes. *Applied Nanoscience, 10*(2), 379–389. https://doi.org/10.1007/s13204-019-01185-0
50. Chauhan, I., Yasir, M., & Verma, M. (2019). Oral delivery of zolmitriptan loaded fast disintegrating film: Formulation development, statistical optimization, in-vitro and in-vivo evaluation. *Journal of Applied Pharmaceutical Sciences and Research., 2*(1), 13–22. https://doi.org/10.31069/japsr.v2i1.3
51. Pardeshi, C.V., Belgamwar, V.S. (2018). N,N,N-trimethyl chitosan modified flaxseed oil based mucoadhesive neuronanoemulsions for direct nose to brain drug delivery. *International Journal of Biological Macromolecules, 120* (Pt. B), 2560–2571. https://doi.org/10.1016/j.ijbiomac.2018.09.032
52. Pardeshi, C.V., Belgamwar, V.S. (2020). Improved brain pharmacokinetics following intranasal administration of N,N,Ntrimethyl chitosan tailored mucoadhesive NLCs. *Materials Technology: Advanced Performance Materials, 35*(5), 249–266. https://doi.org/10.1080/10667857.2019.1674522
53. Kapadia, R., Jain, M., Patel, D., Devkar, R., & Sawant, K. (2023). Nano-emulsomes for back of the eye delivery of Ganciclovir: Formulation optimization, characterization & in vitro/in vivo evaluation. *Pharmaceutical Development and Technology, 28*(6), 535–546. https://doi.org/10.1080/10837450.2023.2216784
54. Babar, A. V., Jain, G., Patel, J., Sharma, R., Khan, S., & Patel, R. (2022). QbD approach in HPLC method development and validation of tamoxifen. *International Journal of Pharmaceutical Sciences and Medicine., 7*(9), 90–105. https://doi.org/10.47760/ijpsm.2022.v07i09.006
55. Patel, K. Y., Dedania, Z. R., Dedania, R. R., & Patel, U. (2021). QbD approach to HPLC method development and validation of ceftriaxone sodium. *Future Journal of Pharmaceutical Sciences, 7*(1), 1–10. https://doi.org/10.1186/s43094-021-00286-4
56. Krishna, M. V., Dash, R. N., Jalachandra Reddy, B., Venugopal, P., Sandeep, P., & Madhavi, G. (2016). Quality by design (QbD) approach to develop HPLC method for eberconazole nitrate: Application oxidative and photolytic degradation kinetics. *Journal of Saudi Chemical Society, 20*, S313–S322. https://doi.org/10.1016/j.jscs.2012.12.001
57. Suleman, S., Vandercruyssen, K., Wynendaele, E., Hondt, D., Bracke, M., Duchateau, N., et al. (2013). A rapid stability-indicating, fused-core HPLC method for simultaneous determination of beta-artemether and lumefantrine in anti-malarial fixed dose combination products. *Malaria Journal, 12*, 23–31.
58. Schmidt, A. H., & Molnar, I. (2013). Using an innovative quality-by-design approach for development of a stability indicating UHPLC method for ebastine in the API and pharmaceutical formulations. *Journal of Pharmaceutical and Biomedical Analysis, 78–79*, 65–74.
59. Murthy, M. V., Krishnaiah, C., Srinivas, K., Rao, K. S., Kumar, N. R., & Mukkanti, K. (2012). Development and validation of RPUPLC method for the determination of darifenacin hydrobromide, its related compounds and its degradation products using design of experiments. *Journal of Pharmaceutical and Biomedical Analysis, 72*, 40–50.
60. Janardhanan, V. S., Manavalan, R., & Valliappan, K. (2012). Chemometric technique for the optimization of chromatographic system: Simultaneous HPLC determination of Rosuvastatin, Telmisartan, Ezetimibe and Atorvastatin used in combined cardiovascular therapy. *Arabian Journal of Chemistry, 9*(S2), S1378–S1387.
61. Awotwe-Otoo, D., Agarabi, C., Faustino, P. J., Habib, M. J., Lee, S., Khan, M. A., et al. (2012). Application of quality by design elements for the development and optimization of an analytical method for protamine sulfate. *Journal of Pharmaceutical and Biomedical Analysis, 62*, 61–67.

62. Grewal, N. K., & Leadbeater, N. E. (2012). The central role of chemistry in 'quality by design' approaches to drug development. *Future Medicinal Chemistry, 4*, 1799–1810. https://doi.org/10.4155/fmc.12.117
63. Yu, L. X., Lionberger, R. A., Johnston, G., Olson, M. C., Buehler, G., & Winkle, H. (2009). Quality by design for generic drugs. *Pharmaceutical Technology, 33*(10), 1–6.
64. Beg, S., Akhter, S., Rahman, M., & Rahman, Z. (2017). Perspectives of quality by design approach in nanomedicines development. *Current Nanomedicine, 7*, 191–197. https://doi.org/10.2174/2468187307666170519122653
65. Rathore, A. S. (2009). Roadmap for implementation of quality by design (QbD) for biotechnology products. *Trends in Biotechnology, 27*, 546–553. https://doi.org/10.1016/j.tibtech.2009.06.006
66. Bhutani, H., Kurmi, M., Singh, S., Beg, S., & Singh, B. (2014). Quality by design (QbD) in analytical sciences: An overview. *The Pharma Times, 46*, 71–75.
67. Panda, S. S., Beg, S., Kumar, B., & Sahu, J. (2016). Implementation of quality by design approach for developing chromatographic methods with enhanced performance: A mini review. *Journal of Analytical & Pharmaceutical Research, 2*(6), 39–43.
68. Orlandini, S., Pinzauti, S., & Furlanetto, S. (2013). Application of quality by design to the development of analytical separation methods. *Analytical and Bioanalytical Chemistry, 405*, 443–450.
69. Reid, G. L., Morgado, J., Barnett, K., Harrington, B., Wang, J., Harwood, J., et al. (2013). Analytical quality by design (AQbD) in pharmaceutical development. *American Pharmaceutical Review*, 144191.
70. Beaman, J. (2010). Stability testing: Doing everything or doing the right thing? *Pharmacological Reviews, 13*, 73–76.
71. Waterman, K. C. (2011). The application of the accelerated stability assessment program (ASAP) to quality by design (QbD) for drug product stability. *AAPS PharmSciTech, 12*(3), 932–937.
72. Karmarkar, S., Garber, R., Genchanok, Y., George, S., Yang, X., & Hammond, R. (2011). Quality by design (QbD) based development of a stability indicating HPLC method for drug and impurities. *Journal of Chromatographic Science, 49*(6), 439–446. https://doi.org/10.1093/chrsci/49.6.439
73. Dickinson, P. A., Lee, W. W., & Stott, P. W. (2008). Clinical relevance of dissolution testing in quality by design. *The AAPS Journal, 10*(2), 380–390.
74. Miroslav, Ž. M., Valentina, D. M., Predrag, S. S., Radosav, M. P., & Dragan, M. M. (2010). An HPLC method for the determination of digoxin in dissolution samples. *Journal of the Serbian Chemical Society, 75*, 1583–1594.
75. Huang, J., Goolcharran, C., & Ghosh, K. (2011). A quality by design approach to investigate tablet dissolution shift upon accelerated stability by multivariate methods. *European Journal of Pharmaceutics and Biopharmaceutics, 78*(1), 141–150.
76. Sangshetti, J. N., Deshpande, M., Zaheer, Z., Shinde, D. B., & Arote, R. (2017). Quality by design approach: Regulatory need. *Arabian Journal of Chemistry, 10*, S3412–S3425.
77. Cook, J. A. (2012). A technique to estimate in vivo dissolution profiles without data from a solution. *The AAPS Journal, 14*(3), 433–436. https://doi.org/10.1208/s12248-012-9355-4
78. Aksu, B., Beer, T. D., Folestad, S., Ketolainen, J., Lindén, H., Lopes, J. A., et al. (2012). Strategic funding priorities in the pharmaceutical sciences allied to Quality by Design (QbD) and Process Analytical Technology (PAT). *European Journal of Pharmaceutical Sciences, 47*, 402–405. https://doi.org/10.1016/j.ejps.2012.06.009
79. Cogdill, R. P. (2008). Case for process analytical technology: Regulatory and industrial perspectives. In S. C. Gad (Ed.), *Pharmaceutical manufacturing handbook: Regulations and quality* (pp. 313–352). John Wiley and Sons.
80. Gaspar, R., Aksu, B., & Cuine, A. (2012). Towards a European strategy for medicines research (2014–2020): The EUFEPS position paper on horizon 2020. *European Journal of Pharmaceutical Sciences, 47*, 979–987.

81. Schmitt, S. (2011). *Quality by design: Putting theory into practice*. Parenteral Drug Association.
82. Quality Risk Management Guideline, Q9. (2005). *International Conference on Harmonization of Technical Requirements for Registration of Pharmaceuticals for Human Use*.
83. Pharmaceutical Quality System (PQS), Q10. (2008). *International Conference on Harmonization of Technical Requirements for Registration of Pharmaceuticals for Human Use*.

Chapter 2
Regulatory Aspects of Quality by Design

Ayca Altay Benetti and Camillo Benetti

Abstract The regulations for Quality by Design (QbD) guidelines clearly state the necessary requirements for effectively managing quality risks, Critical Process Parameters (CPAs), and Critical Quality Attributes (CQAs), implementing control strategies, and relentlessly striving for continuous improvement. Scientific principles, statistical tools, and risk management techniques MUST be utilized to design and enhance the manufacturing process if you aim to implement Quality by Design (QbD) in the development process. In order to gain approval for a drug, regulatory agencies require the submission of a QTPP, a design space, and a control strategy as part of the approval process. This is a mandatory requirement that cannot be overlooked or ignored. The impact of QbD regulations on the development, approval, and monitoring of pharmaceuticals will be thoroughly examined in this section.

Keywords Regulatory perspective · QbD · Quality control attributes · Design models

2.1 Introduction

Background and motivation for the Quality by Design (QbD) approach in drug development comes with regulatory framework, key terminology in the QbD context, perspective of the QbD approach for the pharmaceutical business area. This chapter topic of choice would be the regulatory aspects of QbD in the pharmaceutical industry. QbD is a method to develop a product, or a process based on the evaluation of all critical its parameters. This way of thinking was developed by the quality pioneer Dr. Joseph M. Juran [1]. In this framework, quality is not a result of testing or inspection, but rather is built into the drug product through its development and manufacturing processes [1].

A. A. Benetti (✉) · C. Benetti
Department of Pharmacy, National University of Singapore, Singapore, Singapore
e-mail: ayca.ben@nus.edu.sg; ben.cam@nus.edu.sg

N. K. Jain, N. Bajwa (eds.), *Introduction to Quality by Design (QbD)*,
https://doi.org/10.1007/978-981-99-8034-5_2

The regulatory agencies such as the US Food and Drug Administration (FDA) and the European Medicines Agency (EMA) have recognized QbD as the best approach for drug development and quality assurance [2, 3]. In this chapter, we will explore the regulatory aspects of QbD and its implications on the development, approval and post-marketing surveillance of pharmaceuticals.

2.2 Regulatory Framework

The first aspect to consider is the regulatory guidance on QbD. The FDA has issued several guidance documents, including International Conference on Harmonization (ICH) Q8, Q9 and Q10, which provide a framework for QbD implementation [4–7]. The European Medicines Agency (EMA) has also issued similar guidance documents. The agencies' guidance documents describe the expectations for quality risk management, critical process parameters and critical quality attributes (CPAs and CQAs), control strategies and continuous improvement (Table 2.1).

2.2.1 EMA Point of View

According to the guidelines approved by EMA (Table 2.1), an enhanced QbD approach to pharmaceutical development provides opportunities for more flexible regulatory approaches, for example:

- Risk-based regulatory decisions with assessment and inspections
- Manufacturing process changes within the approved Design Space without further regulatory review
- Reduction of post-approval submissions

Table 2.1 Guidelines and views of the authorities

Authority	Guideline (QbD upstream shifting concepts already introduced)
EMA	**EMA/CHMP/ICH/167068/2004—ICH Q8 (R2)**—Pharmaceutical Development [4]
	EMA/CHMP/ICH/24235/2006—ICH Q9—Quality Risk Management [5]
	EMA/CHMP/ICH/214732/2007—ICH Q10—Pharmaceutical Quality System [6]
	CPMP/QWP/3015/99—Note for Guidance on Parametric Release [8]
	EMA/CPMP/ICH/367/96—ICH Q6A—Test Procedures and Acceptance Criteria for New Drug Substances and New Drug Products: Chemical Substances [9]
	EUDRALEX GMP Volume 4 ANNEX 15—Section 5, Process Validation [10]
FDA	**Process Validation: General principles and Practises FDA** [11]
	ICH Q8(R2) Pharmaceutical Development [12]
	ICH Q9(R1) Quality Risk Management [13]
	ICH Q10 Pharmaceutical Quality System [14]

- Real-time quality control, leading to a reduction of end-product release testing, namely Real Time Release using Process Analytical Technology (PAT)

In this process, the quality, safety and effectiveness of the products were evaluated, thus QbD was represented and well-defined by common technical document (CTD) section called Mock p.2 created by European Federation of Pharmaceutical Industries Association (EFPIA) in order to support with public debate between industry and regulators in pursuit of this objective [15–18]. This assessment may be repeated where different types of variations are required, however, the guidelines consider the change in design space as a type of changes not requiring any prior approval by the authorities such as EMA [19].

On the other hand, CTD, is a global standard for regulatory submissions and is recognized by FDA and EMA, is a set of specifications that outline the information required for the submission of regulatory information for pharmaceutical products. It includes five modules, each with specific information requirements:

- Module 1: Administrative information and prescribing information.
- Module 2: Common Technical Document Summaries.
- Module 3: Quality.
- Module 4: Nonclinical Study Reports.
- Module 5: Clinical Study Reports.

The Module 3 is the quality description of the pharmaceutical product which should fulfil the requirements of QbD within complemented by a 'mock' section P.3.3 aside from describing the proposed process control strategy for routine production, that is based on available insight and perception linked to a review of the quality risk management regarding the available information [17]. QbD can be summarized within CTD implementation, as well as, can be identified as QbD Dossier (Fig. 2.1).

2.2.2 *FDA Point of View*

According to the guidelines approved (Table 2.1) by FDA, QbD can be explained within three terms:

- Design of Experiment (DoE) (ICH Q8)
- Risk assessment (ICH Q9)
- Process analytical technology (PAT) [20]

A correct implementation can be illustrated from sequential to cyclic approach in the Fig. 2.2. The first step is to define the desired clinical product performance and, in turn, the quality target product performance (QTPP). On the basis of the QTPP, the next step is to identify appropriate limits of the critical quality attributes (CQAs) of the product using preliminary experiments and prior knowledge. The third step is process characterisation. This involves thorough risk assessment (failure mode and effects analysis—FMEA, Ishikawa diagrams etc.) and prior knowledge to identify

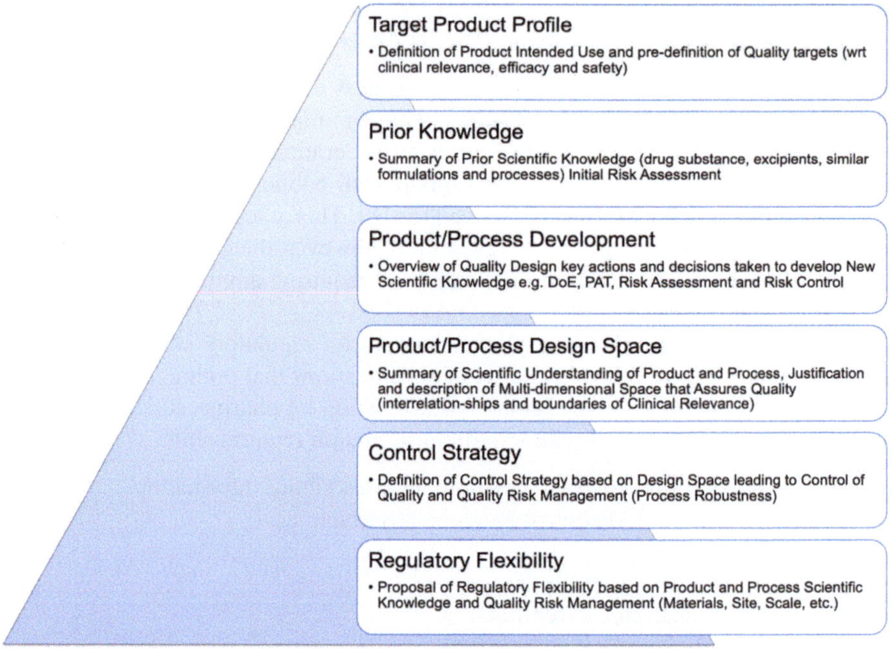

Fig. 2.1 QbD dossier

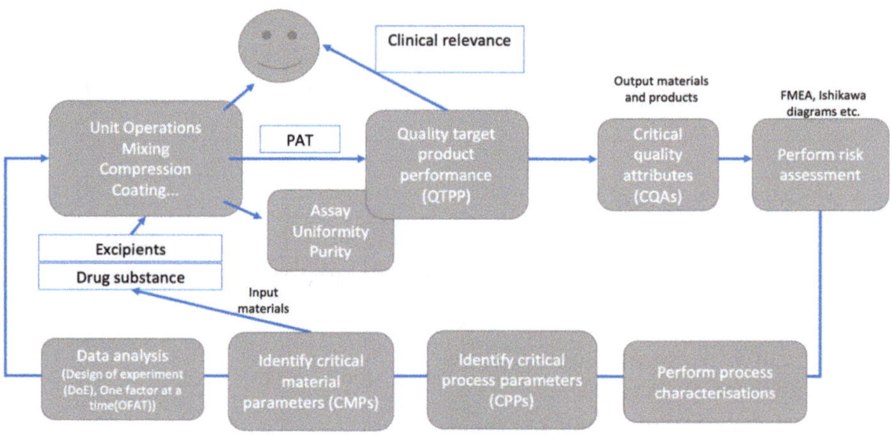

Fig. 2.2 An overall implementation from sequential to cyclic approach [21]

the CPPs and critical material parameters (CMPs). These parameters require monitoring and control the process which have a material impact on a CQA. In order to analyse these parameters we need to use a sort of data analysis which might be involved by design of experiment [21].

2.3 Development and Approval of QbD

The development of QbD has been driven by frequent quality issues, delays in product development and a lack of regulatory compliance. To address these challenges, the International Conference on Harmonization (ICH) issued Q8(R2), Q9 and Q10 guidelines that paved the way for QbD [22, 23]. Q8(R2) provides guidance on pharmaceutical development, Q9 provides guidance on quality risk management and Q10 provides guidance on pharmaceutical quality management systems.

Furthermore, QbD has been developed through the implementation of the ICH quality guidelines and the scientific principles of pharmacopoeias, including the United States Pharmacopeia (USP) and the European Pharmacopeia (EP) [1, 24, 25]. It promotes designing quality into the product by identifying and assessing CQAs. This ensures that quality is built into the product during its development, from the formulation design phase to the final manufacturing process. The approval of QbD has also led to a shift from a traditional trial-and-error approach to a more risk-based approach in the manufacturing process [7, 26]. With QbD, manufacturers can anticipate potential problems and design innovative solutions to them, improving the quality of the product. This approach involves a thorough understanding of the product's characteristics, the manufacturing process and the potential risks to the final product.

2.3.1 International Conference on Harmonization (ICH) Q8 (R2): Pharmaceutical Development

The development of a pharmaceutical product involves a complex process that requires the integration of a number of factors such as chemistry, manufacturing and controls. The main objective of the pharmaceutical industry is to produce safe and effective products that comply with regulatory standards. To achieve this objective, the industry has developed several quality control frameworks, with ICH Q8 R2 being one of them.

ICH Q8 R2 (Pharmaceutical Development) provides guidance for the pharmaceutical industry to improve the design and development of products. One of the key parameter of the guideline is the use of QbD principles, which emphasizes designing and developing products with a focus on quality and reducing the risks of failure and variability.

QbD means a whole product lifecycle; pool of techniques to be applied 'transversally' to the sectors:

- The techniques can be of organization, data collection and quantitative determination of the associated risk (Risk Analysis) and statistical-mathematical (uni- and multi-variate data analysis).

Table 2.2 Regulatory framework of ICH Q8

Section 1: Introduction	This section provides an overview of the regulatory framework of ICH Q8. It outlines the purpose and scope of the framework and provides definitions of key terms that are used throughout the document.
Section 2: Pharmaceutical Development	This section outlines the key principles of pharmaceutical development. It emphasizes the need to understand the CQAs of a product and the importance of designing a product that meets these CQAs via DoE. The section also provides guidance on the development of a product from early stages through to commercialization.
Section 3: Manufacture	This section outlines the key principles of manufacturing a pharmaceutical product. It emphasizes the need to develop a process that is robust and that can consistently produce a product that meets the required quality standards. The section also provides guidance on the design and development of manufacturing facilities, as well as on the control of the manufacturing process.
Section 4: Product Lifecycle Management	This section outlines the key principles of product lifecycle management. It emphasizes the importance of monitoring product performance throughout its lifecycle and the need to make adjustments to the manufacturing process as necessary to maintain product quality. The section also provides guidance on the management of any changes to the manufacturing process and the need to communicate these changes to regulatory authorities.

- Statistical-mathematical techniques are divided into active (design of experiment (DoE), Latin squares, etc...) or passive (control charts, multivariate statistical process control (MSPC), corrective and preventive actions (CAPA) analysis, etc...)
- As specified by ICH Q8 R2, they are implemented 'gradually' and maintaining a 'flexible' approach referring to the product/process in question and in any case to the pharmaceutical field.
- They represent a (potentially incontestable) way to make a defined test or model objective.
- The regulatory framework of ICH Q8 is divided into four main sections, as discussed below at Table 2.2.

2.3.2 International Conference on Harmonization (ICH) Q9

The ultimate goal is to identify factors (few survivors) that it makes sense to actively investigate and define control conditions for other potential influencing factors. In order to establish the control conditions, ICH Q9 is settled to be a guidance document on quality risk management by the International Conference of Harmonisation (ICH) of Technical Requirements for Registration of

Fig. 2.3 Tools for risk identification and analysis

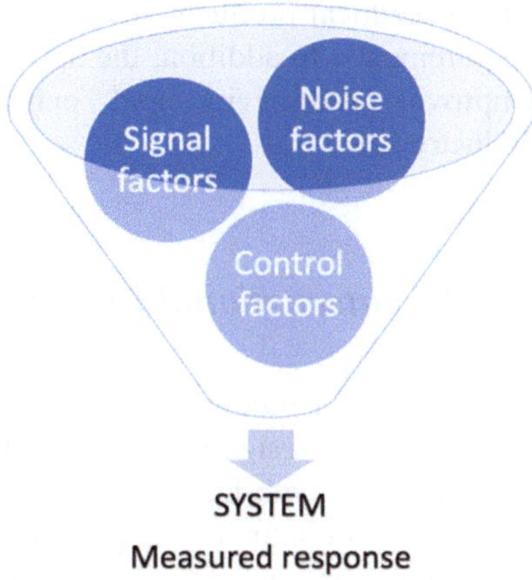

Pharmaceuticals for Human Use. To overpass the risk monitoring, there are several methos in this guideline, extract from ICH Q9 (QRM); Section 5 (Fig. 2.3):

- Basic risk management facilitation methods (flowcharts, check sheets etc.)
- Failure Mode Effects Analysis (FMEA)
- Failure Mode, Effecrs and Criticality Analysis (FMECA)
- Fault Tree Analysis (FTA)
- Hazard Analysis AND Critical Control Points (HACCP)
- Hazard Operability Analysis (HAZOP)
- Preliminary Hazard Analysis (PHA)
- Risk ranking and filtering
- Supporting statistical tools

The principles of ICH Q9 can be described with some terms as the following [27]:

- Risk assessment: It is important to identify, assess and analyze the potential risks associated with the quality of pharmaceutical products or clinical trials.
- Risk control: Once risks have been identified, measures should be taken to mitigate or control them by designing appropriate systems and processes.
- Risk communication: Risks should be effectively communicated between the stakeholders involved in the pharmaceutical industry, including manufacturers, suppliers, regulators and patients.
- Risk review and monitoring: The risks identified and controlled should be periodically reviewed and monitored to ascertain the effectiveness of the risk management system put in place.
- Risk management throughout the life cycle: The process of risk management should be an ongoing activity and should be applied throughout the life cycle of the pharmaceutical product or clinical trial.

The principles of ICH Q9 are relevant to the pharmaceutical industry as they promote a proactive approach to managing risks associated with the quality of the products or trials. The risk associated management creates an environment that the pharmaceutical products or trials are safe, effective and compliant with regulatory requirements. In addition, the application of these principles can help reduce costs, improve productivity and enhance the reputation of the pharmaceutical industry [28].

2.3.3 International Conference on Harmonization (ICH) Q10

ICH Q10 is a guideline for quality management systems in the pharmaceutical industry. It is designed to ensure that pharmaceutical products are developed and manufactured with high levels of quality and safety. The guideline is based on the principles of continuous improvement and risk management and provides a framework for companies to develop robust quality management systems that meet regulatory requirements.

Such that quality is a critical component and is essential to ensure that products are safe, effective and meet the intended specifications by the pharmaceutical companies. Thus, quality management systems play a key role in achieving and maintaining high levels of quality throughout the product life cycle, from development to manufacturing, distribution and post-market surveillance [28, 29] .

The Implementation of ICH Q10 requires a collaborative approach between industry, regulators and other stakeholders. It involves the establishment of processes to identify, assess and manage risks, as well as the implementation of measures to monitor and control quality throughout the product life cycle.

Overall, ICH Q10 by adhering to the principles and guidelines set forth, the pharmaceutical companies can ensure that their products are of high quality, safe and effective and meet the needs of patients worldwide. There are several benefits for this implementation of ICH Q10 who is followed by the regulatory guidelines. Those benefits of adhering to the principles of ICH Q10 in ensuring the quality of pharmaceutical products can be summarized with some terms as the following:

- Consistent quality: The adherence to the principles of ICH Q10 ensures that the quality of pharmaceutical products is consistent throughout the manufacturing process.
- Improved efficiency: ICH Q10 emphasizes the need for continuous improvement of the manufacturing process, leading to enhanced efficiency and reduced costs.
- Reduced variability: By following ICH Q10 principles organizations can minimize variability in their processes, leading to improved quality and safety of products.
- Regulatory compliance: Adhering to ICH Q10 guidelines ensures regulatory compliance, which is critical for the pharmaceutical industry.
- Risk-management: ICH Q10 focuses on proactive risk management, allowing organizations to identify and mitigate potential risks before they become critical issues.

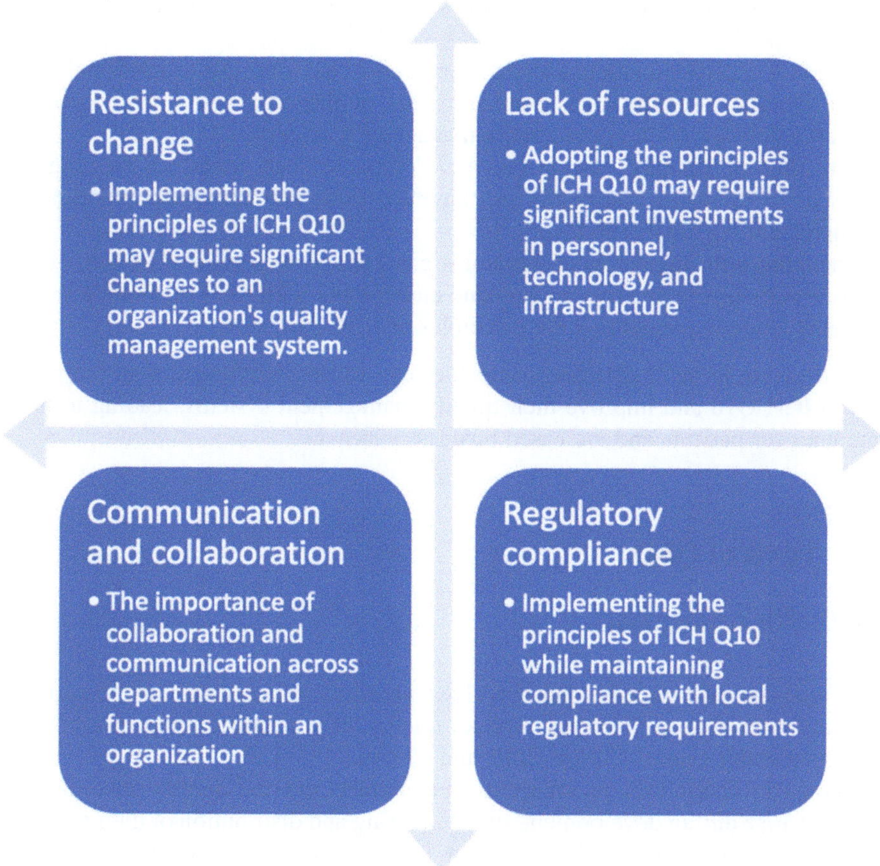

Fig. 2.4 The challenges and barriers for the pharmaceutical companies

- Enhanced customer satisfaction: Quality pharmaceutical products result in increased customer satisfaction, leading to customer loyalty and increased revenue.
- Stronger business reputation: By following ICH Q10 principles organizations build a strong reputation for manufacturing high-quality products in a compliant and safe manner.

Although, there are several challenges and barriers that companies may face when trying to adhere to the principles of ICH Q10. These include in Fig. 2.4 [30, 31]:

To address these challenges and barriers, companies can take several steps, including [30]:

1. Building a culture of quality: Companies can foster a culture of quality by promoting a shared understanding of the importance of quality across the

organization and incentivizing employees to prioritize quality in their day-to-day work.

2. Building a cross-functional team: Companies can form cross-functional teams that include representatives from all areas of the organization to ensure that everyone has a voice in the quality management system.
3. Investing in training and resources: Companies can invest in training and resources to ensure that employees have the skills and knowledge necessary to implement ICH Q10 effectively.
4. Engaging with regulators: Companies can engage with regulators early in the process to ensure that they understand and comply with local regulatory requirements while implementing the principles of ICH Q10.

By addressing these challenges, companies can successfully implement the principles of ICH Q10 and improve their quality management systems, leading to better outcomes for patients and increased business success.

2.4 Tools of QbD

2.4.1 Risk Analysis

Risk analysis is a process that involves identifying, assessing and controlling potential risks that may impact product quality. It is a critical step in QbD as it helps to identify critical CQAs, CPPs and CMAs.

Through risk analysis, pharmaceutical manufacturers can identify potential risks that may arise during development, manufacturing and distribution of their products. This enables them to take the necessary steps to mitigate these risks and ensure product quality and safety. Risk analysis also helps to optimize process design and control, which ultimately leads to increased efficiency, reduced costs and improved product quality [28, 32].

The risk analysis should be considered with the final product in market, thus, product understanding is depending on the desired level of 'deepening', the applicable techniques become more and more complex (Fig. 2.5). By identifying potential risks and controlling them throughout the product lifecycle, manufacturers can optimize their processes and ensure that their products meet the required quality standards in the research and development process [33].

Overall, this implementation can be illustrated with a couple of examples as the following Fig. 2.6.

Tablet dissolution can be given as an example of risk analysis using the fishbone diagram (also known as Ishikawa diagram or cause-and-effect diagram) which is already mentioned at ICH Q9, is graphical tool used to identify and analyze the potential risks and its causes. Even though, this diagram is often used in quality management and process improvement, it has a structure that resembles a fishbone,

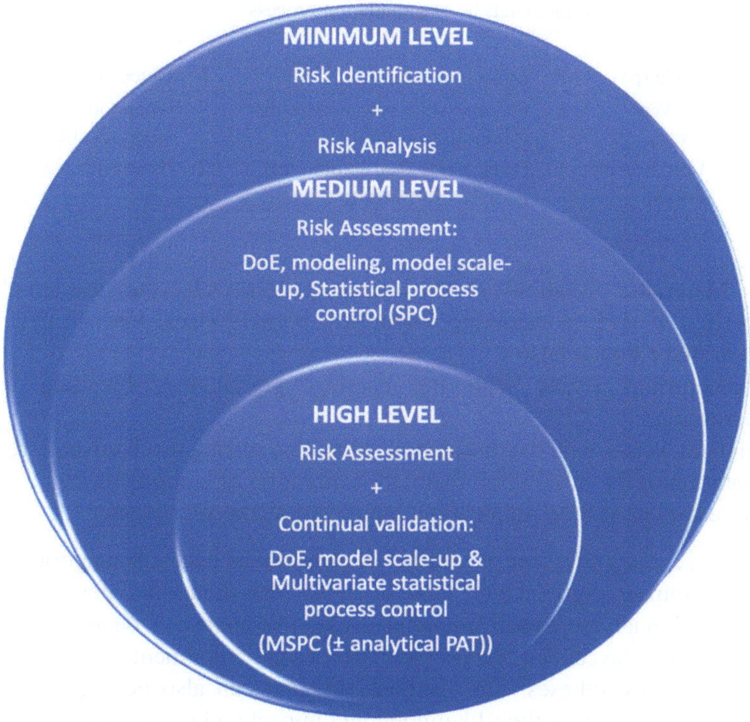

Fig. 2.5 The desired level of "deepening"

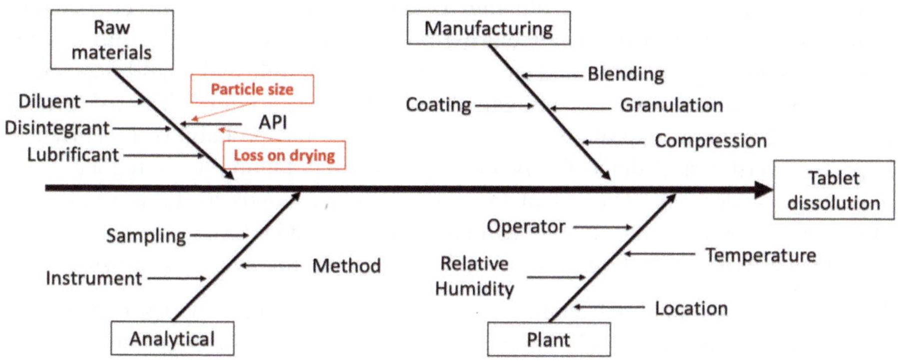

Fig. 2.6 An example of Fishbone (Ishikawa) Diagram for the tablet dissolution

with the "head" representing the problem or effect and the "bones" representing the potential causes.

In the case of tablet dissolution at Fig. 2.5, the problem or effect might be that the tablets are not dissolving properly or at a consistent rate. The fishbone diagram can be used to identify the potential causes of this problem, which could include:

Table 2.3 Failure mode effect analysis of granulation process

Variable	Failure mode (p)	P	Impact	I	Detectability	D	RPN
Mesh size	Broken screen	1	Compression, particle size distribution, flowability	5	Operator, batch record	4	20
Loss on drying	Low loss on drying	1	Particle size distribution, fine particles, flowability	2	Near infra-red	1	2

- Raw Materials: variations in the excipients, especially the physico-chemical properties of API (considering impurities, particle sizes, loss on drying etc.) used to make the tablets.
- Manufacturing: issues with the coating, granulating, blending and the compression.
- Analytical: issues with the dissolution apparatus, stirring mechanism, errors in the preparation of the dissolution medium, calibration of equipment, etc.
- Plant: errors in sample preparation, temperature control, pH, etc.

Each of these categories can be further broken down into subcategories, such as specific equipment components, process steps, ingredients or operator tasks. By systematically analyzing each potential cause, the team can develop hypotheses about what might be causing the problem and design experiments or tests to confirm or refute those hypotheses. The fishbone diagram can also be used to prioritize potential causes based on their likelihood and impact and to guide the development of corrective actions.

On the other hand, failure mode and effects analysis (FMEA) is a systematic method for identifying and evaluating potential failures and their effects on a pharmaceutical product or manufacturing process. It involves identifying possible failures, evaluating the likelihood and severity of each failure and identifying actions or controls that can prevent or mitigate the effects of each failure. The risk priority number (RPN) is a quantitative score that assesses the probability (occurrence), impact (severity) and detectability of each potential failure, with higher scores indicating greater risk. The calculation of RPN corresponds to the multiplication of the score of probability, impact and detectability (Table 2.3).

In the case of granulation process, an FMEA might involve the following steps:

1. Form a cross-functional team to identify potential failure modes in the tablet compression process. This team might include a process engineer, a quality control specialist, a chemist and a production supervisor.
2. Brainstorm potential failure modes and their effects on the granulation process. For example, a failure mode variable might be "broken screen of a mesh used in the granulation process" with the impact of "reduced tablet compression, particle size and flowability."
3. Use a quantitative approach to evaluate the probability, impact and detectability of each failure. Impact refers to the failure on product quality or safety;

probability refers to the likelihood of the failure occurring; and detectability refers to the likelihood of the failure being detected before it causes harm.

4. Calculate the RPN for each failure by multiplying the probability, impact and detectability scores. A high RPN indicates a higher risk failure mode that requires more attention and controls.
5. Prioritize and address high RPN failure modes with appropriate actions or controls.

An example of a high RPN failure mode in granulation process could be "broken screen of a mesh" with an impact score of 5, a probability score of 1 and a detectability score of 4, giving an RPN of 20. Addressing this failure mode may require steps such as evaluating and improving punch and die alignment, improving tablet hardness testing and quality control or implementing improved maintenance procedures on the batch record.

2.4.2 Design of Experiment

As describe before, design of experiments (DOE) is a powerful statistical tool that is recommended in the ICH Q8 R2 guidelines. It is a structured and systematic approach to identifying and optimizing CQAs, CPPs and CMPs to achieve the desired product quality.

DOE provides a framework to systematically vary one or more variables at a time to establish cause-effect relationships and determine the optimal conditions for product quality and performance [34]. This methodology helps in reducing the number of experiments required by enabling the identification of significant factors and interactions that have a major impact on the product quality and performance.

The use of DOE in pharmaceutical development helps to optimize the formulation, manufacturing process and analytical method development. It is important to note that DOE is not a singular approach but rather a set of methods that can be adapted to the specific needs of a particular product or process [35, 36]. Therefore, it is highly recommended to involve statistical experts in the design and analysis of DOE studies.

Overall, the use of QbD principles and DOE can significantly improve the efficiency, quality and robustness of the pharmaceutical development process. The tools of QbD can be illustrated with Fig. 2.7.

Parametric and nonparametric univariate tests are applicable for all departments in pharmaceutical companies. The aim of using these tools is mostly to obtain a data processing and significance of comparison tests. Parametric tests make assumptions about the distribution of the data, such as normality and equal variances and are typically used when the data are continuous and normally distributed. Examples of parametric tests commonly used in pharmaceutical companies include t-tests, ANOVA and linear regression analysis. These tests are often preferred when the data meet the assumptions of normality and homogeneity of variance because they

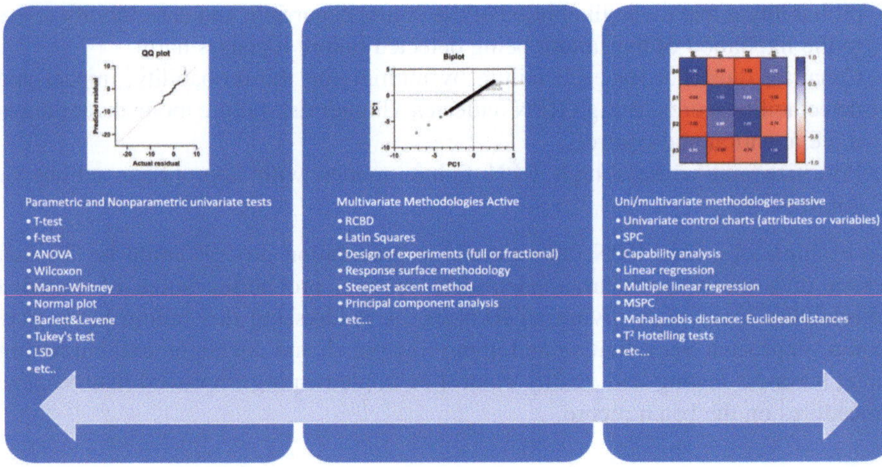

Fig. 2.7 Tools of QbD

generally have greater statistical power and are more sensitive to detecting differences or associations when they exist. However, for cases where these assumptions are not met, the results of a parametric test may not be valid and nonparametric tests may need to be used instead.

Nonparametric tests, on the other hand, do not make assumptions about the distribution of the data and may be used with data that are not normally distributed or when the sample size is small. These tests typically use rank-based methods and frequency counts rather than arithmetic means and therefore are less sensitive to outliers and distributional shape. Examples of nonparametric tests commonly used in pharmaceutical companies include the Wilcoxon rank-sum test and Kruskal-Wallis test for comparing groups and the Mann-Whitney test for comparing two unpaired groups. These tests can be more robust when the assumptions of the parametric test are not met, but may have less statistical power than parametric tests in detecting differences or associations when they exist.

While multivariate methodologies are commonly meant for the product and process development or analytical process departments. The aim of these tools is process understanding and modelling analytical methods. These methods allow for the identification of patterns, relationships and interactions among variables in a single analysis. One of the widely used multivariate methodology is Principal component analysis (PCA). PCA is a technique used to reduce the dimensionality of high-dimensional data by identifying underlying patterns and factors. This method is used to identify the important variables that contribute the most to the variability in the data. PCA is often used in analytical methods and drug discovery to identify active compounds and compound libraries that have similar structures.

The traditional univariate methods analyzes one variable at a time, as opposed to multivariate methods that analyze multiple variables simultaneously. The aim of these tools is to understand the process and modelling continual validation (periodic review). These tools are commonly used in quality assurance to keep the track data processing in manufacturing side.

Regression and survival analysis are the most important tests for the quality assurance. Regression analysis is a method used to assess the relationship between two variables, typically a dependent variable and one or more independent variables. In pharmaceutical companies, regression analysis might be used to assess the impact of dosage or treatment duration on a clinical outcome. While, survival analysis is a statistical method used to analyze the time-to-event data, such as the time to drug failure or the time to disease progression. This method is widely used in pharmaceutical companies to analyze clinical trial data.

2.4.2.1 Full Factorial of Factorial DoE

A full factorial design of experiment is a statistical technique used to investigate the effects of multiple factors or variables on a response. It allows researchers to systematically vary the levels of each factor and observe the resulting changes in the response variable.

In a full factorial design, all possible combinations of factor levels are tested, resulting in a complete and balanced experimental design. This means that each factor is tested at every possible level in combination with every other factor. The design is called "factorial" because it involves multiplying the number of levels for each factor to determine the total number of experimental conditions [37].

Let's consider an example to illustrate the concept. Suppose you want to investigate the effects of two factors, A and B, on a response variable, Y. Factor A has two levels (low and high) and factor B has three levels (low, medium and high). To create a full factorial design, you would need to test all possible combinations of the factor levels.

The experimental conditions would be as follows:

- A1 (low level of factor A) and B1 (low level of factor B)
- A1 and B2
- A1 and B3
- A2 (high level of factor A) and B1
- A2 and B2
- A2 and B3

For each of these conditions, you would measure the response variable Y. By testing all possible combinations, a full factorial design allows you to assess the main effects of each factor (the average effect of each factor across all levels of the other factors) as well as the interaction effects (how the combination of factors influences the response variable beyond their individual effects).

The advantages of a full factorial design include its ability to estimate and quantify the effects of factors and their interactions accurately. However, as the number of factors and levels increases, the number of experimental conditions grows exponentially, which can become impractical or costly. In such cases, fractional factorial designs or other design strategies may be employed to reduce the number of experimental conditions while still capturing the essential information [38].

2.4.2.2 Full Factorial of Factorial DoE

A response surface design is a statistical experimental design used to explore and optimize the relationship between multiple factors and a response variable. It allows researchers to model and analyze complex nonlinear relationships between the factors and the response.

In a response surface design, the factors of interest are typically continuous variables and the goal is to identify the optimal factor settings that maximize or minimize the response variable. The design involves systematically varying the levels of the factors within a predefined range and collecting response measurements at each combination of factor levels.

The design is called “response surface” because it aims to estimate the shape of the underlying response surface, which represents the relationship between the factors and the response. The response surface can take different forms, such as linear, quadratic or higher-order polynomial models, depending on the complexity of the relationship being investigated.

The design typically involves three main steps [38, 39]:

Initial Design: A set of initial experimental conditions is selected based on a design criterion, such as a central composite design or Box-Behnken design. This initial design is often a subset of the possible combinations of factor levels and is chosen to cover a wide range of factor settings while efficiently estimating the response surface.

Data Collection: The response variable is measured at each combination of factor levels in the initial design. Additional data points may be collected near the region of interest to refine the response surface model and identify the optimal factor settings.

Response Surface Modeling and Optimization: Statistical techniques such as regression analysis or design of experiments (DOE) methods are used to fit a mathematical model to the collected data. The model describes the relationship between the factors and the response and allows for prediction and optimization. Based on the model, the optimal factor settings can be determined to maximize or minimize the response variable.

Response surface designs are useful in various fields, including engineering, chemistry and manufacturing, where the relationship between factors and responses is nonlinear or involves interactions. They provide a systematic and efficient approach to explore and optimize the factor settings, leading to improved processes, products or outcomes.

It’s worth noting that response surface designs are often conducted in multiple stages, refining the model and factor settings based on the initial results. This iterative approach helps to better understand the underlying relationships and enhance the accuracy of the response surface model.

2.4.2.3 Mixture Design

A mixture design is a type of experimental design used to study and optimize the composition of a mixture or blend of different components. It is commonly employed when the factors of interest are proportions or percentages of the components, rather than discrete levels.

In many real-world scenarios, products or processes involve combining different ingredients or components to achieve a desired outcome. Mixture designs allow researchers to systematically vary the proportions of the components and study the effects on the response variable of interest.

There are several types of mixture designs, including simplex designs, simplex-centroid designs and extreme vertices designs. Each type has its own characteristics and advantages, but they all share the fundamental principle of representing the composition of the mixture as a vector of proportions or percentages.

The key steps in conducting a mixture design are as follows:

Define the Components: Identify the components or ingredients that make up the mixture. Each component should be measurable and have a proportion or percentage associated with it.

Determine the Constraints: Specify any constraints or restrictions on the proportions of the components. For example, the proportions may need to sum up to a certain value or there may be lower or upper limits on individual component proportions.

Choose a Design: Select an appropriate mixture design based on the specific goals and constraints of the study. The design determines the number and combinations of experimental conditions to be tested.

Set the Proportions: Assign the proportions or percentages of the components for each experimental condition based on the chosen design. These proportions can be randomized or systematically determined.

Collect Data: Perform the experiments according to the defined conditions and measure the response variable of interest for each mixture. Multiple replicates are often conducted to increase the reliability of the results.

Analyze the Data: Analyze the collected data using statistical techniques appropriate for mixture designs. Regression models, such as the linear or nonlinear mixing models, are commonly used to estimate the effects of the components and their interactions on the response variable.

Optimize the Mixture: Use the fitted model to optimize the composition of the mixture. This involves finding the proportions of the components that maximize or minimize the desired response variable or meet certain criteria.

Mixture designs are valuable in industries such as pharmaceuticals, food and beverage, cosmetics and materials science. They provide insights into the optimal combinations of components, allowing for improved formulations, cost-effective mixtures and enhanced product performance.

2.4.2.4 Case Study

2.4.2.4.1 First Example

An example of DoE applied to the pharmaceutical filed could be the degradation of a tablet in storage at different level of relative humidity (three levels) and temperature (two levels) over time.

A full factorial design with a central point was employed. The factors and their levels were chosen based on prior knowledge and preliminary experiments (Fig. 2.8).

Three tablets for each batch were analysed at any given time points for the amount of degradation products on the two different drugs in each tablet. A regression model was then calculated using the degradation product against temperature and time (Fig. 2.9).

The results were then assayed using a specific formulation in order to verify the predictability if the model. The values used in this step should not be the same used in the drafting of the Design Of Experiments.

Predicted shelf life is approximately less than 30% of the actual one if the storing conditions, became harsher. Using this method would have allowed to set an 8 months storage time in lieu of the actual 12 months using only a 2 weeks stability test.

2.4.2.4.2 Second Example

A full factorial design with three factors and two levels was employed. The factors and their levels were chosen based on prior knowledge and preliminary experiments. The two levels for each factor were: binder type (A1 = PVP, A2 = HPMC), compression force (B1 = 10 kN, B2 = 20 kN) and lubricant concentration (C1 = 1%, C2 = 2%). This resulted in a total of 2^3 = 8 experimental conditions.

		Independent Variable 2	
		Level 1	Level 2
Independent Variable 1	Level 1	Dependent Variable	Dependent Variable
	Level 2	Dependent Variable	Dependent Variable
	Level 3	Dependent Variable	Dependent Variable

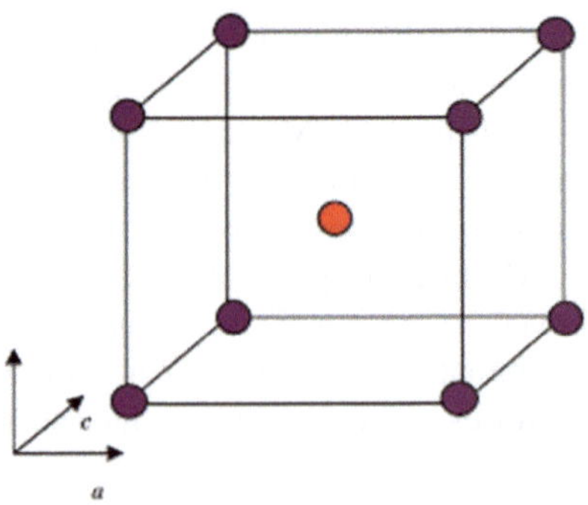

Fig. 2.8 Case study on full factorial design

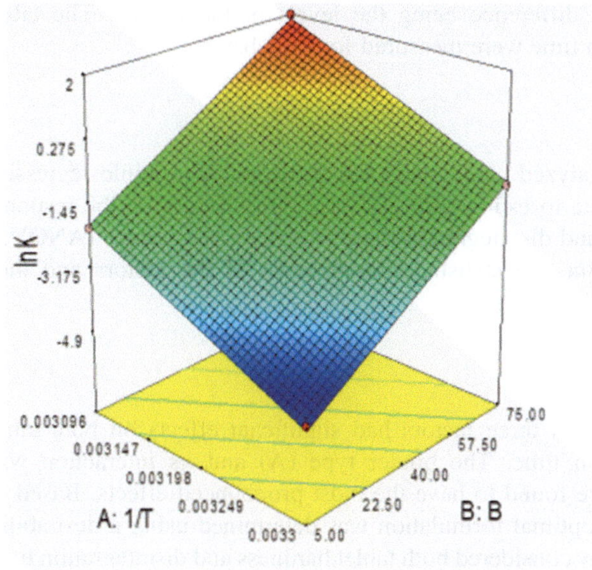

Response 1 lnK

ANOVA for selected factorial model

Analysis of variance table [Partial sum of squares - Type III]

Source	Sum of Squares	df	Mean Square	F Value	p-value Prob > F	
Model	22.51	2	11.26	351.07	0.0003	significant
A-1/T	*7.87*	*1*	*7.87*	*245.32*	*0.0006*	
B-B	*14.65*	*1*	*14.65*	*456.82*	*0.0002*	
Curvature	0.39	1	0.39	12.27	0.0394	significant
Residual	0.096	3	0.032			
Lack of Fit	*0.039*	*1*	*0.039*	*1.39*	*0.3594*	*not significant*
Pure Error	*0.057*	*2*	*0.028*			
Cor Total	23.00	6				

Fig. 2.9 Case study on ANOVA

Experimental Procedure

Eight batches of tablets were prepared according to the experimental conditions defined by the DOE design. Each batch was manufactured using the same process parameters, with the only difference being the levels of the factors. The tablet hardness and disintegration time were measured for each batch.

Data Analysis

The collected data was analyzed using statistical software. A multiple regression model was fitted to the data to estimate the effects of the factors on the response variables (tablet hardness and disintegration time). Analysis of variance (ANOVA) was conducted to determine the statistical significance of the factors and their interactions.

Results and Optimization

The analysis revealed that all three factors had significant effects on both tablet hardness and disintegration time. The binder type (A) and its interaction with compression force (B) were found to have the most pronounced effects. Based on the regression model, the optimal formulation was determined using a desirability function that simultaneously considered both tablet hardness and disintegration time.

Conclusion

This case study demonstrated the application of Design of Experiments (DOE) in optimizing the tablet formulation in pharmaceutical development. By systematically varying the levels of three factors and analyzing the response variables, valuable insights were gained regarding the effects of the factors and their interactions on tablet quality attributes. The optimized formulation obtained through DOE can contribute to improved product quality, manufacturing efficiency and patient satisfaction in the pharmaceutical industry.

2.5 Conclusion

In conclusion, QbD is a systematic approach to drug development that emphasizes the design and development of quality into the product. Regulatory guidance on QbD describes the expectations for quality risk management, CPAs and CQAs, control strategies and continuous improvement. The incorporation of QbD in the development process requires the use of scientific principles, statistical tools and risk management techniques to design and optimize the manufacturing process. The regulatory agencies require the submission of a QTPP, a design space and a control strategy in the drug approval process. Post-marketing surveillance of pharmaceuticals produced using QbD requires the monitoring of the manufacturing process to ensure the product remains within its design space.

References

1. Yu, L. X., et al. (2014). Understanding pharmaceutical quality by design. *The AAPS Journal, 16*(4), 771–783. https://doi.org/10.1208/s12248-014-9598-3
2. Djuris, J., Ibric, S., & Djuric, Z. (2013). *Quality-by-design in pharmaceutical development*. Woodhead Publishing Limited.
3. Beg, S., Rahman, M., & Kohli, K. (2019). Quality-by-design approach as a systematic tool for the development of nanopharmaceutical products. *Drug Discovery Today, 24*(3), 717–725. https://doi.org/10.1016/j.drudis.2018.12.002
4. International Council for Harmonisation. (2009). *EMEA/CHMP, 2009, ICH Topic Q 8 (R2) pharmaceutical development, step 5: note for guidance on pharmaceutical development* (Vol. 8). European Medicines Agency.
5. International Council for Harmonisation. (2015). *ICH guideline Q9 on quality risk management* (Vol. 44 (1), pp. 1–20). European Medicines Agency.
6. International Council for Harmonisation. (2015). *ICH guideline Q10 on pharmaceutical quality system* (Vol. 44, pp. 1–20). European Medicines Agency.
7. Korakianiti, E., & Rekkas, D. (2011). Statistical thinking and knowledge management for quality-driven design and manufacturing in pharmaceuticals. *Pharmaceutical Research, 28*(7), 1465–1479. https://doi.org/10.1007/s11095-010-0315-3
8. The European Agency for the evaluation of medicinal product. (2001, March). Note for guidance on parametric release. *October* (pp. 0–5).
9. International Council for Harmonisation. (2010). *ICH topic Q6A specifications: Test procedures and acceptance criteria for new drug substances and new drug products: Chemical substances* (Vol. 375(9728), p. 1753). European Medicines Agency. https://doi.org/10.1016/S0140-6736(10)60785-4
10. Stuggins, G., Sharabaroff, A., & Semikolenova, Y. (2015). *EUDRALEX GMP Volume 4—Annex 15* (Vol. 4, pp. 47–70). European Medicines Agency. https://doi.org/10.1596/9780821398036_bm
11. Baseman, H. (2012). Commentary on the US Food and Drug Administration's 2011 'Guidance for industry, process validation general principles and practices'. In *Process validation in manufacturing of biopharmaceuticals* (3rd ed., pp. 11–50). Taylor & Francis. https://doi.org/10.1201/b12013-6
12. International Council for Harmonisation. (2009). *ICH Q8(R2) pharmaceutical development* (Vol. 8, p. 28). U.S. Department of Health and Human Services Food and Drug Administration Center for Drug Evaluation and Research (CDER) Center for Biologics Evaluation and Research.
13. Rick Friedman, F. (n.d.). *Q9(R1)—Quailty risk management* (Vol. 9). U.S. Department of Health and Human Services Food and Drug Administration Center for Drug Evaluation and Research (CDER) Center for Biologics Evaluation and Research.
14. H. Services. (2009). *Guidance for Industry—Q10-pharmaceutical quality system* (Vol. 51, pp. 20852–21448). U.S. Department of Health and Human Services Food and Drug Administration Center for Drug Evaluation and Research (CDER) Center for Biologics Evaluation and Research.
15. Sprenger, K., Nickerson, D., Meeker-O'Connell, A., & Morrison, B. W. (2013). Quality by design in clinical trials: A collaborative pilot with FDA. *Therapeutic Innovation & Regulatory Science, 47*(2), 161–166. https://doi.org/10.1177/0092861512458909
16. Costa, E., Girotti, S., Pauro, F., Leufkens, H. G. M., & Cipolli, M. (2022). The impact of FDA and EMA regulatory decision-making process on the access to CFTR modulators for the treatment of cystic fibrosis. *Orphanet Journal of Rare Diseases, 17*(1), 1–14. https://doi.org/10.1186/s13023-022-02350-5
17. Fischer, G., et al. (2006) A guide to EFPIA's Mock P. 2 document.
18. Bloom, S., & Mock, T. (2014). Sakura bloom tablets P2 mock (pp. 1–57), 2015.

19. Dickinson, P. A., et al. (2008). Clinical relevance of dissolution testing in quality by design. *The AAPS Journal, 10*(2), 380–390. https://doi.org/10.1208/s12248-008-9034-7
20. Food and Drug Administration. (2004). *Guidance for industry, PAT-A framework for innovative pharmaceutical development, manufacturing and quality assurance*. U.S. Department of Health & Human Services.
21. Altay-Benetti, A. (2020). *Mannitol polymorphs as carrier in dpis* [PhD thesis]. University of Parma.
22. Allison, G., et al. (2015). Regulatory and quality considerations for continuous manufacturing May 20-21, 2014 continuous manufacturing symposium. *Journal of Pharmaceutical Sciences, 104*(3), 803–812. https://doi.org/10.1002/jps.24324
23. Charoo, N. A., & Rahman, Z. (2020). Integrating QbD tools for flexible scale-up batch size selection for solid dosage forms. *Journal of Pharmaceutical Sciences, 109*(3), 1223–1230. https://doi.org/10.1016/j.xphs.2019.12.007
24. Tang, P., Kwok, P. C., Tong, Z., Yang, R., Raper, J. A., & Chan, H. K. (2012). Does the United States Pharmacopeia throat introduce de-agglomeration of carrier-free powder from inhalers? *Pharmaceutical Research, 29*, 1797–1807. https://doi.org/10.1007/s11095-012-0703-y
25. Politis, S. N., Colombo, P., Colombo, G., & Rekkas, D. M. (2017). Design of experiments (DoE) in pharmaceutical development. *Drug Development and Industrial Pharmacy, 43*(6), 889–901. https://doi.org/10.1080/03639045.2017.1291672
26. Sallam, A. S. A. N. (2019). *Quality by design considerations for product development of dry-powder inhalers*. Elsevier.
27. O'Connor, T., Yang, X., Tian, G., Chatterjee, S., & Lee, S. (2017). *Quality risk management for pharmaceutical manufacturing: The role of process modeling and simulations*. Elsevier.
28. Tian, G., Koolivand, A., Arden, N. S., Lee, S., & O'Connor, T. F. (2019). Quality risk assessment and mitigation of pharmaceutical continuous manufacturing using flowsheet modeling approach. *Computers and Chemical Engineering, 129*, 106508. https://doi.org/10.1016/j.compchemeng.2019.06.033
29. Arden, N. S., Fisher, A. C., Tyner, K., Yu, L. X., Lee, S. L., & Kopcha, M. (2021). Industry 4.0 for pharmaceutical manufacturing: Preparing for the smart factories of the future. *International Journal of Pharmaceutics, 602*, 120554. https://doi.org/10.1016/j.ijpharm.2021.120554
30. Panzitta, M., et al. (2017). The strategic relevance of manufacturing technology: An overall quality concept to promote innovation preventing drug shortage. *International Journal of Pharmaceutics, 516*(1–2), 144–157. https://doi.org/10.1016/j.ijpharm.2016.11.018
31. Orpana, I. F. (2019). Manufacturing pharmaceutical medicines in a regulated environment—An auditor's perspective. *Procedia Manufacturing, 39*(2019), 1773–1782. https://doi.org/10.1016/j.promfg.2020.01.262
32. Yang, Kai, and Basem S. El-Haik. (2009). Design for Six Sigma: A roadmap for product development. 2nd ed. New York: McGraw-Hill. (pp 387–410).
33. Schuhmacher, A., Brieke, C., Gassmann, O., Hinder, M., & Hartl, D. (2021). Systematic risk identification and assessment using a new risk map in pharmaceutical R&D. *Drug Discovery Today, 26*(12), 2786–2793. https://doi.org/10.1016/j.drudis.2021.06.015
34. Zagalo, D. M., Sousa, J., & Simões, S. (2022). Quality by design (QbD) approach in marketing authorization procedures of non-biological complex drugs: A critical evaluation. *European Journal of Pharmaceutics and Biopharmaceutics, 178*, 1–24. https://doi.org/10.1016/j.ejpb.2022.07.014
35. Ashworth, J. L., et al. (1999). Fibrillin degradation by matrix metalloproteinases: Implications for connective tissue remodelling. *The Biochemical Journal, 340*(1), 171–181. https://doi.org/10.1042/0264-6021:3400171
36. Zlota, A. A. (2021). Recommendations for effective and defendable implementation of quality by design. *Organic Process Research and Development, 26*, 899–914. https://doi.org/10.1021/acs.oprd.1c00265

37. Nelsen, T. C. (2023). Chapter Fourteen: Efficient design of experiments (DOE). In *Probability and statistics for cereals and grains: A guide to measurement and analysis* (pp. 201–214). Elsevier. https://doi.org/10.1016/B978-0-323-91724-7.00011-3
38. Armstrong, N. A. (2006). *Pharmaceutical experimental design and interpretation* (2nd ed.). Routledge.
39. Shariare, M. H., De Matas, M., York, P., & Shao, Q. (2011). The impact of material attributes and process parameters on the micronisation of lactose monohydrate. *International Journal of Pharmaceutics, 408*(1–2), 58–66. https://doi.org/10.1016/j.ijpharm.2011.01.047

Chapter 3
Selection of Design of PAT

Suchitra Nishal, Virender Kumar, Parmita Phogat, Vandana Garg, and Harish Dureja

Abstract Process analytical technology (PAT) is a blanket term that refers to a variety of instruments and is frequently mentioned as being a crucial part of other innovative projects, such as real-time release, continuous manufacturing, and quality by design. It is crucial in understanding and controlling processes in various industries. A suitable PAT design for a particular application is selected based on several considerations and methodologies described in this abstract. A focus on PAT in the introduction emphasizes its importance for optimizing manufacturing processes and ensuring product quality. This study provides recommendations to enable real-time monitoring and analysis of PATs.

Using effective PAT design selection, researchers, practitioners, and industry professionals can optimize manufacturing processes by contributing to the growing knowledge regarding PAT applications.

Keywords Process · Analytical · Quality by design · Methodologies · Design · Optimize

---

S. Nishal
College of Pharmacy, Pandit Bhagwat Dayal Sharma University of Health Sciences, Rohtak, India

V. Kumar (✉)
College of Pharmacy, Pandit Bhagwat Dayal Sharma University of Health Sciences, Rohtak, India

Department of Pharmaceutical Sciences, Maharishi Dayanand University, Rohtak, India
e-mail: virenderkumar.copharmacy@uhsr.ac.in

P. Phogat
School of Medical and Sciences, G D Goenka University, Gurugram, India

V. Garg · H. Dureja
Department of Pharmaceutical Sciences, Maharishi Dayanand University, Rohtak, India

N. K. Jain, N. Bajwa (eds.), *Introduction to Quality by Design (QbD)*,
https://doi.org/10.1007/978-981-99-8034-5_3

3.1 Introduction

3.1.1 Overview of Process Analytical Technology (PAT)

Historically, the mainstay of quality control in the pharmaceutical industry has been statistical process control, which serves to understand the process along with anticipated specification boundaries as well as to uphold a stable process by removing allocable causes of variation. In the pharmaceutical manufacturing process, unit operations are evaluated for performance suitability techniques, such as control charts and run charts. Additionally, the majority of offline studies and monitoring carried out during the production batch course to assess attributes of the intermediate as well as accomplished products. Process verification for final batches is carried out using the conventional method under predetermined process conditions. The inability to confirm the quality attributes of intermediate goods all through industrial manufacturing process is a downside of this technology. As a result, locating and fixing issues that crop up during the process takes a long time and produces significantly amplified quality errors. Additionally, there needs to be more guarantee that whole lot complies with necessary criteria, and procedure is inapplicable universally as a solution for each quality issue. To address limitations of statistical process control, assure process control, and enhance process as well as product quality, ICH introduced continuous process verification. This continuously monitors and evaluates manufacturing process performance. In addition, it offers more details about process variability and control, increasing statistical confidence, enhancing evaluation of pharmaceutical production processes, and ensuring a better level of continuous control status. Quality by design (QbD) is a further tactic utilized by pharmaceutical industries to enhance process together with quality control understanding. Generally speaking, in the course of pharmaceutical production, QbD tools such as design of experiments (DoE), pragmatic modelling along with response surface study can generate a design space and identify process inconsistency. QbD paves ways for technology and science built drug quality management, in contrast to current quality by testing (QbT), which primarily uses quality test of finished product. PAT seeks to advance effectiveness and innovation in development, manufacturing, and quality assurance of pharmaceutical products. Centre for Drug Evaluation and Research (CDER) of USFDA conferred requirement on behalf of FDA advice to assist PAT adoption [1–4]. To foster creative along with risk-dependent regulatory assessments by industry together with Agency, agenda is built on process understanding. Major anticipations were raised for greater implementation of PAT *via* pharmaceutical production throughout all stages of development by FDA's drive on cGMP for twenty-first century and FDA PAT direction that followed in 2004. These initiatives aimed to encourage the pharmaceutical industry to adopt new technological advancements. Although one may claim that the pharmaceutical sector has greatly increased since overall implementation of PAT. The outline of the processes under PAT are described in Fig. 3.1 [6, 7].

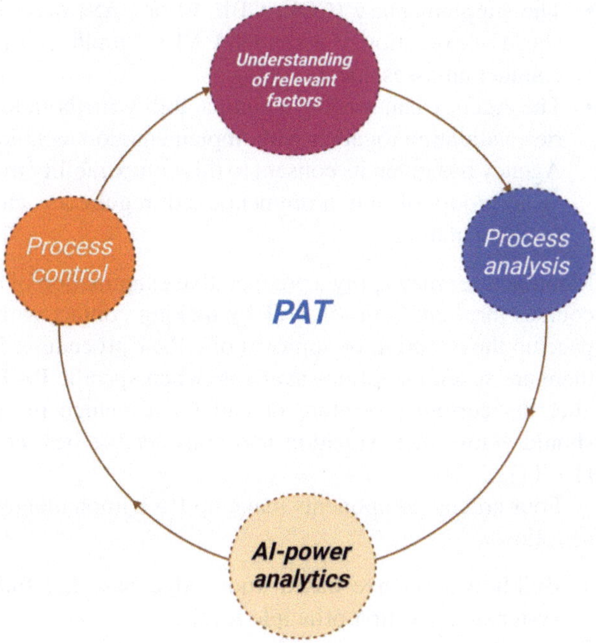

Fig. 3.1 Outline of PAT process [5, 6]

3.1.2 Implementation of PAT

In automotive, food and beverage, semiconductor, and petrochemical industries, PAT is process knowledge and control based concept that has been in use for a while. These sectors of industry have always seen investments made to raise quality and lower costs. Use of PAT tools can be advantageous in three different situations, such as, processes with lengthy periods of processing, inadequate efficiency, or excessive waste; novel products whereby PAT can offer additional data for development as well as scale-up; practices that need to be sufficiently amended to contend with generic drug producers one the period of market exclusively ends. However, it needs to be remembered that every PAT area is not suitable for each product or business. PAT is impracticable for low-strength items as well as small businesses are probably unable to afford it. PAT, however, makes strong commercial sense for pharmaceutical businesses that produce numerous batches every year in specific locations or on specific lines [8, 9]. PAT implementation plan should, in general, be risk-based. Following potential implementation strategies are being suggested by the agency, if necessary:

- The resource's inherent quality system may be used to adopt PAT. PAT implemented may be preceded or followed by cGMP reviews conducted by Team from PAT or a PAT licenced investigator.

- The supplementary (CBE, CBE-30 or PAS) may be presented to the authority ahead of execution, if essential, PAT Committee or PAT trained investigator may conduct an assessment.
- The Agency can receive a comparability methodology summarizing PAT inquiries, validation together with implementation techniques, and timelines. After the Agency has given its consent to this comparability methodology, either individual or a group of the aforementioned regulatory channels may be chosen for implementation

Manufacturer may apply a preoperative examination of a PAT developed facility and courses through Team of PAT by making contact with FDA's PAT Team so as to pace up the adoption or approval of a PAT procedure. It needs to be mentioned that there are several solutions available when specific PAT implementation plans don't alter the current procedure or call for a change in specifications. Manufacturers should assess their situation and consider the best course of action with Agency [10, 11].

Four crucial components make up PAT implementation. These can be described as follows:

1. Building a science based knowledge base i.e. fully grasping all processes at systematic and first principle levels.
2. Process monitoring including control: retrieving of CPPs and CQAs and assortment of analysis, measurement along with control means to modify processes to produce anticipated product quality.
3. PAT system's validation
4. Regulatory methods [12, 13]

3.2 Background

3.2.1 Framework and Tools of PAT

PAT strives to enhance understanding and control of production processes, which is consistent with present quality system: Products cannot be assessed for quality; instead, it should be incorporated into the design. Designing and creating clear processes that reliably guarantee a specific quality at conclusion of production is one of the anticipated outcomes of PAT framework. Such practices would be in-line with the fundamental principles of QbD, and they might lessen quality hazards and regulatory concerns while increasing efficiency. The framework for PAT described under headings: Process understanding, Principles & Tools, Risk-based approach, Integrated Systems Approach, and also Real-time Release (Fig. 3.2) [10].

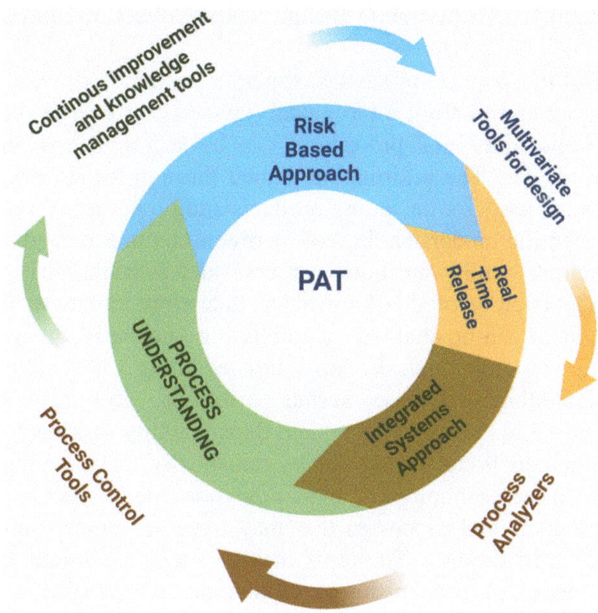

Fig. 3.2 Illustration of PAT framework and tools [5, 14]

3.2.2 *Process Understanding*

A process in general regarded as being favourably understood when all significant variables have been labelled and rationalized, variability has been controlled by process. Product quality attributes can be predicted with high accuracy and reliability over the design space established for the materials used, the process parameters, and manufacturing, environmental, and other conditions. Prediction skills demonstrate a deep level of process knowledge. Even while retrospective process capability statistics show a level of control, they may not be enough to measure or convey process understanding. It is possible to gain more knowledge and understanding of processes allowing process advancement, optimisation, technology transfer, and scale-up through systematized product along with process advancement on a small scale [10, 15].

3.2.3 *Tools for Implementing PAT*

Using a number of methodologies, it is feasible to comprehend all aspects of the development, production, and quality control of pharmaceuticals in a way that is scientifically sound. When integrated into a system, such tools could offer quick and easy ways to collect data that will aid in process understanding, on-going improvement, and the development of risk-reduction strategies. These tools fall into the following categories within PAT frameworks:

3.2.3.1 Multivariate Design, Data Collection, and Analysis Tools

Pharmaceutical processes together with products are intricate multi-factorial programmes from a biological, physical, or chemical standpoint. Finding the best formulations and procedures can be accomplished via a variety of development strategies. The information gained through these development programmes serves as the basis for designing products and processes. When a knowledge base includes scientific understanding of pertinent multidimensional exchanges (for instance, among quality attributes, process and formulations), as well as a way to judge applicability of this knowledge in various contexts, it can be most helpful. This benefit can be attained by combining knowledge management systems with multivariate mathematical approaches including statistical experimental design, process modelling, response surface methods, and pattern identification algorithms (Fig. 3.3). The usefulness and dependability of knowledge contained in mathematical models can be assessed by statistically analysing model predictions. With the aid of these technologies, it is now possible to recognise and evaluate aspects of products and processes that may have an impact on effectiveness and calibre of finished product. The tools may be used to estimate the effects of various failure modes and mechanisms on product quality [5, 10, 11].

3.2.3.2 Process Analysers

Major improvements in process analysis have been backed by industrial drivers of productivity, quality, and environment impact. Tools are now available that assess biological, chemical, and physical properties in addition to univariate process metrics like pH, pressure, and temperature. It's true that some process analysers offer non-destructive tests with data on physical, chemical, and biological characteristics of materials being processed. These analyses could consist of:

At-line: involve obtaining sample, isolating it from process' flow, and interpreting it there.

On-line: sample is acquired from process stream, and then may be added back in.

In-line: measurements that maintain sample moving through the process without being disruptive or non-disruptive.

In PAT, batch records should include scientific and procedural information that demonstrates good process quality and product conformance. For quality control and real-time manufacturing control, accessibility to such data ought to be simple. Installed information technology systems ought to facilitate such functions. It is not necessary for the measurements obtained from these process analysers to be absolute values of the relevant property. When different quality elements along with additional process facts are used to generate the process, flexible process can be constructed and justified. Process control will benefit from the capacity measure relative variations in materials both before and during processing. It is possible to

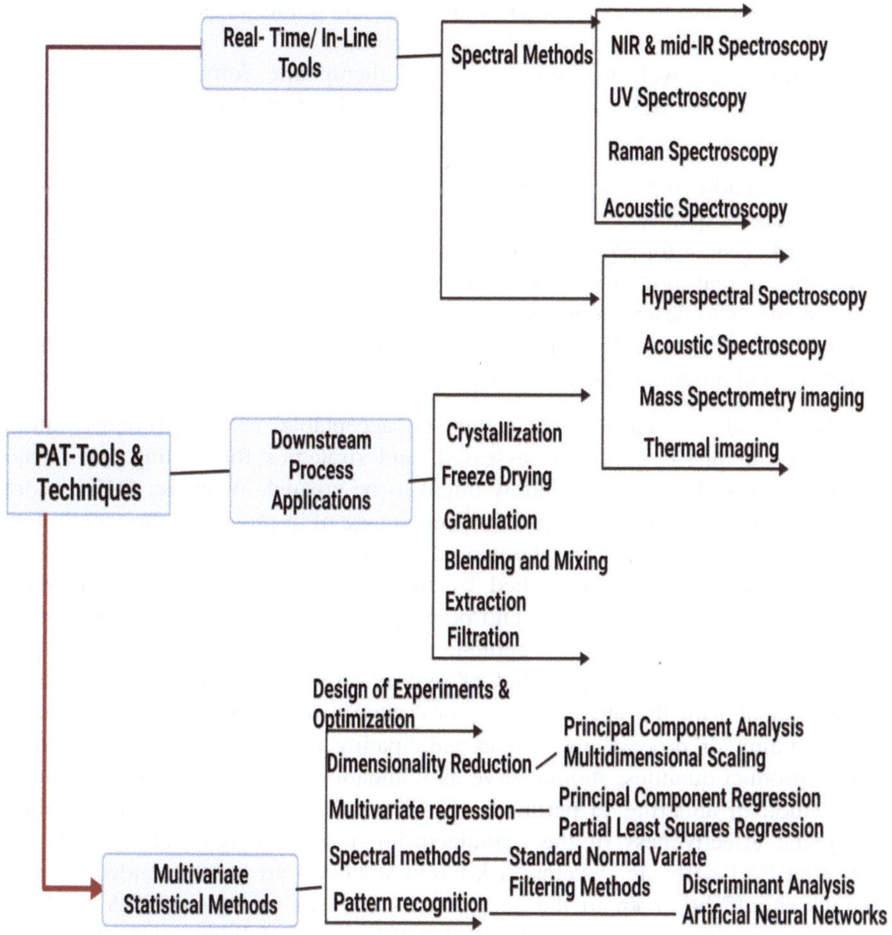

Fig. 3.3 Depiction of tools and techniques of PAT [14, 16]

create a flexible process to handle the variety of materials being processed. However, to extract crucial process knowledge for in-the—moment control and QA, multivariate techniques are frequently required. To ensure that this installation won't negatively impact the process or product quality, risk analysis should be performed before installing process analysers on current production equipment [17–19].

3.2.3.3 Process Control Tools

Process control and surveillance techniques aim to keep an eye on process's status and actively intervene to sustain the anticipated state. Strategies should consider input material qualities, ability and dependability of process analysers to assess critical features, and the achievement of process end points so as to sustain uniform standards for resulting elements and finished product.

The design as well as optimization of therapeutic formulations including manufacturing processes can be done using following techniques within the framework of PAT:

1. Ascertain and compute chief material as well as manufacturing attributes affecting product's quality.
2. Establish system for process assessment that enables monitoring of all important qualities in real-time or almost real-time.
3. Adjustments to design controls that guarantee monitoring all crucial aspects.
4. Establish mathematical connections among essential material including process metrics and product quality features.

The likelihood of achieving a range of acceptable process timings during manufacturing phase should be assessed, and strategies for dealing with major departures from the process window ought to be created. Wherever PAT encompasses complete manufacturing process, percentage of in-process components along with end products tested throughout production may be far higher in contrast to whatever is presently accomplished by means of laboratory evaluation. When creating acceptance standards for end point qualities that take measurement along with sampling procedures into account, rigorous statistical principles should be applied. Realising the full potential of real-time measurements may be possible and beneficial with the aid of process envisaging multivariate statistical control. Understanding of process, as well as anticipating and management of pertinent process/product qualities, should serve as foundation for quality decisions. This is one method of adhering to pertinent cGMP standards, as these control methods certify the effectiveness of the manufacturing process (21CFR211.110(a)). For validating PAT software systems, risk-based strategies are recommended. Considerations should be given to the suggestions made by additional FDA guideline documents, for instance, General Principles of Software Validation [10, 20, 21].

3.2.3.4 Continuous Improvement and Knowledge Management Tools

To promote on-going learning, data should be obtained and assessed over a product's life cycle. These data can be used to prove that proposals for post-approval changes are legitimate. Aside from being useful for the manufacturers, processes together with IT systems allowing acquirement of knowledge from such databases can also make it easier for scientists to communicate with the Agency. The technical knowledge of pertinent multi-faceted interactions and method to assess how useful this knowledge is in various contexts are the two components of a knowledge base that can be of the most use.

Any one of these, or complete production course comprising quality assurance, may benefit from application of a suitable combination of some, all, or, none of these tools [22, 23].

3.2.4 Risk-Based Approach

The risk of producing an inferior product inside a well-established quality system and for a particular manufacturing system would be expected to be inversely related to the level of process understanding. Possibilities are there to create simpler regulatory procedures to manage change for well-understood processes. As a result, placing an emphasis on process awareness can help with risk-based regulatory decisions and innovation [2, 10, 24].

3.2.5 Real-Time Release

It refers to the capacity to quantify and ensure acceptable quality of final product or in-process product established on process information. Element of PAT for real-time release typically consists of reliable set of evaluated material properties with process controls. Analytical techniques that focus on direct or indirect processes can be used to evaluate material properties (Fig. 3.3). Material characteristics as well as process factors are monitored and regulated during real-time release. Real-time release of items that are part of market appliances or licences should be done with the Agency's consent. Real-time release complies with 21 CFR 211.165 [10, 25].

3.2.6 Integrated System Approach

In order to evaluate and implement effective tools and systems that meet demands of patients and business in a timely manner; integrated systems thinking is required due to rapid speed of innovation in modern era. The development, manufacturing, QA, and information management roles are becoming increasingly intertwined as a result of recent developments, and it is important that each region be synchronized in an holistic way [22, 26, 27].

3.3 Assessment of Process Needs

3.3.1 Importance of Selecting the Right Design for PAT

The term "PAT" refers to a wide range of tools that are usually mentioned as an essential constituent of additional cutting-edge initiatives, including QbD, continuous manufacturing together with real-time release [28]. Up until approximately the time the FDA guidance was published, the pharmaceutical industry had been sluggish to adopt and fully utilise the variety of technologies and methodologies that support PAT. Because of its reliance on proven unit operations, it may be argued that the pharmaceutical business entered the scene late. PAT's objective is to comprehend and manage the manufacturing process, which is congruent with our current drug quality system. Quality cannot be tested into products; it should be integrated or purposefully created. Pharmaceuticals are made with quality in mind by thoroughly comprehending:

- The patient group, the administrative route, pharmacological, toxicological, and pharmacokinetic properties of a medicine, as well as its intended therapeutic goals.
- Properties of a medicine that are chemical, physical, and biopharmaceutical.
- Choosing a product's design, elements, packaging centred on aforesaid qualities.
- Material science, engineering, and QA ideologies are worked in the design of manufacturing processes to assure acceptable and repeatable product quality and performance for the course of a product's shelf life.

By incorporating quality into goods, this guidance emphasises the need for process comprehension and prospects for increasing manufacturing efficiency through innovation and improved scientific communication between manufacturers and the Agency. More attention may be given to the important multi-factorial interactions between the materials used, the manufacturing process, environmental factors, and their influence on quality as a result of the increased emphasis on incorporating quality into products. This sharpened focus serves as a foundation for finding and comprehending linkages between various crucial formulation and process aspects as well as for creating efficient risk mitigation plans.

Pharmaceutical production is still among the last to mostly use batch methods, which are ineffective and can lead to drug shortages due to their lengthy lead times or quality issues. Because of the many benefits of continuous manufacturing, pharmaceutical businesses are moving away from outmoded batch lines [2, 29–31]. Reduced process variability is the main goal of PAT. Reduced product variability should result from the integration of on-line measurement and/or modelling of essential quality attributes with automated feedback control of the process parameters affecting these attributes. Consequently, lowering product variability will lower the chance of introducing off-spec products into market. A real quality control strategy, where the process is continuously modified to preserve product quality, is being pushed in place of the traditional quality assurance technique. Eli Lilly's tenfold increase in throughput made possible by bringing HPLC assays online serves as an example of how cycle time may be decreased and process efficiency can be raised by getting rid of the delays brought on by off-line assays. With PAT in place throughout the process development and product production life cycle, it should be considerably easier to identify critical-to-quality process parameters and their ideal operating ranges. Genentech mentioned this as a benefit of using online ion chromatography to test amino acids, even if it would be challenging to quantify [32].

Continuous manufacturing has been acknowledged as a key invention with the potential to enhance resilience, agility, and robustness in the production of polymeric thin films and solid oral dosage forms. The development of continuous manufacturing is also being applied to the fabrication of solid oral dosage forms after production of pharmaceutical active components. FDA is aware that "lack of agility, robustness, and flexibility in the pharmaceutical manufacturing sector poses a potential public health threat as failures within manufacturing facilities that result in poor product quality can lead to drug shortages". Pharmaceutical companies are increasingly looking into the implementation of continuous processes because it could speed up the time it takes for a product to reach the market by removing the need for process scale up. A process could be optimized with designed tests in just a few days, saving both time and materials. Numerous investigations have been made to fully comprehend the feeding and mixing processes in continuous manufacturing [33, 34].

3.3.2 Evolution of PAT

Designing and creating clear processes that reliably guarantee a specific level of quality at the conclusion of the manufacturing process is one of the anticipated outcomes of the PAT framework. Such practices could lower quality hazards and regulatory concerns while increasing efficiency and would be in line with the fundamental principles of quality by design. Gains in efficiency, safety, and/or quality will depend on the procedure and the final product and are most likely to result from:

- By means of on/in/at-line valuations along with controls to shorten production cycle time preventing reprocessing, trash, and rejection.
- Real-time release
- Increase operator security and decrease manual mistakes, automation should be increased.
- Enhancing capacity, improving material and energy utilization.
- Enabling continuous processing to manage unpredictability and increase efficiency.

For decades, methods have been used in the production of API, which were formerly thought to be a component of process analytical chemistry (PAC). The traditional functional zone for PAC in production of API has been process control, which involves factors like pressure, temperature, moisture, solvent drying, redox probes, pH, and spectroscopic techniques like Raman, NIR, IR, UV, etc. However, chemometrics is being utilised in process modelling, control, and supervising. For the goal of process comprehension in development, workstations, automated reactors, and synthesis robots for programmed process optimisation using statistical experimental design along with multivariate data analysis are currently in use. PAT goes much beyond PAC methods, and process analysers. "Transfer of laboratory methods to on-line /in-line /at-line methods may not always be PAT", according to guidance of FDA on PAT. When researching the history of PAT philosophy, one can find examples such as the usage of online analysers, which are used to impose measurement results from which one can draw conclusions about the quality of the product. Through this method, one is in a position to make early process changes to guarantee the required product quality will be achieved [9, 13].

PAT programme, which was launched a few year ago by FDA, contains many of the instruments for which applications in the production and development of APIs have been found. But this endeavour is mostly focussed on the production of medicinal products, not just API. The sophistication of analytical techniques has reached a point where they can now provide data in a meaningful manner while navigating the complex matrix various dose forms. For determining moisture, radio frequency, microwaves, near-infrared spectroscopy (NIR), and other technologies have been used [9]. Various instruments under PAT, there diverse applications, and specific attributes analysis are summarized in Tables 3.1, 3.2, and 3.3, respectively.

3.3.3 Define the Scope and Objectives of the PAT Application

To design, assess, and manage manufacturing processes, the chemical and pharmaceutical industries use a variety of methods and technologies together stated as PAT. Physical, chemical, microbiological, risk analysis, and mathematical that is carried out in an integrated manner are all considered to fall under the broad definition of "analytical" in PAT. To assure that predefined critical quality attributes (CQAs) of ultimate product are regularly met, critical process parameters (CPPs) and key performance indicators (KPIs) are prudently specified, watched continually as well as absolutely figured out. PAT is a vital constituent of the QbD philosophy, implying that quality ought to be ingrained from beginning and not just tested in the finished product [39–41]. Basic goal in incorporating PAT into flow analysis is to transform continuous-flow chemistry from a forefront of science to a different manufacturing infrastructure for use in industrialized processes. The objectives of PAT can be expressed with the help of Fig. 3.4.

PAT recommends features for in-, at-, on-, and off-line assessment. In-line alongside online techniques impart incessant real-time tracking of processes that bring about easier implementation of in-process regulators at vital units of operation rather than relying on manual sampling and end-product testing [42]. The goal of PAT is to manufacture pharmaceuticals by employing most modern risk management and quality systems approaches while ensuring product quality. In PAT, utilisation of most recent technological and scientific advancements in pharmaceutical manufacture is encouraged (Fig. 3.5). For PAT, the Agency's inspection and submission review programmes function in a coordinated and synergistic way. The Agency and the manufacturer apply laws and production requirements consistently. The risk-based approach used by the agency to manage risk promotes innovation in the pharmaceutical manufacturing industry [5].

PAT emphasises the need for process awareness and prospects for improving production efficiencies through innovation and improved scientific communication between manufacturers and the Agency using the strategy of building quality into

Table 3.1 An overview of PAT's major instruments [5, 14, 24, 35]

S. no.	Technology and instrument used	Operational procedures	Applications in PAT
1	Near Infra-Red (NIR)	The transmission, reflectance, and transflectance of a sample are used to obtain NIR spectral information, enabling the measurement process to be adapted to the sample and avoiding the need to prepare the sample before analysis	Preparation of pharmaceutical products and identification of raw materials. Physical parameters are determined. Moisture measurement. A pharmaceutical ingredient is determined by its active pharmaceutical ingredient (API).
2	Process Raman Spectroscopy	The Raman scattering photons are detected by using a monochromatic light source, such as a laser. In a Raman spectrum, stokes portions are generally on the left, while anti-stokes portions are on the right, if plotted on a wavelength axis. The relative intensity difference between stokes and anti-stokes, however, distinguishes them from each other	There is no need to prepare, condition, or destroy samples for this test. A variety of sample forms can be tested, such as solids, liquids, gases, or any combination between them. The sample size is not restricted by its nature. When laser wavelength, laser power, and optics are selected appropriately, macro- and micro-samples can be measured. As a result, it produces spectra that have attractive spectral properties
3	Process Infra-Red	A spectrograph based on diffraction gratings is used to analyse the infrared beam coming from the sample. An aperture above the surface of a rotating encoder disk enables the dispersed radiation from the grating to be imaged. One detector generates a discrete interferogram by bringing the reflected beams to a common image point	In addition to reaction monitoring, dryer monitoring-vapor blending control, composition monitoring, and raw material screening quality control, it is also used for quality control of raw materials screening.
4	On-Line UV-Vis Analysis	US-Vis colour measurements are based on transmittance and reflectance data. Liquid colour is measured with transmittance spectra, while powders and opaque liquids like paints are measured with reflectance spectra	Dissolution testing. Liquid analysis-vessel cleaning.
5	NIR Chemical Imaging	Two-dimensional optical detectors are used for chemical imaging to reproduce spatial information from a scene. A wide range of wavelengths is used to collect grayscale or colour images	Measuring content uniformity in high-throughput systems, Detecting counterfeits, measurement of blend homogeneity

(continued)

Table 3.1 (continued)

S. no.	Technology and instrument used	Operational procedures	Applications in PAT
6	Process MS	Multiple ions, fragmentation, isotopic abundances, etc. in mass spectra lead to a more complex spectrum that provides useful information about unknown chemicals in the spectrum	It is easy to multiplex (up to 64 streams per analyser) and provides a wide dynamic range, which makes it useful for process diagnostics.

Table 3.2 Industrial applications of PAT tools [26, 36, 37]

S. no.	Analysing technique	Applications	Outcomes
1.	NMR spectroscopy	The presence of organic matter in waste-water is analysed.	Increase in time for analysis, increase in productivity.
2.	NIRS	A simultaneous quality control and raw material detection process.	Increase in recognition time, cost effective.
3.	In-line NIRS	An example of a catalysis reaction is that which transforms acetone into methyl isobutyl ketone.	Impacts the production and selectivity of methyl isobutyl ketone.
4.	Raman spectroscopy, attenuated total reflectance (ATR) and FTIR	Analysing crystal polymorphs and solute concentration at the same time.	A quantitative description of the effects of reactant addition on crystal polymorphic states.
5.	DIA	Characteristics of particle reduction.	It can be applied to milling processes.
6.	PSD analyser or SFT	An analysis of particle size distributions.	The size and characteristics of particles are quantified and qualitatively described.
7.	XRD, FTIR	Identification of APIs, drugs, or excipients.	During intermediate reactions, polymorphs can be classified qualitatively.
8.	HPLC, RP-HPLC	Analysing developed products for bioactives.	In a dosage form, what are the chances of quantifying raw materials or APIs.
9.	MRI, CLSM	Topographies near and on the surface.	Provides improved clarity of surface alteration in complexes and blends.

goods. A greater importance on product quality enables for more attention to be paid to the important multi-factorial interactions between materials used, the manufacturing process, environmental factors, and the influence these have on quality. This sharpened focus serves as a foundation for recognising and comprehending the connections between many crucial formulation and process aspects as well as for creating efficient risk mitigation plans. Preformulation programmes, development

Table 3.3 Application of PAT in processes with specific attributes [5, 21, 38]

PAT	Process	Attribute analysed	On/in/off-line
FT-IR with ATR probe	Pharmaceutical salt formation process	End point monitoring	In line
Image probe (CCD camera and high energy XE lighting system)	High shear granulation	Particle shape and size	In line
Raman spectroscopy	Compression	API analysis in tablets	Off-line
NIR spectroscopy-transmission	Compression	Quantification of active ingredient	Off-line
Temperature sensor and increase	Granulation	Granulation end point	In line
NIR spectroscopy-reflectance	Raw material	Active ingredient quantification	Off-line
NIR spectroscopy-reflectance	Packing line	Identification	On-line
NIR spectroscopy-reflectance	Granulation	End point of wet granulation	On-line
NIR spectroscopy-reflectance	Packing component identification	Blister PVC-films identification	Off-line
NIR spectroscopy-reflectance	Compression-capsules and tablets	Assay and content uniformity	Off-line
NIR spectroscopy-reflectance	Powder	Moisture content	Off-line

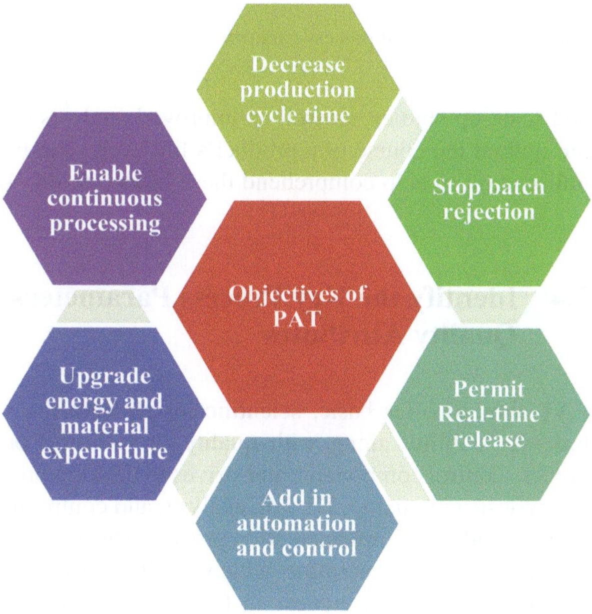

Fig. 3.4 Overview of objectives of PAT [2, 36]

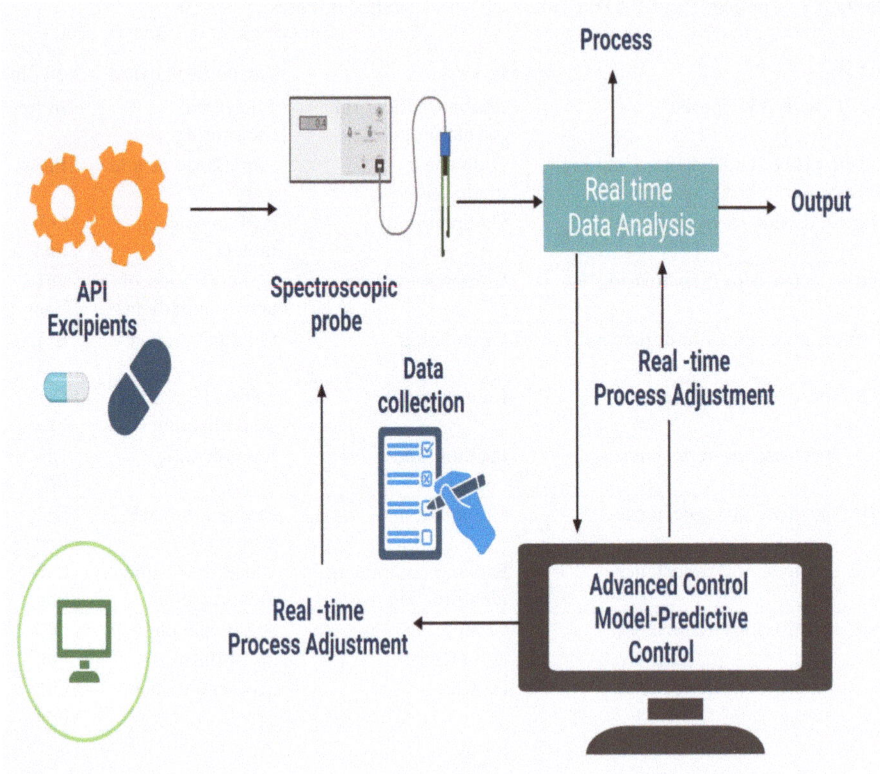

Fig. 3.5 Outlining the process parameters and assessment tools through PAT [1, 43]

and scale-up studies, as well as improved analysis of manufacturing data gathered throughout the course of a product's life cycle, can be used to harness the data and information to help comprehend these links [10, 44, 45].

3.4 Identify the Key Process Parameters and Critical Quality Attributes

PAT aims to offer basic, scientific insight into the variables that are essential to process' stability along with quality of resulting product. PAT alludes to QbD process control components and strives to develop procedures that generate products with consistent quality. Design, analyse, and control are the three steps that make up a successful PAT implementation process. During the design phase, experimentation is carried out to determine which quality attributes are relevant to certain unit operation and which process factors and raw material attributes have the greatest influence. Using this information, it is then possible to pinpoint the qualities, process variables, and raw material characteristics that should be taken into account when designing an efficient PAT-based control scheme for the process. In the analyse step, a suitable analytical tool is chosen to examine the chosen quality characteristic, as well as the selected process parameters and the attributes of raw materials, if necessary. Finally, the control stage entails assimilating the analysis results into a suitable control scheme to determine what to do in the event that the process' performance deviates from the ideal course [46].

In-line (exclusive of taking off sample), on-line (sample gets removed from central process, examined, and potentially be returned back), at-line (by taking away sample while carefully analysing process), along with off-line (taking away sample and analysing independently of the process) are the four different types of processes that PAT can be used to measure and analyse effectively. While at-line measures are discovered to be drawn-out and time-consuming, online or in-line measuring techniques give the quickest results [26, 36, 47].

Several practices, involving magnetic resonance imaging (MRI), light and electron microscopy, positron emission tomography (PET), confocal laser scanning microscopy, Raman spectroscopy (RS), NIRS, and laser prompted break-down spectroscopy, have been exercised to perceive various characteristics including quality of film coating [36, 48, 49]. Each methodology has been determined to be important and offers essential information on the steps taken during coating, but each technique has its own limitations. In order to tackle multiple layer coating problems with a single measurement, many of these procedures or methods have been discovered to be damaging. As a result, a suitable multivariate analysis model may be needed for data analysis [37, 50].

By efficiently controlling the potential CPPs, such as time, pH, and temperature, CQAs, such as product yield, should be regulated so as to increase quality in product. Additionally, PAT instruments are used at four crucial steps throughout biopharmaceutical processes. Online detection, omics-centered analysis, and CQAs assessment are the three strategies used in early phase to gather quality related information. The information gathered in first phase is examine utilising assessment or any additional particular vigorous channel in the second step. With the use of preprogrammed control devices, the third step of procedure is controlled. The final phase involves making human decisions about whether to discharge the batch or not depending on data generated to create a precise set of product standards limitations. A suitable multivariate assessment model may be needed for analysis because several among these procedures are proven to be damaging as well as incapable of resolving multiple layer coating with just one measurement [26, 27, 51]. Due to their ability to control manufacturing process, PAT appliances have also been identified by manufacturers as an important and promising tool in the production of injectable [26]. The density and viscosity of solvents used to prepare injectables were assessed using the acoustic resonance spectroscopy method [52]. Drugs or medicinal substances' moisture content and dissolution profiles were found using the calorimetry approach [53]. The ratio of salt or conjugates of salt in solvent media expended in injectable production was assessed using dielectric spectroscopy [54]. For measuring lysine clonixinate in injectable, NIRS technique was successfully devised

[55]. Amount of baicalin, luteoloside, chlorogenic acid, caffeic acid, ursodesoxycholic acid located in a traditional Chinese medicine were significantly quantified using methods like FT-NIRS and fractional least-square regression. Another study employed HPLC method to determine concentration of 5-fluoruracil [52].

PAT minimised batch loss, met essential standards with just minimal variations, enhanced product safety issues for hot melt extrusion. The importance of in- and on-line NIRS processing for hot melt extrusion was originally proven by Apruzzese et al., in 2003. Using a VIS-NIRS sensor coupled to an extruder to detect the colour dye, the residence time distribution was determined in their work [56]. Another work employed in-line NIRS to evaluate content of drug together with plasticizer simultaneously in a mix of polymer met copolymer, with various opacity. During HME process, NIRS is utilised to quantity in-line process constraints such as concentration of API, polymer-drug conjugate behaviour, and molecular interfaces. NIRS was used as a tool to monitor HME operations and demonstrated its value as a PAT tool for keeping an eye on key elements of HME process as a whole. In order to assess the various configurations of polymer-polymer composited, the in-line RS for HME used a number of probes, including Raman probes that were enclosed with the extruder die section of a specific screw extruder [57–59].

Focussed beam reflectance measurement (FBRM), spatial filtering technique (SFT), and dynamic image analysis (DIA) are among non-disruptive PAT techniques to quantify particle-size scattering [60, 61]. Main benefits of the non-intrusive procedures were that they didn't disrupt on-going practices and endorsed real-time process monitoring and regulation. These methods also did away with the extra phases for formulating sample required by traditional off-line methods. Because the typical stream procedures were greatly focussed with flow rates, which caused the detectors to get entangled, a lateral stream of particles was frequently needed in these techniques to become separated. Instantaneous observations using laser diffraction to do correct analysis and modify the classifier mill's uneven feeding rate [62, 63]. Similar to this, Chan et al. created at-line along with in-line laser diffraction setups then used them to determine the PSD of microspheres. According to the findings, at-line processes were suggested as being more appropriate and reliable for assessing particle size because, when using in-line laser diffraction techniques, a significant amount of agglomeration disintegration was seen. In a further investigation, the researchers compared the behaviour of the decrease of particle size in conical mill using different DIA modes, embracing on-line, at-line, and in-line [64]. Additionally, feedback regulating practices could be linked to conical mill concurrently having DIA sensors for improved comprehension together with control of the dry milling process in the pharmaceutical sectors [65, 66].

Manufacturing related crystallization process is seen as an undesirable phase, hence it is crucial to purpose, optimize, and regulate discreet phases in order to control crystallization. There have been several PAT appliances with methods reported that can be used to monitor as well as control crystallization process in pharmaceutical companies, to illuminate crystalline components, in addition to better understand the processes involved in order to ensure product safety and higher

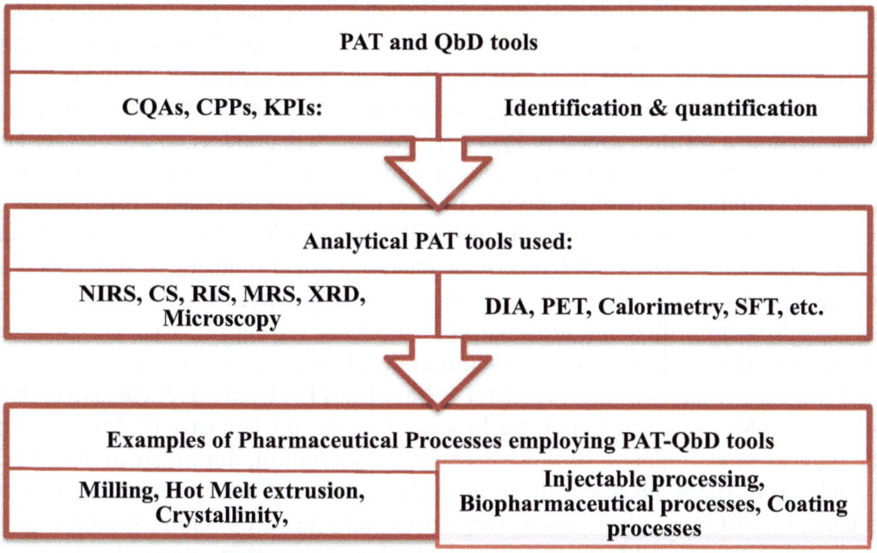

Fig. 3.6 Description of QbD combined PAT tools for identification of CQAs and CPPs for various pharmaceutical processes [69]

quality [67, 68]. Innumerable novel PAT practices that can be exercised to satisfy regulatory plans should efficiently supervise and control CPPs, like polymorphic change along with phase transformations, throughout the manufacturing processes, supervise and regulate CQAs, like particle size, recognise API's ultimate products by expending spectroscopic tools or XRD, compute polymorphs, along with crystallinity (Fig. 3.6) [6, 15].

3.5 Understand the Regulatory Requirements and Guidelines

Centre for Drug Evaluation and Research (CDER) of FDA created programme with goal of enticing businesses to implement contemporary PAT in pharmaceuticals production as well as quality control. PAT's effort was subsequently conferred for the very first time in public at an FDA Advisory Committee for Pharmaceutical Sciences (ACPS) gathering in July of 2001 that also included business along with scientific representatives [9]. A subcommittee of ACPS was established by the FDA to investigate online pharmaceutical surveillance technologies. The subcommittee is made up of around 30 professionals from business and academia, including FDA. The experts' expertise ranged from production skills and development to academics, chemometrics, pharmaceutics, chemical engineering, and information technology. The committee's second job was to consider broad regulatory applications of PAT principles, such as those governing method validations and specifications together with validation of chemometric tools, as well as the viability of employing parametric release. The third assignment required the group to implement PAT and consider PAT rules. Before execution, a supplement may be presented to the agency, and when needed, a certified group of PAT investigators will conduct an inspection. The agency can receive a comparable protocol summarising PAT research, implementation, and validation techniques along with timelines. FDA introduced programme in August of 2002 called "Pharmaceutical cGMP for the Twenty-First Century: A Risk-based approach". FDA's PAT effort for pharmaceutical industry is anticipated to increase production productivity and drug product quality, as well as decrease or perhaps do away with some finished product release testing [2]. Ajaz Hussain, the Deputy Director, CDER's Office (Pharmaceutical Science), who serves as head of PAT steering group, was driving factor behind FDA'PAT initiative. He also collaborated with Mel Koch of Centre for Process Analytical Chemistry (CPAC), one of the pioneer in assimilation of process analysis and control, during creation of PAT guidelines. In 1980s, an academic group at University of Washington developed idea to launch CPAC, an industry-university cooperative research centre, similar to this, FDA and Pfizer agreed to work together on chemical imaging research when they signed a "Cooperative Research and Development Agreement" in 2003 [9].

The pharmaceutical sector created in US had potential to exercise PAT- allied process control practices; nevertheless they were unsuccessful in doing so since it was expensive in terms of equipment costs, costs to develop quality assurance systems along with costs to validate all systems. Although PAT had improved production efficiency rate, this might not be viewed as a business perspective as a convincing reason for modification as a result of fact that implementation expenses might recompense turnover for investments in some cases, especially for micro pharmaceutical enterprises that manufacture drug products that were previously in competition with constricted boundaries. Industry experts are now very concerned about how the regulatory organisations will react to examination of a specific equipment services, thus the regulatory agencies have also been able to overcome their role as a significant contributor to the decline in PAT implementations. The lack of international regulatory prospectus relating to PAT instruments that are unified is a key concern among the industry. To provide a unified approach to PAT, FDA has participated in worldwide symposiums. The ICH has a special role to play in enabling PAT framework's implementation due to its organisational goal of harmonising regulatory approaches and regulations in the pharmaceutical industries. The ICH recommendations Q7 to Q11 exposed the real-world challenges for all other parties-regulatory, professional, academia, business, and pharmaceutical firms-to collaborate and create PAT methodologies, technologies, and implementations [8, 70].

3.6 Assessment of Process Needs

3.6.1 Review the Existing Process and Analytical Methods

The core of PAT entails process comprehension together with real-time scrutinising of raw materials or in-process product features in order to restrain process by modifying process related parameters and guarantees quality of the finished product. At designated assessment points and based on certain decision criteria, it is decided whether to move on to the next process step or alters a parameter. With the aim of effectively accomplishing anticipated material qualities of end product, this approach shifts product quality assurance moves from rigid controls (for example, batch testing of finished product) to adaptive controls (supervised as well as controlled during manufacturing process) [21]. The conventional approaches to pharmaceutical production have numerous shortcomings that obstruct innovations, on-going development, QbD, and productivity gains. These flaws not only cause pharmaceutical items to be expensive, but they also have an impact on our customers' health. In conventional samples were taken at regular intervals and processed in batches before being analysed in the lab. Even when personnel consistently adhere to an established formula with pre-set set points, the end products vary because the raw ingredients are changeable and the batch operation has its own variability in equipment behaviour. To guarantee the calibre of end result, this approach needed a system approaches used [3]. The main issues with contemporary manufacturing model are inconsistent quality of final products; emphasis placed more on compliance than on application of science, and high rates of beyond specifications along with rejections in conventional manufacturing, which result in pricey production. Additionally, a lot of samples are taken from final products, intermediate products, and raw materials, which extends the testing process and delays batch release. Processes used in production nowadays validate external elements and provide ensured products with predestined quality. The advantages of applying PAT include improved processes, and safer products [71].

3.6.2 Identify the Gaps or Limitations in the Current System

Implementing PAT involves a number of difficulties, both technical and organizational-cultural. A corporation must alter its development, manufacturing, and quality. Applying PAT poses many difficulties, for instance, integration of technology, expensive starting price, need for specialized analysers, changes in corporate culture, sophisticated chemometric models, system integration, PAT data administration, development of PAT systems, and requirement for regulatory approval. The costs associated with implementing PAT cover the purchase of new technology, adjustments to current infrastructure, ongoing maintenance, and employee training. To achieve a successful deployment without operational impediments, corporate culture is essential. PAT increases organizational effectiveness and promotes the creation of secure and long-lasting procedures, which quickens the pipeline for new products. PAT needs to be viewed as a top-down strategic objective from the standpoint of the company culture. This means that for PAT to be implemented successfully and without operational obstacles, senior management, technical team, along with production people must cooperate it [72, 73]. The majority of PAT might not be stability indicating, thus conventional assessment for stability will still be required. Despite FDA's PAT proposal has not yet valid to biotechnology goods, it is conceivable that its success in drug manufacture will have an impact on the production of biologics. As they work to gain greater insight of their progressions, as well as to increase quality as well as manufacturing efficiency, makers of biotechnology products might take a look at a number of technologies [2].

3.6.3 Define the Specific Benefits and Applications of Implementing PAT

PAT aims to offer basic, scientific insight into the variables that are essential to a process' stability and the quality of resulting output. With aim of enhancing quality of final product alongside dropping manufacturing costs, PAT is used, which greatly benefits the pharmaceutical industry. PAT provides immediate analysis of CQAs. PAT's benefits include cost cutting and improved manufacturing process understanding and control (Fig. 3.7). Scientific principles along with PAT tools that may be exercised to better understand in addition control processes are provided in the FDA guidance. In addition to advancing manufacturing efficiency and strengthening the scientific foundation for generating regulatory criteria, PAT is anticipated to maintain or advance product quality assurance already in place. A risk-related assessment procedure that concentrates on problems that represent most threat to safety alongside quality will be implemented as a result of PAT, which will also provide consistency to facility inspections. PAT also facilitates statistical process control encompassing raw materials through in-process quality control of finished product. It also helps to design quality into processes. Additionally, PAT takes into account statistical variances, which are rarely used in development of novel procedures and products in pharmaceutical industry [3]. With aforementioned objectives PAT has been implemented in diverse pharmaceutical sectors, the applications of PAT in industries can be understood with the help of Fig. 3.8.

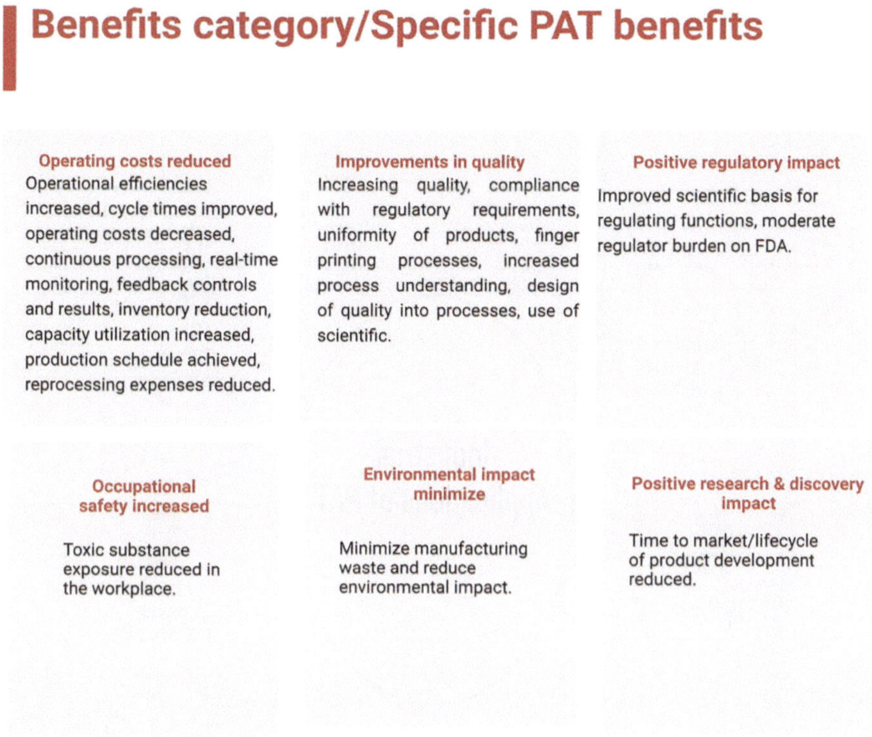

Fig. 3.7 Overview of specific benefits of PAT [74, 75]

3.7 Selection Criteria

3.7.1 Reliability and Accuracy

Some of the crucial components for validation activities include consistency, reproducibility, and reliability to make sure the product satisfies the release specification and instils quality into it. It might be difficult to validate new technologies for monitoring and controlling the crystallization process scientifically; in that scenario, first-principle modelling might offer a different strategy. Accuracy along with precision is key component of a chromatographic test procedure. Accuracy is degree to which experimental values closely approaches actual concentration of chemical in matrix. Measures of precision used to determine how well two measurements match. An analytical method's accuracy is impacted by a number of things. These can typically be improved throughout the technique advancement and optimisation stage of study and range from extraction efficiency to analyte stability to adequate chromatographic separation. Assumptions established during assay setup and executions are crucial but frequently disregarded aspects that affect accuracy. Purity of the reference materials worked to isolate analyte, build calibration curve, further

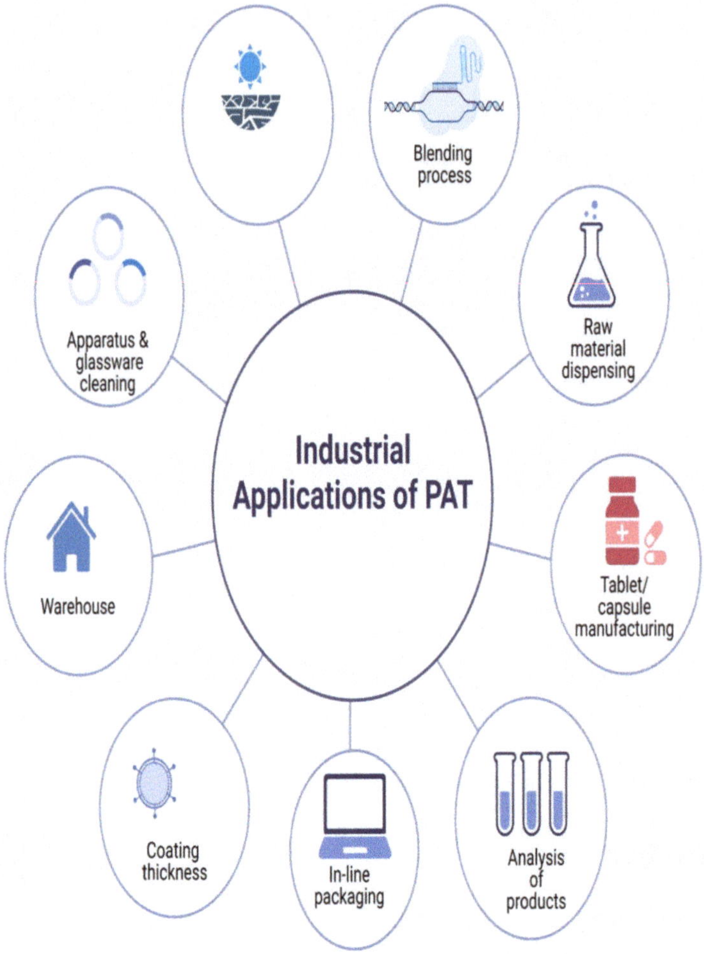

Fig. 3.8 Describing the wide applications of PAT [36, 76, 77]

produce a quantitative analytical result is the subject of the first assumption. Accuracy can also be impacted by the reliability of the references. Precision can be decreased by using improper or inaccurate tools, for instance, pipettes or analytical balances, by failing to manage light or moisture whenever necessary, or by using analysts who aren't properly trained. As data management systems strive to complete assimilations against uneven baselines, incompetent chromatographic resolution, descending peaks, and challenges to measure many analytes round an extensive dynamic range could also reduce precision. The issue becomes more serious when analysing complex natural products for both low and high quantities of analytes at the same time. In multi-laboratory investigations, the absence of homogeneity between parts might provide apparent imprecision. This may be appropriate for UV detection if the target compound's particular extinction coefficients are analogous; less analogous they are, the less accurate the results. Concern has been raised about the precision and dependability of multivariate calibration models for processes that are affected by fluctuations in both process variables and physical features. Such fluctuations may have a no-linear impact on spectra, resulting in subpar calibration performance and perhaps subpar closed loop process management. As a result, it becomes more difficult to extract from spectroscopic observations the pertinent chemical information and, ultimately, a trustworthy understanding of the process needed for process modelling, control, and optimisation. For PAT/ QbD including closed loop process control to be fully implemented, vigorous as well as transferable calibration models are required [6, 78]. A suitable calibration model built on log-polynomial expansion was designed to decouple impacts of solute concentration and also temperature on dielectric constant for the given solution so as to increase accuracy in cloud point estimation employing measurements of dielectric constant. The solute concentration profile was created by converting dielectric constant profile using calibration model. Turbidity measurements demonstrate that cloud points calculated from solute concentration profile are more accurate than those calculated straight from dielectric constant profile [79]

3.7.2 Scalability, Flexibility and Cost-Effectiveness of PAT

Manufacturing offers a unique chance to get current information for prompt process rectification decisions owing to simplicity of doing process analytical measurements. Without requiring complex process control interfaces up front, at-line solutions deliver near-time information. When process decisions can be made during a defined operation, such as batch processing applications or unit activities, this technique works effectively. Measuring raw materials, mixing premix batches or monitoring quality check points before sending final goods to the finished limits are effective at-line approaches. Although the 'realised cost reductions' are frequently used as rationale for implementing PAT, other PAT benefits have been found that are frequently much more crucial to the bottom line. This category contains the knowledge acquired about products and processes that, in the end, can increase an organisation's quality plans efficacy while lowering costs. Early in the product development, employment of automated reaction vessels in conjunction with in-situ monitoring techniques, statistical and chemometric modelling, as well as other factors, can offer valuable insight and understanding of chemistry of final product as well as manufacturing process. A greater understanding of chemistry before scaling up and product launch is achieved, which could ultimately decrease time to market new products. The additional information obtained when PAT tools are employed during scaled-up activities, such as in a pilot-plant, is frequently utilised to optimise the production of items in greater quantities (Fig. 3.9).

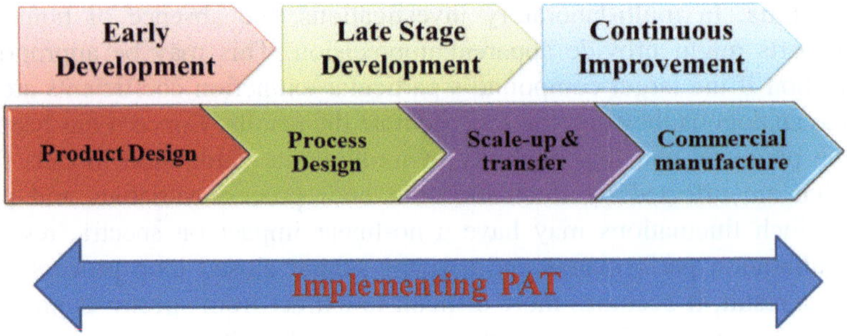

Fig. 3.9 Role of PAT in scalability [8, 76, 80, 81]

The industries fundamental components include projects' expenses, benefits, risk assessments, and techniques for assessing the project's performance in relation to its return on investment. PAT doesn't immediately improve effectiveness of manufacturing process. Benefits of PAT rely on how well it is integrated into lean manufacturing and the quality system. Costs associated with QC laboratory testing are frequently reduced as a result of the project, and the real-time process monitoring may hasten product release. PAT is similarly predicted on the idea that the likelihood of generating a subpar product diminishes as process expertise grows. PAT could be used to ensure proper blending or prevent over drying of APIs. PAT can therefore aid in the decrease of issues with quality. As a result, PAT helps to reduce manufacturing process costs and expedite product release by doing away with post-production testing. It also cuts down on the amount of energy needed to run the machinery and the cycle time. The entire cost of labour is decreased by automating the procedures [76, 79, 82–87].

3.8 Discussion

PAT can be worked and used in the pharmaceutical industry to improve the quality, safety, and stability of products while delivering superior benefits in terms of usage and economic values. By describing raw materials' physicochemical characteristics and distinct elements crucial for manufacturing procedures, PAT has also contributed to a better knowledge of raw materials. Overall, PAT aid in the development of more reliable processes, higher-quality goods, improved process control, and time savings, all of which contribute to the creation of a distinctive brand image for organization. Objective of PAT ought to implement reliable techniques that are adaptable enough to provide a specific amount of inconsistency in CQAs by modifying contexts of processes. These resources might be effectively employed in a variety of unit operations for industries beyond pharmaceutical. Although there are significant restrictions and difficulties in their implementations, these issues can be resolved with the aid of rules established by regulating bodies, FDA. PAT makes an effort to change approach pharmaceutical drugs are developed from an artistic to scientific method. In order to communicate with agency using quantitative approach for medication processing and specification setting activities, this will require practice of more modern process design concepts and risk management tools [2, 6, 76, 80]. PAT's appropriateness and applicability will undoubtedly aid researchers and producers in the near future in reducing undesirable difficulties and improving product quality.

References

1. Kim, E. J., Kim, J. H., Kim, M.-S., Jeong, S. H., & Choi, D. H. (2021). Process analytical technology tools for monitoring pharmaceutical unit operations: A control strategy for continuous process verification. *Pharmaceutics, 13*, 919.
2. Munson, J., Freeman Stanfield, C., & Gujral, B. (2006). A review of process analytical technology (PAT) in the US pharmaceutical industry. *Current Pharmaceutical Analysis, 2*, 405–414.
3. Scott, B., & Wilcock, A. (2008). Process analytical technology and validation. In *Validation of pharmaceutical processes* (3rd ed.). Taylor & Francis.
4. Wu, H., & Khan, M. A. (2010). Quality-by-design (QbD): An integrated process analytical technology (PAT) approach for real-time monitoring and mapping the state of a pharmaceutical coprecipitation process. *Journal of Pharmaceutical Sciences, 99*, 1516–1534.
5. Mali, A. S., Jagtap, M., Karekar, P., & Maruška, A. (2016). A brief review on process analytical technology (PAT). *International Journal of Current Pharmaceutical Research, 8*, 10–15.
6. Simon, L. L., Pataki, H., Marosi, G. R., Meemken, F., Hungerbühler, K., Baiker, A., Tummala, S., Glennon, B., Kuentz, M., & Steele, G. (2015). Assessment of recent process analytical technology (PAT) trends: A multiauthor review. *Organic Process Research & Development, 19*, 3–62.
7. Chanda, A., Daly, A. M., Foley, D. A., Lapack, M. A., Mukherjee, S., Orr, J. D., Reid, G. L., III, Thompson, D. R., & Ward, H. W. (2015). Industry perspectives on process analytical technology: Tools and applications in API development. *Organic Process Research & Development, 19*, 63–83.
8. Chew, W., & Sharratt, P. (2010). Trends in process analytical technology. *Analytical Methods, 2*, 1412–1438.
9. Hinz, D. C. (2006). Process analytical technologies in the pharmaceutical industry: The FDA's PAT initiative. *Analytical and Bioanalytical Chemistry, 384*, 1036–1042.
10. Food & Drug Administration. (2004). *Guidance for industry, PAT—A framework for innovative pharmaceutical development, manufacturing and quality assurance*. Retrieved from http://www.fda.gov/cder/guidance/published.html
11. Watts, C. (2004). *PAT—A framework for innovative pharmaceutical development manufacturing and quality assurance*. FDA/RPSGB Guidance Workshop.
12. Casian, T., Nagy, B., Kovács, B., Galata, D. L., Hirsch, E., & Farkas, A. (2022). Challenges and opportunities of implementing data fusion in process analytical technology—A review. *Molecules, 27*, 4846.
13. Pomerantsev, A. L., & Rodionova, O. Y. (2012). Process analytical technology: A critical view of the chemometricians. *Journal of Chemometrics, 26*, 299–310.
14. Misra, N., Sullivan, C., & Cullen, P. (2015). Process analytical technology (PAT) and multivariate methods for downstream processes. *Current Biochemical Engineering, 2*, 4–16.
15. Wu, H., Dong, Z., Li, H., & Khan, M. (2015). An integrated process analytical technology (PAT) approach for pharmaceutical crystallization process understanding to ensure product

quality and safety: FDA scientist's perspective. *Organic Process Research & Development, 19*, 89–101.
16. Gao, Y., Zhang, T., Ma, Y., Xue, F., Gao, Z., Hou, B., & Gong, J. (2021). Application of PAT-based feedback control approaches in pharmaceutical crystallization. *Crystals, 11*, 221.
17. Korasa, K., & Vrečer, F. (2018). Overview of PAT process analysers applicable in monitoring of film coating unit operations for manufacturing of solid oral dosage forms. *European Journal of Pharmaceutical Sciences, 111*, 278–292.
18. Lundsberg-Nielsen, L., Schlindwein, W. S., & Berghaus, A. (2018). Process analytical technology (PAT). In *Pharmaceutical quality by design: A practical approach* (pp. 227–255). Wiley.
19. Methfessel, J., & Roy, J.-R. (2006). A practical approach to PAT implementation. *Pharmaceutical Technology Europe, 18*, 24–31.
20. Mercier, S. M., Diepenbroek, B., Wijffels, R. H., & Streefland, M. (2014). Multivariate PAT solutions for biopharmaceutical cultivation: Current progress and limitations. *Trends in Biotechnology, 32*, 329–336.
21. Read, E., Park, J., Shah, R., Riley, B., Brorson, K., & Rathore, A. (2010). Process analytical technology (PAT) for biopharmaceutical products: Part I. Concepts and applications. *Biotechnology and Bioengineering, 105*, 276–284.
22. Clark, G. (2004). FDA's PAT initiative. Pharmaceutical Technology Europe, 16.
23. Scott, B., & Wilcock, A. (2006). Process analytical technology in the pharmaceutical industry: A toolkit for continuous improvement. *PDA Journal of Pharmaceutical Science and Technology, 60*, 17–53.
24. Rathore, A. S., Bhambure, R., & Ghare, V. (2010). Process analytical technology (PAT) for biopharmaceutical products. *Analytical and Bioanalytical Chemistry, 398*, 137–154.
25. Drakulich, A. (2011). Real time release testing. *Pharmaceutical Technology, 35*, 42–49.
26. Glassey, J., Gernaey, K. V., Clemens, C., Schulz, T. W., Oliveira, R., Striedner, G., & Mandenius, C. F. (2011). Process analytical technology (PAT) for biopharmaceuticals. *Biotechnology Journal, 6*, 369–377.
27. Wasalathanthri, D. P., Rehmann, M. S., Song, Y., Gu, Y., Mi, L., Shao, C., Chemmalil, L., Lee, J., Ghose, S., & Borys, M. C. (2020). Technology outlook for real-time quality attribute and process parameter monitoring in biopharmaceutical development—A review. *Biotechnology and Bioengineering, 117*, 3182–3198.
28. Clegg, I. (2020). *Process analytical technology. Specification of drug substances and products*. Elsevier.
29. Aksu, B., De Beer, T., Folestad, S., Ketolainen, J., Lindén, H., Lopes, J. A., De Matas, M., Oostra, W., Rantanen, J., & Weimer, M. (2012). Strategic funding priorities in the pharmaceutical sciences allied to quality by design (QbD) and process analytical technology (PAT). *European Journal of Pharmaceutical Sciences, 47*, 402–405.
30. Hu, C. (2021). Reactor design and selection for effective continuous manufacturing of pharmaceuticals. *Journal of Flow Chemistry, 11*, 243–263.
31. Mandenius, C.-F., Derelöv, M., Detterfelt, J., & Björkman, M. (2007). PAT and design science. *European Pharmaceutical Review, 12*, 74–80.
32. Cooley, R. (2007). On-line process control: Automating the control of process-scale purification columns using on-line liquid chromatography. *BioPharm International, 2007*.
33. Pauli, V., Roggo, Y., Pellegatti, L., Trung, N. Q. N., Elbaz, F., Ensslin, S., Kleinebudde, P., & Krumme, M. (2019). Process analytical technology for continuous manufacturing tableting processing: A case study. *Journal of Pharmaceutical and Biomedical Analysis, 162*, 101–111.
34. Vargas, J. M., Nielsen, S., Cárdenas, V., Gonzalez, A., Aymat, E. Y., Almodovar, E., Classe, G., Colón, Y., Sanchez, E., & Romañach, R. J. (2018). Process analytical technology in continuous manufacturing of a commercial pharmaceutical product. *International Journal of Pharmaceutics, 538*, 167–178.

35. Murphy, T., O'Mahony, N., Panduru, K., Riordan, D., & Walsh, J. (2016). Pharmaceutical manufacturing and the quality by design (QBD), process analytical technology (PAT) approach. In *27th Irish Signals and Systems Conference (ISSC), 2016* (pp. 1–7). IEEE.
36. Das, S. S., Alkahtani, S., Nayak, A. K., & Hasnain, M. S. (2021). Process analytical technology (PAT) tools: Uses in pharmaceutical manufacturing. In *Advances and challenges in pharmaceutical technology*. Elsevier.
37. Knop, K., & Kleinebudde, P. (2013). PAT-tools for process control in pharmaceutical film coating applications. *International Journal of Pharmaceutics, 457*, 527–536.
38. Pérez-Beltrán, C. H., Jiménez-Carvelo, A. M., Torrente-López, A., Navas, N. A., & Cuadros-Rodríguez, L. (2023). QbD/PAT—State of the art of multivariate methodologies in food and food-related biotech industries. *Food Engineering Reviews, 15*, 24–40.
39. Marvaniya, H. M., Vanparia, D. J., Chauhan, R., & Sen, D. J. (2010). Process analytical technology (PAT): An initiative objective for pharmaceutical C-GMP for the 21st century. *Research Journal of Science and Technology, 2*, 129–133.
40. Narang, A. S., & Dubey, A. (Eds.). (2023). *Continuous pharmaceutical processing and process analytical technology*. CRC Press.
41. Ralbovsky, N. M., & Smith, J. P. (2022). Process analytical technology and its recent applications for asymmetric synthesis. *Talanta, 252*, 123787.
42. Fonteyne, M., Vercruysse, J., De Leersnyder, F., Van Snick, B., Vervaet, C., Remon, J. P., & De Beer, T. (2015). Process analytical technology for continuous manufacturing of solid-dosage forms. *Trends in Analytical Chemistry, 67*, 159–166.
43. Chun, M.-H., Kim, J. Y., Park, E.-S., & Choi, D. H. (2021). Development of a robust control strategy for fixed-dose combination bilayer tablets with integrated quality by design, statistical, and process analytical technology approach. *Pharmaceutics, 13*, 1443.
44. Afnan, A. (2004). *PAT—A framework for innovative pharmaceutical development, manufacturing and quality assurance*. Guidance for Industry.
45. FDA. (2004). *Guidance for industry: PAT—A framework for innovative pharmaceutical development, manufacturing, and quality assurance*. FDA.
46. Rathore, A. S. (2014). QbD/PAT for bioprocessing: Moving from theory to implementation. *Current Opinion in Chemical Engineering, 6*, 1–8.
47. Ündey, C., Ertunç, S., Mistretta, T., & Looze, B. (2010). Applied advanced process analytics in biopharmaceutical manufacturing: Challenges and prospects in real-time monitoring and control. *Journal of Process Control, 20*, 1009–1018.
48. Metz, H., & Mäder, K. (2008). Benchtop-NMR and MRI—A new analytical tool in drug delivery research. *International Journal of Pharmaceutics, 364*, 170–175.
49. Richardson, J. C., Bowtell, R. W., Mäder, K., & Melia, C. D. (2005). Pharmaceutical applications of magnetic resonance imaging (MRI). *Advanced Drug Delivery Reviews, 57*, 1191–1209.
50. Seo, K.-S., Bajracharya, R., Lee, S. H., & Han, H.-K. (2020). Pharmaceutical application of tablet film coating. *Pharmaceutics, 12*, 853.
51. Streefland, M., Martens, D. E., Beuvery, E. C., & Wijffels, R. H. (2013). Process analytical technology (PAT) tools for the cultivation step in biopharmaceutical production. *Engineering in Life Sciences, 13*, 212–223.
52. Henriques, J., Sousa, J., Veiga, F., Cardoso, C., & Vitorino, C. (2019). Process analytical technologies and injectable drug products: Is there a future? *International Journal of Pharmaceutics, 554*, 21–35.
53. Marabi, A., Mayor, G., Raemy, A., Bauwens, I., Claude, J., Burbidge, A. S., Wallach, R., & Saguy, I. S. (2007). Solution calorimetry: A novel perspective into the dissolution process of food powders. *Food Research International, 40*, 1286–1298.
54. Fagan, C. C., Everard, C., O'Donnell, C. P., Downey, G., & O'Callaghan, D. J. (2005). Prediction of inorganic salt and moisture content of process cheese using dielectric spectroscopy. *International Journal of Food Properties, 8*, 543–557.

55. Lopez-Arellano, R., Santander-García, E., Andrade-Garda, J., Alvarez-Avila, G., Garduno-Rosas, J., & Morales-Hipolito, E. (2009). Quantification of lysine clonixinate in intravenous injections by NIR spectroscopy. *Vibrational Spectroscopy, 51*, 255–262.
56. Apruzzese, F., Pato, J., Balke, S., & Diosady, L. (2003). In-line measurement of residence time distribution in a co-rotating twin-screw extruder. *Food Research International, 36*, 461–467.
57. Hitzer, P., Bäuerle, T., Drieschner, T., Ostertag, E., Paulsen, K., Van Lishaut, H., Lorenz, G., & Rebner, K. (2017). Process analytical techniques for hot-melt extrusion and their application to amorphous solid dispersions. *Analytical and Bioanalytical Chemistry, 409*, 4321–4333.
58. Maniruzzaman, M., & Nokhodchi, A. (2017). Continuous manufacturing via hot-melt extrusion and scale up: Regulatory matters. *Drug Discovery Today, 22*, 340–351.
59. Saerens, L., Dierickx, L., Lenain, B., Vervaet, C., Remon, J. P., & De Beer, T. (2011). Raman spectroscopy for the in-line polymer–drug quantification and solid state characterization during a pharmaceutical hot-melt extrusion process. *European Journal of Pharmaceutics and Biopharmaceutics, 77*, 158–163.
60. Silva, A. F., Burggraeve, A., Denon, Q., Van Der Meeren, P., Sandler, N., Van Den Kerkhof, T., Hellings, M., Vervaet, C., Remon, J. P., & Lopes, J. A. (2013). Particle sizing measurements in pharmaceutical applications: Comparison of in-process methods versus off-line methods. *European Journal of Pharmaceutics and Biopharmaceutics, 85*, 1006–1018.
61. Tok, A. T., Goh, X., Ng, W. K., & Tan, R. B. (2008). Monitoring granulation rate processes using three PAT tools in a pilot-scale fluidized bed. *AAPS PharmSciTech, 9*, 1083–1091.
62. Harvill, T. L., & Holve, D. J. (1993). In-process particle size distribution measurements and control. *Particle & Particle Systems Characterization, 10*, 262–265.
63. Harvill, T. L., Hoog, J. H., & Holve, D. J. (1995). In-process particle size distribution measurements and control. *Particle & Particle Systems Characterization, 12*, 309–313.
64. Chan, L., Tan, L., & Heng, P. W. (2008). Process analytical technology: Application to particle sizing in spray drying. *AAPS PharmSciTech, 9*, 259–266.
65. Nalluri, V. R. (2011). *Novel process analytical technological approaches of dynamic image analysis for pharmaceutical dry particulate systems*. University of Basel.
66. Nalluri, V. R., Schirg, P., Gao, X., Virdis, A., Imanidis, G., & Kuentz, M. (2010). Different modes of dynamic image analysis in monitoring of pharmaceutical dry milling process. *International Journal of Pharmaceutics, 391*, 107–114.
67. Samad, N. A. F. A., Sin, G., Gernaey, K. V., & Gani, R. (2013). A systematic framework for design of process monitoring and control (PAT) systems for crystallization processes. *Computers & Chemical Engineering, 54*, 8–23.
68. Wu, H., & Hussain, A. S. (2005). Use of PAT for active pharmaceutical ingredient crystallization process control. *IFAC Proceedings Volumes, 38*, 147–152.
69. Yu, L. X., Amidon, G., Khan, M. A., Hoag, S. W., Polli, J., Raju, G., & Woodcock, J. (2014). Understanding pharmaceutical quality by design. *The AAPS Journal, 16*, 771–783.
70. Price, G. A., Mallik, D., & Organ, M. G. (2017). Process analytical tools for flow analysis: A perspective. *Journal of Flow Chemistry, 7*, 82–86.
71. Layloff, T. (2004). An overview of process analytical technology in the pharmaceutical industry. *American Pharmaceutical Review, 7*, 30–34.
72. Nagy, B., Galata, D. L., Farkas, A., & Nagy, Z. K. (2022). Application of artificial neural networks in the process analytical technology of pharmaceutical manufacturing—A review. *The AAPS Journal, 24*, 74.
73. Owczarek, D. (2021). *Process analytical technology (PAT) and artificial intelligence: The expanding impact of ML in drug manufacturing*. Retrieved June 20, 2023 for future prospects of AI in PAT, from https://nexocode.com/blog/posts/process-analytical-technology-pat-with-ai-in-biopharmaceutical-manufacturing/
74. Cullen, P. J., O'Donnell, C. P., & Fagan, C. C. (2014). Benefits and challenges of adopting PAT for the food industry. In *Process analytical technology for the food industry* (pp. 1–5). Springer.
75. Rathore, A., Jesubalan, N., & Thakur, G. (2022). Perspectives on process analytical technology. *BioPharm International, 35*(5), 31–35.

76. Cooley, R. (2010). *What industry needs does Process Analytical Technology meet?* Retrieved June 23, 2023, from https://www.europeanpharmaceuticalreview.com/article/3643/process-analytical-technology-pharma-industry/
77. Lawrence, X. Y., Lionberger, R. A., Raw, A. S., D'Costa, R., Wu, H., & Hussain, A. S. (2004). Applications of process analytical technology to crystallization processes. *Advanced Drug Delivery Reviews, 56*, 349–369.
78. Betz, J. M., Brown, P. N., & Roman, M. C. (2011). Accuracy, precision, and reliability of chemical measurements in natural products research. *Fitoterapia, 82*, 44–52.
79. Hermanto, M. W., He, G., Tjahjono, M., Chow, P. S., Tan, R. B., & Garland, M. (2011). Calibration of dielectric constant measurements to improve the detection of cloud and clear points in solution crystallization. *Chemical Engineering Research and Design, 89*, 2613–2619.
80. Bakeev, K. A. (2010). *Process analytical technology: Spectroscopic tools and implementation strategies for the chemical and pharmaceutical industries*. John Wiley & Sons.
81. Ganguly, J., & Vogel, G. (2006). Process analytical technology (PAT) and scalable automation for bioprocess control and monitoring—a case study. *Pharmaceutical Engineering, 26*, 8.
82. Das, P., Mishra, S. R., & Nayak, B. S. (2017). Process analytical technique (PAT): An integral part of pharmaceutical process automation. *International Journal of Pharmaceutics and Drug Analysis, 5*, 31–35.
83. Fontalvo-Lascano, M., Méndez-Piñero, M., & Romañach, R. (2020a). Design and development of a cost model for the implementation of process analytical technology in the pharmaceutical industry. In *9th Annual World Conference of the Society for Industrial and Systems Engineering* (pp. 17–18).
84. Fontalvo-Lascano, M. A., Méndez-Piñero, M. I., & Romañach, R. J. (2020b). Development of a business case model for process analytical technology implementation in the pharmaceutical industry. In *Proceedings of the 5th NA International Conference on Industrial Engineering and Operations Management* (pp. 1144–1152). IEOM Society International.
85. Graham, L. (2019). *Empowering an effective PAT methodology*. Retrieved June 25, 2023, from https://www.isa.org/intech-home/2019/march-april/features/empowering-an-effective-pat-methodology
86. Peru, D. (2008). Pat-reducing the Cost of quality in consumer products manufacturing. *European Pharmaceutical Review, 13*, 84.
87. Vasava, S., Patel, D., Patel, G., & Meshram, D. (2021). QBD and PAT tools in drug product life cycle. *International Journal of Pharmaceutical Research and Applications, 6*, 818–827.

Chapter 4
Analytical Quality by Design

Falguni B Tandel, Madhusmruti Khandai, Pranav Shah, Dibya Sundar Panda, and Vipul P. Patel

Abstract QbD applied to analytical procedure development is termed as Analytical Quality by Design (AQbD). ICH Q14 encompasses the approach of AQbD. Appropriate information from analytical procedure risk assessment and development studies to support the proposed strategy of lifecycle management should be submitted in summarized form in the regulatory submission sections for analytical procedure validation. AQbD elements are Analytical Target Profile (ATP) or Quality Target Method Profile (QTMP), Risk Assessment, Critical Method Parameters (CMP), Critical Quality Attributes (CQA), Method Operable Design Region/Space (MODR) & Method Validation. AQbD being systematic approach develops analytical method by multivariate Design of Experiment (DoE) rather that tedious univariate One Factor at a time (OFAT) concept which gives better understanding of the method and robustness. MODR thus obtained is used to set control strategy of the analytical method which is further validated and proven as fit for use and continuously monitored for improvement. Analytical methods are key part of control strategy with reference to pharmaceutical quality system. Thus implementation of analytical QbD in manufacturing process as control strategy will ensure predetermined performance and product quality. The chapter describes the concept, importance, elements, systematic steps etc. of AQbD. The chapter in depth explains

F. B. Tandel
Indus Institute of Pharmacy & Research, Indus University, Ahmedabad, Gujarat, India

M. Khandai
Royal College of Pharmacy and Health Sciences, Berhampur, Odisha, India

P. Shah
Maliba Pharmacy College, Uka Tarsadia University, Bardoli, Gujarat, India

D. S. Panda
Department of Pharmaceutics, College of Pharmacy, Jouf University, Sakaka, Kingdom of Saudi Arabia

V. P. Patel (✉)
Sanjivani College of Pharmaceutical Education and Research (Autonomous), Kopargaon, Maharashtra, India

N. K. Jain, N. Bajwa (eds.), *Introduction to Quality by Design (QbD)*,
https://doi.org/10.1007/978-981-99-8034-5_4

implementation of AQbD with an insight in to the practical aspects giving examples of chromatographic method development.

Keywords Analytical quality by design · AQbD · DoE · Risk management · QbD

4.1 Introduction to AQBD

Analytical Quality by Design (AQbD) is the concept of QbD applied to analytical procedure development.[1] Since 2002, a lot has happened in Pharmaceutical QbD area resulting in issuance of various regulatory guidelines. Analytical method development and product development go parallel throughout the life cycle of Pharmaceutical product. The classical approach of analytical procedure development is based on OFAT (one factor at a time) concept which is very tedious and leads to method failure in many cases. Hence to smoothen the method development, AQbD approach has slowly been adopted by scientist.

4.1.1 QbD and AQbD

As per ICH Q8, Quality by Design (QbD) is a systematic approach to development that begins with predefined objectives and emphasizes product and process understanding and process control, based on sound science and quality risk management [2]. A systematic approach to development can include incorporation of prior knowledge, results of studies using design of experiments, use of quality risk management and use of knowledge management throughout the lifecycle of the product. The approach can enhance achieving the desired quality of the product and help the regulators to better understand a company's strategy. Product and process understanding can be updated with the knowledge gained over the product lifecycle.

QbD extended to Analytical method development is termed as AQbD, Analytical Quality by Design (Table 4.1).

Table 4.1 QbD vs. AQbD

QbD approach	AQbD approach
QTPP identification	ATP identification
Risk assessment, critical process parameters, critical material attributes and CQA identification	Risk assessment, CMP and CQA identification
Formula and process development and optimization with DoE	Method development and optimization with DoE
Design space	MODR
Control strategy and risk assessment	Control strategy and risk assessment
QbD formulation validation	AQbD method validation
Continuous process monitoring	Continuous method monitoring

AQbD is the systematic approach to development that begins with predefined objectives and emphasizes analytical method understanding and method control based on sound science and quality risk management as described in ICH Q9 with method lifecycle management and continual improvement as described in ICH Q10. AQbD ensures that the method is working as anticipated.

This AQbD approach leads to Improved understanding of the procedure and control of sources of variability linked to the intended use of the procedure as described in the analytical target profile (ATP). AQbD approach develops method that is more robust, resulting in fewer failures during use and during qualification in a new laboratory leading to reduction of overall resources required for a new or revised procedure, the levels of effort, formality and documentation.

AQbD gives flexibility to analytical method for movement within method operable design region (MODR). Out-of-trend (OOT) results and out-of-specification (OOS) results can be reduced when analytical method is developed by AQbD approach due to the robustness of the method within MODR.

Key Benefits adopting QbD and AQbD concepts are as follows:

- Facilitation of continuous improvement and technological innovation through more advanced regulatory approaches to change management
- More effective use of industry and regulatory resources
- Enhanced method robustness and ruggedness through the product life cycle
- Fewer investigations related to analytical-method performance
- Reduction in the costly and time consuming investigations.
- Reduced Out of specification (OOS) and Out of trend (OOT) results during analysis
- Elimination of batch failures due to analytical method variations.
- More robust method knowledge transfer as a result of enhanced analytical-method understanding as well as improved knowledge management (Fig. 4.1).

AQbD elements are Analytical Target Profile (ATP) or Quality Target Method Profile (QTMP), Risk Assessment, Critical Method Parameters (CMP), Critical Quality Attributes (CQA), Method Operable Design Region/Space (MODR) & Method Validation. ATP is based on and performance characteristics as per ICH Q2(R1) guideline and should be aligned with acceptance criteria of the method to meet the CQA of the method. Based on the analytical method, various procedural parameters are required to be studied to select best suited method which is done by risk assessment. Analytical method parameters are identified and assessed for risk associated with it using various risk assessment tools like Ishikawa diagram, Failure Mode Effect Analysis etc. Critical Method parameters selected based on medium/ high risk using risk assessment study. AQbD being systematic approach develops analytical method by multivariate Design of Experiment (DoE) rather that univariate One Factor at a time (OFAT) concept which gives better understanding of the method and robustness. MODR or design space is obtained for CMP which are used to set control strategy of the analytical method. Analytical method is further validated and proven as fit for use. During routine analysis, the method is continuously monitored for improvement.

Fig. 4.1 AQbD process

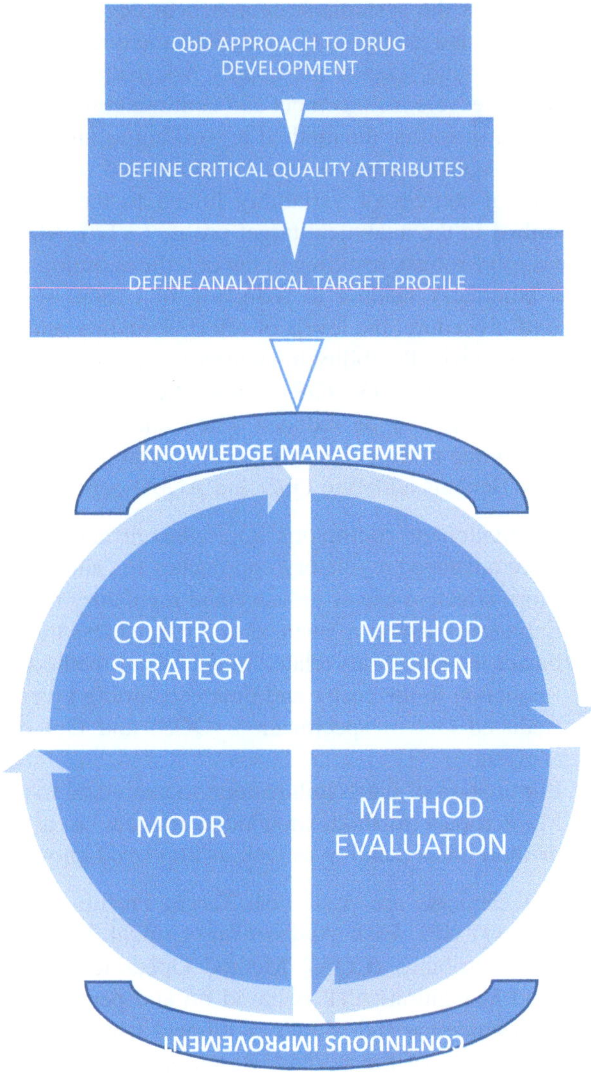

4.2 Traditional Approach Versus AQBD and Regulatory Considerations

The traditional approach is based on 'trial and error' for analytical method development. Traditional approach optimizes one factor at a time (OFAT) using prior knowledge. Traditional approach may not give analytical method with optimal conditions and may have robustness related issues. Traditional vs. AQbD approach is well explained by comparison as shown in Table 4.2.

Table 4.2 Traditional vs. AQbD approach

Aspect	Traditional approach	AQbD approach
Method development	Selection of analytical techniques and method conditions using empirical and OFAT approach	Method attributes to be measured are based on structured risk assessment. Method conditions are selected based on a systematic and multivariate strategy that is aligned with the end-user's preferences
Method validation	One time exercise as per ICH Q2 (R1) guideline for required performance characteristics	Start with the end in mind i.e. predefined analytical target profile, based on the intended used of the method. Identifies and considers all variables that have a potential to impact the method performance criteria and uses risk assessment tools to identify potentially critical and critical variables for study
Method transfer	One time exercise using limited numbers of operators/instruments that demonstrate the receiving site can run the method	Considered as a change control process where variables likely of change as a consequence of operating the method at the receiving site are risk assessed and, if appropriate, included in a measurement systems analysis study to assess conformance to the analytical target profile
Method operation	Relies on general system-suitability checks to indicate how well the method is operating. Occasional method failures and investigations into method performance as consequence of a typical, suspect, or out of specifications results	Robust methods with system suitability tests specifically derived from a formal assessment of the risk of method factor variation on the method performance criteria
Life-cycle management	Introduction of method changes or improvements are constrained by the costs of regulatory action, methods only changed if major issues are identified	Changes are made with reference to the accumulated knowledge generated during method development, validation, technology transfer and from performance data obtained during routine use of the method. Changes are made without the need for prior regulatory approval, including the introduction of more appropriate technology as long as the analytical target profile is met

Traditional approach, based on One Factor at a Time concept, utilises study of one method parameter at a time for the expected response whilst others remain constant hence study of effect of multifactor at a time cannot be studied resulting in narrow robust behaviour of the method. OFAT leads to high risk of method failure henceforth revalidation or alternative method development. AQbD is the systematic approach which sets goal of method to be achieved and identify critical method parameters through risk assessment. The interaction among critical method parameters affecting the responses are studied through DoE which gives method understanding and define the MODR within which method is flexible. Working within MODR is not considered as change and do not raise Post Approval Change. AQbD is a preferred and recommended approach for analytical method development to attain regulatory flexibility and reduce OOS, OOT, OOC, and OOSC results, high degree of robustness and cost effective analytical method [3].

4.2.1 AQBD and Regulatory Considerations

Analytical methods are key part of control strategy with reference to pharmaceutical quality system (ICH Q 10 guidelines) [4]. Thus implementation of analytical QbD in manufacturing process as control strategy will ensure predetermined performance and product quality. AQbD includes processing parameters and material/analytical attributes related to drug substance and drug product which includes facility, instrumental conditions, finished pharmaceutical product specification and the associated procedures and frequency.

Designing or building the quality into products is necessary in pharmaceutical development. A lot has happened in pharmaceutical QbD area since 2003 which results in issuance of many regulatory guidelines by regulatory agencies like USFDA, EMA, WHO, ICH. It formally started with release of a progress report by USFDA in 2003 [5]. Final report of the same was published in 2004. In 2004. USFDA published a guidance document introducing PAT in 2004 [6]. The European Medicines Agency: published Road Map to 2010 for the quality product development in 2005 [7].

Quality Risk Management (ICH Q9) was published in November, 2005, It states that quality risk management is a valuable component of an effective quality system. A proactive approach to quality risk management is beneficial, as it facilitates robust product design and continual improvement. The approach to quality risk management is very important to achieve an effective pharmaceutical quality system [8]. In same year, 2005, ICH published Q8 guideline: Pharmaceutical Development and then published "Annex to the Parent Guideline: Pharmaceutical Development" introducing Quality by Design concept for product development which includes Risk assessment as one of the element [2]. The guideline shows how parent Q8 document could be put into practice for all dosage forms. To enhance science- and risk-based regulatory approaches, quality by design and quality risk management are linked to an appropriate pharmaceutical quality system. Pharmaceutical development should include, at a minimum, the elements like Defining the quality target product profile (QTPP), Identifying potential critical quality attribute (CQAs) of the drug product and the drug substance, selecting an appropriate manufacturing process and defining a control strategy. While an enhanced QbD approach would additionally include a systematic evaluation, understanding and refining of the formulation and manufacturing process through identification of material attributes and process parameters that can have an effect on product CQAs and determination of relationship of them with CQAs; to establish an appropriate control strategy. Thus QbD approach facilitates continual improvement throughout the product lifecycle. Elements of QbD as per Q8 are QTPP, CQA, Risk Assessment which links Material Attributes and Process Parameters to Drug Product CQAs, Design Space, Control Strategy and Product Lifecycle Management and Continual Improvement.

USFDA published "Guidance for Industry Q8, Q9, and Q10 Questions and Answers (R4)" in May 2010 then revise in November 2010 and re revision on November, 2011 to clarify key issues faced after implementing the guidelines in the ICH regions. In December 2011, "ICH-Endorsed Guide for ICH Q8/Q9/Q10 Implementation" was published by ICH working group covering topics relevant to the implementation of ICH Q8(R2), Q9 and Q10, which supplement the existing Questions and Answers.

Guideline on Real Time Release Testing (RTRT) published by European Medicines Agency in March, 2012. RTRT recognises that under specific circumstances an appropriate combination of critical process parameters with pre-defined material attributes may provide greater assurance of product quality than end product testing and an integral part of the control strategy.

QbD implementation to Analytical development serves the same purpose as for formulation development. ICH published draft guideline "Q14 Analytical Procedure Development-step 2" encompassing the approach of AQbD in March, 2022 [9]. This guideline is intended to complement ICH Q2 Validation of Analytical Procedures. This guideline describes development of analytical procedures based on science and risk-based approaches suitable for the assessment of the quality of drug substances and drug products. The systematic approach suggested in ICH Q8 Pharmaceutical Development together with principles of ICH Q9 Quality Risk Management can also be applied to the analytical procedures development and lifecycle management. The guideline describes principles supporting change management of analytical methods which are based on risk management, adherence to predefined criteria for performance characteristics and comprehensive understanding of the analytical procedure using the tools described in ICH Q12 [10]. The guideline also describes submission of analytical procedure development and related lifecycle information in the Common Technical Document (CTD) format (ICH M4Q).

Together ICH Q14 and ICH Q2(R2) describe the development and validation activities suggested during the lifecycle of an analytical procedure used for the assessment of the quality of drug substances and drug products. ICH Q14 describes the scientific principles for development, change management and submission requirement of analytical procedures for the minimal and enhanced approach. ICH Q2(R2) provides guidance for establishing, submitting and maintaining evidence that an analytical procedure is fit for purpose (assuring drug quality). EMA published overview of the comments received on ICH Q14 in August, 2022.

Current approach of Analytical development includes, at a minimum, the elements like Identifying attributes of the drug substance or drug product which need to be tested by the analytical procedure, Selecting an appropriate analytical procedure technology and related instruments or suitable apparatus. Conducting appropriate development studies to evaluate analytical procedure performance characteristics and Defining an appropriate analytical procedure description including the analytical procedure 64 control strategy.

An enhanced approach offers a systematic way of developing and refining knowledge of an analytical procedure. In addition to the minimal approach, should include the following elements: (a) An evaluation of the sample properties and the expected variability of the sample, (b) Defining the analytical target profile, (c) Conducting risk assessment and evaluating prior knowledge to identify the analytical procedure parameters, (d) Conducting uni- or multi-variate experiments, (e) Defining an analytical procedure control strategy, and (f) Defining a lifecycle change management plan. AQbD approach can lead to better understanding of the impact of procedure parameters, lead to more robust method, more flexibility to life cycle management like wider operating range.

The analytical procedure description(s) should be included in the ICH M4Q CTD. Validation data (As per ICH Q2) and any supportive information needed to justify the analytical procedure control strategy should be included in the CTD for drug substance or drug product [9]. The criteria used in the validation study should be included in the submission. In some cases, depending on the intended use it may be appropriate to submit development data. Performance characteristics and acceptance criteria (as per ATP) and other elements of the enhanced approach (e.g., MODRs), should be described for analytical procedure description in the dossier sections. In the submission, Risk-based categorization of changes and corresponding reporting categories should be included. Appropriate information from analytical procedure risk assessment and development studies to support the proposed strategy of lifecycle management should be submitted in summarized form in the regulatory submission sections for analytical procedure validation.

4.3 AQBD and Risk Management

Risk is defined as severity harm and the combination of the probability of occurrence of that harm. The attributes that are important to the quality of the drug product are maintained and the product remains safe and effective when product quality is assured based on appropriate risk based decision-making throughout the product lifecycle. Quality risk management is a valuable component of an effective quality system [8]. An effective quality risk management approach provides a proactive means to identify and control potential risks to quality during development to ensure the quality of the drug product to the patient. Quality risk management includes an appropriate application of root cause analysis that can identify and address the root cause(s).

The tool of quality risk management is used in the method development to produce robust analytical procedure and reduce risk of poor method performance and reporting incorrect results. Risk assessment can be formal or informal and can be supported by prior knowledge. Risk assessment tools as described in ICH Q9 can be

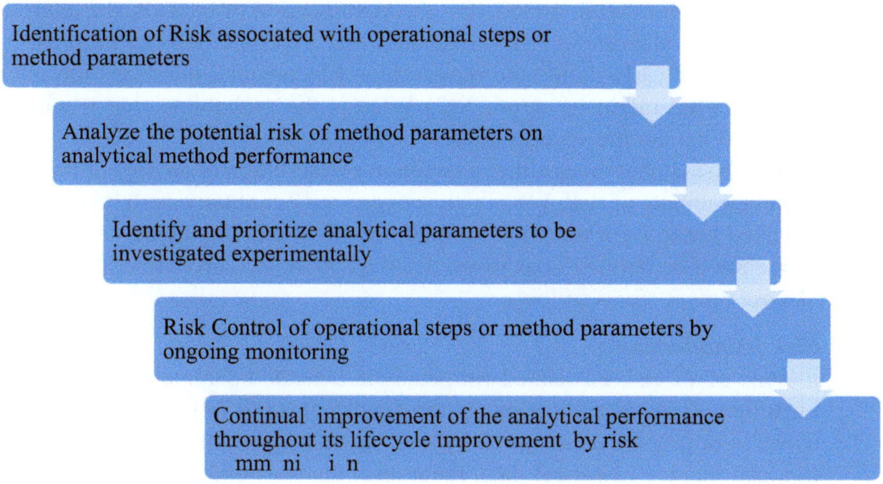

Fig. 4.2 Risk assessment process of analytical procedure development

used to: (a) identify analytical procedure parameters (factors and operational steps) termed as Critical method parameters (CMP) with potential impact on its performance. (b) Assess the potential impact of analytical procedure parameters on the analytical procedure performance. (c) Identify and prioritise analytical parameters to be investigated experimentally. Risk control can be used to establish an analytical procedure control strategy. As part of risk review, ongoing monitoring is recommended. To support continual improvement of the analytical procedure performance throughout its lifecycle, Risk communication should be used. When implementing changes to analytical procedures, QRM can be used to evaluate the impact of the changes. The risk assessment outcome informs the design and extent of the studies required to support the change which includes an appropriate bridging strategy to demonstrate that the changed procedure is fit for purpose.

Risk Management process of analytical procedure development is as follows (Fig. 4.2):

Risk Management is performed as per following steps:

1. Risk Assessment: It is performed in three steps as follows.
    (a) Risk Identification
    (b) Risk Analysis
    (c) Risk Evaluation
2. Risk Control
3. Risk Communication
4. Risk Review

1. ***Risk Assessment*:** Once Analytical target profile is established, risk assessment process starts with a well-defined problem description or risk question. Risk

assessment is performed in three steps: the identification of hazards, the analysis of hazards and evaluation of risks associated with exposure to those hazards.

As an aid to clearly define the risk(s) three fundamental questions are often helpful: (1) What might go wrong? (2) What is the probability that it will go wrong? (3) What are the severity levels?

The risk assessment gives either a qualitative description of a range of risk or a quantitative estimate of risk. When risk is expressed quantitatively, a "risk score" is calculated based on a numerical probability. When risk is expressed qualitatively, it should be defined in as much detail as possible. Three descriptive steps of risk assessment are as follows.

(a) ***Risk Identification***:

First of all, risk associated with analytical method parameters needs to be identified. Risk identification is to identify hazards referring to the risk question or problem description by a systematic use of information. Information can include historical data, theoretical analysis and informed opinions. Hazard identification is done to find "What might go wrong?" including identifying the possible consequences. Generally Fish-Bone/Ishikawa diagram is used for risk identification.

(b) ***Risk Analysis***

Once risks are identified, they are analysed. Risk Analysis is the qualitative or quantitative process of linking and the severity of harms, probability of occurrence and also the detectability of harm (in some cases. After analysis the risk is categorised as Low, Medium or High.

(c) ***Risk Evaluation***

Factors/method parameters with medium/high risks are evaluated practically to study and understand the impact of on performance characteristics. Risk evaluation compares the identified and analysed risk against given risk criteria. Risk evaluations consider the strength of evidence for all three of the fundamental questions.

2. ***Risk Control***

The purpose of risk control is to reduce the risk to an acceptable level. Risk control includes decision-making to reduce (Risk reduction) and/or accept risks (Risk acceptance). Risk reduction focuses on processes for mitigation or avoidance of risk when it exceeds a specified/acceptable level. Risk reduction might include actions taken to mitigate risk severity and risk occurrence probability or to improve the detectability of hazards. Risk acceptance is a decision to accept risk. For some types of harms, even the best quality risk management practices might not entirely eliminate risk, in such cases, it might be agreed that an appropriate quality risk management strategy has been applied and that quality risk is reduced to an acceptable level.

3. ***Risk Communication***

The output of the risk management process should be appropriately communicated and documented among interested parties, e.g. within a company, industry or regulatory authority, etc. Communication need not be carried out for each and every risk acceptance.

4. ***Risk Review***

   Risk review is for continuous monitoring of the method after transfer. It is done for events that might impact the original quality risk management decision. The output of the risk management process should be reviewed to get new knowledge and experience. The risk review frequency should be based upon the level of risk. In some case, after risk review, reconsideration of risk acceptance decisions required.

4.3.1 Quality Risk Management Methods and Tools [8]

Traditionally, risk management is done in a variety of ways based on compilation of observations, trends and other information etc. Additionally, the pharmaceutical industry and regulators can assess and manage risk using recognized risk management tools listed below.

- Basic risk management facilitation methods (flowcharts, check sheets etc.)
- Failure Mode Effects Analysis (FMEA);
- Failure Mode, Effects and Criticality Analysis (FMECA);
- Fault Tree Analysis (FTA);
- Hazard Analysis and Critical Control Points (HACCP);
- Hazard Operability Analysis (HAZOP);
- Preliminary Hazard Analysis (PHA);
- Risk ranking and filtering;
- Supporting statistical tools.

It might be appropriate to adapt these tools for use in specific areas on case to case basis. Ishikawa (fishbone/cause-effect) diagram and failure mode and effects analysis (FMEA) are widely used risk assessment tool. Fish-bone diagram is a qualitative tool which allows one to identify and classify risk by answering the question "What might be wrong?". FMEA method is a quantitative risk assessment tool that provides a risk priority number which estimated based on the occurrence probability (P), severity (S), and likelihood of detection (D). FMEA answers the questions "What is the probability of it will be wrong?" and "What is the severity?" Based on risk priority number the risk is categorised as low, medium or high.

4.4 AQBD and DOE

DoE is a structured and organized method for determining the relationships between input factors (x—independent variables) affecting one or more output responses (y—dependent variables), through the establishment of mathematical models. In the

DoE approach, the controlled input factors are systematically varied to determine their effects on the output responses, which enables the identification of input factors leading to optimised output responses, the determination of the most important input factors, and the clarification of interactions between input factors.

4.4.1 Experimental Design Selection

A successful experimental design should take into account various criteria, including established objectives, the amount of input components and interactions to be researched, and the statistical validity and effectiveness of each design. Experimental designs can be classified into two groups in order to provide a better understanding of DoE applications: Screening designs (a) and optimisation designs (b).

The parameters of screening and optimisation designs, such as the quantity of experiments needed, the number of levels of input components, and the quantity of factors to be examined, are summarised in Table 4.3.

4.4.1.1 Screening Designs

The most popular screening designs, which have the best value for money, include two-level full factorial designs, fractionate factorial designs, and Placket-Burman designs. These experimental setups make it possible to explore a variety of input variables with fewer experiments. To provide a better understanding of the effects of input elements on output responses, they also have some limitations that need be taken into account. The most effective screening designs are two-level full factorial ones because they enable estimation of the primary effects of input variables and their interactions with output responses. When compared to fractionate factorial and Plackett-Burman designs, the fundamental constraint of two-level full factorial designs is the enormous number of experiments required. For two-level full factorial designs, the number of experiments required is 2k, where k is the number of input factors to be explored. As a result of its ability to evaluate numerous input components with a minimal number of tests, fractionate factorial designs are among the

Table 4.3 Summary of the characteristics, number of experiments, levels, and parameters for screening and optimisation designs

Applications	Experimental design	Experiments	L levels	Factors
Screening	Plackett-Burman	N	2	<N−1
	Factionate factorial	2k-pRes	2	k > 4
	To-level full factorial	2k	2	2 < k < 5
Optimization	B Box-Behnken	2k(k − 1) + C	3	3 < k < 5
	Central composite design	2k + 2k + C	5	2k < 5
	3-level factorial	3k	3	2 < k < 3

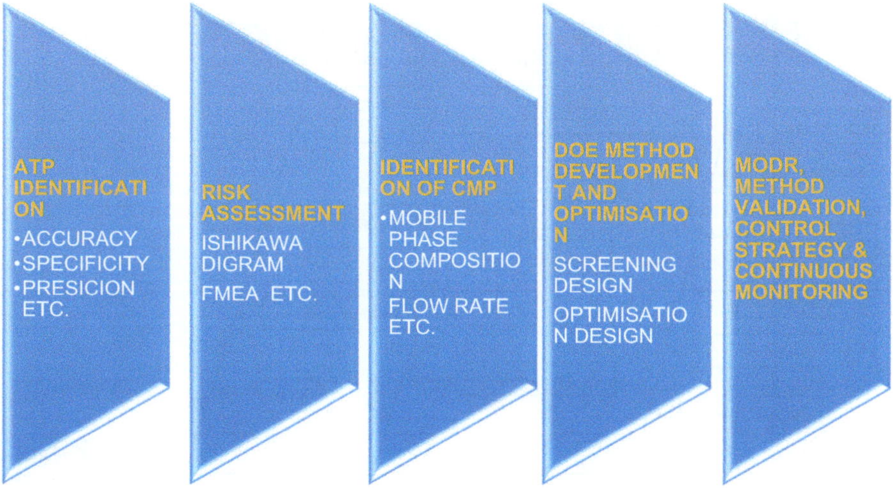

Fig. 4.3 Steps of AQbD

most popularly utilised for screening purposes. Given that some of them are confounded, attention should be taken while estimating main effects and interaction effects using fractionate factorial designs. The fractionate factorial designs might not be appropriate for determining how different elements interact. The selection of generators determines the resolution of fractionate factorial designs. Plackett-Burman designs are particular kinds of two-level fractionate factorial designs that provide the investigation of up to N − 1 input factors with respect to N experiments (N should be a multiple of 4). In the initial stage of DoE, screening designs are frequently used to pick the most important input elements and eliminate the less important ones. Pareto charts are important tools for accomplishing this goal since they allow you to determine the order of the input factors and their interconnections. When using a fractionate factorial design, pay special attention to the Pareto chart. In other words, the interaction of variables X1 and X2 aliased the main effect of input component X3 (Fig. 4.3a). Consider the situation as follows: while developing a chromatographic method, one wants to describe the resolution between two peaks as a function of the pH of the mobile phase (X1), the amount of organic solvent in the mobile phase (X2), and the flow rate (X3). The flow rate is a significant analytical parameter, according to the Pareto chart in Fig. 4.3a. However, the interaction between pH and the quantity of organic solvent in the mobile phase ($X3 = X1 \times X2$) causes the flow rate to be aliased (or confounded). This conclusion was incorrectly drawn since the Pareto graphic in Fig. 4.3a was created using a fractionate factorial design with resolution III. The Pareto chart in Fig. 4.3b demonstrates that when utilising a full factorial, flow rate will not be aliased (confounded) by the interaction between pH and the amount of organic solvent in the mobile phase. Pareto charts do not, however, show how changing the degree of an input element affects the output responses. A main effects and interactions graph

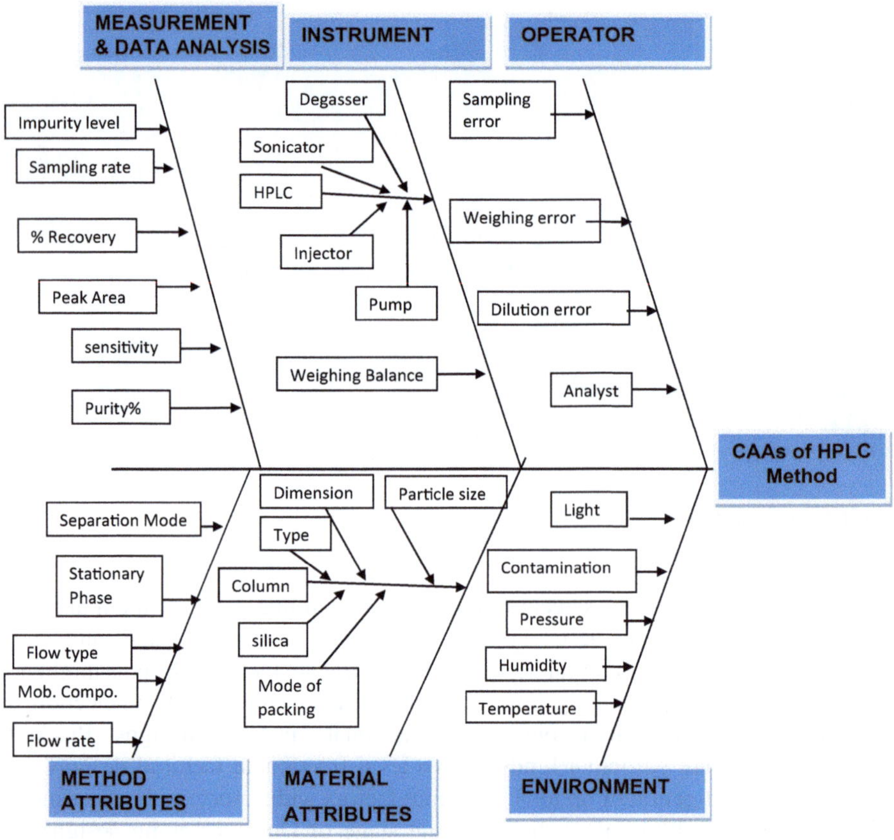

Fig. 4.4 Ishikawa Diagram illustrating influence of factors on CAAs of LC method

(Fig. 4.4) may offer this information. For instance, if the input factor X1 is changed from low (−1) to high (+1), the output response increases, and if the input factor X2 is changed from low (−1) to high (+1), the output response reduces. The advantage of the DoE approach over OFAT testing is the identification of interactions between input components. Varying the input factor X1 from low (−1) to high (+1) level with X2 fixed at low level (−1—grey dots and line), while varying the input factor X1 from low (−1) to high (+1) level with X2 fixed at high level (+1—black dots and line), the output response increases significantly. Interaction effects plots can help you identify synergies or antagonism between input elements and output responses.

4.4.1.2 Designs for Optimisation

The most popular optimisation designs are three-level full factorial, central composite, and Box-Behnken because they enable modelling of complex response surfaces.

One of the biggest drawbacks of screening designs is that because there are only two levels for each input factor, they can only be used to represent first order (linear) response surfaces. The modelling of the second order (quadratic) response surface is made possible by the employment of optimisation designs that use 3–5 levels of each input factor [11]. They are typically employed to analyse a smaller number of input parameters, however, as a result of the demand for an increased number of experiments (Table III).

The three-level full factorial design is frequently only employed when two or three input components need to be investigated because it necessitates more tests. When k is the number of input elements to be examined, the necessary number of experiments can be computed as 3k. For example, 27 experiments are needed for a three-level full factorial design with three input factors.

One of the most popular optimisation designs is the central composite design (CCD), which uses five levels of each input component while requiring fewer experiments than the three-level complete factorial design.

Box-Behnken designs are unique types of three-level fractionate factorial designs that enable modelling of first and second order response surfaces. When there are many input factors, these designs are more economical than three-level complete factorial designs. The full factorial design is used to pick all of the points for Box-Behnken designs.

4.4.1.2.1 Method Operable Design Region (MODR)

Method Operable Design Region (MODR), also known as analytical design space in analytical Quality by Design (QbD), is the multidimensional combination and interaction of analytical conditions (input factors, such as chromatographic conditions) that have been shown to provide assurance performance of analytical method (output response, such as asymmetry of peak). Graphical optimisation can be used to determine the Operable Design Region from counter plots of output responses (Ys) as functions of input components (Xs) that are overlaid.

In order to ensure that critical method parameters (CMP) are within predicted limits, a control strategy is required. Obviously, control space must be contained within the Modr. Process Analytical Technology (PAT) is a significant instrument in the implementation of control strategies because it enables real-time release testing and provides a higher level of quality assurance than traditional end-product testing. In analytical QbD, the control strategy is developed from the data obtained during method development and validation, allowing the technique's capacity to satisfy the analytical target profile (ATP) to be predicted. Continuous monitoring of method performance enables the detection, identification, and correction of out-of-trend analytical method performance.

The use of screening designs in pharmaceutical AQbD allows for the identification of critical method parameters (CMPs) (independent variables) influencing critical quality attributes (CQAs) (dependent variables) and, thus, the Analytical target profile (ATP). Furthermore, optimising design and surface response methods, as well as multiple response optimisation, enable the definition of a design space region where CQAs and ATP are attended. MODR can be defined using DoE results. Working within MODR is not considered a change, and hence does not require method revalidation.

4.4.1.2.2 Advantages of MODR

1. Methods will be more robust and rugged, resulting in fewer resources spent investigating out-of-specification results and greater confidence in analysis testing cycle times
2. Resources currently invested in performing traditional technology transfer and method validation activities will be redirected to ensuring methods are truly robust and rugged
3. The introduction of new analytical methods—from research and development to quality control laboratories—using a QbD approach will lead to a higher transfer success rate than with traditional technology-transfer approaches
4. A specified process will help the systematic and successful implementation of the QbD methodology and fosters a team approach
5. A true continuous learning process is established through the use of a central corporate knowledge repository that can be applied across all methods.
6. By registering only a commitment to ensure method changes meet the registered method performance criteria, flexibility to continuously improve methods can be achieved.

4.4.1.2.3 Limitations of MODR

1. Same facility is required at development laboratory and QC laboratory.
2. Acceptance must be gained for registration of the method performance criteria rather than the method conditions.
3. Analysts must learn new tools and skills.

DoE has recently been employed in the logical development and optimisation of analytical methods. Examples of input components (independent variables) that can be screened and optimised using DoE include the composition of the culture media, the composition of the mobile phase, flow rate, and the time of incubation. There are a number of output responses (dependent variables), including microbial growth, resolution between peaks, and retention time.

Dry lab (soft ware for method development) [12].

The revolutionary HPLC method development and optimisation tool ‘Dry Lab’ predicts chromatograms under a far wider variety of experimental conditions than is ever possible in a lab.

4.5 Implementation of AQBD: An Insight in to the Practical Aspects

Analytical procedures can be developed using a minimal (traditional) approach or elements of an enhanced approach. The enhanced approach is based on the Quality by Design approach, which must be implemented for the development and upkeep of analytical processes appropriate for the evaluation of the quality of drug substances and drug products. An analytical procedure's knowledge can be developed and improved in a systematic manner using the AQbD technique. It is practically applicable to analytical methods for assay, stability indicating, impurity estimation etc. It is applied to different analytical techniques like spectrophotometric, chromatographic etc. Steps of AQbD are described in Fig. 4.3.

4.5.1 Step 1: ATP Identification

The first step in AQbD (Fig. 4.1) is a predefine analytical target profile (ATP) which is also known as Quality target method profile (QTMP) (analogues to QTPP of QbD). It is the requirement for the method which is the combination of all performance criteria (specificity, accuracy, precision, range etc.) required to adequately describe what a method has to achieve by measurement means it is the target profile of CQAs. ATP defines requirements of measurement of CQA which needs to be met by an analytical procedure. The ATP is used to define and assess the fitness of the method in the development phase and during all changes across the analytical lifecycle. The ATP help to focus on the method parameters with impact quality decisions rather than method descriptions thus enable regulatory flexibility [13].

Pharmaceutical products undergo examination to make sure they operate as intended. Drug safety and effectiveness are both part of product performance. Pharmaceutical products are frequently evaluated for drug release and assay to determine the efficacy of the medicine. Impurities in medicinal products are also measured for safety purposes. As a result, while developing an analytical method, the most typical goals are assay estimate, drug release determination, and impurity measurement in pharmaceutical products [14].

The examples of ATP are given in following table (Table 4.4).

It should be established before starting the development of the method, defining its intended purpose but without specifying the analytical technique. Therefore, an example of an ATP for determination of related substances and degradation products of an API in a drug product with a specification limit of not more than 0.20% could be the following: "The analytical procedure should be able to quantitatively determine all known/specified related substances and degradation products of an API in the presence of excipients and other components of a drug product from a reporting limit to 120% of their specification limit with an accuracy within ±20.0% of the true value." Once it is approved by the regulatory authorities, an ATP facilitates greater

Table 4.4 Analytical target profile

Test	Critical quality attribute	Regulatory requirement	Analytical target profile
Assay method	Tailing factor	NMT 2.0	NMT 1.5
	Precision of peak area	NMT 2.0	NMT 2.0
	Theoretical plates	NLT 2000	NLT 4000
	Accuracy	97.0–103.0%	98.0–102.0%
	Run time	–	Less than 8 min
Drug release method	Tailing factor	NMT 2.0	NMT 1.5
	Precision of peak area	NMT 2.0	NMT 2.0
	Theoretical plates	NLT 2000	NLT 4000
	Accuracy	90.0–110.0%	95.0–105.0%
	Run time	–	Less than 7 min
Impurity estimation method	Tailing factor	NMT 1.5	NMT 1.5
	Precision of peak area	NMT 10	NMT 5
	Theoretical plates	NLT 2000	NLT 4000
	Accuracy	85.0–115.0%	85.0–115.0%
	Resolution	NLT 1.5	NLT 2.0
	Run time	–	Less than 28 min

continuous improvement of the method by supporting regulatory flexibility and should be considered in all stages of the analytical procedure lifecycle. Next step after defining ATP, is selection of an analytical technique capable of fulfilling the ATP requirements. There are many analytical techniques with same principle are available, out of which, the most suitable techniques can be considered.

4.5.2 Step 2: Risk Assessment

Risk assessment is the process where risk is identified, analyzed, evaluated and controlled. Various tools are used for risk assessment from which Ishikawa diagram or Fish-Bone diagram and Failure Mode Effect Analysis are very widely used.

Ishikawa diagram identifies potential factors and distributes factor-related risks into categories related to instrumentation, materials, methods, measurements, laboratory environment, and human factors. Ishikawa diagram is also known as cause and effect diagram as well as Fish-bone diagram.

Potential categorical factors associated with LC method are given as Ishikawa Diagram in following figure.

4.5.2.1 Cause Effect Relationship of CMP and CAA

Effect of various method parameters on Critical Analytical Attributes is depicted in following Table 4.5.

Factors are further assessed and prioritized by failure mode effect analysis (FMEA) or other prioritization matrices to assess the severity of risk each factor presents for a method's performance. Once failure modes are established, risk reduction can be used to eliminate, contain, reduce or control the potential failures. FMEA methodically breaks down the analysis of complex processes into manageable steps [8]. It is a powerful tool for summarizing the important modes of failure, factors causing these failures and the likely effects of these failures. The Risk Priority Number, or RPN, is a numeric assessment of risk assigned to a method as part of Failure Modes and Effects Analysis (FMEA in which a team assigns each failure mode numeric values that quantify likelihood of occurrence, likelihood of detection, and severity of impact). Each failure mode gets a numeric score that quantifies (a) Occurrence (O): likelihood that the failure will occur, (b) Delectability (D): likelihood that the failure will *not be* detected, and (c) Severity (S): the amount of harm or damage the failure mode. The product of these three scores is the Risk Priority Number (RPN) for that failure mode. The risk priority number is calculated as per the formula below,

$$\text{RPN} = \text{S} \times \text{P} \times \text{D}.$$

where S is the severity of the effect of failure, Occurrence probability, and D is the ease of detection.

Rating scales for S, P and D usually range from 1 to 5 or from 1 to 10, with the higher number representing the higher seriousness or risk. For example, on a ten-point Probability scale, ten indicates that the failure is very likely to occur and is worse than one, which indicates that the failure is very unlikely to occur. As an example, following table shows a generic scale for Severity.

Severity The severity of the failure mode is ranked on a scale from 1 to 10.

A high severity ranking indicates severe risk.

Table 4.5 Cause effect relationship of method parameters and CAA

Sr. no.	Method parameters	CQA
1	Column flow rate, pH of mobile phase buffer, concentration of organic modifier in mobile phase, column oven temperature	Retention time, tailing factor and plate counts
2	pH of mobile phase buffer, organic modifier and its concentration in mobile phase, gradient program, column stationary phase, dimension of HPLC column	Resolution between adjacent peaks
3	Diluents, sample extraction methodology i.e. shaking or sonication, shaking/sonication time, temperature during sonication	Drug recovery from sample matrix

Qualitative ranking and criteria	Quantitative ranking
Very high (the failure will affect product)	9, 10
High (there will be a high degree of failure)	7, 8
Moderate (the failure will generate some effect)	4, 5, 6
Low (the failure will only cause minor effects)	2, 3
Minor (may not even become aware of the failure)	1

Probability The potential of failure occurrence is rated on a scale from 1 to 10.

A high Probability rank reflects high failure occurrence potential.

Qualitative ranking	Criteria	Quantitative ranking
Very high (the failure is very likely to occur)	1 in 2	10
Very high	1 in 8	9
High (the failure will occur often)	1 in 20	8
High	1 in 40	7
Moderate (the failure will occur occasionally)	1 in 80	6
Moderate	1 in 400	4, 5
Low (the failure will rarely occur)	1 in 4000	3
Low	1 in 20,000	1, 2

Detection The capability of failure detection is ranked on a scale from 1 to 10.

A high detection rank reflects low detection capability.

Qualitative ranking and criteria	Quantitative ranking
Non-detection inevitable (potential design problems cannot be detected by the program)	10
Very low (program probably will not be able to detect a potential design problem)	9
Low (the program is unlikely to detect a potential design problem)	7, 8
Moderate (program may detect a potential design problem)	5, 6
High (there is a good chance that the program will detect a potential design problem)	3, 4
Very high (it is almost certain that the program will detect a potential design problem)	1, 2

The RPN value for each failure ranges between 1 and 1000.The sum of the RPNs for the failure modes is the overall RPN for the method. The RPN value for each potential problem can then be used to compare the issues identified within the analysis. Typically, if the RPN falls within a pre-determined range, corrective action may be recommended or required to reduce the risk (i.e., to reduce the likelihood of occurrence, increase the likelihood of prior detection or, if possible, reduce the severity of the failure effect). A failure mode with a high *RPN* number should be

Table 4.6 Ranking of risk priority number

Risk priority number	Risk category	Nature of impact	Acceptance criteria
<50	Low	Nonimpact	Risk acceptable, no action required
<200	Medium	Indirect impact	Acceptable with control
≥200	High	Major impact	Requires control measure before acceptance

given the highest priority in the analysis and corrective action. Following is the ranking of RPN (Table 4.6).

Risk analysis by FMEA is illustrated for HPLC method development in following table (Table 4.7).

4.5.3 Step 3: Identification of CMP

Depending on the analytical method, various parameters may affect the method's performance, the variability of which may have a significant impact on the selected Critical Method Attributes. Parameters which affect the method's performance significantly are called, Critical Method Parameters (CMPs). Critical method parameters are identified based on risk analysis. Analytical method parameters with high RPN is considered as Critical Method Parameters (CMPs). Following table describes categorization of Critical Method Parameters based various factors affecting Critical Analytical Attributes(CAAs) (Table 4.8).

More specifically Critical Method Parameters for HPLC, GC and TLC methods is described in following table (Table 4.9).

CMPs are the risk parameters which are further evaluated practically as a part of method development to study their individual as well as interactive effects on each other. DoE is applied for the same purpose.

4.5.4 Step 4: DOE Method Development and Optimisation

The Critical Method DOE evaluates the parameters that have been shortlisted by risk assessment. In DoE, CMPs are input variables (factors, independent variables). CMPs can be either qualitative or quantitative. Different stationary phases with different make/model and lot, the quality/grade of the buffer for the mobile phase are examples of qualitative variations. Flow rate, oven temperature and concentration of organic modifier in the mobile phase, etc. are quantitative variables. After the selection of input variables, responses (Dependent variables) are finalized. Dependent variables are resolution between adjacent peaks, tailing factor, recovery of drug

Table 4.7 Failure Mode Effects Analysis for HPLC method development

Sr no	Feature	Failure mode	Impact	Cause	Current controls	S	P	D	RPN	Risk classification	Action to be taken
1	pH of mobile phase	Mobile phase	Peak shape and retention time	pH not adjusted properly or pH not selected based on the property of analyte	Calibarte the pH meter properly and then adjust the pH of mobile phase. Consider the properties of analyte and then select the pH	5	6	6	180	High	There should be 1 or near to 1 value difference between the pH of mobile phase and pKa value. Also calibrate the pH meter and ensure it's maintenance regularly
2	Column length	Column	Theoretical plate, retention time	Wrong column selection	Consider the size of analyte and then select the column	1	2	1	2	Low	Column of appropriate length should be selected based on the particle size of the analyte
3	Type of column	Column	Retention time, peak area, peak shape	Column of appropriate type not used	Consider the properties of the analyte while selecting the type of column	1	2	2	4	Low	Appropriate column must be selected based on the analyte
4	Mobile phase ratio	Mobile phase	Retention time, peak shape	Analyte properties not taken into consideration.	Select the mobile phase by considering the solubility and polarity of analyte	6	4	5	120	High	Increasing the polarity of mobile phase leads to shorter retention time. Thus the ratio of the mobile phase should be selected accordingly

5	Buffer concentration	Chemical	May affect the column life because very high concentration causes precipitation in the column and blockage of channels and line filters	Very high concentration of buffer	Before placing the mobile phase into the HPLC system, check if the precipitates are forming or not	4	4	1	16	Low	During the selection of mobile phase, check if the selected mobile phase containing buffer is forming precipitates or not
6	Light	Light	Reduce peak height, peak area	Exposure of light sensitive API to light may degrade the drug	Check if the API is light sensitive or not	3	1	3	9	Low	Check the light sensitivity of API and if it is light sensitive then take proper care and do not allow the API to come in contact with light for more period of time
7	Balance	Weighing balance	It may affect the concentration of buffer and sample concentration	Error in weighing or weighing balance not calibrated.	Calibrate the weighing balance	1	1	1	1	Low	Calibrate the weighing balance and ensure its maintenance regularly so that there is no problem in the weighing balance
8	Sample vials contamination	Vials	Reduced resolution	Vials not cleaned properly	Ensure that the vials are properly cleaned	2	2	2	8	Low	Before sampling clean all the vials properly and ensure that they are properly dried and there are no

(continued)

Table 4.7 (continued)

Sr no	Feature	Failure mode	Impact	Cause	Current controls	S	P	D	RPN	Risk classification	Action to be taken
											traces of previous material
9	Ambient temperature	Temperature	Detection problem, resolution problems	Stability of drug at ambient temperature not checked	Check that the drug is stable at which temperature range	1	2	1	2	Low	Check the stability of drug and store it appropriately at the suitable temperature range

Table 4.8 Categorization of critical method parameters (CMP)

Sr. no.	Category of CMP	CMP
1.	Material attributes	• Solvents and chemicals used for buffer preparation used in mobile phase • Purity of reference standard • Different make/model/lots of stationary phase • Glassware Type e.g. amber colored or clear, Class A or B • Sample and mobile phase filter paper: type and make
2.	Instrument related aspects	• Column: packing and dimensions • HPLC detectors: UV/PDA, RI, fluorescence • HPLC: make, model • LC system configuration: size of injector loop and diameter of tubing
3.	Instrument operating parameters	• Flow rate, wavelength of detection, oven temperature, gradient program, sampling rate, needle wash
4.	Method parameter	• Buffer pH, buffer concentration, organic modifier, diluents for sample preparation, time of sonication

Table 4.9 Critical method parameters for HPLC, GC and TLC methods

Critical method parameters			
Sr. no.	HPLC method	GC method	TLC method
1	Stationary phase (make, model, dimensions, packing, life)	Stationary phase (make, model, dimensions, packing, life)	Stationary phase and plate coating thickness
2	Column flow	Column flow	Development distance
3	Oven temperature	Oven temperature	Temperature of mobile phase
4	Mobile phase pH	Carrier gas	Solvent mixture composition
5	Mobile phase composition	Split flow	Solvent mixture pH
6	Mobile phase gradient	Temperature program	Sample solution spotting volume
7	Organic modifier in mobile phase	Injector temperature	Size and shape of spot
8	–	–	Drying time and conditions of TLC plate
9	–	–	Technique used for visualizing the spot e.g. by spraying reagent, detection under UV light

from the sample matrix, etc. After selecting independent variables, their levels are selected. Selection of the levels depends on normal operating ranges of the factors.

In HPLC method development, some of the independent variables areas follows:

- HPLC Column (dimensions, stationary phase, make, ageing)
- Flow Rate
- Column oven temperature

- Buffer for mobile phase
- Buffer concentration
- Concentration of additives (ion pair etc.)
- pH of mobile phase buffer
- Mobile phase gradient
- Organic modifier in mobile phase etc.

The dependent variables (Responses) for HPLC method are as follows:

- Asymmetry
- Theoretical Plate
- Resolution etc.

Method development is performed using screening designs like Placket-Barman, full factorial design etc. After screening, method parameters are optimized using response surface methodology. Generally Box-Behnken design, Central composite design, fractional factorial design are used for method optimization. The soft ware then gives contour plots and 3D plots. Interpretation of the ANOVA table, charts, plots etc. gives better understanding of the effect of different parameters on method performance. Various softwares are available for DoE study like Minitab, Design Expert etc. The study gives Method Operable Design Region.

4.5.5 Step 5: MODR, Method Validation, Control Strategy and Continuous Monitoring

MODR (Analytical Design Space) is the design space in which the analytical procedure has to meet the Analytical target profile (ATP). MODR gives optimised ranges of Critical Method Parameters proving robustness of the method. MODR ranges should be always broader than normal operating ranges. The final analytical method with optimized Critical Method Parameters is then validated within the MODR.

Based on MODR, control strategy can be decided. Basically, the control strategy is finalised using all the knowledge that was acquired during the various QbD techniques, such as risk assessment, experiment design, and method operable design range. It aids in implementing the necessary controls in the analytical process to prevent analysis failures. The control strategy is in the form of limits specification of system suitability, specifying any grade of materials, setting mobile phase composition limits or any other specific precaution.

Few examples of control strategy are given below

- In case of separation criticality, resolution criteria must be added as part of system suitability.
- In case of reagent/chemical criticality, suitable make/grade of reagent must be specified.

- In methodology, if sonicator time and temperature found to affect the response control of temperature during sonication can be incorporated in the STP.

Finally, the analytical method is implemented for routine analytical testing. Due to Laboratory transfer, some challenges may arise due to different operating environment, instrument models etc. These challenges are monitored very carefully and accordingly; adjustments are to be made in methodology within a Design Space. Due to continuous improvement, the analytical method may need some changes or improvements. As a part of continuous improvement, the performance of the analytical method is monitored by trend analysis, out of specification (OOS), and out of trend (OOT) occurred during a specified time. If trend analysis shows out of trend, it is investigated and necessary changes are done. If the changes are within the defined MODR, Re-method validation is not required as it is not considered as change but when they are out of the MODR, revalidation required.

4.6 Conclusion

Analytical methods play a key role in the development and production of a drug product. One of the most important features of a method is its robustness, which should be evaluated during method development. Robust methods are able to provide credible results of constant quality. The traditional OFAT approach to method development usually results in an unoptimized method without any knowledge of its robustness, often leading to method failure during its routine use. AQbD is a systematic approach to method development which comprises defining the Analytical Target Profile and Critical Analytical attributes; performing risk assessment and optimization of critical method parameters using DoE. Experimental design makes it possible to investigate the effects of factors and their interactions on particular responses (CMAs). MODR and control strategy is defined based on the outcome of the DoE.

AQbD is emerging gradually and a lot more analytical methods are being developed with the AQbD approach. The recent ICH Q14 guideline facilitates fast implementation of AQbD. AQbD can offer more flexibility for analytical methods which derives from analytical method understanding. Moreover, it helps in gaining regulatory flexibility. The most common benefit is methods are more robust and rugged, and thus sustain through the challenges of long-term usage in the product life cycle. By designing quality into the method, AQbD completes the overall lifecycle concept of analytical methods.

References

1. Bhutani, H., Kurmi, M., Singh, S., Beg, S., & Singh, B. (2014). Quality by design (QbD) in analytical sciences: An overview. *Pharma Times, 46*(8).
2. ICH harmonised tripartite guideline, pharmaceutical development, Q8(R2).
3. Peraman, R., Bhadraya, K., & Reddy, Y. P. (2015). Analytical quality by design: A tool for regulatory flexibility and robust analytics. *International Journal of Analytical Chemistry, 2015*, 1–10.
4. ICH harmonised tripartite guideline, pharmaceutical quality system, Q10.
5. USFDA. (2003). *Pharmaceutical cGMP for the 21st century—A risk-based approach: Second progress report and implementation plan introduction*. Food and Drug Administration.
6. Guidance for industry: PAT—A framework for innovative pharmaceutical development, manufacturing, and quality assurance. (2004).
7. The European Medicines Agency road map to 2010: Preparing the ground for the future. (2005).
8. ICH harmonised tripartite guideline, quality risk management, Q9 (R1).
9. ICH harmonised tripartite guideline, analytical procedure development-step 2, Q14.
10. ICH harmonised tripartite guideline, product life cycle management, Q 12.
11. Fukuda, I. M., Pinto, C. F. F., dos Santos Moreira, C., Saviano, A. M., et al. (2018). Design of experiments (DoE) applied to pharmaceutical and analytical quality by design (QbD). *Brazilian Journal of Pharmaceutical Sciences, 54*(Special), e01006.
12. Retrieved from https://molnar-institute.com/drylab/#how-it-works
13. Jackson, P., Borman, P., Campa, C., Chatfield, M., Godfrey, M., Hamilton, P., Hoyer, W., Norelli, F., Orr, R., & Schofield, T. (2019). Using the analytical target profile to drive the analytical method lifecycle. *Analytical Chemistry, 91*(4), 2577–2585.
14. Nitin Kumar, D., & Sangeetha. (2020). Analytical method development by using QbD—An emerging approach for robust analytical method development. *Journal of Pharmaceutical Sciences and Research, 12*(10), 1298–1305.

Chapter 5
API Quality by Design

Madhusmruti Khandai, Dibya Sundar Panda, Pranav Shah, Vipul P. Patel, and Falguni Tandel

Abstract The pharmaceutical sector is always adapting, and so must the idea and practise of quality by design (QbD). In regulatory submissions, active pharmaceutical ingredients (API) production procedures are often detailed; however, there are few and restricted instances of the actual practise of adding QbD evaluations. We are aware that there are important distinctions between creating QbD strategies for API and medicinal product manufacture. When compared to the criteria provided in ICH Q8, which are largely focused on the use of QbD for pharma product manufacture, the creation of appropriate process knowledge for API synthesis may be relatively difficult. This study provides an illustration of desirability analysis by combining a risk assessment with instances of design and a suggested criticality evaluation. The present study provides an insight into the pros and cons of design space along with roadblocks and remedies thereof after critical evaluation.

Keywords Active pharmaceutical ingredients · ICH · API synthesis · QbD · Critical evaluation

M. Khandai
Royal College of Pharmacy and Health Sciences, Berhampur, Odisha, India

D. S. Panda
Department of Pharmaceutics, College of Pharmacy, Jouf University, Sakaka, Kingdom of Saudi Arabia

P. Shah
Maliba Pharmacy College, Uka Tarsadia University, Bardoli, Gujarat, India

V. P. Patel (✉)
Sanjivani College of Pharmaceutical Education and Research (Autonomous), Kopargaon, Maharashtra, India

F. Tandel
Indus Institute of Pharmacy & Research, Indus University, Ahmedabad, Gujarat, India

N. K. Jain, N. Bajwa (eds.), *Introduction to Quality by Design (QbD)*,
https://doi.org/10.1007/978-981-99-8034-5_5

5.1 History of QBD

The history of Qbd is as important as its need in the current pharmaceutical industry, and also understanding of the conventional terms can lead to the improvement, in terms of present challenges. Perspective and proportion are missing from the QbD argument. In healthcare, mistakes can be fatal, therefore understanding of quality history minimises the danger of repeating mistakes.

Mass production methods became well known throughout the industrial revolution. Over the course of several years, the final product was the only gauge of the product's quality. By 1930, American statistician Shewart adopted the technique that emphasises process quality to produce a quality product. Deming, Juran, and Shewart introduced various concepts of quality management.

The notion of Quality by Design (QbD) was initially introduced by Dr. Joseph M. Juran, a well-known name in the field related to quality management [1]. According to Dr. Juran's philosophy, the incorporation of quality into a product's design is crucial, as many quality issues and predicaments stem from the initial design of the product. According to Woodcock's definition [2], a high-quality pharmaceutical product is one that consistently delivers the therapeutic benefit to the user that was stated on the label and is free of impurities as well as contaminants. QbD encourages risk-based drug product development, manufacture, and regulation by the FDA. After discovering additional testing does not always improve product quality, FDA prioritised QbD based products which are designed with excellent parameters.

With the release of ICH Q8 (R2) for pharmaceutical manufactured formulations, ICH Q9 for management of quality aspects based on risk analysis and ICH Q10 for pharmaceutical quality system over course of several years, QbD in the pharmaceutical industry has seen significant development [3–5]. Furthermore, the ICH Q1WG has released a set of questions and answers on various guidelines like Q8 to Q10, along with the ICH Q8/Q9/Q10 road maps. In addition, the ICH Q11, which deals with the design, formulation, development and production of drug substances, has now been made available to the public as well. Furthermore, outcomes related to the FDA-EMA's joint valuation of QbD components and marketing applications thereof have also been made available [6–9]. The aforementioned documents offer comprehensive guidance regarding the extent and characterization of quality benchmarkings in relation to pharmaceuticals manufacturing units. However, several aspects of implementation remain unaddressed in the aforementioned guidelines and documents. Despite recent publications, there exists a state of perplexity among industry scientists, academicians, and regulators [10–13].

The focus of this chapter is aimed to define the objectives of Quality by Design (QbD) in the pharmaceutical industry, elucidate its fundamental principles and components, and clarify the tools and investigations employed in its implementation.

5.2 Pros and Cons of QBD [14] Continuous Improvement

The application of the Quality by Design (QbD) methodology can ensure the quality of pharmaceutical goods in terms of both their safety and their effectiveness, in addition substantially enhancing manufacturing process's quality performance. The underlying concept is that of ongoing enhancement, which stems from the escalating demand of manufacturing entities to augment their comprehension of finished products as well as leverage expanding ideas repository which evolves as a active agent traverses its life cycle. The technique lets organisations enhance and modify plans. This ensures that manufacturing difficulties are adequately addressed and remedied as a project moves from the lab through the pilot plant to commercial manufacture. Each development step increases knowledge, and the risk assessment may be revised to fix any weaknesses. After a product is commercialised, fresh data and experience may be leveraged to enhance process and product. These stable and predictable performance conditions are sought for via the employment of statistical approaches in the final procedures. In the developmental stage, the utilisation of DoE proves beneficial in determining the consequences of significant elements or criteria and their reciprocations. Identifying critical factors beforehand facilitates the execution of risk mitigation measures and thereby prevents batch rejection. Furthermore, the statistical design methodology is employed to ascertain the most favourable operational conditions and establish constraints analysis diverse process parameters.

5.3 Alteration Control

The QbD strategy excels when it improves the ability to support change. If quality is integrated throughout the product development process, revising regulatory documentation following any modifications to the manufacturing process may simply call for minor adjustments. This is due to the fact that a thorough understanding of the operational conditions and its limitations features are already been developed and data accumulated to support any modifications. These are different from the previous strategy, which avoided change because it might influence reversely on the quality of the final product. Knowing the product and how it works thoroughly will make it simpler to decide nearest changes that are allowed and among these which can be averted. The International Conference on Harmonisation related to technical requirements for registration of pharmaceuticals for human use, pharmaceutical development Q8 (R2) document defines an example of a design domain that always produces a product with specific characteristics. A design space can be created by combining all the data points and know-how gathered during the processing phase. It is potentiated to ensure that the process persists robustness and that a final product developed from active ingredient satisfies the required standards is still attainable in the midst of production within framed as well as established planned space and it is requisite to progress out with minor changes or deviances in terms of the intended production method because of technological constraints. In demarcation to the primitive "Quality by QC" processing plans, this permits some latitude for process change as long as it will not affect the qualities of a product. Here, the change's potential effects are unclear, and it can't be ruled out that it might produce a product with unfavourable characteristics. This could lead to a significant regulatory change, necessitating the collection of new data, which could have an impact on both time and cost.

5.4 Failure Prevention

Manufacturing teams may have a more beneficial awareness about the constraints of the ontogenesis process thanks to QbD methodology. This means that if specific process elements were expected throughout development and documented with the proper regulative body, the squad may be able to calibrate them. Results may be continually fine-tuned and guided to improve accuracy and proximity to the target value. It may also be less likely for batch failure to happen due to unforeseen factors if the process parameters are thoroughly understood and connected together, as all potential interactions have already been considered and are recognised. The additional benefit of lower overall costs may result from this. QbD centres on the practise of using multivariate analysis. To better recognise and understand critical material features and production process parameters, this is generally utilised in tandem with modern process-analytical chemical methods and knowledge-management technology. Using multivariate tools to monitor batch performance in real time and displaying the findings in multivariate control charts, process engineers may determine whether a batch is operating as expected or if it is straying early. Early detection of deviations allows the process to be returned to normal before it is too late. With this newfound knowledge, manufacturers will be able to incorporate quality checks into their operations and produce ever-better goods. To reduce failed batches, the manufacturing economics and environment will benefit.

5.5 Consistency

Quality design improves batch uniformity. When quality is designed into the process from the beginning, it leads to higher consistency from lot to lot. These demonstrated consistency will assist to instill higher assurance in the regulatory body regarding the consistency of both the method and the finished good. This may make it possible for regulators to exercise reduced stringent control in the course of the enrolment process and require fewer post-approval submissions and requests. QbD improves pharmaceutical consistency and reduces recalls, boosting consumer confidence in manufacturers.

5.6 Right First Time

The QbD strategy aims to improve the likelihood that products will be registered "right the first time." Statistical designs ensure all facts of variability affecting a course of experiment are recognised, analysed, effectively accomplished. On doing such analysis, items are guaranteed to continuously meet predetermined criteria and be "right the first time." Lean asset management ensures that a company's resources are being utilised effectively and helps ensure first-time success. Since, the registration and approval processes are streamlined, this also results in a reduced amount of time needed to bring the product to market.

5.7 Reduced Controls

Having an in-depth understanding of the procedures involved is an excellent predictor of the quality of the final goods before they are even evaluated. This is because the assurance of quality has already been integrated into the processes. This means that controls over the process's intermediates and outputs are unnecessary, since they are being implemented in real time inside the process itself. This not only cuts costs but also shortens the amount of time required for manufacturing, testing, and product release for manufacturing organisations and their respective clients.

As forward-thinking businesses strive to increase their production capacities, decrease their throughput times, and implement more continuous manufacturing processes, the relevance of quality by design (QbD) will only grow. QbD uses data to improve manufacturing process knowledge, minimise batch failures, improve change management, and boost return on investment, time, and cost savings in this increasingly non-stop production environment. The utilisation of QbD will become increasingly crucial as regulative bodies start to expect that manufacturers, design quality into final designed products at each level of the product's development, unit operation and manufacturing process.

5.8 QBD in Association with ICH

In May of 2006, the International Conference for Harmonisation released the guideline known as Q8 for pharmaceutical development. This guideline was then amended by the guidelines known as Q9 which is for quality risk management [15], Q10 for pharmaceutical quality system [16], and Q11 for development and manufacture of drug substances [17] between 2009 and 2012. Specifically, the Quality Target Product Profile (QTPP), Critical Quality Attributes (CQA), Quality Risk Management, Design Space and Control Strategies are principles of QbD which are described in ICH guideline that was published in 2009. It is generally accepted within the pharmaceutical industry that the primary benefits of quality by design are an enhanced awareness of the processes involved, a better ability to maintain control over product manufacture, and the incorporation of the intended level of quality into the product's design. However, the manufacturer's who begin the process of implementing QbD report higher beginning expenses and a greater number of perceived regulatory impediments as some of the disadvantages [18]. Pharmaceutical companies needs to maintain adequate time to utilise QbD for the ontogenesis of their medicines from the time of its inception in to the ICH guidelines Q8, Q9, Q10, and Q11 (more than 10 years ago). However, the European Medicines Agency (EMA) stated in 2014 that application dossiers containing QbD data are still far to be implemented as normal strategy, only with a tiny fraction of commercialising approval applications or MAA submitted in Europe contain supporting QbD based reports [19].

The present research examined QbD utilisation in regulatory dossiers for EU-approved medical product development across a 6-year period starting 5 years after ICH Q8 acceptance (2014–2019).

5.9 QbD and Its Components

QbD components	Explanation
Control strategies	A premeditated collection consisting measures, obtained from present comprehension of products and processes, which guarantees the performance of processes and quality of products.
Critical process parameters (CPP)	To achieve the desired quality, processes must be monitored or controlled if one or more of their parameters is volatile and has a major impact on a key quality metric.
Critical quality attributes (CQA)	A characteristic of the product that must be within a certain range, limit, or distribution in order to preserve the desired quality. This characteristic might be physical, chemical, biological, or microbiological.
Design space	Quality assurance via a tried-and-true multidimensional combination and interplay of process parameters and input elements like material characteristics.
Proven acceptable range (PAR)	A range of a process parameter that, when utilised in isolation, will provide products that meet the required quality requirements while all other variables are held constant.
Quality risk management	A systematic procedure used to assess, handle, communicate, and evaluate quality risks of pharmaceutical goods at every stage of the product life cycle.
Quality target product profile (QTPP)	An anticipated abstract of a drug product's quality features that must be met to guarantee the intended quality, bearing in mind the medicine's safety and effectiveness.

5.10 Challenges in Application of QBD in Formulation Design

The following are some of the obstacles encountered during the creation of generic and novel drugs while using the QbD methodology [20–24].

- Submissions based on QbD are expected to meet customary standards.
- Regulatory bodies and industrial units together should lead the training initiative necessary to put the QbD idea into practise.
- Setting the standard for the depth of information included in regulatory filings (the kind and scope of data to be included in future CMC filings).
- Gaining regulatory leeway while maintaining high-quality output; Finding a happy medium between QbD and more conventional approaches of proving quality.
- Exchange of confidential data with regulatory bodies; Variations in strategy to account for wide range of drugs.
- Comparison of small chemical vs. huge biological.
- Oral solid pharmaceutical products versus novel or modified dosage forms, Active ingredients versus combination therapy products.
- CMC Regulatory Agreement, a possible approach to regulation.
- The hindrances to implementing statistical or QbD in course of product development and manufacturing unit operations include insufficient comprehension and confidence, as well as the expenses associated with such a decision, which are subject to developing and marketing considerations.
- Various regulatory procedures such as biologics licence application (BLA), new drug application (NDA), abbreviated new drug application (ANDA), and follow-on practises are implemented in the pharmaceutical industry. These procedures are accompanied by distinct regulatory practises and cultural norms. The amalgamation of review and inspection techniques.
- The current research is to investigate the existing evaluation tools and the necessary cultural shifts in the business world and the FDA.

5.11 Current Approaches of QBD in Formulation and Manufacturing

5.11.1 Plackett-Burman (PB) Design

The Plackett-Burman designs (PBDs), which was introduced by both renowned scientists Plackett and Burman, are widely recognised as non-regular designs. PBDs are frequently employed for screening purposes due to their ability to explore multiple elements in a relatively small number of empirical runs [25].

5.11.2 Central Composite Designs

Central composite designs (CCDs) was first coined by Box and Wilson (1951) for sequential testing [26]. F factorial points, NC centre runs, and 2k axial points all play a role in the construction of a CCDs, as do two-level factorial points (full or factorial). The progressive aspect of the design is immediately apparent. Adding axial points to the system's curvature allows for efficient estimation of pure quadratic components. CCD is based on sequential testing, however it may be utilised well in non-sequential batch response surface experiments [27].

5.11.3 Face Cantered Design

Under specific conditions a CCD is defined to be a FCD type when $\alpha = 1$, where FCD is a three level study design [28].

5.11.4 Box-Behken Design

A series of three-level designs known as the Box-Behken designs (BBDs), created by two eminent scientists Box and Behnken in 1960, is effective in accommodating 2°-order response surfaces. It is supported on building unfinished block patterns that are balanced.

Researchers in many fields of science that use RSM need three levels that are evenly spaced, therefore BBD serves as a highly effective substitute for CCD [29, 30].

5.11.5 Star Designs

A simple method for fitting a quadratic model is a star pattern. The formula for the number of trial runs required in this design is 2k + 1 [31, 32].

5.11.6 Center of Gravity Designs (CGDs)

Designs keeping gravity in due consideration are variations on CCD. Using these designs, the total trial runs is reduced to 4k + 1 [33, 34].

5.11.7 *Equiradial Designs*

The first degree of response surface designs are known as equiradial designs, and they are comprised of N points that are arranged in a regular polygon that is centred on a circle [35, 36].

5.11.8 *Mixture Designs*

There are two sub-categories of mixture designs: standard mixture designs and limited mixture designs. Both of these sub-categories are included under the umbrella term “mixture designs” [32].

5.11.9 *Taguchi Designs*

Experiential design is referred to “off-line quality control” in Taguchi design since it’s a way for assuring excellent functioning in the creation of goods or processes. The intention of these reliable trails is to identify the control factor configurations that will result in satisfactory responses and to categprise the variability of the system according to the different sources. The use of indicative components and noise factors are distinguishing characteristics of this method.

The constrained criteria or inputs that govern the system are the signal factors. It is often impossible or impractical to regulate the components that contribute to noise [37, 38].

5.11.10 *Optimal Designs*

Postulating an exact model, defining the changeable space, and determining the no. of trail points must be done as part of the optimum design technique. These steps must be taken in such a way that the model coefficients may be found with the highest possible probability. These prominent designs allow for continual design point addition and staged experimentation [39, 40].

5.11.11 *Rechtschaffner Designs*

In models with major effects and first-order interactions, these designs matter [41–43].

5.11.12 Cotter Designs

In majority of the experiments, these patterns are used for screening [11].

5.12 QBD and API Manufacturing [37]

The first phase of the process is when the QbD requirements for NDA goods and ANDA products diverge the most significantly from one another. In the case of an NDA, the TPP is still in the process of being developed, but in the case of an ANDA product, the TPP has already been thoroughly established based to the labelling and clinical studies that were carried out to support approval of reference product. The following phases of generic medication development include quality-based drug development:

Detailed description of the product used as a reference This provides all the specifics surrounding the evidence of the structure of the substance.

5.12.1 Roadblocks 1

Denoted product complexity, biological product origin, polydisperse admixture, and supramolecular complexity are the challenges. The company filing the ANDA may need more detailed product definition than the innovator. The present method and the QbD based method for developing generic products are both summarised in the excogitation of generic entity and process.

5.12.2 Roadblocks 2

Dosage form critical quality attributes (CQAs) are identified, and a empirical space is created by varying a few key parameters this also is a challenging factor.

In case product development: present vs. QbD Current method, QbD as a data-driven specification setting systemic, linking mechanistic stimulation material qualities and operational factors to medicinal products, CQAs, Clinical (safety and efficacy) as well as uncertain clinical linkage guaranteed. Negotiating specification due to inadequate data and absence of systematic product development methodology, design space specifications may not represent product quality, sometimes resulting in out-of-spec results that may lead to

- Violations of compliance and probes thereof.
- Market withdrawal, quarantine, or recall of goods, as appropriate;
- Instances of market-wide drug scarcity.

'True' product quality is always reflected in a design space that has been authorised. To comprehend a product or process, developmental research often uses multivariate trials with one variable at a time. Regulatory roadblock to ongoing progress Regulations are changed to allow for ongoing progress within the design space. Different review cycles for chemistry, processing and controls a solitary CMC critical review cycle and risk-based requirements. The production procedures are set in constrained conditions, which are sometimes defined as a "frozen process," and they oppose any flexibility. Initial full-scale runs validate production method. Life cycle glide path to fabricating process validation and uninterrupted confirmation emphasises optimisation, repeatability, robustness, and control strategy. Product requirements form an integral element of the gross quality sum-up and are derived from batch data that is readily accessible throughout the registration process. The performance goals and supporting data used to create a product's specifications are determined by those goals. Method of control based on checking and checking again Control scheme by design; control scheme via web based testing and investigation; real-time release conceivable; risk-based approach to risk management. Management of product quality throughout its life cycle entails a combination of proactive measures (such as preventing issues from occurring in the first place) and reactive measures (such as fixing them when they do). The Statistical Process Control Unit Method is under utilised, overutilized, or used primarily.

5.12.3 Root Cause Analysis

- In-case of majority of the experiments, the reason behind root cause cannot be determined.
- Insufficient comprehension of real-world variations, such as those associated with products, operator of the manufacturing process, fundamental to the formulation are less understood.
- Variability in the measurement system. Methods of analysis (like the USP calibrator table) for dissolving.
- Lack of or inadequate knowledge hinders drug development.

5.12.4 Raw Material Properties

- Manufacturing processes (unit operations) influenced by the qualities of formulation components. Manufacturing's impact on medicine product CQAs
- Connection between active pharmaceutical ingredients and excipients and drug product quality features
- Concerns about the impact on product quality
- Tools from the field of Process Analytical Technology (PAT) are crucial in the creation of modern pharmaceuticals.

- Process Control Analysis Offline is also essential.
- In-process testing are the major means of determining if a process is good to go or not after it has been approved.
- Process control PAT devices with feed forward and feedback control may help identify CQAs in oral suspension dose forms.
- Key biological sample set.
- Bioequivalence research.

5.12.5 Roadblocks

Bioequivalence for drugs with systemic or local effects; expensive, time-consuming, or unanticipated bioequivalence testing; few companies develop medicines requiring bioequivalence studies of clinical end points.

5.12.6 QbD for Bio-equivalence Trails

The approach of QbD helps with product design and production for bioequivalence. Modelling of formulation In vivo-in vitro correlation and dissolution manufacturing of commercial goods.

5.12.7 Roadblocks

Scale-up issues, wasted commercial batches, failure to fulfil requirements, process variability, and sometimes the necessity for process or product reformulation on the part of the sponsor. A generic sponsor scales up the operation from the show lot given at the ANDA to the manufacture of commercial products using a control approach, a component of QbD.

5.13 Application of QBD in API Design [44]

Robust procedures provide high-quality products on a consistent basis when using a QbD strategy. A QbD strategy is defined as a strategy that small drug manufacturers are attempting to implement. It aims to experimentally develop an entity and work out with the features that will match certain standards specified following rigorous examination of the designated therapeutic effect. The formulated drug's (drug substance's) physicochemical qualities and performance behaviour are examples of key quality attributes CQAs of both the finished drug product and the potential pharmaceutical ingredient. Multiple intermediates are commonly needed in the many methods used to create small-molecule active pharmaceutical ingredients. Each of these procedures must be solid if these CQAs are to be met on a regular basis. To ensure the API process reliably produces product meeting the specified CQAs, it is necessary to establish and maintain control over a number of critical process parameters (CPPs).

5.14 Linking CQAs with CPPs

Process factors impact various features of a formed medication, primarily the API, even if real CQAs are in the final specification. According to Chris Senanayake, vice-president of chemical development at Boehringer Ingelheim Pharmaceuticals, the features like (total purity, impurity levels, polymorph(s), solvates, pH, water content, particle size) must be connected to the CPPs for each API production process.

Gert Thurau, group head, Technical Regulatory at Roche, says the most prevalent CQAs for APIs are chemical process impurities and drug substance physical features (e.g., particle size, polymorphism). Controlling impurities and potential active substance solid-state characteristics ensures product safety and bioavailability. He adds that once these traits are identified, it's feasible to identify weather is vital to the operation and develop a product that is closer to quality by design as well as very resilient.

According to Dr. Reddy's Custom Pharmaceutical Services (CPS) assistant director of quality assurance Vishwanath Nadig, a high assay indicates that the medicine is operating effectively and a low impurity level signifies that patients are safe. Nadig also says that the amount of moisture is important because, in some cases, moisture can speed up the breakdown of the drug substance.

If the finished product includes these CQAs, it will operate as planned. According to Senanayake, CPPs should be chosen because they relate to the final CQAs and, when regulated, ensure product quality.

Cambrex's director of regulatory relations, quality, and compliance Mark TePaske defines CPPs as "those process parameters that have an impact on the CQAs of the API." An acceptable range must be determined for each essential process parameter before it can be used in routine production and process control. Nadig emphasises the crucial nature of establishing suitable Critical Process Parameters (CPPs) for each specific step in order to ensure the drug's quality, which subsequently has the potential to influence its efficacy.

According to TePaske, the establishment and fulfilment of Critical Process Parameters (CPPs) and Critical Quality Attributes (CQAs) enables integration of quality into process and provides a level of resilience that extends beyond the scope of release testing. The author additionally observes that clearly defined Critical Process Parameters (CPPs) are essential for maintaining a process in a controlled and validated state. This is of utmost importance as process validation is a mandatory requirement under current Good Manufacturing Practises (cGMP). According to Thurau, a complete risk-based approach to product and process development begins with defining of CQAs for final medicinal commodity and API. Critical quality features for both the finished product and the medication ingredient are determined from the analysis's starting point, which is often the quality target product profile. It's crucial to interpret the chemistry of the distinct experimental stages, he says, since the process-related impurities that come from the beginning and raw materials are the most significant CQAs of the final API. Senanayake argues that risk assessment is only one factor among many in establishing CQAs and CPPs, along with previous information, intended trials, and regulatory expectations.

5.15 Risk-Assessment Approaches

Once Critical Quality Attributes (CQAs) have been put forward, usually during the initial phase of scaling up, it becomes imperative to determine the suitable Critical Process Parameters (CPPs). According to Senanayake, it is the belief of Boehringer Ingelheim that the utilisation of a DoE based approach is the most effective methodology to illustrate interrelationship between critical process parameters and critical quality attributes in the context of individual process steps. According to Nadig, in addition to conducting common risk assessments, the usage of design failure mode and effects analyses (FMEA) proves to be efficacious in delineating crucial parameters pertaining to therapeutically potential active substances and fabrication processes. According to the speaker, when employed in a suitable manner, these two methodologies prove to be adequate for delineating the significant characteristics.

According to Senanayake, although orthogonal experiments can also be employed, the use of Design of Experiments (DoE) provides valuable insights into the conception space for each operation that cannot be acquired through blocking out experiments. However, screening experiments still are beneficial for ascertaining specific elements of unit operations, for instance identifying optimal solvents or catalysts or promoters of a reaction. According to TePaske, it is necessary to experimentally affirm critical attributes once they have been defined, and establish control limits. In order to achieve this, processes, procedures, equipment, and/or facilities may need to be revised as necessary. It is important to acknowledge that contemporary risk-assessment methodologies exhibit a significantly higher level of rigour compared to their historical counterparts. TePaske (year) notes that historically, the responsibility for defining the process and establishing criteria for critical quality attributes (CQA) was primarily utilized by the formulation and product design development chemist. Furthermore, extent as well as comprehensiveness of the researches conducted are contingent upon the expertise and capabilities of the chemist involved. Consequently, the perception of process development was predominantly centred around its execution within a laboratory setting. The author proceeds to observe that contemporary methodologies exhibit a greater degree of rigour, with research endeavours being characterised by the involvement of project teams comprised of individuals from various disciplines. The aforementioned teams engage in the evaluation and harmonisation of discoveries and risk evaluations, while simultaneously incorporating the apprehensions of various departments such as production, materials, quality, engineering, and others that are implicated in the manufacturing process. The definition and development process of CQA have undergone a transformation from a mere laboratory exercise to a worldwide endeavour aimed at preparing for process validation and commercial manufacturing in accordance with Good Manufacturing Practises (GMP). The author arrives at the determination that adopting a teamed-up approach is the optimal method, as this effectively incorporates requirements of all branches engaged in mercantile manufacturing. Moreover, this approach significantly reduces the likelihood of determining CQAs and CPPs that cannot be executed due to limitations existing in the facilities and equipment.

5.16 Colligating CPPs to CQAs

Thurau suggests investigating alternative reaction and work-up settings and thoroughly analysing the products from each process stage for impurity profiles and other attributes based on the quality risk assessment. For instance, impurity structures can be determined, allowing experimental conditions to be lined up so as to minimise impurities or the course in downstream steps. He says contaminants are eliminated or set at safe levels. He emphasises the need of being methodical and linking CPPs under examination to the drug product or active agent CQAs.

5.17 Obstructions for Developing CPPs

QbD and CPPs related to end product and drug substance CQAs have numerous advantages, but adopting them is difficult. Senanayake says time and money are the primary obstacles to process knowledge. The author asserts that it is imperative to overcome these obstacles, as a comprehensive understanding of a process in its commercial manufacturing stage is crucial for achieving success. Consequently, it's requisite to allocate sufficient time to thoroughly comprehend the response space. Thurau concurs that primary constraints are predominantly practical in essence and generally be surmounted. The author acknowledges the existence of unresolved inquiries regarding the explanations and variations in the suitability of the idea as determined through different federal agency. When the level of detail and specificity in qualification of products developed and manufactured through the Quality by Design (QbD) increases, it is anticipated that the enhanced understanding will facilitate more adaptable change management practises (i.e., "do and tell") rather than the prevailing rigid approach (i.e., "tell and do").

QbD and CQAs/CPPs need skilled team members from throughout the organisation, according to Nadig and TePaske. TePaske believes the interdisciplinary approach needs direct engagement of various spheres of studies whose major task responsibility has traditionally been manufacturing and demands fresh thinking. He argues that risk-assessment might hinder innovation by focusing on what can be done.

5.18 Applicative Use of CPPs

Despite few challenges as well as constraints, several businesses recognise usefulness of QbD strategy and need of integrating essential parameters for specific trail stages with crucial quality features of small-molecule APIs and formulated medicinal products. According to Senanayake and Boehringer Ingelheim evaluating process robustness throughout development and utilising batch histories to determine how it converts to commercial scale. The corporation is also conducting DOEs on key production phases. "Our validation efforts and control charts for individual steps show our success," he adds. According to Nadig, while developing Critical Performance Parameters (CPPs) for each process step at Dr. Reddy's CPS, a cross-functional responsible group of scientists meet to analyse the crucial parameters of the process parameters and material characteristics. These critical sensitivities are first measured in significance, and then tolerances on the variables are set. After that, DOE screening is carried out in order to determine which parameters are most important, and this is followed by optimisation and the development of the experimental space in order to comprehend what will happen to the CQAs based on the various parameter values. During the pre-planning discussions, he says, "it is important for us to try to engage senior members from cross-functional teams so that an exhaustive risk assessment can be conceptualised and a mitigation plan can be worked out." Throughout the whole procedure, the organisation places a strong emphasis on maintaining open lines of communication, cooperating with customers, and providing due consideration to the interests of all relevant stakeholders.

5.19 Potential Consequences

The consequences of failing to connect CPPs to CQAs might be severe. If the right controls aren't in place at a certain stage of production, "a material may not meet criteria for reasons we don't know." If the intermediate goes through the commercial process, it may be recalled. Senanayake adds that systems without sufficient controls can produce low-quality or inconsistent products. As per Thurau, whenever unsuitable process parameters or in-process material characteristics are set, then the end product might look high, but it's due to the consequence of screening quality into it. At last, TePaske emphasises that, process validation and performance in a validated condition need control.

5.20 Conclusions

- An improved understanding of the Target Product Quality Profile and its role in establishing a numeric performance target for QbD.
- For determination of the important material characteristics that serve as a mechanistic connection between the product quality and production process.
- It is important to note that crucial process parameters refer to unit operation parameters and need to be considered in conjunction with vital material properties to effectively describe relationship between experimental unit operation input and output.
- A comprehensive classification system for production parameters and in-process material properties, encompassing non-critical, unclassified, and critical categories.
- The control strategy plays a crucial role in gradually implementing QbD elements into practical applications.
- A strategic approach to navigating a design space by identifying process variables that do not interact and subsequently excluding them from formal experimental designs.

References

1. Juran, J. M. (1992). *Juran on quality by design: The new steps for planning quality into goods and services*. The Free Press.
2. Woodcock, J. (2004). The concept of pharmaceutical quality. *American Pharmaceutical Review, 47*(6), 1–3.
3. U. S. Food and Drug Administration. (2009). Guidance for industry: Q8 (2) pharmaceutical development.
4. U. S. Food and Drug Administration. (2006). Guidance for industry: Q9 quality risk management.
5. U. S. Food and Drug Administration. (2009). Guidance for industry: Q10 pharmaceutical quality system.
6. U. S. Food and Drug Administration. (2011). Guidance for industry: Q8, Q9, and Q10 questions and answers.
7. ICH Quality Implementation Working Group. (2011). Points to consider. ICH-endorsed guide for ICH Q8/Q9/Q10 implementation.
8. U. S. Food and Drug Administration. (2012). Guidance for industry: Q11 development and manufacture of drug substance.
9. U. S. Food and Drug Administration. (n.d.). FDA-EMA parallel assessment of quality-by-design elements of marketing applications. Retrieved from http://www.fda.gov/Drugs/DrugSafety/ucm365524.html

10. Yu, L. X. (2008). Pharmaceutical quality by design: Product and process development, understanding, and control. *Pharmaceutical Research, 25*, 781–791. https://doi.org/10.1007/s11095-007-9511-1
11. Lionberger, R., Lee, S., Lee, L., Raw, A., & Yu, L. X. (2008). Quality by design: Concepts for ANDAs. *The AAPS Journal, 10*, 268–276. https://doi.org/10.1208/s12248-008-9026-7
12. Raw, A. S., Lionberger, R., & Yu, L. X. (2011). Pharmaceutical equivalence by design for generic drugs: Modified-release products. *Pharmaceutical Research, 28*, 1445–1453. https://doi.org/10.1007/s11095-011-0397-6
13. Rathore, A. S., & Winkle, H. (2009). Quality by design for biopharmaceuticals. *Nature Biotechnology, 27*, 26–34. https://doi.org/10.1038/nbt0109-26
14. The advantages of a 'Quality by Design' approach in pharma drug development, December 2019, https://doi.org/10.13140/RG.2.2.21090.04802. Report number: Pharma Manufacturing: https://www.pharmamanufacturing.com/articles/2019/the-advantages-of-a-quality-by-design-approach-in-clinical-and-commercial-pharma-development/Affiliation:Recipharm
15. ICH Q9. (2006). Quality risk management. ICH harmonized tripartite guidelines. In *International Conference on Harmonization of Technical Requirements for Registration of Pharmaceuticals for Human Use*.
16. ICH Q10. (2007). Pharmaceutical quality system, ICH tripartite guidelines. In *International Conference on Harmonization of Technical Requirements for Registration of Pharmaceuticals for Human Use*.
17. ICH Q11. (2012). Development and manufacture of drug substances, ICH tripartite guidelines. In *International Conference on Harmonization of Technical Requirements for Registration of Pharmaceuticals for Human Use*.
18. Vishwasrao, S. S., & Singh, S. (2016). Current perspective on opportunities and adoption challenges of QbD implementation in pharmaceutical product development. *Invent Rapid, 2*, 1–8.
19. Milmo, S. (2014). Quality by design: Bridging the gap between concept and implementation. *BioPharm International, 27*, 40–45.
20. Chen, C.-V., Moore, C., & ONDQA/CDER/FDA. *Role of statistics in pharmaceutical development using quality by design approach—An FDA perspective*. FDA/Industry Statistics Workshops.
21. Nasr, M.M. (2006). Implementation of Quality by Design (QbD)—Status, challenges and next steps. ONDQA, OPS, CDER, Advisory Committee of Pharmaceutical Science (ACPS), 5 October.
22. Nasr, M. M. (2007). Quality by Design (QbD)—A modern systemic approach to pharmaceutical development and manufacturing—FDA *(3) (PDF) opportunities and challenges to implementing quality by design approach in generic drug development*.
23. Winkle, H. N. (2007). Consumer healthcare products association. In *39th Manufacturing Controls Seminar*; 4 October, New Brunswick, NJ: OPS,CDER, FDA.
24. (2007). *Annual Meeting Programming: RS (Regulatory Science) Highlights and discussion* (pp. 1–4).
25. Zhang, L., & Mao, S. (2017). Application of quality by design in the current drug development. *Asian Journal of Pharmaceutical Sciences, 12*(1), 1–8. *The paper covers the applications of quality-by-design in drug development.
26. Box, G. E. P., & Wilson, K. B. (1992). On the experimental attainment of optimum conditions. In S. Kotz & N. L. Johnson (Eds.), *Breakthroughs in statistics: Methodology and distribution* (pp. 270–310). Springer New York.
27. Collins, A. (1992). Toward a design science of education. In *New directions in educational technology* (NATO ASI series) (Vol. 96, pp. 15–22). Springer.
28. Yang, Y. (2008). *Multiple criteria third-order response surface design and comparison*. Florida State University.
29. Box-Behnken designs: NIST/SEMATECH. (2003). Retrieved January 23, 2018, from http://www.itl.nist.gov/div898/handbook/pri/section3/pri3362.htm

30. Box, G. E. P., Hunter, W. G., & Hunter, J. S. (1978). *Statistics for experimenters: An introduction to design, data analysis, and model building* (Wiley series in probability and statistics) (Illustrated ed., Vol. 154). Wiley.
31. Araujo, P. W., & Brereton, R. G. (1996). Experimental design II. Optimization. *Trends in Analytical Chemistry, 15*(2), 63–70.
32. Doornbos, D. A., & de Haan, P. (2001). Optimization techniques in formulation and processing. In J. Swarbrick & J. C. Boylan (Eds.), *Encyclopedia of pharmaceutical technology* (Vol. 20 (Suppl. 3), pp. 77–160). Marcel Dekker.
33. Pinto, J., Podczeck, F., & Newton, J. (1997). Investigations of tablets prepared from pellets produced by extrusion and spheronisation. II. Modelling the properties of the tablets produced using regression analysis. *International Journal of Pharmaceutics, 152*(1), 7–16.
34. Podczeck, F. (1996). The development and optimization of tablet formulations using mathematical methods. In G. Alderborn & C. Nyström (Eds.), *Pharmaceutical powder compaction technology* (pp. 561–593). Informa Healthcare.
35. Chatchawalsaisin, J., Podczeck, F., & Newton, J. M. (2004). The influence of chitosan and sodium alginate and formulation variables on the formation and drug release from pellets prepared by extrusion/spheronisation. *International Journal of Pharmaceutics, 275*(1), 41–60.
36. Lewis, G. A., Mathieu, D., & Phan-Tan-Luu, R. (1998). *Pharmaceutical experimental design*. CRC Press.
37. Scheffé, H. (1958). Experiments with mixtures. *Journal of the Royal Statistical Society, Series B (Statistical Methodology), 2*(2), 344–360.
38. Taguchi, G., & Organization AP. (1986). *Introduction to quality engineering: Designing quality into products and processes*. Unipub/Quality Resources.
39. Lewis, G. (1991). Non classical experimental designs in pharmaceutical formulation. *Drug Development and Industrial Pharmacy, 17*(12), 1551–1570.
40. Wehrlé, P., Palmieri, G., & Stamm, A. (1994). The Taguchi's performance statistic to optimize theophylline beads production in a high-speed granulator. *Drug Development and Industrial Pharmacy, 20*(18), 2823–2843.
41. Chariot, M., Lewis, G., Mathieu, D., et al. (1988). Experimental design for pharmaceutical process characterisation and optimisation using an exchange algorithm. *Drug Development and Industrial Pharmacy, 14*(15–17), 2535–2556.
42. Nielloud, F., Mestres, J. P., Fortuné, R., et al. (2003). Formulation of oil-in-water submicron emulsions in the dermatological field using experimental design. *Polymer International, 52*(4), 610–613.
43. Rechtschaffner, R. (1967). Saturated fractions of 2n and 3n factorial designs. *Technometrics, 9*(4), 569–575.
44. Challener, C. A. (2014). QbD in API manufacturing. *Pharmaceutical Technology, 38*(9).

Chapter 6
Quality by Design in Formulation Development

Camillo Benetti and Ayca Altay Benetti

Abstract Formulation development moved to Quality-by-Design in recent years under the aegis of the FDA. Quality-by-Design allow manufacturers to build the quality into the process instead of simply testing the final product and it also allow to manage small changes of the final formulation or dosage form without a complete redesign of the production line.

Quality-by-Design require the planning of experiments using Design-of-experiments in order to calculate the contribution of each factors and their interactions for the final products and what happened inside the space described by those experiments so that the outcome of every single variations will be known before doing the experiment itself. Besides Design-of-Experiments allows to include statistics inside the experiments rather than at end.

Quality-by-Design require planning, rigorous testing and, if possible, prior knowledge of the subject to better guide the process. All those procedures need to be thoroughly documented and subjected to continuous improvement and discussion.

Keywords Quality by Design · Formulation development · Design of experiments · Quality control

6.1 Introduction

Formulation development is the process of designing and developing a stable and effective dosage form that contains one or more active pharmaceutical ingredients (APIs) and other excipients in a suitable form for administration to patients. The formulation should have a suitable kinetics of release, be stable over time (shelf life) and improve patient compliance, either by reducing the number or size of administrations or making the formulation itself more palatable.

C. Benetti (✉) · A. A. Benetti
Department of Pharmacy, National University of Singapore, Singapore, Singapore
e-mail: ben.cam@nus.edu.sg; ayca.ben@nus.edu.sg

N. K. Jain, N. Bajwa (eds.), *Introduction to Quality by Design (QbD)*,
https://doi.org/10.1007/978-981-99-8034-5_6

Formulation development involves a series of steps, including selecting appropriate excipients, determining the optimal dosage form, developing a formulation that provides the desired release profile and evaluating the stability of the formulation over time. The process also includes assessing the compatibility of the formulation with packaging materials and designing appropriate packaging for the product. Formulation development is a critical part of the drug development process, as it can significantly impact the efficacy, safety and patient acceptance of a drug product. A well-designed formulation can improve the bioavailability and therapeutic efficacy of an API, while a poorly designed formulation can lead to adverse effects, reduced efficacy or even product failure [1].

In recent years, FDA pushed for the use in Quality by Design (QdB) in formulation development thanks to its regulatory framework and guidance documents, which encourage the implementation of QbD principles throughout the product lifecycle [2, 3].

Quality by design is a systematic and holistic approach that emphasizes the importance of understanding the impact of formulation and process variables on product quality and properties; it covers the entire product lifecycle, from development to commercialization and focuses on the identification of the targeted product profile (TPP) and control of critical quality attributes (CQAs) that are important for product performance and safety.

Quality by Design is an idea of Joseph M. Juran [4] who thought quality could be arranged and any issue can be identified with the manner by which quality was arranged. The concept has been received with FDA that provided a framework for the QdB in pharmaceutical development. The QbD approach to formulation development involves a series of steps that are designed to ensure that product quality is built into the formulation so that the product can be free of contaminants and deliver the therapeutic effect [3, 5].

The QdB has also the advantage of reducing product variability that may lead to rejection and waste [6].

In order to help the implementation of the QbD in to the pharmaceutical development, FDA has taken some steps, which are:

1. Guidance Documents: FDA has issued several guidance documents that provide detailed instructions on implementing QbD principles in pharmaceutics; those documents are available for researcher and academics to align their research and this helps the transition between laboratories to manufacturing.
2. Lifecycle Approach: FDA encourages the application of QbD principles from early drug development through commercial manufacturing covering the whole product lifecycle; this allow to increase the knowledge of the product in very stage and it also help for some minor modifications.
3. Design Space and Control Strategy: it is a multidimensional space that defines the acceptable ranges of process parameters and material attributes for the manufacturing process together with a control system; this is the space covered by the experiments and it is where the relationship between factors and final product are calculated.

4. Process Analytical Technology (PAT): it is a combination of advanced analytical techniques, real-time monitoring and control strategies to understand and control critical process parameters and quality attributes. This continuous control is necessary to ensure that all the parameters work inside the limits that ensure the final quality of the product.
5. Quality Risk Management (QRM): it is a systematic approach to quality control based on risk assessment and reduction; this quality control method was extended throughout the product lifecycle thanks to the QbD framework .
6. Regulatory Flexibility in order to facilitate the implementation of QbD in the regulatory processes. This framework consist of:

   - ICH Q8 for pharmaceutical development;
   - ICH Q9 for quality risk management;
   - ICH Q10 for quality systems;
   - ICH Q11 for development and manufacture of drug substance.

6.2 Quality by Design in Formulation Development

The aim of QdB is to ensure the consistent production of a safe and effective product among all the stages of development, minimizing the number of out-of-specifics products [7, 8].

In QbD, three key elements are considered: Design, Understanding, and Control.

Design in QbD refers to a profound and comprehensive understanding of the product and process: every parameters that may affect the final outcome need to be investigated, properly described and quantified. It is not enough to identify what can affect the finall product but also how much the parameters can change before the expected outcome is not reached. In order to do that, several techniques ranging from risk assessment to the modelling are used to study the relationship between those parameters and the product.

Understanding is the acquisition of the necessary knowledge through experiments, data analysis and process characterization in order to identify the potential risks and the source of the possible variability into the final product. This understanding is necessary to define the control strategies need.

Those control strategies are the way to monitor the production and the intermediate by-products to ensure consistent product quality: reducing any risk identified during the previous step or preventing any deviation from the established protocols, it is possible to achieve and maintain the desired quality attributes.

QbD promotes a proactive and science-based approach to quality assurance, reducing the reliance on end-product testing and facilitating continuous improvement [8, 9].

In order to incorporate QbD in formulation development, the usual quality control system, based on the testing of the final product need to be completely revamped to accommodate real-time process controls and design of experiments [10–12].

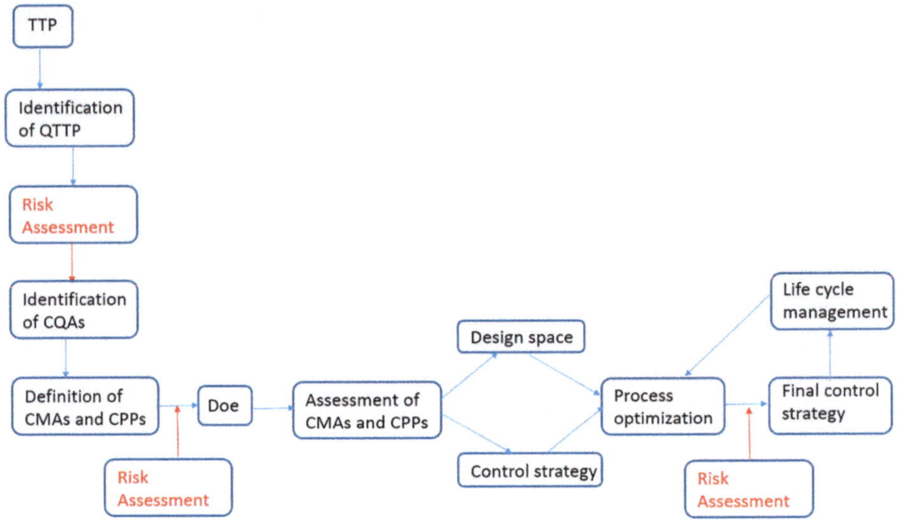

Fig. 6.1 Quality-by-Design in formulation development

This process can be summarized in the image below (Fig. 6.1).

6.2.1 *Target Product Profile and Quality Drug Profile*

The target product profile (TPP) is a strategic document that defines the desired attributes of the final product, including its intended use, dosage form, stability, clinical data and technical details of the final form like packaging. The TTP is a key document for regulatory bodies and it needs to be update regularly to reflect any changes in the science related to the final product [13, 14].

The TPP describes the final product but it doesn't contain any information on how to manufacture that product. The TPP need to be converted in the quality target product profile (QTPP) which describe the quality attributes of a drug product. The QTPP, according to ICH8, translates TPP in quantitative tests that needs to be fulfilled to meet the quality standards of the final product.

6.2.2 *Critical Quality Attributes*

Once QTPP is defined, the next step is to identify critical quality attributes (CQAs): CQAs are the product attributes that have a significant impact on its safety, efficacy and performance. The identification of CQAs is based on a comprehensive understanding of the product, its intended use and the patient population [3].

CQA refers mostly to raw materials and intermediate by-products and are a subset of the QTPP because some characteristics of the final product may not be defined as CQA, like strength and dosage form.

Identification of QCA is part of the ICH9 guidance and it is performed via risk assessment, together via prior knowledge stemming from laboratory practice and preclinical trials.

Critical material attributes (CMAs) and critical process parameters (CPPs) are also part of CQAs via risk assessment and may cause variation in the CQAs.

6.2.2.1 Critical Material Attributes

Critical material attributes (CMAs) are the specific characteristics of a raw material or component that can significantly affect a pharmaceutical product.

CMAs play a vital role in the manufacturing process and are considered critical because they have a direct impact on product performance and if the product can actually meet the predefined acceptance criteria. In order to determine if a property is really critical for quality and performance of a pharmaceutical product, not only a risk assessment is performed, but it is also necessary to have knowledge of the materials and the processes where those are involved.

Because of the effects of CMAs on the process, any variation need to be monitored to ensure that both the material and the final product stay inside the specifications. Those variations may not only depends from batch to batch variations but also on the processes in which the material is involved.

Regulatory authorities, such as the FDA, require the manufacturers to properly identify and describe in details the effects of CMA and their impact on the final product during all stages of the product lifecycle.

6.2.2.1.1 Critical Material Process Parameters

Critical process parameters (CPPs) are the key factors that affect the manufacturing process and the final product. Like CMAs they are identified via risk assessment and prior knowledge and they do also require a proper control plan to assess if they specifications are met. The use of CPPS allow manufacturers to improve the manufacturing process reducing variability and waste and increasing efficiency.

The difference between CQA, CMA and QPP is summarized in the table below (Table 6.1).

In summary, CQAs represent the desired quality characteristics of the final product, CMAs are the critical attributes of the materials used in manufacturing and CPPs are the key process variables that need to be controlled to ensure consistent product quality. All three elements are interconnected, as variations in CMAs and CPPs can directly impact CQAs and controlling CMAs and CPPs is crucial for achieving the desired product quality.

The selection process of CQA is shown in the Fig. 6.2 below:

Table 6.1 Summary of CQA, CMA and QPP

	Critical quality attributes	Critical materials attributes	Critical process parameters
Definition	Measurable properties of a pharmaceutical product that are critical to its role or use	Characteristics of a raw material or component that can significantly impact the quality, safety, and efficacy of a pharmaceutical product	Factors that have a significant impact on the manufacturing process and can influence the quality
Role	CQAs are the quality characteristics that determine the performance and acceptability of the final product from the patients	CMAs focus on the characteristics of the materials or components used in the manufacturing process that can affect the product's quality and performance	CPPs are the specific parameters within the manufacturing process that need to be controlled within defined ranges to ensure consistency and reproducibility of product quality
Examples	Potency, purity, content uniformity, dissolution rate, particle size distribution, stability, sterility	Particle size, purity, chemical composition, solubility, moisture content, polymorphic form or bioburden	CPPs can include parameters such as temperature, pressure, time, mixing speed, drying time, pH, agitation rate or sterilization conditions

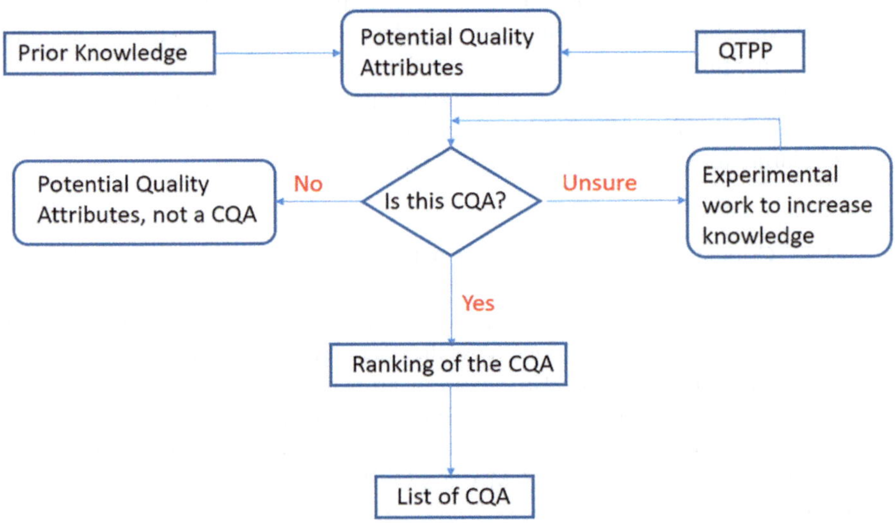

Fig. 6.2 Selection process for CQA. (Adapted from Shoaeb [5])

6.2.2.1.2 Role of Pharmaceutical Excipients

A major source of variability that should be addressed via risk assessment and QCA are the pharmaceutical excipients [3, 5, 13].

Pharmaceutical excipients are components of a drug product other than the active pharmaceutical ingredient, which may affect drug delivery, safety and identification of the final dosage form.

They are classified by their functions and their role in pharmaceutical risk assessment can be summarized as follows:

- Safety Evaluation: excipients undergo rigorous safety evaluations to ensure they do not pose any significant risks to patients. These evaluations include assessing their toxicological profiles, potential interactions with the API and possible adverse effects. Excipients should comply with regulatory guidelines and be approved for use in pharmaceutical formulations.
- Compatibility Assessment: excipients must be compatible with the API and other components of the formulation. Compatibility studies examine the physical, chemical and functional interactions between excipients and the API to ensure stability, efficacy and safety. Incompatibilities may lead to degradation of the API, decreased efficacy or adverse reactions.
- Manufacturing and Quality Control: excipients contribute to the manufacturability and quality control of pharmaceutical products. They aid in processes such as tablet compression, granulation and coating, ensuring proper formulation characteristics and dosage uniformity. Quality control measures involve testing excipients for identity, purity, strength and other specifications to maintain consistent product quality.
- Pharmacokinetic Considerations: excipients can influence drug absorption, distribution, metabolism and excretion (ADME) characteristics. In risk assessment, excipients are evaluated for their potential impact on the bioavailability and pharmacokinetic behaviour of the API. Any alterations in drug release, dissolution or solubility caused by excipients need to be assessed for their impact on therapeutic efficacy.
- Allergenicity and hypersensitivity: some excipients may have the potential to cause allergic reactions or hypersensitivity in certain individuals. Risk assessment involves evaluating the allergenic potential of excipients through testing, literature review and monitoring adverse events. Information on potential allergenic excipients should be provided to healthcare professionals and patients.
- Regulatory Compliance: excipients are subject to regulatory oversight and their use in pharmaceutical products must comply with applicable regulations and guidelines. Risk assessment ensures that excipients meet the necessary regulatory requirements, including their safe use, appropriate labelling and documentation of specifications.

Interaction between different drugs are also a possible source of variability that need to be addressed via compatibility studies. These studies are crucial to ensure that the components used in the formulation or manufacturing process do not negatively interact with each other, leading to potential degradation, loss of efficacy or safety concerns [15].

There are several mechanisms that may lead to issues with the compatibility between excipients, drugs, solvent or even the container.

- Physical compatibility relates to the ability of the formulation components to blend uniformly and maintain their physical characteristics throughout the manufacturing process. For example, in solid dosage forms like tablets or capsules, it is important to ensure that the API and excipients can be effectively mixed and compressed into a solid mass without segregation or stratification. In liquid formulations, physical compatibility includes the ability of the components to dissolve, disperse or suspend uniformly.
- Chemical compatibility refers to the interactions between the formulation components that could lead to chemical degradation or unwanted reactions. Components should be evaluated for potential incompatibilities such as hydrolysis, oxidation or degradation caused by pH, temperature, light or other factors. Chemical compatibility studies are conducted to ensure that the API and excipients remain stable and do not undergo significant chemical changes that may impact the therapeutic efficacy or cause harmful effects.
- Container compatibility is critical to maintain product stability and prevent interactions that could compromise safety or efficacy. Container compatibility considerations include evaluating the potential for leaching or adsorption of the formulation components into the container material, as well as interactions between the container and the formulation that may affect stability, such as moisture permeation, light exposure or gas permeability. Compatibility testing is performed to ensure that the container is appropriate for the formulation and does not introduce any adverse effects or changes.
- Process compatibility with the manufacturing process involves assessing how the formulation components interact with the processing conditions and equipment. The API and excipients should be compatible with the chosen manufacturing techniques such as blending, granulation, compression or coating. The processing parameters, such as temperature, pressure or shear forces, should not cause significant degradation or alter the properties of the formulation components. Compatibility studies help identify any process-related risks and ensure that the manufacturing process does not compromise the quality or integrity of the product.

6.2.3 Risk Assessment

Risk assessment is also known as process mapping, where all the processes or unit operations are sequentially listed in the form of a flow chart. The oldest one is the Ishikawa diagram (Fig. 6.3), also known as fishbone diagram for its shape: it is a risk identification and analysis tool that provides a systematic way of looking at effects and causes [9, 16, 17].

There are several methods for risk assessment according to ICH guidelines and are:

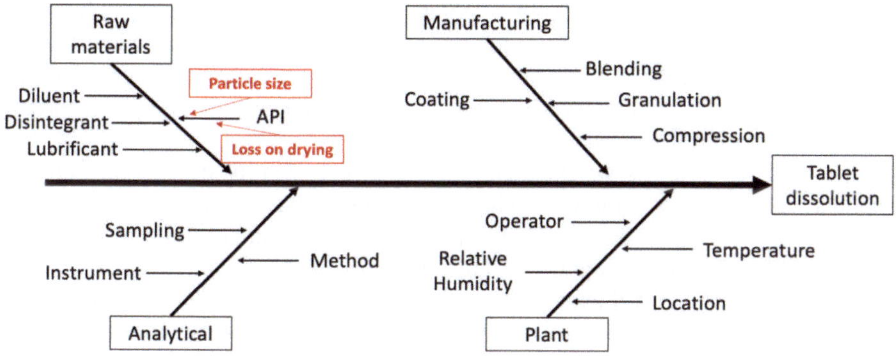

Fig. 6.3 An example of Ishikawa diagram for the tablet dissolution

- Failure mode effect analysis (FMEA);
- Failure mode effect and critical analysis (FMECA);
- Fault tree analysis (FTA);
- Hazard analysis and critical control points (HACCP);
- Hazard operability analysis (HOP/HAZOP);
- Preliminary hazard analysis (PHA);
- Relative risk-based matrix assessment (RRMA);
- Risk ranking and filtering.

FMEA is a quantitative tool to identify the actual critical material and process parameters that affect the critical quality attributes of a pharmaceutical product together with a risk priority number that is obtained by multiplying the score for probability of occurrence, severity and detectability. There is limit for RPN set for the process, which point out if the parameter is critical [18].

FMECA is an evolution of FMEA and it involves breaking down a system or process into its constituent components and identifying the potential failure modes for each component. Once the failure modes are identified, the next step is to analyze the effects of each failure mode. This analysis involves determining the consequences or impacts that the failure mode may have on the system or process, including the potential harm to people, the environment, equipment or overall system performance. The effects analysis helps prioritize the most critical failure modes for further investigation and mitigation [19].

Fault Tree Analysis (FTA) is a systematic and graphical method used to analyze the various combinations of events or failures that can lead to a specific undesired or critical outcome using a logical diagram that represents the relationships between events and failures that can contribute to the occurrence of the undesired event. The fault tree visually depicts the cause-and-effect relationships and shows how different failures can combine to cause the top-level event using what are known as gates. In order to build a fault tree, it is necessary to identify the relevant events and their relationships based on expert knowledge, historical data, system documentation and analysis of failure modes [20].

Hazard analysis and critical control points (HACCP) is a systematic approach to safety management that identifies and controls potential hazards after identification of critical points and their limits. This method also involves a continuous monitoring and established verification and corrective actions, together with a proper recording and documentation system [21].

The HAZOP analysis systematically explore and identify deviations from the intended operation that could lead to hazardous situations, considering past incidents that had likelihood for catastrophic consequences, human-controlled factors and consequences of failure of applied control measures [22].

The preliminary hazard analysis (PHA) is usually the first attempt in the system safety process to identify potential hazards associated with the running operation.

Relative risk-based matrix assessment is a method used to evaluate and compare the risks associated with different hazards or scenarios by assigning relative risk scores and prioritizing them based on their severity and likelihood, creating therefore a matrix which provides a structured approach to allocate resources effectively [23].

Risk ranking and filtering is a process to assess potential risks based on their severity and likelihood in order to prioritize the most critical ones so that resources (that are often limited) can be allocated where they are more effective [16].

6.2.4 Design Space

The design space is the range of formulation and process variables that have been determined to be safe and effective in producing a product with the desired quality attributes. It is defined through a series of experiments and analyses that explore the impact of various variables and their interactions on product quality and it is subject to regulatory approval.

The design space include the control space, which is the space inside which variables, process parameters and excipients ensures consistent production within established limits. The design space is a part of the bigger knowledge space which is the summa of all the information available on the product, its manufacturing and its life-cycle as explained in Fig. 6.4.

The design space is linked to the results of the risk assessment and it describes mathematically all the relationship between QCA and the critical control points [3, 14, 17, 24].

Design space may be scale and equipment dependent and therefore the scale-up of a process require the revamping of the design space due to differences in the mechanical and physical properties of the equipment itself.

In order to help during this assessment, the experiments can be designed in order to highlight the contribution of the single variables and how those variables interact between each other's. This methodology is called Design of Experiment or DoE.

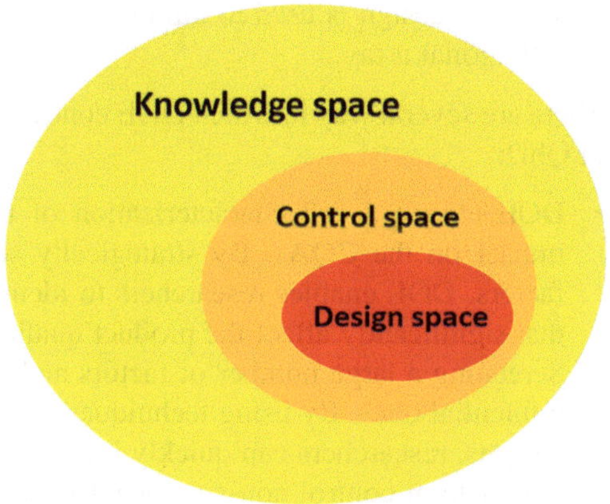

Fig. 6.4 Control, design and knowledge space

6.2.4.1 Design-of-Experiments

Design of experiments was first introduced by Sir Ronald Fisher in the 1930s and it postulates the application statistical analysis during the planning phase rather than at the end: the input factors are varied systematically to study their effect on the responses, connecting CPPs with the CQAs via mathematical equations [25, 26].

The building of a design is composed of four parts: objective, input variables with their levels, measured response variables and design.

- The type design depends on the issue and the aim (screening vs. optimization designs), number of factors and their interaction which will affect time and resources needed for the study and prior knowledge about the subject.
- Factorial designs are used for process development when the factors that are independent of each other and the responses depend on the levels of the factors. Process parameters are varied based on a predetermined matrix of factor level combinations [26, 27].
- Fractional factorial design is a part of the full factorial design, used for screening; its main limitation is the aliasing of the main effects and the interactions, which affects the resolution of the design [27].
- Plackett–Burman design is mostly used for screening, because the main effects are heavily confounded with two-factor interactions: therefore this design is viable for a quick study of the main effects when the interactions between factors can be considered negligible [28].
- A central composite design is a full quadratic model used when sequential experimentation is required and it is also known as response surface design. This model is suitable in case of continuous variables and it can describe different kinds of relation between variables in order to minimize or maximize the response [28].
- Box–Behnken design is a small response surface design with three levels for each factor and it can estimate up to second order coefficients [29].

- Taguchi design is used to minimize the noise from uncontrollable factors via an orthogonal array.

There are several way in which DOE contributes to the formation of the design space in QbD:

- DOE allows a full characterization of the input variables and their potential impact on the CQAs. By strategically selecting and varying the levels of the factors, DOE enables researchers to identify the critical factors and their ranges that significantly affect the product quality.
- Screening a large number of factors and their combinations to identify the most influential ones. By using techniques like factorial designs or fractional factorial designs, researchers can quickly identify the key factors affecting the CQAs and eliminate or control non-significant factors, reducing the number of experiments needed.
- Determining optimal operating conditions thanks to the optimal operating conditions within the design space. By using response surface designs, researchers can model the relationship between the input variables and the CQAs and then optimize the process by identifying the factor levels that result in the desired product quality.
- Robustness: DOE helps in assessing the robustness of the design space by evaluating the effects of process parameter variations and their interactions. By conducting experiments under different conditions and analyzing the response variations, researchers can determine the tolerances for the input variables to maintain product quality within the desired range.
- Regulatory compliance: DOE provides a systematic and scientific approach to understand and control the critical factors affecting product quality. By employing DOE in the design space formation, pharmaceutical companies can meet the regulatory requirements and demonstrate a thorough understanding of their processes, ensuring consistent product quality.

Using DOE techniques, pharmaceutical companies can design robust processes, reduce development time and costs and enhance the quality and reliability of their products [28, 30, 31].

6.2.5 Control Strategy

The control strategy, according to ICH10, outlines the measures that will be taken to ensure that the product consistently meets the defined quality attributes throughout its lifecycle. This includes process controls, specifications and testing protocols.

The finished products are tested to assess if they meet the expected requirements, but the quality depends on manufacturing and controlling variables.

Control strategy includes three levels:

- Level 1: automatic engineering controls to monitor the critical quality attributes of the output variables in real-time using process analytical technology (PAT) or predictive models [3].
- Level 2: reduced quality control of the final product with flexibility in the material attributes and process parameters within the design space.
- Level 3: traditional level of control used in the pharmaceutical industry, depending on end-product testing and constrained material attributes and process parameters.

A hybrid approach combining levels 1 and 2 can also be used according to ICH Q8. This hybrid system need to be derived from current product and process understanding that ensures process performance and product quality.

6.2.5.1 Process Analytical Technology

PAT provide continuous monitoring of CPPs and CMAs and it can therefore indicate if the process is remaining onto the design space and provide adjustment needed for the process [2, 3, 32, 33].

By implementing PAT, manufacturers can move away from traditional quality control methods that rely on post-production testing of samples and instead adopt a proactive approach to quality assurance and process improvement.

PAT incorporates various analytical techniques, including spectroscopy, chromatography, microscopy, near-infrared (NIR) spectroscopy, Raman spectroscopy and others. These techniques provide real-time measurements and analysis of critical quality attributes and process parameters. By monitoring CQAs and process parameters continuously or at frequent intervals, manufacturers can identify deviations, trends or potential issues during the production process.

PAT also utilizes advanced data analysis and multivariate statistical techniques to analyze the vast amount of data generated in real-time. This analysis helps establish relationships between process variables and quality attributes and facilitates the detection of correlations, patterns and potential causes of variability. Statistical process control tools and multivariate data analysis techniques such as principal component analysis and partial least squares regression are commonly employed in PAT applications.

The benefits of implementing PAT in pharmaceutical manufacturing include:

- Real-time process monitoring: By monitoring critical parameters and quality attributes in real-time, manufacturers can detect process deviations and take corrective actions promptly, minimizing the potential for quality issues or product failures.
- Improved process control: PAT facilitates real-time feedback control mechanisms, allowing manufacturers to adjust process parameters in response to real-time data, thereby maintaining product quality within predefined specifications.
- Reduced production variability: By continuously monitoring and controlling critical process parameters, PAT helps reduce variability in manufacturing

processes, leading to increased product consistency and batch-to-batch reproducibility.

The U.S. Food and Drug Administration (FDA) has been a strong proponent of PAT and encourages its adoption in pharmaceutical manufacturing. The FDA's PAT guidance provides regulatory expectations and encourages the use of advanced technologies and quality risk management principles to achieve continuous process improvement and ensure product quality throughout the lifecycle.

6.2.6 Continuous Improvement Program

The continuous improvement program is an ongoing process that monitors the product throughout its lifecycle and makes adjustments to the formulation and process as necessary to maintain product quality through the life cycle or to reduce the possible sources of variability.

A continuous improvement program is a structured approach that aims to identify, analyze and implement incremental improvements in processes, products or services within an organization.

A continuous improvement program typically include:

1. Systematic evaluation: regularly assessing existing processes, systems and practices allow to identify areas that could be improved. This evaluation can be based on various factors, including customer feedback, performance metrics, benchmarking or internal audits.
2. Detailed objectives: clear and measurable improvement goals are established to provide a direction for the program. These goals are aligned with the organization's strategic objectives and are specific for the project and should be possible inside a specific time frame.
3. Employees at all levels are actively engaged and encouraged to participate in the improvement process. This can be achieved through training, workshops, cross-functional collaboration or suggestion systems that allow employees to contribute ideas and solutions.
4. Data-driven approach: data and specific metrics are used to drive decision-making and measure progress. Quantitative and qualitative data are collected, analyzed and utilized to identify improvement opportunities, track performance and monitor the impact of implemented changes.
5. Continuous learning and innovation thanks to a culture of learning and innovation by encouraging experimentation, knowledge sharing and the adoption of best practices. Lessons learned from both successes and failures are used to refine processes and drive further improvement.
6. Process standardization: standardizing processes and workflows helps establish a baseline for improvement and facilitates consistent application of best practices across the organization. Standard operating procedures (SOPs), guidelines and quality management systems are often utilized to ensure process consistency.

7. Continuous improvement tools, such as Lean, Six Sigma, Kaizen, Plan-Do-Check-Act cycles, process mapping, root cause analysis are employed to systematically identify and address improvement opportunities.
8. Strong leadership support is crucial for the success of a continuous improvement program. Leaders provide the necessary resources, create a supportive environment and communicate the importance of continuous improvement to motivate and empower employees.
9. Regular monitoring and review of the program's progress, effectiveness and impact are essential. Key performance indicators and performance metrics are used to assess the success of implemented improvements and to identify areas that require further attention.

By implementing a continuous improvement program organizations can foster a culture of innovation, efficiency and quality. Over time, this iterative and proactive approach to improvement leads to enhanced customer satisfaction, increased operational efficiency, reduced costs and a competitive advantage in the marketplace.

In a QbD development process, the product and process understanding during pharmaceutical development should lead to an early identification and reduction of potential sources of variation thanks to the control strategy; in a non-QbD environment instead, those issues were usually discovered during stability testing or early commercialization, leading to a stop of the production for the necessary redesign and testing [3, 13, 34].

6.2.7 *Case Study*

6.2.7.1 Diclofenac Tablets

This case study focus on a QbD formulation development for a diclofenac dispersible tablet [34].

After setting up the TPP according to the FDA guidelines, a QTPP was outlined in order to describe which methods and manufacturing processes should be used to meet the requirements of the product profile.

The QTPP defined the limit for the disintegration time, dissolution test (85% of the drug released in less than 30 min) and it also contained buffer and all the technical details necessary to perform the tests.

Then a risk assessment was performed in order to assess CQAs from the QTPP; after the identification of CQA, CMAs were then identified using a combination of the all the methods listed previously in order to generate a RPN that depends on the characteristics of the product.

Given that diclofenac is a BCS class II drug, the main CMAs where solubility and particle size because they slow down the dissolution of the drug itself. All the excipients involved, like microcrystalline cellulose and the disintegrating agents and the diluent may also affect the disintegration time of the final dosage form and therefore the drug release.

The physical form of the drug is also a CMA: the salt form, despite being more soluble, has a lower patience compliance due to a worse palatability; the use of the drug as a free acid lead to a specific set-up for the dissolution test involving phosphate buffer instead of simulated gastric fluid to overcome the solubility issue.

The most critical CPP were lubrication and compression because they can slow down the disintegration of the tablet and the dissolution of the drug: higher compression forces slow down the water uptake from the tablet, while a long mixing time with the magnesium stearate may reduce the wettability and the dissolution of the drug [35, 36].

Other process, like mixing, may affect the manufacturing parameters more than the characteristics of the final product and therefore are rated as less critical.

A drug-excipients compatibility study was also performed in order to ensure that there were no issue with the final formulation.

All the CMAs, CQAs and QPPs were then investigated using full factorials designs because they allow to estimate factors and interactions without influences from the other factors.

The DoE performed were:

- Study 1: effect of drug particle size and microcrystalline cellulose on flow characteristics, disintegration time and dissolution.
- Study 2: effect of superdisintegrants and tablet hardness on disintegration time and friability.
- Study 3: effect of lubricant level on disintegration time and dissolution.
- Study 4: effect of packaging material on the stability.
- Study 5: risk analysis of critical process parameters.
- Study 6: effect of blending process on blend homogeneity.
- Study 7: effect of lubrication time on disintegration time.
- Study 8: effect of tablet compression process on CQAs.

All those studies provided the knowledge necessary to set the limits for all the critical attributes in order to ensure that the final product will fit the specifics required. But the range of the specifics may requires additional testing and costs if not already covered by the preliminary experiments and therefore a large design space, albeit being desirable because it provides a lot of regulatory flexibility, is not always achievable if the process should remain cost-effective.

The results of all the DoE were:

- Study 1: particles size of drug and excipients may reduce the drug dissolution but not below the acceptable limit (95% in 30 min), so there are no issues with the materials used in the formulation development.
- Study 2: disintegrating agents and hardness affect the disintegration time but not to the extent of going over the limit (2 min); friability for all the batches was below 1%, so not a CQA.

- Study 3: hydrophobic lubricants may affect the hardness and the friability, but the limits were never reached during the DoE.
- Study 4: dispersible tablets need to be carefully protected from moisture; tablets packed with PVC/PVDC films showed a lower amount of impurities after storage.
- Study 5: particles size may not affect the dissolution time but it can negatively influence the blend homogeneity while lubrication and compression time may also be considered CPP for their effect on friability and hardness.
- Study 6: blends with smaller drug particles showed comparatively better homogeneity and less demixing; there were no PAT available, so a suitable range of speed and time was identified for this process.
- Study 7: there was no discernible effect of lubricant content and lubrication time on the disintegration of the tablets; also in this case there were no PAT available because near-infrared-spectrometry was not sensitive enough to prevent overmixing.
- Study 8: content uniformity and friability were not an issue (as also already stated above) but it was possible to identify the range of compression force that fulfil the requirements; an online feed-back system can be used to reduce the number of tablets outside the specifications because it can adjust the compression force in relation to the filling of the die.

The outcome of those experiments lead to the design space and the control strategy in which all the parameters described in the space were classified according to a FMEA analysis to determine which ones are critical.

The design space was then validated to ensure that all the materials and processes described can actually deliver a final product inside the specifications but it was also necessary to understand the effect of the control strategy on the process, i.e. if the controls in place (both PAT and testing) can actually mitigate the risk of a product outside specifications.

The QTPP can guarantee a tablet with the intended behaviour without the testing of the final product, but the testing remain necessary for the stability and the impurities. Besides this work never shows any planning for the down-time of instruments or equipment that need to be factored in.

6.2.7.2 Lipid Crystalline Nanoparticles

This case study focus on the QbD manufacturing of liquid crystalline nanoparticles [29].

Lipid based crystalline nanoparticles are a mesophase where the components may behave both as solid and as a liquid and, thanks to their properties, are suitable for different routes of administration.

The preparation of those nanoparticles is a sensitive process and there is a lack of data for scale-up. The QbD manufacturing can help in that regard, providing the knowledge necessary to build the model and to extend it.

After the layout of the QTPP, CQAs were identified using a combination of likelihood and severity: if a parameters (related to the material and a process) can push the final product outside specification, it is considered as QCA.

This risk management was done via a multisets process:

1. Risk identification and assessment via Ishikawa diagram.
2. Process mapping to assess possible points of failure or detect areas that can be subject to improvement.
3. Risk analysis using FMEA to quantify the effect of the single parameters on the QTPP, the process was then repeated via a RRMS to visually assess the contribution of QMAs to the final product.

Several DoE were selected according to the aim (screening or optimization) of the single study and the importance of possible interactions versus a larger area or a curvature included into the model. This parameters will affect the design space but the evaluation of the single experiments needs to be based on prior knowledge to determine if the lacking information may be relevant or not.

The design space need to be wide enough so that minor modifications still allow the final product or process to be inside so that the change cannot will not trigger a revision and revamping of the process to abide the regulations, which increase the costs.

Control strategy is based on a combination of automatic engineering controls together with controls during the intermediate steps to reduce the testing of the final product.

The main difference compared to case study 1 is that the lack of a defined scope for the QbD leave the process open to revamping according to the route of administration and drug applied to the dosage form: crystalline lipid nanoparticles for intravenous administration have very different requirements compared to those for oral administration (sterility for example).

6.2.7.3 Immediate and Sustained Release Tablets

This case study focus on the formulation and manufacturing development of a multikinetic (immediate or IR and sustained or SR) tablet using sarpogelate as a model drug [37].

For this dual release tablet, the QTPP involves both physical properties of the excipients and mechanical properties of the layers that constitute the tablets because their dissolution and swelling (in case of the sustained release layer) may affect the release of the drug.

In order to create a comprehensive model, several physical properties and the test used to measure them were included into the design itself:

- Disintegration of the IR layer, because it may affect the gelification of the SR layer and therefore the release.
- Contact angle to determine the wettability of the tablet.

- Size and strength of the polymeric gel layer in the SR section of the tablet.
- Water uptake.

After the definition of the CMAs and CQAs, a FMEA risk assessment was performed considering not only the effect and the probability of occurrence but also the capability of quantify that effect: if the effect of a parameter cannot be assessed or quantified, there is no reason to include that in to the DoE because it can generate background noise that mask the effect or the interactions between parameters.

D-optimal mixture design was used for both IR and SR layer: mixture design is a surface response design that is suitable for formulation development but not for compression parameters that were not therefore included into the design space.

The tablets were evaluated in terms of:

- assay and content uniformity;
- hardness and friability;
- dissolution.

The CQAs were evaluated using multivariate analysis and principal components analysis in order to find the optimal formulation inside the design. CMAs where then related to CQAs using a placebo as a control to confirm that the CQAs were inside the limits when the swelling ad water uptake of the excipients matched the ones in the original study.

The mechanical properties of each layers and the interactions between them were used to find the optimal behaviour of the tablet that was then tested in vivo using healthy human volunteers.

The pharmacokinetics parameters in human volunteers of the optimized formulation were superimposable with the ones of the reference formulation showing how the QdB worked both in vitro and in vivo ensuring the bioequivalence of a formulation developed upon the original study.

6.3 Conclusion

By following a QbD approach to formulation development, manufacturers can ensure that their products are safe, effective and consistently meet the needs of patients but this doesn't means the possibility to reduce the testing of the final product. This approach also reduces the risk of product failures and regulatory issues, which can have significant financial and reputational impacts on a company.

References

1. Razinkov, V. I., & Kleemann, G. R. (2017). 1—High-throughput methodology in biopharmaceutical formulation development. In V. I. Razinkov & G. R. Kleemann (Eds.), *High-throughput formulation development of biopharmaceuticals* (pp. 1–10). Woodhead Publishing.
2. Haleem, R. M., et al. (2015). Quality in the pharmaceutical industry—A literature review. *Saudi Pharmaceutical Journal, 23*(5), 463–469.
3. Yu, L. X., et al. (2014). Understanding pharmaceutical quality by design. *The AAPS Journal, 16*(4), 771–783.
4. Malik, A., et al. (2019). Quality by design: A new practice for production of pharmaceutical products. *Journal of Drug Delivery and Therapeutics, 9*, 416–424.
5. Syed, S.M. and More, R.I., (2020). Quality by design: An approach for formulation development. Inventi Rapid: Pharm Analysis & Quality Assurance, 1, pp.1–6.
6. Arden, N. S., et al. (2021). Industry 4.0 for pharmaceutical manufacturing: Preparing for the smart factories of the future. *International Journal of Pharmaceutics, 602*, 120554.
7. Patil, H., et al. (2023). An overview on quality by design. *Asian Journal of Research in Pharmaceutical Sciences, 13*(1), 49–55.
8. Vivekanandan, Sundaramurthy. (2019). Quality by Design in Pharmaceutical Formulation (pp. 221–239). https://doi.org/10.4018/978-1-5225-7326-5.ch010.
9. Schuhmacher, A., et al. (2021). Systematic risk identification and assessment using a new risk map in pharmaceutical R&D. *Drug Discovery Today, 26*(12), 2786–2793.
10. Bhattacharyya, S., Adhikari, H., & Regmi, D. (2019). A brief review on Qbd approach on liposome and the requirements for regulatory approval. *Research Journal of Pharmacy and Technology, 12*, 4057.
11. Aksu, B., et al. (2013). A quality by design approach using artificial intelligence techniques to control the critical quality attributes of ramipril tablets manufactured by wet granulation. *Pharmaceutical Development and Technology, 18*(1), 236–245.
12. Kumar, V., & Gupta, N. V. (2015). A review on quality by design approach (QBD) for pharmaceuticals. *International Journal of Drug Development & Research, 7*, 52–60.
13. Zagalo, D. M., Sousa, J., & Simões, S. (2022). Quality by design (QbD) approach in marketing authorization procedures of non-biological complex drugs: A critical evaluation. *European Journal of Pharmaceutics and Biopharmaceutics, 178*, 1–24.
14. Zlota, A. A. (2022). Recommendations for effective and defendable implementation of quality by design. *Organic Process Research & Development, 26*(3), 899–914.
15. Rodrigues, A. D. (2019). *Drug-drug interactions*. CRC Press.
16. Haimes, Y. Y., Kaplan, S., & Lambert, J. H. (2002). Risk filtering, ranking, and management framework using hierarchical holographic modeling. *Risk Analysis, 22*(2), 383–397.
17. Tian, G., et al. (2019). Quality risk assessment and mitigation of pharmaceutical continuous manufacturing using flowsheet modeling approach. *Computers & Chemical Engineering, 129*, 106508.
18. Kiran, D. R. (2017). Chapter 26—Failure modes and effects analysis. In D. R. Kiran (Ed.), *Total quality management* (pp. 373–389). Butterworth-Heinemann.
19. Kalathil, M. J., Renjith, V. R., & Augustine, N. R. (2020). Failure mode effect and criticality analysis using dempster Shafer theory and its comparison with fuzzy failure mode effect and criticality analysis: A case study applied to LNG storage facility. *Process Safety and Environmental Protection, 138*, 337–348.
20. Wreh, E. (2023). Chapter 17—Risk management. In E. Wreh (Ed.), *Medical device regulation* (pp. 455–478). Academic.
21. Featherstone, S. (2015). 9—Hazard analysis and critical control point (HACCP) systems in food canning. In S. Featherstone (Ed.), *A complete course in canning and related processes (fourteenth edition)* (pp. 215–234). Woodhead Publishing.
22. Dunjó, J., et al. (2010). Hazard and operability (HAZOP) analysis. A literature review. *Journal of Hazardous Materials, 173*(1), 19–32.

23. Zheng, Q., Liu, X., & Wang, W. (2023). A consensus model-based risk matrix for human error factors risk analysis in medical devices by considering risk acceptability. *Reliability Engineering & System Safety, 238*, 109446.
24. Orpana, I. F. (2019). Manufacturing pharmaceutical medicines in a regulated environment—An auditor's perspective. *Procedia Manufacturing, 39*, 1773–1782.
25. Box, J. F., & R. A. (1980). Fisher and the design of experiments, 1922-1926. *The American Statistician, 34*(1), 1–7.
26. Politis, S. N., et al. (2017). Design of experiments (DoE) in pharmaceutical development. *Drug Development and Industrial Pharmacy, 43*(6), 889–901.
27. Gujral, G., Kapoor, D., & Jaimini, M. (2018). An updated review on design of experiment (DOE) in pharmaceuticals. *Journal of Drug Delivery and Therapeutics, 8*, 147–152.
28. Fukuda, I., et al. (2018). Design of experiments (DoE) applied to pharmaceutical and analytical quality by design (QbD). *Brazilian Journal of Pharmaceutical Sciences, 54*, e01006.
29. Waghule, T., et al. (2021). Quality by design (QbD) in the formulation and optimization of liquid crystalline nanoparticles (LCNPs): A risk based industrial approach. *Biomedicine & Pharmacotherapy, 141*, 111940.
30. Nelsen, T. C. (2023). Chapter fourteen—Efficient design of experiments (DOE). In T. C. Nelsen (Ed.), *Probability and statistics for cereals and grains* (pp. 201–214). Woodhead Publishing.
31. Tavares Luiz, M., et al. (2021). Design of experiments (DoE) to develop and to optimize nanoparticles as drug delivery systems. *European Journal of Pharmaceutics and Biopharmaceutics, 165*, 127–148.
32. Yu, L. X., et al. (2004). Applications of process analytical technology to crystallization processes. *Advanced Drug Delivery Reviews, 56*(3), 349–369.
33. Simão, J., Chaudhary, S. A., & Ribeiro, A. J. (2023). Implementation of quality by design (QbD) for development of bilayer tablets. *European Journal of Pharmaceutical Sciences, 184*, 106412.
34. Charoo, N. A., et al. (2012). Quality by design approach for formulation development: A case study of dispersible tablets. *International Journal of Pharmaceutics, 423*(2), 167–178.
35. Gunawardana, C. A., et al. (2023). Magnesium stearate surface coverage on tablets and drug crystals: Insights from SEM-EDS elemental mapping. *International Journal of Pharmaceutics, 630*, 122422.
36. Virtanen, S., Salokangas, H., & Yliruusi, J. (2008). The effect of mixing time of the magnesium stearate on crushing strengths of tablets. *European Journal of Pharmaceutical Sciences, 34*(1, Supplement), S27.
37. Han, J. K., et al. (2022). A formulation development strategy for dual-release bilayer tablets: An integrated approach of quality by design and a placebo layer. *International Journal of Pharmaceutics, 618*, 121659.

Chapter 7
Quality by Design (QbD) Concept for Formulation of Oral Formulations for Tablets

Mohammed Tahir Ansari, Teejan Ameer Abed Alahmed, and Farheen Sami

Abstract Pharmaceutical scientists use the Quality by Design (QbD) approach to develop robust oral dosage forms, such as tablets and pellets, by considering critical quality attributes and critical material attributes. The critical material attributes (CMAs) of raw materials are crucial for determining the critical quality attributes (CQAs) of a tablet. It is essential to identify and control the properties of active pharmaceutical ingredients (APIs), excipients, lubricants, and other materials to ensure optimal tablet quality and performance. The Risk Priority Number (RPN) method is used to conduct risk assessments of critical process parameters (CPPs) and critical material attributes (CMAs) in oral dosage forms. The RPN method ranks risks based on severity, occurrence, and detectability, and provides a numerical value that evaluates the risk of physical and chemical attributes affecting CQAs in oral dosage forms. To establish the design space, an in-depth understanding of the manufacturing process is crucial, including the effects of raw material properties, process parameters, and intermediate properties. Accomplishing this requires experimental investigations, data analysis, and a comprehensive understanding of the scientific and engineering principles. Multivariate modelling techniques, such as partial least squares (PLS) and response surface methodology (RSM), are utilised to establish connections between the input formulation parameters, process parameters, and tablet attributes. These models help comprehend the impact of variables on tablet properties and aid in the optimisation process. Additionally, Design of Experiments (DoE) is a systematic method employed to study the impact of multiple variables simultaneously and determine critical process parameters. It allows for the examination of a wide range of formulation and process variables to optimise the tablet formulation.

M. T. Ansari (✉) · T. A. A. Alahmed
School of Pharmacy, Faculty of Science and Engineering, University of Nottingham Malaysia, Semenyih, Selangor, Malaysia
e-mail: Tahir.Ansari@nottingham.edu.my

F. Sami
Hygia Institute of Pharmaceutical Education and Research, Lucknow, India

School of Pharmacy, CT University, Ludhiana, Punjab, India

N. K. Jain, N. Bajwa (eds.), *Introduction to Quality by Design (QbD)*,
https://doi.org/10.1007/978-981-99-8034-5_7

Keywords Tablets · Quality by design · Critical material attribute · Critical process parameters · Design of experiments · Critical quality attributes

7.1 Introduction

The pharmaceutical industry has seen a significant shift in recent years, with a growing focus on ensuring drug quality throughout the entire development lifecycle. Quality by Design (QbD) is an approach that integrates quality into the drug development process from the beginning. This chapter discusses the application of QbD principles in the formulation of oral dosage forms, including tablets and pellets. Oral dosage forms are crucial for ensuring drug safety, efficacy, and patient compliance. However, conventional forms such as tablets and capsules can have wide fluctuations in drug concentration, leading to toxicity and poor drug delivery. Therefore, ensuring the quality of oral dosage forms is essential to optimize drug delivery and minimize fluctuations [1].

The controlled release of drugs from oral dosage forms is a crucial aspect of ensuring quality, as it enables sustained and consistent drug delivery, leading to improved patient outcomes and reduced adverse effects [2]. By optimizing the formulation and delivery system, the release of the drug can be controlled, resulting in a more dependable and consistent drug concentration within the body.

The development of a child-friendly dosage forms, such as orally disintegrating tablets or flavored liquids, also requires ensuring accurate dosing and improving patient adherence in children. Children often have difficulty swallowing conventional tablets and capsules, which can lead to medication non-compliance and ineffective treatment outcomes. Therefore, it is crucial to develop oral dosage forms that are appropriate for children's needs. Overall, quality in oral dosage formulation is essential to ensure the safety, efficacy, and patient compliance of pharmaceutical products. By controlling drug release, developing child-appropriate dosage forms, and applying QbD principles, manufacturers can optimize the quality of oral dosage formulations and improve patient outcomes [2–4].

7.2 Principles of Quality by Design

Pharmaceutical Quality by Design (QbD) is a systematic and scientific method to pharmaceutical development that concentrates on understanding the product and process variables that impact product quality. This approach includes designing and developing pharmaceutical products with predetermined quality attributes, while also considering the manufacturing process and control strategies. QbD highlights the significance of recognizing and comprehending critical quality attributes (CQAs) of the product, which are the features that are vital to guaranteeing the safety, efficacy, and quality of the product. By comprehending the relationship between the formulation, manufacturing process, and CQAs, manufacturers can optimize the formulation and process parameters to consistently produce high-quality products [5]. The QbD approach entails utilizing risk assessment and management tools to identify and mitigate potential risks throughout the product's lifecycle. This encompasses identifying probable sources of variability and implementing control strategies to reduce their impact on product quality [6]. QbD in pharmaceutical development has advantages. It streamlines the process by focusing on controlling critical product and process aspects, leading to shorter development time, lower cost, and better process and product quality [7]. Quality by Design (QbD) improves regulatory compliance through a systematic and science-based approach to product development. This approach helps better understand and control the product and process, leading to more reliable manufacturing. QbD is a valuable tool in pharmaceutical development that promotes a proactive and science-based approach to product quality. Manufacturers can optimise processes and produce high-quality pharmaceuticals by focusing on controlling critical quality attributes [8, 9].

7.3 Defining the Quality Target Product Profile (QTPP) for Tablets

The QbD approach starts with the Quality Target Product Profile (QTPP), which is specific to oral solid dosage forms, such as tablets. This QTPP outlines the critical product characteristics that are essential for optimal quality, encompassing dosage form, strength, route, particle size, rheology, drug concentration, homogeneity, pH, drug release, and microbial limits [10]. The patient's acceptance of oral solid dosage forms is a crucial factor in determining QTPP. Taste, ease of swallowing, and visual appearance are the key factors that influence patient acceptance of the dosage form [11]. User acceptability studies, such as visual preference surveys, can inform the development of age-appropriate medicines [12]. The Quality by Design (QbD) approach, which includes the Quality Target Product Profile (QTPP), emphasises the understanding and management of critical quality attributes throughout the product lifecycle. QbD systematically develops and refines the formulation and manufacturing processes to guarantee consistent product quality. Defining the QTPP allows manufacturers to identify and optimise critical quality attributes within the QbD process [13]. The Quality Target Product Profile (QTPP) is tied to the Analytical Target Profile (ATP), which outlines the product's desired analysis performance. Together, they direct the creation of a precise, accurate, and economical quality control analytical method [14]. The QTPP serves as a guiding document throughout the development process, ensuring that a product's quality attributes align with the intended therapeutic outcome (Fig. 7.1).

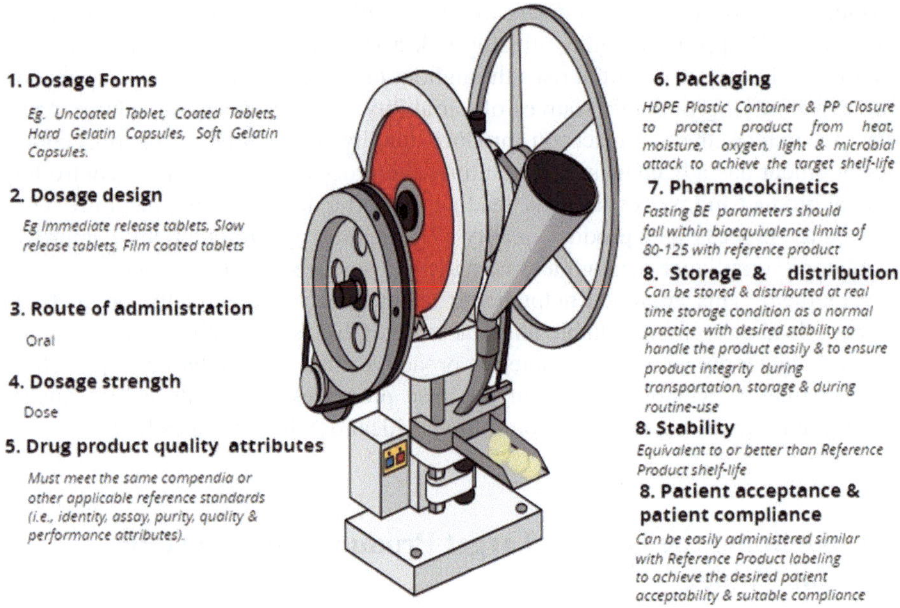

Fig. 7.1 QTPP elements for oral tablets

7.4 Identifying Critical Quality Attributes (CQAs)

The Critical Quality Attributes (CQAs) are the key characteristics of a product that affect safety, efficacy, and quality. Porosity is a CQA for tablets that influences their mechanical properties, mass transport, and dissolution. CQAs for tablets also include other factors that affect performance, quality, and efficacy. Porosity is essential for tablet disintegration and drug release. A high porosity leads to faster drug release and improved bioavailability. It also affects tablet hardness and friability. Tablets with suitable porosity are hard enough for handling and packaging, yet disintegrate to release the drug [15].

The disintegration time is essential for tablets, as it shows how long it takes them to break down in the GI tract. Drug dissolution rate and disintegration time are not always linked, as other factors can affect drug release. Disintegration time is still important, as it indicates the tablet‘s ability to quickly release the drug [16]. Tablet breaking force is a CQA that reflects compaction and bonding [17]. Alyami et al. found a direct link between the tablet disintegration time and hardness. Taste-masking agents and lubricants can slow down dissolution due to increased tablet hardness and prolonged disintegration time. The study also showed that taste-masking agents, superdisintegrants, and lubricants can significantly increase tablet hardness [18]. A study reported that both stress tablet hardness as a key physical-mechanical quality attribute. Dave et al. note that density, porosity, breaking force, viscoelastic properties, and any structural defects or irregularities are important physical-mechanical tablet qualities [18, 19].

Dissolution of tablets is a critical quality attribute. Dissolution testing evaluates drug release and oral solid dosage forms' performance and effectiveness. Altaş et al. [20] stress the importance of dissolution as an official test to predict in vivo performance and assess solid dosage forms. Dissolution studies are regularly conducted in quality control and R&D [20] .

Tablet content uniformity is essential for accurate dosing and efficacy. Dissolution testing is used to assess content uniformity and guarantee product quality. Madathilethu et al. [21] studied hydrocortisone tablets and mini-tablets for kids. They evaluated the weight and content uniformity of quartered tablets and mini-tablets according to pharmacopoeial standards. Over half of the quartered tablets failed the hydrocortisone content requirements, while the mini-tablets met the content uniformity criteria [21]. Su et al. [22] note that tablet quality depends on component segregation and blending, including dosage, content uniformity, and composition [22].

Other CQAs for tablets may include particle size, friability, and mass variation [23–25]. These attributes are crucial for ensuring consistent drug release, dosage accuracy, and product performance. Identifying CQAs is a crucial step as it allows formulators to focus their efforts on controlling and optimizing the factors that truly matter in achieving a high-quality product.

7.5 Understanding Critical Material Attributes (CMAs)

The critical material attributes (CMAs) of raw materials are important for determining the critical quality attributes (CQAs) of a tablet. It is crucial to identify and control the properties of APIs, excipients, lubricants, and other materials in order to ensure optimal tablet quality and performance.

7.5.1 Risk Assessment

Risk assessment is a key part of Quality by Design (QbD) in formulation development. It identifies critical factors and parameters that may influence CQAs, allowing for the optimisation and control of formulation and process variables to ensure product quality and performance. Several studies have applied risk assessment in the QbD approach for formulation development. For tablet formulations, this may involve assessing risks related to excipients, drug-substance properties, and manufacturing processes. Zarmpi et al. studied the effect of excipients, such as superdisintegrants, on drug solubility. Risk assessment was used to assess the impact of excipient physicochemical properties on the product quality [26]. Ikeda et al. studied 3D printing of orally disintegrating tablets and fast-dissolving oral dosage

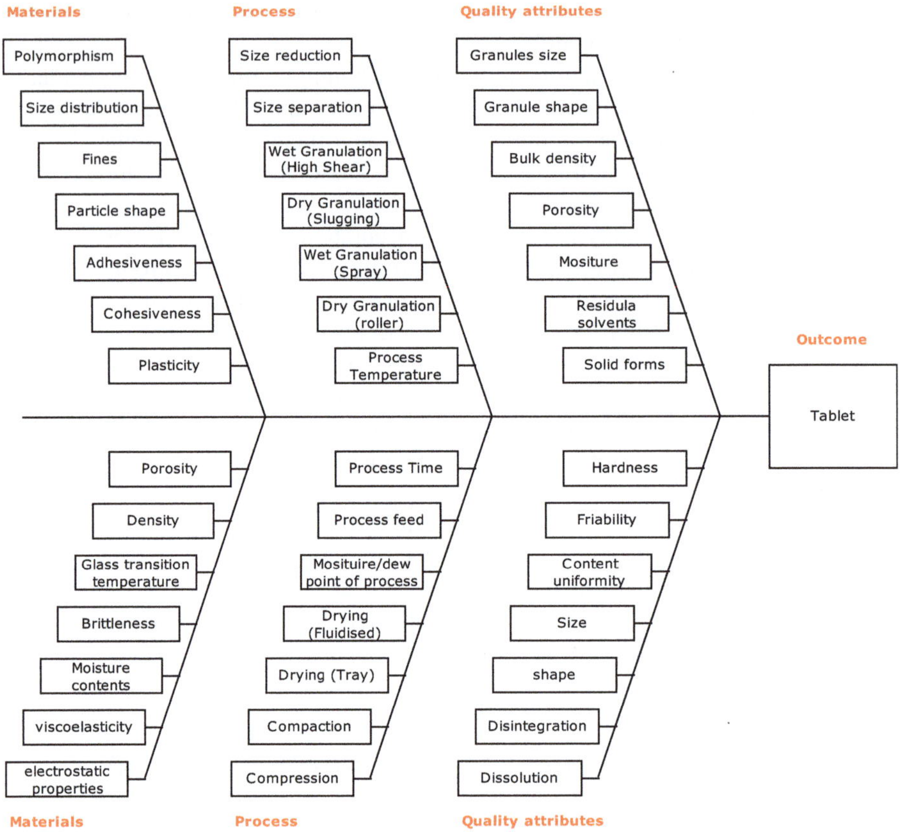

Fig. 7.2 Fishbone diagram showing the CMAs, CPPS and CQA for Tablets

forms. They discussed the application of QbD principles and risk assessment to optimise the formulation and process parameters for rapid drug release and in vivo oral absorption [27]. Tefas et al. used QbD to create liposomes encapsulating doxorubicin and curcumin. They identified critical factors and parameters to assess risks and ensure product quality [28].

Risk assessment helps formulators prioritise by focusing on the most important factors, such as raw material variability or process deviations.

This involves three stages:

1. Risk identification: The material attributes and process parameters are identified and process map or fish bone diagram (Fig. 7.2).
2. Risk analysis methods, such as Risk Ranking and Mitigation Analysis (RRMA) and Failure Mode and Effects Analysis (FMEA) (Table 7.1), Analytic Hierarchy Process (AHP) or Process Failure Mode and Effects Analysis-Value Stream Mapping (PFDA-FMEA-VSM) in the Quality by Design (QbD) concept for

Table 7.1 Risk Ranking and Mitigation Analysis

Finished product CQAs	Co-sifting	High-shear wet granulation	Sieving	Lubrication	Compaction
Shape	Low	High	Medium	High	Low
Size	Medium	High	Medium	High	Low
Hardness	Low	High	Low	Medium	Medium
Disintegration	Low	High	Low	Medium	High
Dissolution	Low	High	Low	High	High

Low	Broadly risk acceptable	Medium	Risk acceptable	High	Risk unacceptable

oral dosage forms, involve evaluating and managing potential risks affecting critical quality attributes (CQAs) [8, 29]

3. Risk Evaluation Risk assessment for critical process parameters (CPPs) and critical material attributes (CMAs) in oral dosage forms is conducted using the Risk Priority Number (RPN) method. The RPN is a numeric value that ranks risks based on severity, occurrence, and detectability, with specific RPN values varying based on the context and scoring system. Higher RPN values suggest greater risk levels, whereas lower values suggest less risk [30]. The RPN score, calculated from the likelihood of failure, impact severity, and detection frequency affecting the CQA, evaluates the risk of physical and chemical attributes affecting CQAs in oral dosage forms.

   Understanding CMAs is essential because it enables formulators to select and source materials that will consistently meet the desired quality standards.

7.5.1.1 Polymorphism

Polymorphism the ability of a substance to exist in different crystal forms or structures, is a critical quality attribute (CQA) in tablet manufacturing that should be carefully controlled and monitored. Different polymorphic forms of a drug substance can exhibit varying physical and chemical properties, such as solubility, stability, and compressibility, which affect the performance and quality of tablets. The crystal packing and bonding strength of polymorphs play a role in their tabletability, and the presence of slip planes in crystals correlates with increased tabletability [31]. The study conducted by Vervaet et al. demonstrated that the polymorphic transition of mannitol during twin screw granulation enhanced the plastic deformability of the material, leading to improved tabletability. The transformation from one polymorphic form to another, caused by moisture, results in a unique granule morphology and better tableting properties. Controlling polymorphism is crucial in tablet manufacturing and requires an understanding of the crystal structure and properties of the drug substance [32].

7.5.1.2 Particle Size

Particle size is a critical quality attribute in tablet manufacturing due to its significant impact on various aspects of tablet performance and quality. Studies have shown that the particle size influences the flowability and packing behaviour of granules, which in turn affects tabletability and compressibility. Smaller particle sizes can lead to improved flow properties and better compaction behaviour, resulting in tablets with higher mechanical strength and uniformity [33, 34]. Particle size distribution plays a crucial role in achieving content uniformity in tablets. Variations in particle size can lead to uneven distribution of the active pharmaceutical ingredient (API) within the tablet matrix, affecting drug dosage and efficacy. The particle size of an API can significantly affect its dissolution rate and subsequent drug release from the tablet. Particle size can affect various downstream processing steps such as blending, granulation, and tablet pressing [22, 30, 35].

7.5.1.3 Fines

The incorporation of fine particles or powders can enhance tablet density and strength. Fines can result in tablets with slower disintegration and dissolution rates because of their compact nature [36, 37]. Furthermore, the presence of fines can result in poor flow properties during tablet compression and potential content nonuniformity [38].Ensuring consistent tablet performance and quality requires controlling the addition of fines as a critical quality attribute in tablet manufacturing. This involves selecting appropriate raw materials, optimising formulation and processing parameters, and implementing effective control strategies to maintain the desired fines content and distribution in the final tablets.

7.5.1.4 Particle Shape

Particle shape can affect tabletability and compaction behaviour, content uniformity, and various downstream processing steps such as blending, granulation, and tablet pressing. When particles have irregular shapes, they can cause difficulties in achieving uniform mixing, granule formation, and tablet compression, which may lead to content non-uniformity and tablet defects. Therefore, controlling particle shape as a critical quality attribute in tablet manufacturing is crucial for ensuring consistent tablet performance and quality [39].

7.5.1.5 Adhesiveness and Cohesiveness

The adhesiveness and cohesiveness of powders play critical roles in tablet manufacturing because of their impact on powder flowability, packing density, and uniformity. The presence of cohesive forces between particles influences their flow behaviour, which is crucial for achieving uniform powder layers and ensuring the overall tablet quality [40–42]. The particle size, surface energy, and mechanical properties of the powder material influence adhesion and cohesion. Fine cohesive particles can result in low-quality powder layers and uneven surfaces. Cohesive interactions between particles can overpower gravity and significantly affect bulk powder behaviour [41–43].

7.5.1.6 Plasticity

Excipients with plastic deformation during compression exhibit strong bonding and tabletability [44]. The plasticity of the co-processed excipients affects their compaction behaviour, as shown by the Heckel yield pressure changes [44, 45]. Higher compaction pressures may be necessary because of reduced plasticity, and the use of excipients such as microcrystalline cellulose (MCC) can impact tablet disintegration [46]. By decreasing the extent of plastic deformation in MCC, a more porous tablet structure can be created, leading to rapid water uptake and disintegration [47].

7.5.1.7 Porosity

The porosity of tablets plays a crucial role in determining the mass transport processes and mechanical changes during disintegration, which ultimately affect the dissolution time of tablets. Tablets with higher porosities tend to disintegrate and release drugs more rapidly, leading to faster dissolution [15]. Porosity and tablet mechanical strength are inversely related, with high porosity associated with low mechanical strength, leading to tablets that are more prone to breakage or damage during handling and packaging. Controlling the porosity of the excipients is crucial for achieving consistent tablet performance, dissolution characteristics, and mechanical strength [48, 49].

7.5.1.8 Density

The density of excipients in tablets directly affects their mechanical properties, including hardness and tensile strength. Tablets with a higher excipient density tended to have lower porosity, resulting in slower disintegration and dissolution rates. Other factors that can affect the structure of tablets include the porosity, thickness, initial particle size distribution, and granule density. The flowability of excipients can also be affected by excipient density, which is important for achieving uniform tablet compression and content uniformity [50].

7.5.1.9 Glass Transition Temperature

The thermal properties of excipients can affect phase transitions that occur during tablet compression [51]. In general, excipients with higher thermal transition temperatures exhibit increased elasticity and a reduced capping tendency, leading to tablets with improved mechanical strength and resistance to damage during handling and storage. On the other hand, excipients with appropriate thermal transition temperatures exhibit desirable compressibility and elasticity, facilitating efficient tablet compression and ensuring consistent tablet quality [52].

7.5.1.10 Brittleness

Tablets containing brittle excipients are more susceptible to fracturing or breaking during compression, handling, and packaging. Conversely, tablets with less brittle excipients displayed increased resilience, reducing the probability of damage during transportation and use [53]. The brittleness of the tablets can affect their disintegration properties. Tables with brittle excipients may experience slower disintegration rates because of their decreased ability to break apart and disintegrate in moisture [44, 54].

7.5.1.11 Moisture Content

An increase in moisture content can have a significant impact on the mechanical properties of tablets, including their crushing and tensile strengths. Additionally, moisture content can play a role in the flowability of pharmaceutical powders, affecting the behavior of the powders and potentially impacting tablet uniformity and content distribution [55, 56]. Moisture content can affect tablet disintegration and dissolution properties [57].

7.5.1.12 Viscoelasticity of Excipients

Viscoelastic materials have elastic and viscous properties that affect tablet compatibility, elastic recovery, and dimensions during compression and decompression. Understanding the viscoelastic behaviour of excipients can help predict tablet changes and properties over time, thereby ensuring tablet stability [58, 59]. Tablets with good plasticity and elastic recovery can be achieved using excipients with appropriate viscoelastic properties, leading to improved tabletability and reduced capping tendency [60].

7.5.1.13 Electrostatic Charge

The blending process can result in electrostatic charges in mixtures, affecting the formulation, manufacturing process, packaging behaviour, and quality of the final product. It has been reported that excipients exhibit increasing negative charge density throughout the blending process. The presence of uncontrolled electrostatic charges can negatively impact the uniformity of the final powder blend [61, 62].

7.6 Understanding Critical Process parameters for Tablets

7.6.1 Size Reduction

Size reduction produces smaller particles, improving flowability and active ingredient distribution. It also ensures a uniform tablet composition and smaller tablets that are easier to swallow [63, 64]. Smaller particles will also have improved bioavailability and compaction properties of tablets [65, 66].

7.6.2 Size Separation

Size separation plays a critical role in controlling the particle size of excipients, which can impact tablet mechanical properties, such as hardness and tensile strength, drug release, ultimately leading to uniform distribution of the active pharmaceutical ingredient (API) within the tablet matrix and ensuring accurate dosage and efficacy [67, 68].

7.6.3 Granulation

Granulation is vital in tablet manufacturing because of its substantial impact on granule and tablet quality. Research has indicated that employing additional kneading elements during granulation can produce higher granule density, resulting in tablets with longer disintegration time [69]. Shaddinger et al. compared the different manufacturing processes of daprodustat tablets and evaluated their impact on bioequivalence and dissolution. The study emphasized the significance of the granulation process in achieving consistent pharmacokinetic properties and dissolution profiles [70]. Optimising the granulation fluid level, wet massing time, granule porosity/density, feed volume, pore size, temperature, and granulation speed is essential for consistent tablet quality. The compression Force, granulation time, roller speed, binder concentration, flow properties, throughput, screw speed, screw configuration, barrel temperature, and binder addition method must be optimised for dry granulation [71–74].

7.6.4 Granulation Temperatures

Azad et al. noted that improper processing temperatures can cause powder flow issues, leading to tablet weight variability, poor content uniformity, and inconsistent tablet properties such as hardness, disintegration, and dissolution. Thus, the temperature is a vital process parameter in tablet manufacturing [75]. Yusof et al. examined how the drying temperature influences pitaya powder colour and water activity, affecting the desired characteristics for tablet production [76]. Singh et al. highlighted the significance of optimising the processing temperature for manufacturing layered tablets. Each layer must demonstrate a sufficient reduction in volume and increase in mechanical strength, which can be influenced by temperature during the compression process [77]. Cho et al. emphasised the importance of proper temperature and humidity control during the manufacturing of tablets containing dapagliflozin propanediol monohydrate. The compound's low melting point and amorphous nature require careful temperature management to ensure successful tablet production [78].

7.6.5 Processing Time

Su et al. highlighted the significance of temperature control during the continuous manufacturing of tablets, emphasising the importance of maintaining proper temperatures during crystallisation and tablet compression to ensure efficient, consistent production, segregation, and homogeneity of the final tablets [22]. Razavi et al. emphasised the significance of processing temperature on the drying process and the properties of the resulting powder. They demonstrated how temperature affects the efficiency of drying and the properties of the powder, which can ultimately affect tablet manufacturing [68, 79].

7.6.6 Moisture and Dew

Yamada et al. investigated orally disintegrating tablets produced by moisture-activated dry granulation. They discovered that a moderate moisture content in the granules prevented agglomeration and facilitated better alignment during tableting, resulting in higher tablet hardness. Additionally, they found that excessive moisture content in the granules could decrease tablet tensile strength, emphasising the significance of moisture control on tablet mechanical properties [80–82].

7.6.7 Drying

Razavi et al. discussed the use of ultrasound measurement as a noninvasive method for forecasting tablet tensile strength, highlighting how the drying process and degree of shear strain during manufacturing can impact tablet hardness and mechanical resilience [68]. Su et al. [22] emphasize the significance of a novel continuous heterogeneous crystallization process, where the drying step plays a crucial role. This drying process enabled the formed product to be dried and shaped into tablets without the need for additional processing steps [22]. To optimise the drying process for tablet manufacturing, key factors such as temperature, residence time, moisture barrier coating, shear strain, drug loading, and processing design should be carefully considered [83].

7.6.8 Compaction and Compression

Compaction and compression processes are crucial in tablet manufacturing because of their considerable impact on tablet quality, manufacturability, and performance. Li and Wu highlight the importance of lubrication on the dynamics of these processes, noting how the presence of lubricants, such as magnesium stearate, impacts the mechanical properties of both compacts and tablets [84]. Haruna et al. investigated the compactability of co-processed excipients and their connection to tabletability. The compactability of a formulation affects its manufacturability and tabletability [45]. Optimising parameters such as the rotor speed, tablet compression force, die lubrication, Heckel analysis, compaction behaviour, and roller compaction can enhance the efficiency of tablet manufacturing [85].

7.7 Understanding Critical Quality Attributes

Adedokun et al. evaluate the physicochemical properties of sustained-release tablets, including granule properties such as size, shape, and density, and their impact on flowability, bulk density, and compressibility. The study determines granule properties by measuring the angle of repose, flow rate, bulk density, tapped density, Carr's index, and Hausner's quotient. The compressed tablets are also evaluated using official and non-official parameters [86]. Yamada et al. investigate the impact of granule density and moisture content on the hardness and disintegration time of orally disintegrating tablets produced by moisture-activated dry granulation. They find that a decrease in granule density and capillary wetting leads to prolonged disintegration time, while an increase in granule density results in increased hardness. The study suggests that adjusting the amount of added water and water activity within specific ranges can produce tablets with desired hardness and disintegration time [87].

7.8 Developing a Design Space

The Design Space is a multidimensional region in which a formulation satisfies all Critical Quality Attributes (CQAs). The design and control space in tablet manufacturing are developed through a combination of process understanding, multivariate modelling, and optimisation techniques. By correlating raw material variability, process parameters, and tablet attributes, a design space consistent with the Quality by Design (QbD) principles can be established. Various approaches can be employed to develop the design space, including building multivariate models that relate raw material attributes and process parameters to final tablet attributes and using optimisation strategies to find the best process parameters that compensate for the effect of incoming raw material variability [88]. Multivariate statistical analysis and modelling techniques, such as partial least squares path modelling, can help understand the interconnections between raw material properties, process parameters, and tablet properties. The design space enables the identification of crucial variables and units that influence process and tablet properties. This provides a structure for controlling and optimising the manufacturing process within defined limits to ensure consistent product quality. The design space is established based on a comprehensive understanding of the process, including the impact of raw material properties, process parameters, and intermediate properties on the final tablet properties [89]. The design-space concept enables flexibility during manufacturing by maintaining product quality, which permits adjustments within predefined boundaries without compromising safety or efficacy.

7.9 Application of QbD in Tablet Formulation

7.9.1 Multivariate Modelling

Multivariate modelling techniques, including partial least squares (PLS) and response surface methodology (RSM), were used to establish relationships between the input formulation parameters, process parameters, and tablet attributes. These models help understand the impact of variables on tablet properties and aid in the optimisation process [90, 91].

7.9.2 Design of Experiments (DoE)

A systematic design of experiments (DoE) is used to study the impact of multiple variables simultaneously and determine critical process parameters. According to Maeda et al. [91] the method allows the examination of a wide range of process conditions to determine the optimal combination of variables that result in tablets with the desired characteristics. Han et al. carried out a thorough investigation using multivariate analysis and the Quality by Design (QbD) approach to control high-shear wet granulation. They identified critical quality attributes (CQAs) such as granule size, granule hardness, flow energy, and Carr's index that have a significant impact on tablet tensile strength and dissolution [92]. Dave et al. discussed emerging technologies for non-invasive characterization of tablet physical-mechanical properties, such as near-infrared spectroscopy and terahertz pulsed imaging. These technologies enable the real-time monitoring and analysis of tablet attributes, supporting the QbD approach [19]. Fayed et al. applied the QbD approach to optimize a green fluidized bed granulation process using DoE to understand the influence of process variables on granule and tablet attributes. This has led to the establishment of a design space [93]. Singh et al. developed a control system that combined feed-forward and feedback mechanisms for a continuous tablet manufacturing process based on Quality by Design (QbD). This approach enables real-time monitoring and control of critical process parameters to ensure consistent product quality. Using QbD principles, manufacturers can optimise the tablet manufacturing process, minimise variability, and enhance product quality and performance. The design space provides a framework for controlling and optimising the manufacturing process within the defined boundaries, resulting in a robust and stable process [74]. Malakar and Nayak used the response surface methodology to optimise the formulation and manufacturing process of bisoprolol fumarate matrix tablets for sustained drug release. They investigated the effects of calcium alginate, HPMC K4M, and Carbopol 943 on tablet properties using a factorial design. The response surface plots helped to estimate the effects of the independent variables on the responses being studied, such as drug release and tablet hardness [94]. Mesut et al. investigated the impact of mannitol particle size on melatonin dissolution and tablet properties using a quality-by-design (QbD) approach. They used response surface methodology to optimise the formulation and process variables. This study highlighted the significance of formulation and process variable variability in achieving the desired product quality [95]. The use of box behnken technique has also been reported for analytical method development and tablet manufacturing process [96, 97].

7.9.3 Statistical Analysis

Statistical analysis methods, such as multiple linear regression analysis and artificial neural networks, are employed to analyse experimental data and determine the significant factors that influence tablet properties. These analyses aid in comprehending the relationships between variables and aid in the optimisation process [91, 98].

7.9.4 Process Understanding

An in-depth understanding of the manufacturing process, including the effects of the raw material properties, process parameters, and intermediate properties, is vital for creating the design space. This knowledge is acquired through experimental investigations, data analysis, and understanding of the underlying scientific and engineering principles [99, 100].

7.9.5 Optimisation Algorithms

Optimisation algorithms, such as genetic algorithms and particle swarm optimisation, can be utilised to determine the optimal combination of process parameters within the design space. These algorithms enable the identification of the most favorable operating conditions that meet the desired tablet attributes [101]. Chen and colleagues emphasized the significance of optimizing powder properties in tablet manufacturing, particularly the role of optimization algorithms in enhancing the tableting properties of active pharmaceutical ingredients (APIs) and excipients to ensure a robust and efficient process [102]. Tuyen et al. devised a scalable process for producing film-coated bilayer tablets. They utilised optimisation strategies to guarantee the stability and success of the manufacturing process, resulting in a consistent product that meets the desired specifications [103].

7.9.6 Continuous Manufacturing

Continuous manufacturing approaches permit real-time monitoring and control, allowing the creation of a design space based on continuous data collection and analysis. Continuous processes provide greater flexibility in adjusting the process parameters and maintaining consistent product quality [104, 105].

To ensure consistent tablet quality and performance, a design space can be established using these optimisation techniques. This design space defines the acceptable ranges of the input variables and process parameters. The design space serves as a framework for controlling and optimising the manufacturing process within defined boundaries, leading to a robust and stable manufacturing process.

Once the process is optimized the approach is repeated three times to confirm reliability of the process. The Process Analytical Technology (PAT) and the application of control charts are beyond the scope of the current chapter.

7.10 Conclusion

Pharmaceutical scientists can use the powerful approach of Quality by Design (QbD) to develop robust oral dosage forms like tablets and pellets by systematically considering critical quality attributes and critical material attributes and employing tools like Design of Experiments.

References

1. Modi, K. A., Modi, M., Mishra, D., Mittal, P., Sorathiya, U., & Shelat, P. (2013). Oral controlled release drug delivery system: An overview. *International Research Journal of Pharmacy, 4*, 70–76. https://doi.org/10.7897/2230-8407.04312
2. Hasnain, M. S., Ahmed, S. A., Beg, S., Ansari, M. T., & Nayak, A. K. (2019). 'Quality by Design' approach for development of multiparticulate drug delivery systems. In *Pharmaceutical quality by design* (pp. 351–365). Elsevier.
3. Sholapur, H. P. N. A., Dasankoppa, F. S., Channabasavaraja, M., Sagare, R. D., Abbas, Z., Sai, L., & Kshatriya, K. (2021). Quality by design approach for design, development and optimization of orally disintegrating tablets of Montelukast sodium. *International Journal of Pharmaceutical Investigation, 11*, 288–295. https://doi.org/10.5530/ijpi.2021.3.51
4. Nsabagasani, X., Hansen, E. H., Mbonye, A. K., Ssengooba, F., Muyinda, H., Mugisha, J., & Ogwal-Okeng, J. (2015). Explaining the slow transition of child-appropriate dosage formulations from the global to National Level in the context of Uganda: A qualitative study. *Journal of Pharmaceutical Policy and Practice, 8*, 19. https://doi.org/10.1186/s40545-015-0039-1
5. Pramod, K., Tahir, M. A., Charoo, N. A., Ansari, S. H., & Ali, J. (2016). Pharmaceutical product development: A quality by design approach. *International Journal of Pharmaceutical Investigation., 6*, 129–138. https://doi.org/10.4103/2230-973x.187350

6. Yu, L. X., Amidon, G., Khan, M. A., Hoag, S. W., Polli, J., Raju, G. K., & Woodcock, J. (2014). Understanding pharmaceutical quality by design. *The AAPS Journal, 16*(4), 771–783. https://doi.org/10.1208/s12248-014-9598-3
7. Swain, S., Parhi, R., Jena, B. R., & Babu, S. M. (2019). Quality by design: Concept to applications. *Current Drug Discovery Technologies, 16*(3), 240–250. https://doi.org/10.2174/1570163815666180308142016
8. Beg, S., Hasnain, M. S., Rahman, M., & Swain, S. (2019a). Introduction to quality by design (QbD): Fundamentals, principles, and applications. In S. Beg & M. S. Hasnain (Eds.), *Pharmaceutical quality by design* (pp. 1–17). Academic. https://doi.org/10.1016/B978-0-12-815799-2.00001-0
9. Beg, S., Swain, S., Rahman, M., Hasnain, M. S., & Imam, S. S. (2019b). Chapter 3 - application of design of experiments (DoE) in pharmaceutical product and process optimization. In S. Beg & M. S. Hasnain (Eds.), *Pharmaceutical quality by design* (pp. 43–64). Academic. https://doi.org/10.1016/B978-0-12-815799-2.00003-4
10. Hildebrandt, C., Gopireddy, S. R., Scherließ, R., & Urbanetz, N. A. (2019). Assessment of material and process attributes' influence on tablet quality using a QbD and DEM combined approach. *Powder Technology, 345*, 390–404. https://doi.org/10.1016/j.powtec.2019.01.015
11. Charoo, N. A., Shamsher, A. A. A., Zidan, A. S., & Rahman, Z. (2012). Quality by design approach for formulation development: A case study of dispersible tablets. *International Journal of Pharmaceutics, 423*(2), 167–178. https://doi.org/10.1016/j.ijpharm.2011.12.024
12. Januskaite, P., Xu, X., Ranmal, S. R., Gaisford, S., Basit, A., Tuleu, C., & Goyanes, A. (2020). I spy with my little eye: A Paediatric visual preferences survey of 3D printed tablets. *Pharmaceutics, 12*, 1100. https://doi.org/10.3390/pharmaceutics12111100
13. Namjoshi, S., Dabbaghi, M., Roberts, M. S., Grice, J. E., & Mohammed, Y. H. (2020). Quality by design: Development of the quality target product profile (QTPP) for semisolid topical products. *Pharmaceutic, 12*, 287. https://doi.org/10.3390/pharmaceutics12030287
14. Marie, A. A., Hammad, S. F., Salim, M. M., Elkhodary, M. M., & Kamal, A. H. (2023). Deduction of the operable design space of RP-HPLC technique for the simultaneous estimation of metformin, pioglitazone, and glimepiride. *Scientific Reports, 13*, 4334. https://doi.org/10.1038/s41598-023-30051-x
15. Bawuah, P., Zeitler, J. A., Farrell, D. P., Evans, M. J., Portieri, A., Anderson, A. E., Goodwin, D. J., Lucas, R., & Zeitler, J. A. (2020). Terahertz-based porosity measurement of pharmaceutical tablets: A tutorial. *Journal of Infrared Millimeter and Terahertz Waves, 41*, 450–469. https://doi.org/10.1007/s10762-019-00659-0
16. Zhao, N., & Augsburger, L. L. (2005). Functionality comparison of 3 classes of Superdisintegrants in promoting aspirin tablet disintegration and dissolution. *AAPS PharmSciTech, 6*, E634–E640. https://doi.org/10.1208/pt060479
17. Markl, D., Sauerwein, J., Goodwin, D. J., Ban, Sander Van Den Ban, & Zeitler, J. A. (2017). Non-destructive determination of disintegration time and dissolution in immediate release tablets by terahertz transmission measurements. Pharmaceutical Research 34 1012–1022 https://doi.org/10.1007/s11095-017-2108-4
18. Alyami, H. S., Khan, I. A., Alyami, M., Dahmash, E. Z., Almeanazel, O. T., Algahtani, T. S., Alanazi, F. K., & Alshora, D. H. (2021). Formulation of sublingual promethazine hydrochloride tablets for rapid relief of motion sickness. *Saudi Pharmaceutical Journal, 29*, 478–486. https://doi.org/10.1016/j.jsps.2021.04.011
19. Dave, V. P., Shahin, H. I., Youngren-Ortiz, S. R., Chougule, M. B., & Haware, R. V. (2017). Emerging Technologies for the non-Invasive Characterization of physical-mechanical properties of tablets. *International Journal of Pharmaceutics, 532*, 299–312. https://doi.org/10.1016/j.ijpharm.2017.09.009
20. Altaş, E., Kayar, G., Gökalp, M., & Gökalp, M. (2021). The development and the validation of a novel dissolution method of Favipiravir film-coated tablets. *Scientia Pharmaceutica, 90*, 3. https://doi.org/10.3390/scipharm90010003

21. Madathilethu, J., Roberts, M. J., Peak, M., Blair, J., Prescott, R., & Ford, J. D. (2018). Content uniformity of quartered hydrocortisone tablets in comparison with mini-tablets for paediatric dosing. *BMJ Paediatrics Open, 2*, e000198. https://doi.org/10.1136/bmjpo-2017-000198
22. Su, Q., Testa, C. J., Perala, S. R. K., Jensen, K., Braatz, R. D., Myerson, A. S., & Trout, B. L. (2017). Continuous heterogeneous crystallization on excipient surfaces. *Crystal Growth & Design, 17*, 3321–3330. https://doi.org/10.1021/acs.cgd.7b00297
23. Azad, M., Capellades, G., Wang, A. B., Klee, D. M., Hammersmith, G., Rapp, K., Brancazio, D., & Myerson, A. S. (2021). Impact of critical material attributes (CMAs)-particle shape on miniature pharmaceutical unit operations. *AAPS PharmSciTech, 22*, 98. https://doi.org/10.1208/s12249-020-01915-6
24. Katayama, H., Itakura, S., Todo, H., Sugibayashi, K., & Takayama, K. (2021). Latent structure analysis of wet-granulation tableting process based on structural equation modeling. *Chemical and Pharmaceutical Bulletin, 69*, 674–680. https://doi.org/10.1248/cpb.c21-00235
25. Willecke, N., Szepes, A., Wunderlich, M., Remon, J. P., Vervaet, C., & Beer, T. D. (2018). A novel approach to support formulation design on twin screw wet granulation technology: Understanding the impact of overarching excipient properties on drug product quality attributes. *International Journal of Pharmaceutics, 545*, 128–143. https://doi.org/10.1016/j.ijpharm.2018.04.017
26. Zarmpi, P., Flanagan, T., Meehan, E., Mann, J., & Fotaki, N. (2020). *Biopharmaceutical understanding of excipient variability on drug apparent solubility based on drug physicochemical properties. Case Study: Superdisintegrants* (Vol. 22, p. 46). *The AAPS Journal.* https://doi.org/10.1208/s12248-019-0406-y
27. Ikeda, S., Kobayashi, M., Aoki, S., Terukina, T., Kanazawa, T., Kojima, H., & Kondo, H. (2023). 3d-printed fast-dissolving Oral dosage forms via fused deposition modeling based on sugar alcohol and poly(vinyl alcohol)—preparation, drug release studies and in vivo oral absorption. *Pharmaceutics, 15*, 395. https://doi.org/10.3390/pharmaceutics15020395
28. Tefas, L. R., Sylvester, B., Tomuta, I., Sesarman, A., Licarete, E., Banciu, M., & Porfire, A. (2017). Development of Antiproliferative long-circulating liposomes co-encapsulating doxorubicin and curcumin, through the use of a quality-by-design approach. *Drug Design Development and Therapy, 11*, –1605, 1621. https://doi.org/10.2147/dddt.s129008
29. Yucesan, M., Gul, M., & Celik, E. (2021). A holistic FMEA approach by fuzzy-based Bayesian network and best–worst method. *Complex & Intelligent Systems, 7*, 1547–1564. https://doi.org/10.1007/s40747-021-00279-z
30. Huang, S., Cheng, H. W., Tsai, J. T., Kuo, C.-Y., Chang, C.-C., Chen, L.-J., Shiau, A.-C., Wang, Y.-J., & Li, M.-H. (2022). Failure mode and effects analysis for errors detected during pretreatment physics plan and chart review in external beam radiotherapy. *Therapeutic Radiology and Oncology, 6*, 11. https://doi.org/10.21037/tro-21-38
31. Khomane, K. S., More, P. K., Raghavendra, G., & Bansal, A. K. (2013). Molecular understanding of the compaction behavior of indomethacin polymorphs. *Molecular Pharmaceutics, 10*, 631–639. https://doi.org/10.1021/mp300390m
32. Vervaet, C., Bekaert, B., Peeters, E., Beer, T. D., Remon, J. P., & Vervaet, C. (2016). Improved Tabletability after a polymorphic transition of Delta-mannitol during twin screw granulation. *International Journal of Pharmaceutics, 506*, 13–24. https://doi.org/10.1016/j.ijpharm.2016.04.025
33. Lakio, S., Ervasti, T., Tajarobi, P., Wikström, H., Fransson, M., Karttunen, A.-P., Ketolainen, J., Folestad, S., Abrahmsén-Alami, S., & Korhonen, O. (2017). Provoking an end-to-end continuous direct compression line with raw materials prone to segregation. *European Journal of Pharmaceutical Sciences, 109*, 514–524. https://doi.org/10.1016/j.ejps.2017.09.018
34. Lakio, S., Siiriä, S., Räikkönen, H., Airaksinen, S., Närvänen, T., Antikainen, O., & Yliruusi, J. (2010). New insights into segregation during tabletting. *International Journal of Pharmaceutics, 397*(1), 19–26. https://doi.org/10.1016/j.ijpharm.2010.06.041

35. Vaithiyalingam, S. R., & Sayeed, V. A. (2010). Critical factors in manufacturing multi-layer tablets--assessing material attributes, in-process controls, manufacturing process and product performance. *International Journal of Pharmaceutics, 398*(1–2), 9–13. https://doi.org/10.1016/j.ijpharm.2010.07.025
36. Castellanos, A. (2005). The relationship between attractive interparticle forces and bulk behaviour in dry and uncharged fine powders. *Advances in Physics, 54*(4), 263–376.
37. Tanaka, R., Osotprasit, S., Peerapattana, J., Ashizawa, K., Hattori, Y., & Otsuka, M. (2021). Complete cocrystal formation during resonant acoustic wet granulation: Effect of granulation liquids. *Pharmaceutics, 13*, 56. https://doi.org/10.3390/pharmaceutics13010056
38. Perez-Gandarillas, L., Perez-Gago, A., Mazor, A., Kleinebudde, P., Lecoq, O., & Michrafy, A. (2016). Effect of roll-compaction and milling conditions on granules and tablet properties. *European Journal of Pharmaceutics and Biopharmaceutics, 106*, 38–49. https://doi.org/10.1016/j.ejpb.2016.05.020
39. Shi, Z., McGhehey, K. C., Leavesley, I. M., & Manley, L. F. (2016). On-line monitoring of blend uniformity in continuous drug product manufacturing process—The impact of powder flow rate and the choice of spectrometer: Dispersive vs. FT. *Journal of Pharmaceutical and Biomedical Analysis, 118*, 259–266.
40. Herbold, E., Walton, O., & Homel, M. (2015). *Simulation of powder layer deposition in additive manufacturing processes using the discrete element method.* Lawrence Livermore National Lab.(LLNL).
41. Meier, C., Weissbach, R., Weinberg, J., Wall, W. A., & Hart, A. J. (2019a). Modeling and characterization of cohesion in fine metal powders with a focus on additive manufacturing process simulations. *Powder Technology, 343*, 855–866.
42. Meier, C. R., Weissbach, R., Weinberg, J., Wall, W. A., & Hart, A. (2019b). Critical influences of particle size and adhesion on the powder layer uniformity in metal additive manufacturing. *Journal of Materials Processing Technology, 266*, 484–501. https://doi.org/10.1016/j.jmatprotec.2018.10.037
43. Agiba, A. M., Abul-Ella, S., & El-Monem, R. A. A. (2021). Pharmacotechnical development and optimization of multilayered tablets: An updated industrial review with emphasis on bilayer tablets. *International Journal of Applied Pharmaceutics, 13*, 55–64. https://doi.org/10.22159/ijap.2021v13i4.41528
44. Bowles, B., Dziemidowicz, K., Lopez, F., Pasqua, O. D., Tuleu, C., Edwards, A. J., & Ernest, T. B. (2018). Co-processed excipients for dispersible tablets–Part 1: Manufacturability. *AAPS PharmSciTech, 19*, 2598–2609. https://doi.org/10.1208/s12249-018-1090-4
45. Haruna, F., Apeji, Y. E., Oparaeche, C., Oyi, A. R., & Gamlen, M. (2020). Compaction and tableting properties of composite particles of microcrystalline cellulose and crospovidone engineered for direct compression. *Future Journal of Pharmaceutical Sciences, 6*, 35. https://doi.org/10.1186/s43094-020-00055-9
46. Thoorens, G., Krier, F., Leclercq, B., Carlin, B., & Evrard, B. (2014). Microcrystalline cellulose, a direct compression binder in a quality by design environment—A review. *International Journal of Pharmaceutics, 473*(1), 64–72. https://doi.org/10.1016/j.ijpharm.2014.06.055
47. Rojas, J., & Kumar, V. (2011). Comparative evaluation of silicified microcrystalline cellulose II as a direct compression vehicle. *International Journal of Pharmaceutics, 416*(1), 120–128. https://doi.org/10.1016/j.ijpharm.2011.06.017
48. Fukami, J., Yonemochi, E., Yoshihashi, Y., & Terada, K. (2006). Evaluation of rapidly disintegrating tablets containing glycine and carboxymethylcellulose. *International Journal of Pharmaceutics, 310*(1–2), 101–109. https://doi.org/10.1016/j.ijpharm.2005.11.041
49. Pabari, R. M., & Ramtoola, Z. (2012). Effect of a disintegration mechanism on wetting, water absorption, and disintegration time of orodispersible tablets. *Journal of Young Pharmacists, 4*, 157–163. https://doi.org/10.4103/0975-1483.100021
50. Zeitler, J. A., Bawuah, P., Ridgway, C. J., Van Den Ban, S., Goodwin, D. J., Ketolainen, J., Gane, P. A., Peiponen, K.-E., & Zeitler, J. A. (2018). Fast and non-destructive pore structure

analysis using terahertz time-domain spectroscopy. *International Journal of Pharmaceutics, 537*, 102–110. https://doi.org/10.1016/j.ijpharm.2017.12.029
51. Juban, A., Briançon, S., & Puel, F. (2016). Processing-induced-transformations (PITs) during direct compression: Impact of tablet composition and compression load on phase transition of caffeine. *International Journal of Pharmaceutics, 501*, 253–264. https://doi.org/10.1016/j.ijpharm.2016.01.079
52. Zhang, J., Cleary, P. W., Pan, X., & Cleary, P. W. (2017). On identification of critical material attributes for compression behaviour of pharmaceutical diluent powders. *Materials, 10*, 845. https://doi.org/10.3390/ma10070845
53. Yin, Z., Hannard, F., & Barthelat, F. (2019). Impact-resistant nacre-like transparent materials. *Science, 364*, 1260–1263. https://doi.org/10.1126/science.aaw8988
54. Jain, S., Kaur, S., Rathi, R., Nagaich, U., & Singh, I. (2023). Application of co-processed excipients for developing fast disintegrating tablets: A review. *Polymers in Medicine, 53*, 59–68. https://doi.org/10.17219/pim/158009
55. Crouter, A., & Briens, L. (2013). The effect of moisture on the flowability of pharmaceutical excipients. *AAPS PharmSciTech, 15*, 65–74. https://doi.org/10.1208/s12249-013-0036-0
56. Khan, F., Pilpel, N., & Ingham, S. (1988). The effect of moisture on the density, compaction and tensile strength of microcrystalline cellulose. *Powder Technology, 54*(3), 161–164.
57. Sun, C. C. (2008). Mechanism of moisture induced variations in true density and compaction properties of microcrystalline cellulose. *International Journal of Pharmaceutics, 346*(1–2), 93–101.
58. Haware, R. V., Tho, I., & Bauer-Brandl, A. (2010). Evaluation of a rapid approximation method for the elastic recovery of tablets. *Powder Technology, 202*(1–3), 71–77.
59. Kachrimanis, K., & Malamataris, S. (2004). "Apparent" Young's elastic modulus and radial recovery for some tableted pharmaceutical excipients. *European Journal of Pharmaceutical Sciences, 21*(2–3), 197–207.
60. Gharaibeh, S. F., & Aburub, A. (2013). Use of first derivative of displacement vs. force profiles to determine deformation behavior of compressed powders. *AAPS PharmSciTech, 14*(1), 398–401. https://doi.org/10.1208/s12249-013-9928-2
61. Rowley, G., & Mackin, L. A. (2003). The effect of moisture sorption on electrostatic charging of selected pharmaceutical excipient powders. *Powder Technology, 135*, 50–58.
62. Vodáčková, P., Vraníková, B., Svačinová, P., Franc, A., Elbl, J., Muselík, J., Kubalák, R., & Solný, T. (2018). Evaluation and comparison of three types of spray dried coprocessed excipient Avicel® for direct compression. *BioMed Research International, 2018*, 2739428. https://doi.org/10.1155/2018/2739428
63. Khan, I. A., Apostolou, M., Bnyan, R., Houacine, C., Elhissi, A., & Yousaf, S. (2020). Paclitaxel-loaded micro or nano transfersome formulation into novel tablets for pulmonary drug delivery via nebulization. *International Journal of Pharmaceutics, 575*, 118919. https://doi.org/10.1016/j.ijpharm.2019.118919
64. Vallet, T., Michelon, H., Jani, Y., Leglise, P., Laribe-Caget, S., Piccoli, M., Fur, A. L., Liu, F., Ruiz, F., & Boudy, V. (2020). Acceptability in the older population: The importance of an appropriate tablet size. *Pharmaceutics, 12*, 746. https://doi.org/10.3390/pharmaceutics12080746
65. Cabiscol, R., Shi, H., Wünsch, I., Magnanimo, V., Finke, J. H., Luding, S., & Kwade, A. (2020). Effect of particle size on powder compaction and tablet strength using limestone. *Advanced Powder Technology, 31*(3), 1280–1289.
66. Jounela, A., Pentikäinen, P., & Sothmann, A. (1975). Effect of particle size on the bioavailability of digoxin. *European Journal of Clinical Pharmacology, 8*, 365–370.
67. Campos-Aldrete, M. E., & Villafuerte-Robles, L. (1997). Influence of the viscosity grade and the particle size of HPMC on metronidazole release from matrix tablets. *European Journal of Pharmaceutics and Biopharmaceutics, 43*(2), 173–178.
68. Razavi, S. M., Callegari, G., Drazer, G., & Cuitiño, A. M. (2016). Toward predicting tensile strength of pharmaceutical tablets by ultrasound measurement in continuous manufacturing.

International Journal of Pharmaceutics, 507, 83–89. https://doi.org/10.1016/j.ijpharm.2016.04.064
69. Vercruysse, J., Díaz, D. C., Peeters, E., Fonteyne, M., Delaet, U. A. C., Assche, I. V., Beer, T. D., Remon, J. P., & Vervaet, C. (2012). Continuous twin screw granulation: Influence of process variables on granule and tablet quality. *European Journal of Pharmaceutics and Biopharmaceutics, 82*, 205–211. https://doi.org/10.1016/j.ejpb.2012.05.010
70. Shaddinger, B., Mahar, K. M., Sprys, M., Andrews, S. M., Chattoraj, S., Israni, R., & Cobitz, A. (2023). Comparison of two manufacturing processes of Daprodustat for bioequivalence and dissolution in healthy volunteers: A randomized crossover study. *Clinical Pharmacology in Drug Development, 12*, 739–748. https://doi.org/10.1002/cpdd.1257
71. De Simone, V., Caccavo, D., Lamberti, G., d'Amore, M., & Barba, A. A. (2018). Wet-granulation process: Phenomenological analysis and process parameters optimization. *Powder Technology, 340*, 411–419.
72. Park, S.-Y., Galbraith, S. C., Liu, H., Lee, H., Cha, B., Huang, Z., O'Connor, T., Lee, S., & Yoon, S. (2018). Prediction of critical quality attributes and optimization of continuous dry granulation process via flowsheet modeling and experimental validation. *Powder Technology, 330*, 461–470.
73. Singh, R., Barrasso, D., Chaudhury, A., Sen, M., Ierapetritou, M., & Ramachandran, R. (2014a). Closed-loop feedback control of a continuous pharmaceutical tablet manufacturing process via wet granulation. *Journal of Pharmaceutical Innovation, 9*, 16–37.
74. Singh, R., Muzzio, F. J., Ierapetritou, M. G., & Ramachandran, R. (2015). A combined feed-forward/feed-back control system for a QbD-based continuous tablet manufacturing process. *Processes, 3*, 339–356. https://doi.org/10.3390/pr3020339
75. Azad, M., Osorio, J. C., Wang, A. B., Klee, D. M., Eccles, M. E., Grela, E., Sloan, R. S., Hammersmith, G., Rapp, K., Brancazio, D., & Myerson, A. S. (2019). On-demand manufacturing of direct compressible tablets: Can formulation be simplified? *Pharmaceutical Research, 36*, 167. https://doi.org/10.1007/s11095-019-2716-2
76. Yusof, Y. A., Salleh, F. S. M., Chin, N. L., & Talib, R. A. (2011). The drying and tabletting of pitaya powder. *Journal of Food Process Engineering, 35*, 763–771. https://doi.org/10.1111/j.1745-4530.2010.00625.x
77. Singh, A., Das, S. K., Gupta, S., & Ghosh, S. (2021). The challenges of producing bilayer tablet: A review. *Journal of Drug Delivery and Therapeutics, 11*, 171–175. https://doi.org/10.22270/jddt.v11i4-s.4922
78. Cho, S., Lee, J. W., Yoo, Y., Cho, M.-Y., Sohn, S., & Lee, B.-J. (2021). Improved manufacturability and in vivo comparative pharmacokinetics of Dapagliflozin Cocrystals in beagle dogs and human volunteers. *Pharmaceutics, 13*, 70. https://doi.org/10.3390/pharmaceutics13010070
79. Kleinebudde, P. (2004). Roll compaction/dry granulation: Pharmaceutical applications. *European Journal of Pharmaceutics and Biopharmaceutics, 58*(2), 317–326.
80. Sun, C. C. (2016). Quantifying effects of moisture content on flow properties of microcrystalline cellulose using a ring shear tester. *Powder Technology, 289*, 104–108.
81. Thapa, P., Lee, A. R., Choi, D. H., & Jeong, S. H. (2017). Effects of moisture content and compression pressure of various deforming granules on the physical properties of tablets. *Powder Technology, 310*, 92–102.
82. Veronica, N., Heng, P. W. S., & Liew, C. V. (2022). Relative humidity cycling: Implications on the stability of moisture-sensitive drugs in solid pharmaceutical products. *Molecular Pharmaceutics, 20*, 1072–1085. https://doi.org/10.1021/acs.molpharmaceut.2c00812
83. Nair, A., Khunt, D., & Misra, M. (2019). Application of quality by design for optimization of spray drying process used in drying of risperidone nanosuspension. *Powder Technology, 342*, 156–165.
84. Li, J., & Wu, Y. (2014). Lubricants in pharmaceutical solid dosage forms. *Lubricants, 2*, 21–43. https://doi.org/10.3390/lubricants2010021

85. Natoli, D., Levin, M., Tsygan, L., & Liu, L. (2017). Development, optimization, and scale-up of process parameters: Tablet compression. In *Developing solid oral dosage forms* (pp. 917–951). Elsevier.
86. Adedokun, M., Onah, B. I., & Attama, A. A. (2018). Physico-mechanical and release properties of sustained release artesunate tablets in hydroxypropyl methylcellulose matrix. *International Journal of Applied Pharmaceutics, 10*, 103. https://doi.org/10.22159/ijap.2018v10i1.22918
87. Yamada, M., Ishikawa, A., Muramatsu, S., Furuishi, T., Onuki, Y., Fukuzawa, K., & Yonemochi, E. (2022). Study of orally disintegrating tablets using Erythritol as an excipient produced by moisture-activated dry granulation (MADG). *Pharmaceuticals, 15*, 1004. https://doi.org/10.3390/ph15081004
88. Muteki, K., Swaminathan, V., Sekulic, S. S., & Reid, G. C. (2011). De-risking pharmaceutical tablet manufacture through process understanding, latent variable modeling, and optimization technologies. *AAPS PharmSciTech, 12*, 1324–1334. https://doi.org/10.1208/s12249-011-9700-4
89. Sun, F., Xu, B., Dai, S., Zhang, Y., & Qiao, Y. (2019). A novel framework to aid the development of design space across multi-unit operation pharmaceutical processes—A case study of Panax Notoginseng Saponins immediate release tablet. *Pharmaceutics, 11*, 474. https://doi.org/10.3390/pharmaceutics11090474
90. Jang, E. H., Park, Y. S., & Choi, D. H. (2021). Investigation of the effects of materials and dry granulation process on the mirabegron tablet by integrated QbD approach with multivariate analysis. *Powder Technology, 382*, 23–39.
91. Maeda, J., Suzuki, T., & Takayama, K. (2012). Design space construction of multiple dose-strength tablets utilizing Bayesian estimation based on one set of design-of-experiments. *Chemical and Pharmaceutical Bulletin, 60*, 1399–1408. https://doi.org/10.1248/cpb.c12-00491
92. Han, J. I., Shin, B. S., & Choi, D. H. (2019). Comprehensive study of intermediate and critical quality attributes for process control of high-shear wet granulation using multivariate analysis and the quality by design approach. *Pharmaceutics, 11*, 252. https://doi.org/10.3390/pharmaceutics11060252
93. Fayed, M., Alalaiwe, A., Almalki, Z. S., & Helal, D. A. (2022). Design space approach for the optimization of green fluidized bed granulation process in the granulation of a poorly water-soluble fenofibrate using design of experiment. *Pharmaceutics, 14*, 1471. https://doi.org/10.3390/pharmaceutics14071471
94. Malakar, J., & Nayak, A. K. (2012). Use of response surface methodology in the formulation and optimization of Bisoprolol fumarate matrix tablets for sustained drug release. *ISRN Pharmaceutics, 2012*, 1–10. https://doi.org/10.5402/2012/730624
95. Mesut, B., Tok, Y. P., Alkan, B., Vefai, M. K., Al-Mohaya, M., & Özsoy, Y. (2023). Effect of mannitol particle size on melatonin dissolution and tablet properties using a quality by design framework. *Dissolution Technologies, 30*, 12–21. https://doi.org/10.14227/dt300123p12
96. Afzal, M., Muddassir, M., Alarifi, A., & Ansari, M. T. (2021). Box-Behnken assisted validation and optimization of an RP-HPLC method for simultaneous determination of Domperidone and lansoprazole. *Separations, 8*(1), 5.
97. Hasnain, M. S., Ansari, S. A., Rao, S., Tabish, M., Singh, M., Abdullah, M. S., & Ansari, M. T. (2017). QbD-driven development and validation of liquid chromatography tandem mass spectrometric method for the quantitation of sildenafil in human plasma. *Journal of Chromatographic Science, 55*(6), 587–594.
98. Kim, J. Y., & Choi, D. H. (2022). Control strategy for excipient variability in the quality by design approach using statistical analysis and predictive model: Effect of microcrystalline cellulose variability on design space. *Pharmaceutics, 14*(11), 2416.
99. Gavi, E., & Reynolds, G. K. (2014). System model of a tablet manufacturing process. *Computers & Chemical Engineering, 71*, 130–140.

100. Matero, S., van Den Berg, F., Poutiainen, S., Rantanen, J., & Pajander, J. (2013). Towards better process understanding: Chemometrics and multivariate measurements in manufacturing of solid dosage forms. *Journal of Pharmaceutical Sciences, 102*(5), 1385–1403.
101. Liu, F., Ranmal, S. R., Batchelor, H., Orlu-Gul, M., Ernest, T. B., Thomas, I. W., Flanagan, T., Kendall, R., & Tuleu, C. (2015). Formulation factors affecting acceptability of oral medicines in children. *International Journal of Pharmaceutics, 492*, 341–343. https://doi.org/10.1016/j.ijpharm.2015.05.013
102. Chen, F.-C., Liu, W.-J., Zhu, W.-F., Yang, L.-Y., Zhang, J.-W., Feng, Y., Ming, L.-S., & Li, Z. (2022). Surface modifiers on composite particles for direct compaction. *Pharmaceutics, 14*, 2217. https://doi.org/10.3390/pharmaceutics14102217
103. Tuyen, N. D., Nghiem, L. T. P., Tuan, N. H., & Le, P. H. (2021). Development of a scalable process of film-coated bi-layer tablet containing sustained-release metoprolol succinate and immediate-release amlodipine Besylate. *Pharmaceutics, 13*, 1797. https://doi.org/10.3390/pharmaceutics13111797
104. Boukouvala, F., Niotis, V., Ramachandran, R., Muzzio, F. J., & Ierapetritou, M. G. (2012). An integrated approach for dynamic flowsheet modeling and sensitivity analysis of a continuous tablet manufacturing process. *Computers & Chemical Engineering, 42*, 30–47.
105. Singh, R., Sahay, A., Muzzio, F., Ierapetritou, M., & Ramachandran, R. (2014b). A systematic framework for onsite design and implementation of a control system in a continuous tablet manufacturing process. *Computers & Chemical Engineering, 66*, 186–200.

Chapter 8
Quality by Design Topical Formulations

Burcu Uner and Ahmet Dogan Ergin

Abstract Pharmaceutical companies are currently investing a significant amount of their resources in developing advanced carrier systems. These systems are designed to deliver drugs over an extended period with minimal side effects, which improves patient compliance and decreases variability that is often associated with intermittent dosing. The utilization of the Quality by Design (QbD) methodology has been shown to improve the efficacy of drug development and regulatory processes, leading to quicker access to high-quality topical dermatological and transdermal products. Moreover, this approach could decrease costs for both the healthcare system and consumers. In order to adopt the QbD approach, it is necessary to establish the Quality Target Product Profile (QTPP) and Critical Quality Attributes (CQAs) for a pharmaceutical product. This entails evaluating potential risks to determine Critical Material Attributes (CMAs) and Critical Process Parameters (CPPs). Subsequently, a Design Space should be defined through the use of Design of Experiments (DoEs). Additionally, it is important to establish a Control Strategy and uphold a commitment to continuous improvement and innovation throughout the product's lifecycle. In this chapter, we will discuss the creation and assessment of Critical Quality Attributes (CQAs) for both topical and transdermal products. These products can range from basic semisolid formulations and patches to more advanced nanotechnological-based technologies.

Keywords Topical formulations · QbD · CQAs · Nasal · Ocular · Transdermal

---

B. Uner (✉)
Department of Pharmaceutical and Administrative Sciences, University of Health Science and Pharmacy in St. Louis, Saint Louis, MO, USA
e-mail: burcu.uner@uhsp.edu

A. D. Ergin
Department of Pharmaceutical Technology, Faculty of Pharmacy, Trakya University, Edirne, Turkey

Department of Neuroscience, University of Turin, Turin, Italy
e-mail: adoganergin@trakya.edu.tr

N. K. Jain, N. Bajwa (eds.), *Introduction to Quality by Design (QbD)*,
https://doi.org/10.1007/978-981-99-8034-5_8

8.1 Introduction

In the realm of pharmaceutical development, there exists a methodology known as Quality by Design (QbD), which is thoughtfully structured with defined objectives. This approach is centered around the principles of science and quality risk management, with a particular emphasis on understanding and managing both the product and process in question. Pharmaceutical QbD aims to achieve the following objectives [1, 2]:

1. According to establish product quality standards that reflect real-world clinical performance.
2. In order to optimize the quality of product manufacturing and mitigate potential defects, our objective is to elevate the level of proficiency in the design, comprehension, and regulation of both the product and manufacturing process.
3. In order to improve the efficiency of product development and manufacturing processes, we need to take certain steps.
4. To enhance the examination of fundamental factors and effective handling of modifications post-endorsement.

QbD is an imperative process that links product quality to its desired clinical performance. It is absolutely crucial to create a reliable formulation and manufacturing process that guarantees consistent achievement of the desired product quality. The U.S. Food and Drug Administration (FDA) has executed great performance regarding the goal of performance-based quality specifications since implementing pharmaceutical QbD. Recent FDA discussions about the potency limits of narrow therapeutic index drugs and the physical characteristics of generic drug products demonstrate this progress [3]. It is important to note that the International Committee of Harmonization (ICH) documents did not explicitly mention clinical performance-based specifications as a QbD goal. However, a scientific paper did acknowledge this as a recent development [4]. QbD is an essential and systematic approach to designing and developing products. It not only enhances development capabilities but also accelerates the process and improves formulation design. Moreover, it shifts resources from a reactive approach to a proactive one, effectively preventing issues before they occur [5, 6].

QbD concepts and applications are highly recommended for pharmaceutical dosage forms, with particular emphasis on oral dosage forms. However, in this review, we will be focusing on the use of QbD approaches in the development of topical dosage forms (TDFs), specifically generic forms [7]. Developing a standardized TDF can often be a complex and time-consuming task. However, by employing QbD methodologies, this process can be streamlined. In this review, we delve into the fundamental QbD approaches and provide insight into their practical applications [8].

Drugs that are applied directly to the specific area where they are required are commonly referred to as topical products. These are specialized drugs meant to be applied on a specific area of the body and have minimal impact on the rest of the body. To obtain regulatory approval, a topical product must closely resemble the commercialized reference product. To ensure consistency in pharmaceutical and therapeutic effects, generic TDFs must imitate the reference listed drug. QbD focuses on creating TDF specifications that are clinically based and meaningful in terms of performance [9, 10].

The FDA has provided guidance on the essential elements for creating generic drugs. These include the requirement to prove clinical endpoint bioequivalence for locally acting TDFs, as outlined in regulation 21CFR 320.24 [11]. In the future, non-oral dosage forms like TDFs will be critical. It is essential to utilize QbD approaches during development to ensure quality. In order to determine the bioavailability of a drug product or establish its bioequivalence, conducting well-controlled clinical trials is an absolute requirement. This can only be achieved through the implementation of 21 CFR 320.34 and the design of appropriately comparative clinical trials. The changes between the traditional and contemporary approaches was given in Fig. 8.1.

Among the methods used to measure bioavailability or show bioequivalence, this particular approach is the least precise, responsive, and consistent. Therefore, this chapter aims to thoroughly explore the specific perspectives of QbD in the production of topical dosage forms.

8.2 Purpose of Using QbD in Topical Route

Developing a nuanced topical nanocarrier can present difficulties due to the need to coordinate multiple materials and processes. To achieve the desired characteristics and quality of a drug, it is essential for formulation scientists to methodically fine-tune all pertinent variables while adhering to the product development guidelines outlined in ICH Q8 (R2) [5, 13].

The outdated method of optimizing drug delivery one variable at a time is simply not enough. In today's world, a more comprehensive and systematic multivariate response surface methodology is the preferred approach, fully aligned with regulatory requirements. The QbD approach is absolutely essential when it comes to formulating, optimizing, and developing dosage forms. It is a systematic approach that establishes predefined objectives and prioritizes understanding product and process, as well as controlling processes based on sound science and quality risk management [8, 14, 15].

Over the past decade, numerous scientists have devised novel delivery techniques utilizing the QbD approach. This approach mandates the identification of a Quality Target Product Profile (QTPP), Critical Quality Attributes (CQAs), Critical Material Attributes (CMAs), and Critical Process Parameters (CPPs) [13, 16]. By utilizing tools such as Design of Experiment (DoE) and Central composite design, scientists can establish the relationship between CQAs and CMAs [1, 15]. QTPP is a summary that outlines the quality characteristics of a pharmaceutical product (Fig. 8.2) [1, 12, 17].

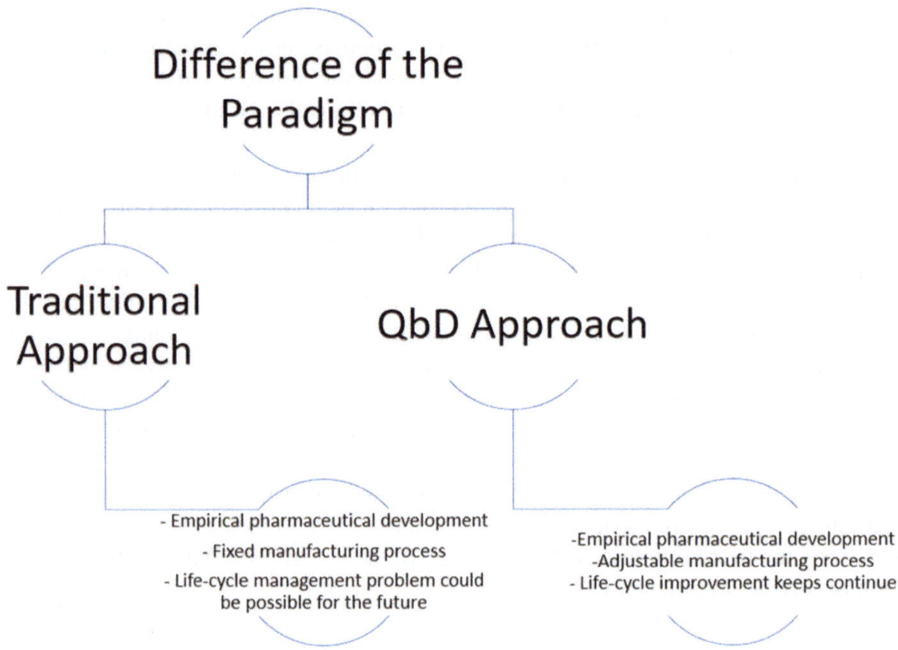

Fig. 8.1 Shifting between the traditional and contemporary approaches [12]

Its purpose is to ensure the safety and effectiveness of the product. QTPP is a crucial aspect that specifies the desired outcome for a pharmaceutical product. It considers critical factors such as the intended use in a clinical setting, the mode of administration, the form of delivery, strength, packaging, properties that may affect it, and quality standards. This essential information is based on rigorous research conducted by different working groups [18–20]. In this chapter, we will explore how to apply a topical dosage form from a QbD perspective with a focus on the criteria for the route of administration. The table used in this chapter was shared with permission from Dr. Andre S. Raw [12] (Table 8.1).

Regarding the incomes of the Table 8.1, one of the studies was investigated by using their outcomes (Table 8.2).

Although there are multiple tools available for QbD, their intended usage varies depending on the application. As a result, it's crucial that the initiator can effectively communicate their purpose to the examiner.

8.3 Preparing Topical Formulations with QbD Tools

8.3.1 Experimental Design

The systematic approach of DoE establishes the correlation between process factors and their outcomes. It is imperative to note that this method encompasses various designs such as Taguchi design, Box-Behnken design, and mixture composite design. In the course of DoE, scientists are confronted with the daunting task of examining numerous input and output variables, which includes process parameters for each unit operation. This can pose a practically impossible challenge. Nevertheless, scientists can draw upon their pre-existing knowledge and risk management expertise to determine the most critical input and output variables as well as process parameters [6]. Through the use of DoE, it is possible to pinpoint the key factors that are linked to CQAs. This includes investigating any interactions and synergies that may exist between these factors. By establishing the allowable limit of CQAs, it becomes possible to determine the design space for CPPs [1, 21].

8.3.2 Risk Management

ICH Q9 quality risk management defines *risk assessment* as the combination of probability and severity of harm. There are several tools available to manage risks

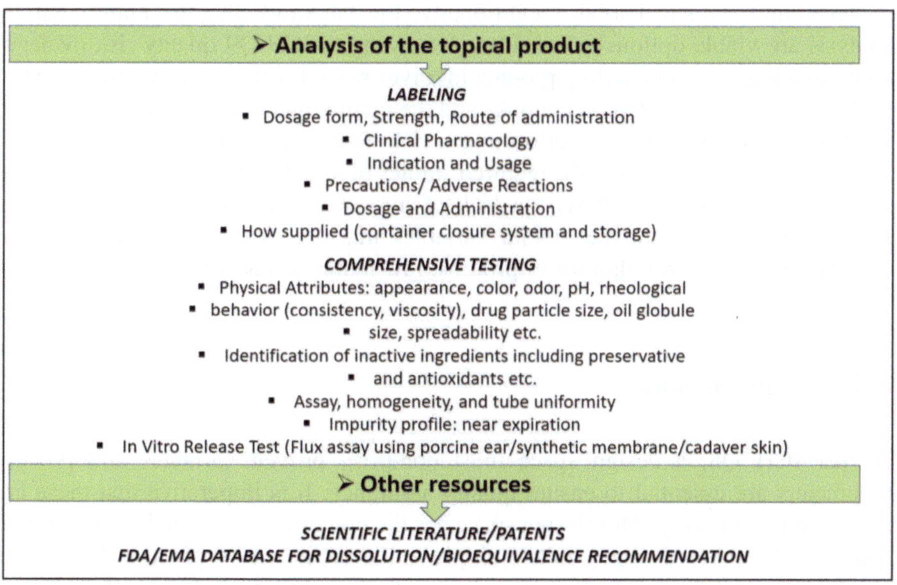

Fig. 8.2 QTTP criteria for topical products

Table 8.1 CQA for one of the samples in topical products

CQA	Purpose	Rationale
Identification	(+) for active	Mandatory for clinical effectiveness
Assay	85–10%	Mandatory for clinical effectiveness
Impurities	Impurity A: NMT 0.25% Impurity B: NMT 0.25%	Mandatory for safety
Homogeneity	90–110% and SD NMT 5%	Mandatory for clinical effectiveness
Physical attributes Rheological behavior Particle/ droplet size	Match USP information/other trusted published references	Mandatory for clinical effectiveness and patient safety
In vitro release	Match USP information/other trusted published references	Mandatory for clinical effectiveness and patient safety
Microbial limitations	USP/EMA	Mandatory for safety
Remained solvents	USP/EMA	Mandatory for safety

NMT Not more than, *SD* Standard deviation

effectively. These range from basic facilitation methods like flowcharts and check sheets to more advanced techniques such as failure mode effects analysis, fault tree analysis, and statistical tools—additionally, the Ishikawa diagram and "what if" analysis are viable options to consider. According to ICH Q9 quality risk management, making and using a drug product involves some level of risk. It's important to evaluate the balance between the risks and the quality and ultimately connect this to the benefits for the patient. Performing a risk assessment can aid pharmaceutical manufacturers in identifying the required studies before developing a new drug. Risk assessments must not be based on lack of information or uncertainty. The study results clearly highlight the crucial variables that are essential in establishing a comprehensive control plan for in-process, raw material, and final testing [22].

8.3.3 Design Space

As per ICH Q8, a design space must comprise proven variables and process parameters are essential to ensure quality assurance. It is imperative that these are not altered in any way. This design space has the potential to vary in dimensions and can be utilized for a single unit, multiple units, or the entire process. It presents regulatory flexibility and is a crucial aspect of ensuring quality in the process (Fig. 8.3).

Table 8.2 Outcomes of QTTP Design Purpose for study of *Rapalli* et al. [17]

CQA	Meet the requirements	Properties
Identification	+	100 mg ketoconazole
Assay	+	Excipients: Poloxamer 407: 2–5% Glycerol monooleate: 2–5% PVA: 2–5% Preservative: Benzalkonium chloride: 0.01%
Impurities	+	DSC thermogram and FTIR spectrums proved that there is no impurities are available in the formulation apart from the excipients' peaks.
Homogeneity	+	No physical separation has been occurred for 3 months
Physical attributes Rheological behavior Particle/droplet size	+	Physical attributes: Except the high temperature (40 °C), the formulations observed as stabile according to the particle size changing, the rheological behaviors, pH, and etc. Rheological behavior: The Newtonian rheological behavior was observed due to the regular topical cream Particle size: 200–250 nm (appropriate for the passing into the other layers of the skin)
In vitro release	+	In vitro release of the cubosomes were passed significantly higher than the regular ketoconazole cream
Microbial limitations	+	No significant increase in microbial activity was observed until the end of the 3-month period.
Remained solvents	+	Cell culture was proved that the remained solvent is not existed in the formulation.

Working within the design space is not regarded as a modification, and it indicates a comprehensive comprehension of the process. Nonetheless, there are drawbacks to the design space, such as the expense of time and money. Also, an unfavorable outcome may occur if any essential parameter is overlooked during the construction of the design space. According to FDA guidelines, making a change within the approved design space of a manufacturing process is not considered a change. However, if the process moves outside of that space, it is considered a change.

If any changes need to be made after a product is approved, a process called regulatory post-approval change is necessary. The design space for making these changes differs depending on the formulation design and equipment used by the sponsors during drug product development.

8.3.4 Implications of the Surface Designs

The screening designs unequivocally identify the process variables that shall subsequently be optimized for the surface. By using response surface designs like

Box-Behnken, central composite, and three-level factorial designs, one can determine the best processing conditions. Factorial designs are a preferred choice because they can handle missing data well. Additionally, they allow for the incorporation of corner and center point trials from past factorial experiments, making them a strong option. However, to simplify the experiment, mixture design is utilized due to the three levels of each factor [7].

The designated design area encompasses the range where a consistently high-quality product can be produced, known as robust operating space. If a deviation from the typical process occurs within the designated design space, it is considered acceptable. Flexibility in regulatory approval can only be achieved through conducting QbD experimentation at a manufacturing scale, as shown in Fig. 8.3.

8.3.5 Application of the Analytical Perspective

PAT is an indispensable monitoring tool that must be utilized to keep track of variable processing parameters and ensure that the process remains within its operating range, regardless of whether it is inline or online. This tool is crucial for identifying processing failures and determining the exact operating parameters that must be controlled or adjusted to guarantee the product's quality. This system is essential for designing, analyzing, and controlling manufacturing processes [19, 22].

It takes timely measurements to ensure the final product's quality, and its implementation is crucial for achieving success in manufacturing. One benefit of using PAT is that it offers greater flexibility, and its ongoing assessment helps to confirm the strength of the methodology. An ideal online process monitoring is the one which should detect any out of control situation during manufacturing. PAT can identify, simulate, and control the manufacturing process. Currently, continuous monitoring with an improved and extensive [24]. For manufacturing, a good online process monitoring system should be able to identify any instances of the process going out of control.

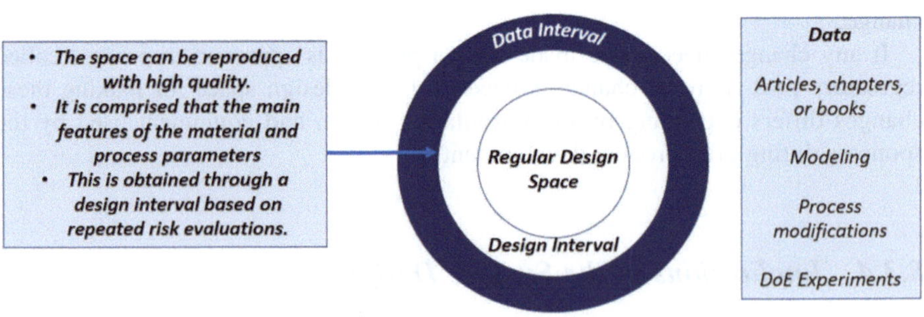

Fig. 8.3 The relationship between the data Interval and design interval (*The image was adapted Roy et al by permission*) [23]

8.4 Applications for Topical Route

8.4.1 *Transdermal Applications*

It's crucial to be aware that the skin is the largest organ in the human body, consisting of three distinct layers: the stratum corneum, the living epidermis, and the dermis. The stratum corneum serves as the primary barrier that effectively controls the diffusion rate through the skin. This complex two-compartment system comprises keratinized cells and a multilamellar lipid matrix consisting of neutral lipids and ceramides. Corneocytes, i.e., keratin-filled dead cells, are surrounded by crystalline lamellar lipid regions. The primary routes for drug delivery to the skin are through intracellular and transcellular pathways, with sweat glands and hair follicles contributing to a lesser extent. It is important to get drugs to the dermal layer; one way to do that is through topical drug delivery using drug transport [9, 25].

When dermatological products are applied to the skin, they are usually meant to have a local effect. However, some products can also have a broader impact on the body and are referred to as transdermal drug delivery systems. Preparations that act locally affect the skin's surface and change the function of the epidermis or dermis. Drug molecules enter the skin mainly by following a complex and unbroken path between cells.

Extensive research has been undertaken to assess the effectiveness of both traditional and unconventional dosage methods, employing the QbD methodology. During the period of 2022–2023, research groups conducted 793 studies utilizing Quality by Design (QbD) to improve transdermal formulations. Table 8.3 displays the most recent studies conducted on transdermal applications that are relevant to the DoE. One of the ground-breaking studies that emphasized the importance was performed by *Bagde* et al. They created 3D-microneedles patch loaded with ibuprofen using three different techniques (stereolithography, photo initiation, and lithography) that incorporated QbD design. QbD assisted them in identifying the optimal technique and mixture for the appropriate mold design [26] (Table 8.3).

On the contrary that ongoing the studies, the sanctions of FDA and EMA keep continuing strictly. A topical dermatological product's abbreviated new drug application (ANDA) is subjected to multiple review cycles by the FDA before receiving approval in most cases. Waiting for a decision can be a lengthy process, and the sponsor must exercise patience. Despite the clear guidelines provided in the ICH common technical document Q8 on how to prepare a product development report, numerous ANDA sponsors continue to struggle with its practical application. A necessary step in addressing this problem is to produce a comprehensive product development report that contains validated and substantiated specifications derived from QbD endeavors. This can aid the Office of Generic Drugs in gaining a deeper comprehension of the generic product that has been developed [8, 15].

Moreover, the in vitro diffusion technique is mandatory as well as the primary means of determining the most effective formulation among multiple options developed for a product. Before getting regulatory permission some of the limitations lead to impact regulatory process. It is important to note that this approach cannot serve as a replacement for determining in vivo bioavailability or bioequivalence. The other crucial limitation is to find appropriate amount and suitable tissue. When faced with the challenge of poor water-solubility of a drug, opting for full thickness skin may not be the best solution. Instead, it is recommended to consider alternative options such as using dermatomed skin or isolated epidermis. Maintaining sink conditions is of utmost importance in effectively dissolving agents in the receptor medium. Utilizing lipophilic solutions in a medium could heighten the percutaneous absorption of lipophilic drugs which poses a potential downside [32, 33]. To fully comprehend the outcomes, it is imperative to examine the influence of diverse elements on the QbD pattern of the formula. By conducting astute evaluations, we can attain a more profound comprehension of these consequences.

8.4.2 Nasal Applications

Nasal drug delivery stands as a vital means of drug administration, capable of producing local, systemic, and central nervous effects. As such, the role of nasal drug formulations in modern pharmaceutical technological advancements is critical. Nasal administration can serve as a viable alternative to intravenous administration, and it is common practice to use local formulations. Nasal delivery has several benefits such as a wide surface area for absorption, numerous pores in the epithelial barrier, an abundance of capillaries, avoidance of liver first-pass metabolism, and easy usage for patients [34, 35]. Nasal gels with nanosized active pharmaceutical ingredients (APIs) can help improve the effectiveness of drugs that don't dissolve well in water. The development of these products relies heavily on particle engineering, selecting suitable excipients, and implementing the appropriate technological process. These factors are crucial for success. In recent decades, more attention has been given to studying Central Nervous System (CNS) diseases through the nasal route, including Parkinson's, Alzheimer's, migraines, and epilepsy [34, 36–38]. Emesis, the forceful expulsion of gastrointestinal contents through the mouth, is a precisely coordinated process that is under the complete control of the vomiting center, situated in a specific area of the medulla [39]. In the years 2022–2023, there were approximately 1750 studies carried out on the subject of CNS order, utilizing the QbD approach [40].

A recent study highlighted the significance of the QbD perspective above all others. A team of researchers, Michels et al. in 2023, created a gel formulation using intranasal nanoemulsion to specifically target the brain. This formulation is utilized for treating brain tumors. The authors used two gelling agents, namely P407 (7.5%) and Lipoid mixture (1%), and the Design Expert software (version 14.4, State Ease, Iowa, IL, US) to achieve their results. The in vitro study showed sustained release of 80% of drug for 21 days. In the study conducted on live subjects, it was observed that the adhesive quality was higher on the 117 nasal mucosa and the retention time in the nasal passage was longer when compared to the control group. Furthermore, the

Table 8.3 The last performed studies of transdermal applications in relevance with the DoE

Formulation	Problem	API	CQA	Optimum formulation	Graphical result
MN Patch [26]	Analgesic	Ibuprofen	• *The process of photoinitiated polymerization* • Activation of a photo initiator • *RoC* -minimum • *Uniformity*- 100% • *Permeation*- high	0.23% (w/v), LAP; 0.25% (w/v). Me HA; 10% (v/v), DW	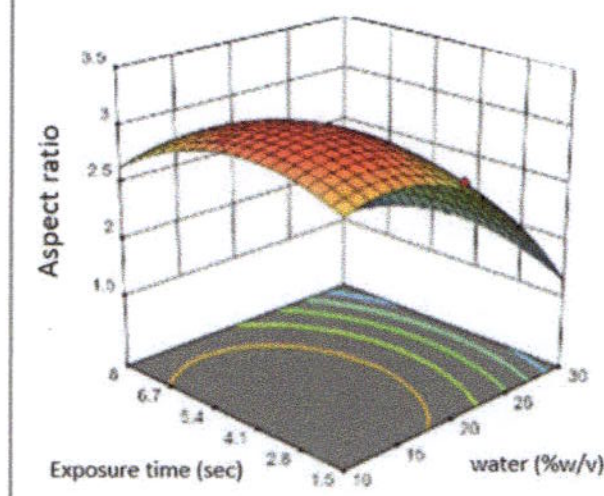 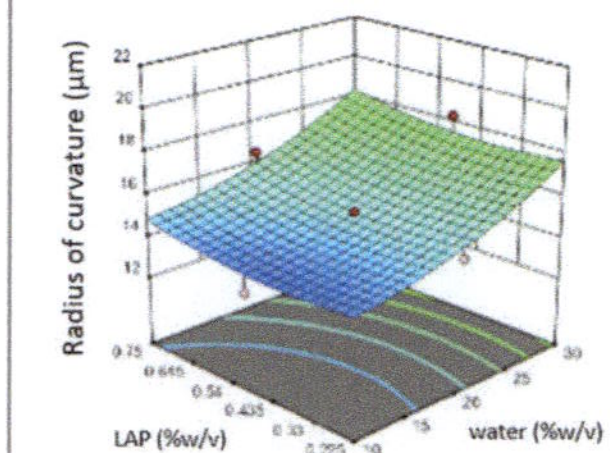
Patch [27]	PPAH	Sildenafil	• *Uniformity*- 100% • *Permeation*- high • *Drug content- high* • Flatness- 97≤ • Endurance- 360≤ • Thickness-minimum	0.1% (w/v), EC; 10% (w/v). TEC; 0.08% (w/v), sildenafil	

(continued)

Table 8.3 (continued)

Formulation	Problem	API	CQA	Optimum formulation	Graphical result

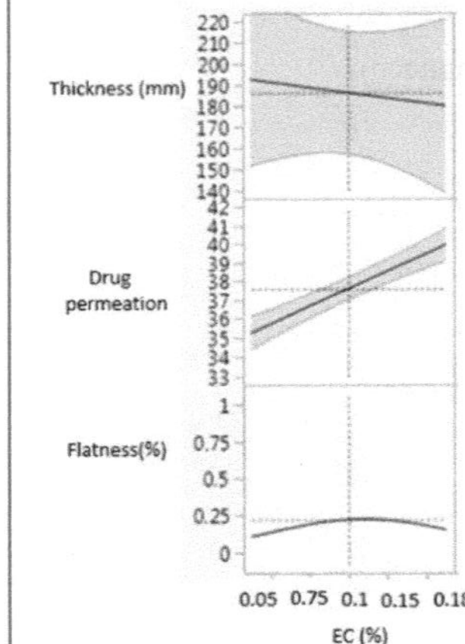

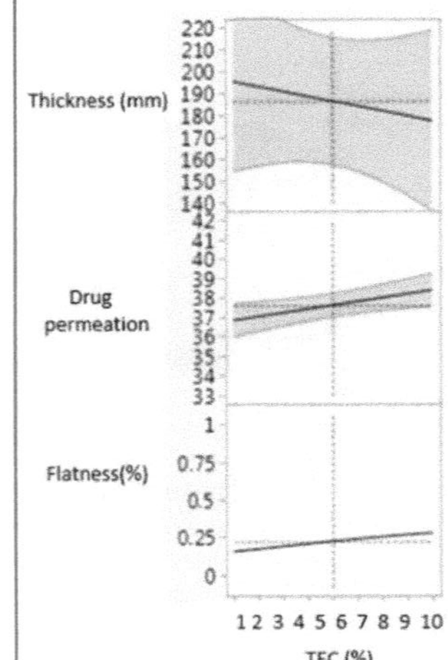

Cubosome [28]	Osteoporosis	Raloxifene	• *Particle size- 90 ≤ ≤100 nm* • *Zeta potential (mV)- + 20≤* • *Entrapment efficiency- high*	100 mg, RLX; 1500 mg. GMO; 1200 mg, F127; 6000 mg, ethanol	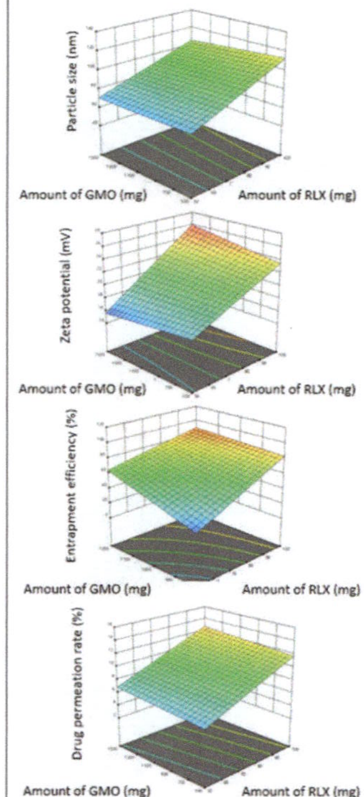
Cream [29]	Fungal infection	Clotrimazole	• Storage modulus (G`)- higher • *Loss modulus (G")-lower* • Tan δ (G"/G`)- lower • *Drug release- high-* • *Content uniformity- 100%*	0.5% (w/w), CA; 5% (w/w), OD	

(continued)

Table 8.3 (continued)

Formulation	Problem	API	CQA	Optimum formulation	Graphical result
			• *Viscosity- high* • *In-vitro permeation- high*		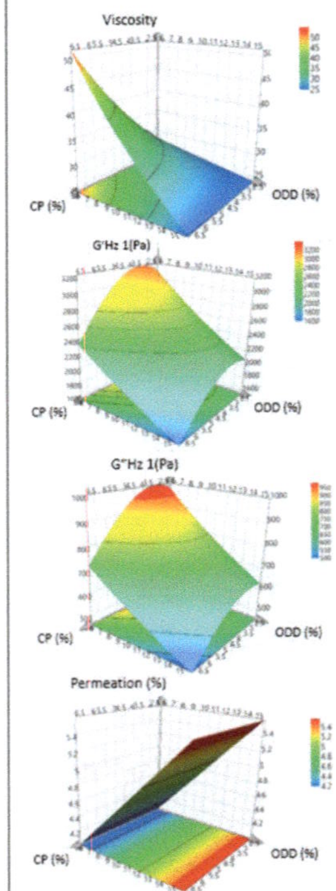

Transethosome [30]	RA	Baicalin (Flavanoid)	• *Drug loading- high* • *Vesicle size-* minimum • *PDI- minimum*	50 mg, P90; 20% (v/v), ethanol; 20 mg, NaCho	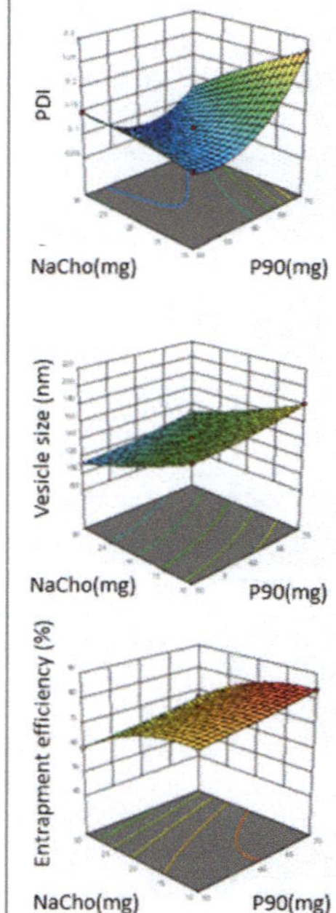
NLC [31]	Osteoarthritis	Diacerein	• *Particle size-* minimum • *Viscosity- high* • *In-vitro permeation- high*	0.2 mg, SA; 0.47 mg, OA	

(continued)

Table 8.3 (continued)

Formulation	Problem	API	CQA	Optimum formulation	Graphical result
					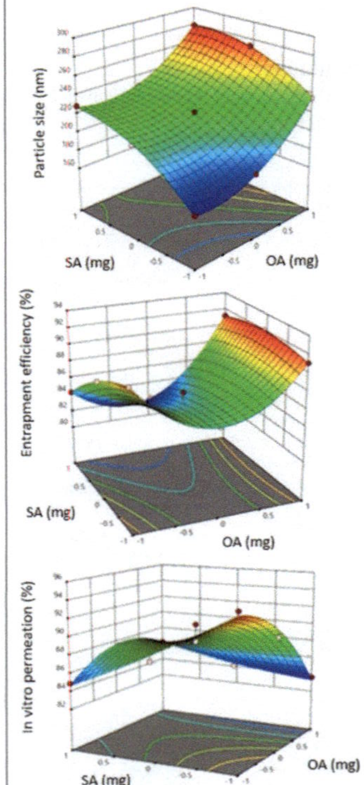

CP cethyl palmitate, *EC* ethyl cellulose, *F127* pluronic F127, GMO glyceryl monooleate, *LAP* lithium phenyl (2,4,6-trimethylbenzoyl) phosphinate, *Me HA* methacrylated hyaluronic acid, *MN* microneedle, *ODD* octyldodecanol, *NaCho* sodium cholate, *OA* oleic acid, *P90* phospholipon 90G, *PEGDAMA* polyethylene glycol dimethacrylate, *PPAH* pediatric pulmonary arterial hypertension, *RA* rheumatoid arthritis, *RLX* raloxifene, *RoC* radius of curvature, *SA* stearic acid, *TEC* Triethyl citrate

Figures were adopted with permissions

in vivo morphological estimation unequivocally confirmed the absence of nasal irritation or damage to the mucociliary system, serving as conclusive proof of the safety of the formulation. The intranasal route of drug administration yielded significantly higher concentration levels in specific tumor regions of the brain when compared to the intravenous route [41]. Table 8.4 highlights some of the studies concerning CNS diseases and emesis by using the QbD perspective.

8.4.3 Ophthalmic Applications

The administration of drugs to the eye can prove to be an arduous task due to the intricate anatomy of the eye and the physiological barriers it poses. Administering drugs systemically is not recommended for minor eye conditions [2, 10]. The blood-aqueous barrier and blood-retina barrier can prevent the drug from reaching the necessary level, making it ineffective. Hence, for overcoming of this problem, high systemic drug dosage is being applied; unfortunately, this approach causes side effects in patients. The most favored way of treating surface-level eye diseases is through topical drug administration, with eye drops accounting for more than half of all ophthalmic preparations. Reducing the number of experiments in literature is essential for obtaining more useful data. There are many different experimental designs that have been described which can be useful. To effectively differentiate between dependent and independent variables in an experimental design, it is highly recommended to utilize first-order designs such as Plackett-Burman or Taguchi from a QbD perspective [47, 48].

Fangueiro et al. successfully created a lipid nanoparticle with a water/oil/water structure that boasts a positive potential. This was achieved through the strategic use of Softisan® 100 (w/w, %), Lipoid S75 (soybean phosphatidylcholine) (w/w, %), and Poloxamer 188 (w/w, %). It is crucial to note that the concentration of lecithin in Lipoid S75 plays a significant role in determining the size of nanoparticles and their PDI ($p < 0.05$), as indicated by the results of the Taguchi factorial design model utilizing Cetyltrimethylammonium bromide (CTAB). The optimized formulation, containing 0.5% CTAB in the lipid phase, demonstrated excellent biocompatibility with the human retinoblastoma cell line [49] (Table 8.5).

On the other hand, for precise results, it is imperative to employ two design models concurrently in order to create formulations. Yerlikaya et al. successfully utilized both Plackett-Burman and Box-Behnken designs to prepare Paclitaxel nanoparticles [50]. To pinpoint potential risk factors, we conducted an analysis on various parameters, including paclitaxel (w/w,%), PLGA (w/w,%), PLGA molecular weight, PLGA terminal group, surfactant concentration (w/w,%), surfactant type, homogenization rate (rpm), and homogenization time. Through the Plackett-Burman experimental design, we were able to study the effects of these factors on critical quality properties of nanoparticles and identify the specific risk factors involved. The amount of PLGA (w/w,%), surfactant concentration (w/w,%), and homogenization rate (rpm) in the emulsion have a significant impact on the particle size, zeta

Table 8.4 the last performed studies of nasal applications in relevance with the DoE

Formulation	Problem	API	CQA	Optimum formulation	Graphical result
ME [42]	Epilepsy	Midazolam	• *Color*- transparent • *Odor*- no unpleasant odor • *Uniformity*- 100% • *Globule size*- ≤200 nm • *Permeation*- high • Viscosity- moderate • %transmittance- high	30% (w/w), CMCM; 40% (w/w), T80:PEG400 (1:1,w/w); 30% (w/w), DW	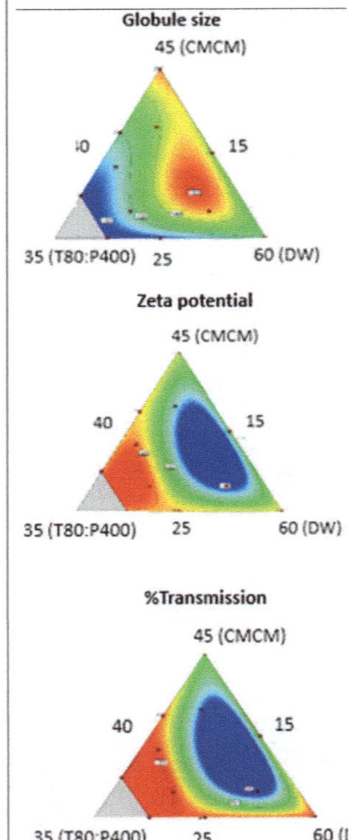

SLN [43]	AD	Rivastigmine tartrate	• *Uniformity*- 100% • *Particle size*- 80 ≤ ≤120 nm • *Entrapment* - ≥80% • *Permeation*- ≥90% • Viscosity-moderate	0.59% (w/w), GMS; 2% (w/w), P80 97.41% (w/w), DW	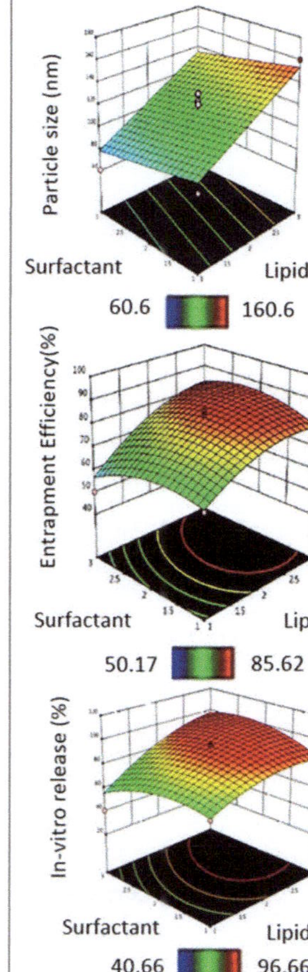

(continued)

Table 8.4 (continued)

Formulation	Problem	API	CQA	Optimum formulation	Graphical result
MCS [44]	AD	Donepezil HCl	• *Yield (%)-* 70≤ • *Swelling ability* • *PDI* - ≤ 0.5 • *Flow-* 0 ≤ ≤20 • *Spray cone angle –* 30 ≤ ≤60 • *Moisture content (%)-* • *Nasal deposition profile-* 60≤	0.4% (w/w), DH; 6% (w/w), mannitol; 30° C, inlet temperature, 100° C, AAH	Not available
ME [45]	Emesis	Domperidone	• *Drug loading- high* • *Globule size-* ≤ 100 nm • *Content uniformity- 100%* • *Mucoadhesion time-* • *In-vitro permeation- high*	8% (w/w), oleic acid; 32.5% (w/w), LAB:PG (1:1,w/w); 59.5% (w/w), DW	

					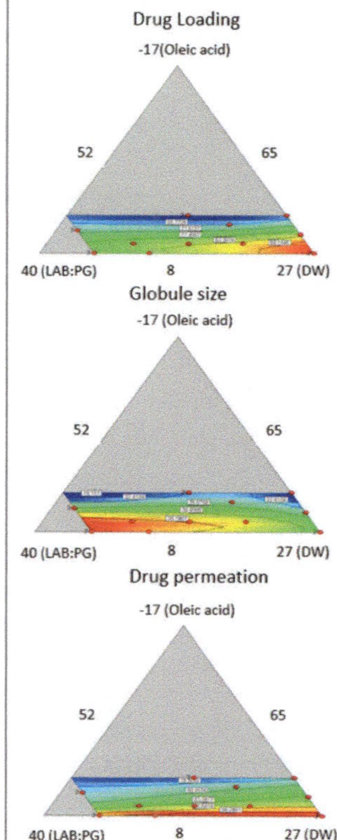 Drug Loading -17(Oleic acid) 52 65 40 (LAB:PG) 8 27 (DW) Globule size -17 (Oleic acid) 52 65 40 (LAB:PG) 8 27 (DW) Drug permeation -17 (Oleic acid) 52 65 40 (LAB:PG) 8 27 (DW)

(continued)

Table 8.4 (continued)

Formulation	Problem	API	CQA	Optimum formulation	Graphical result
Cubosome [46]	Emesis	Granisetron	• *Drug loading- high* • *Vesicle size-* ≤ 250 nm • *Zeta potential (mV)-* -20 ≤ ≤ − 30	5% (w/w), granisetron; 10% (w/w), P407 3% (w/w), T80	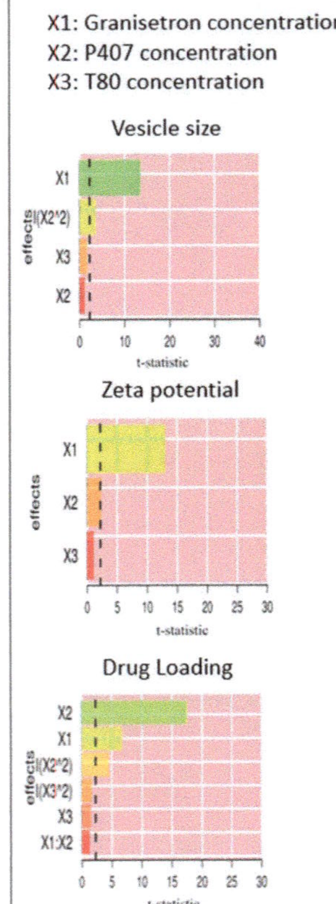

NE [41]	Glioblastoma	Temazolamide	• *Drug content- high* • *Vesicle size-* ≤ 250 nm • *Zeta potential (mV)-* high • *PDI* – Low • Ph- between 5.8–6	2% (w/w), TMZ; 7.5% (w/w), P407; 1% (w/w), MCT: LIPOID (1:1,w/w)	Not available

AAH administration angle in relation to horizontal plane, *AD* Alzheimer's disease, *CMCM* Capmul™ MCM, *CCD* central composite design, *CNS* Central nervous system, *CQA* critical quality attribute, *DH* Donepezil HCl, *DW* Distilled water, *GMS* glyceryl monostearate, *MCS* microsphere, *MCT* capryl/caprylic triglycerides, *ME* microemulsion, *P407* poloxamer® 407, *P80* polysorbate 80, *T80* Tween 80, *TMZ* Temozolomide, *SLN* solid lipid nanoparticle
Figures were adopted with permissions

Table 8.5 The last performed studies of ocular applications in relevance with the DoE

Formulation	Problem	API	CQA	Optimum formulation	Graphical result
LN [49]	Cytotoxicity study	NA	• *Color*- white and homogeneous • *Uniformity*- 100% • *Particle size*- minimum • *Permeation*- high • Zeta potential- close to 0	5% (w/w), S100; 0.5% (w/w), lecithin; 1% (w/w), P188	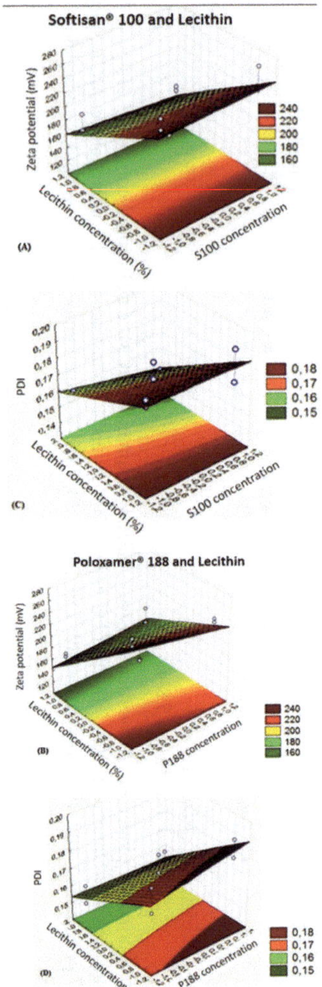
PLGA NP [50]	OST	Paclitaxel	• *Uniformity*- 100% • *Particle size*- minimum • *Entrapment* - ≥60% • *Permeation*- ≥90% • Zeta potential- higher	2 mg, paclitaxel; 40 mg, PLGA; 3% (w/w), PVA ; 11,000 rpm, homogenization rate; 3 mins; homogenization duration, 3 min	

(continued)

Table 8.5 (continued)

Formulation	Problem	API	CQA	Optimum formulation	Graphical result
					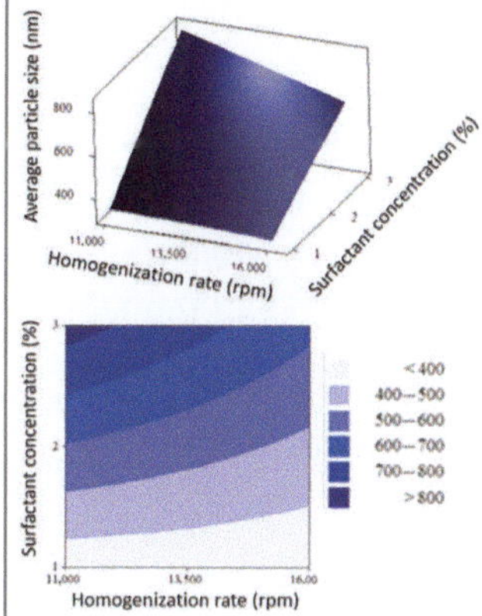 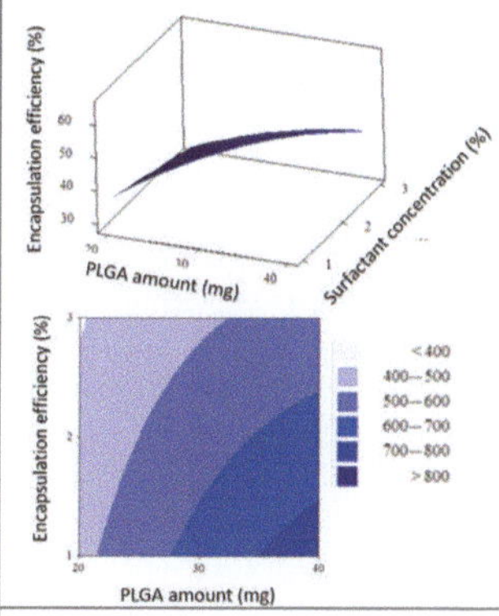
NLC [51]	PCS	Pranoprofen	• *Uniformity*- 100% • *Viscosity- high* • *PDI* - ≤ 0.5 • *Zeta potential*- -20 ≤ ≤ − 30 • *Entrapment* - ≥90%	1.5% (w/w), PF; 3.6% (w/w) Lanette® 18:, 1.4% (w/w, 50: 50), PEG 8 and Castor oil; 2.5% (w/w), TW80	

(continued)

Table 8.5 (continued)

Formulation	Problem	API	CQA	Optimum formulation	Graphical result
					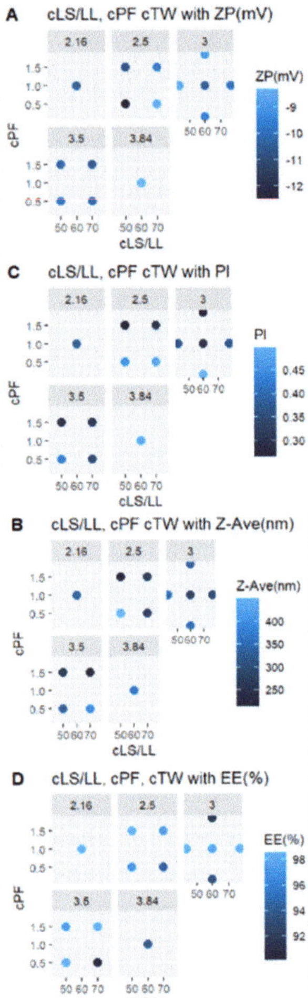
SNES [52]	FC	Terconazole	• *Drug loading- high* • *Globule size-* ≤ 100 nm • *Content uniformity- 100%* • *Mucoadhesion time-* • *In-vitro permeation- high*	10 mg, terconazole; 100 mg, span 80; 20 mg, cholesterol; 5 mg, SDC	

(continued)

Table 8.5 (continued)

Formulation	Problem	API	CQA	Optimum formulation	Graphical result
					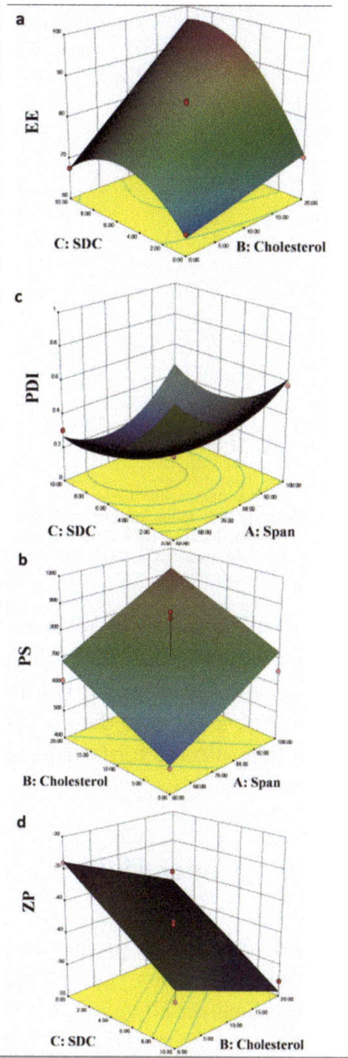
NSLMN [53]	Anti-inflammatory	TA	• *Zeta potential (mV)*-high • *Vesicle size*- ≤ 250 nm • *PDI* – Low • *Endurance*-high • *Thickness*-minimum • *Uniformity*- 100%	100 mg TA; 15% w/v PVA (31–50 kDa) and 20% w/v PVP-taken from this mixture 3% w/v; Volume of beads: Liquid, 0.33	

(continued)

Table 8.5 (continued)

Formulation	Problem	API	CQA	Optimum formulation	Graphical result
			• *Perme-ation*- high		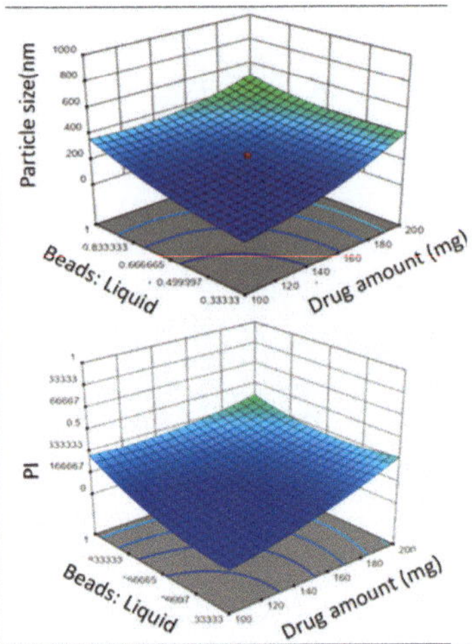

FC fungal keratitis, *LN* lipid nanoparticle, *LS/LL* liquid surfactant/liquid lipid concentration, *NSLMN* Nanosuspension-loaded dissolving bilayer microneedle, *OST* Ocular solid tumor, *P188* poloxamer® 188, *PCS* postoperative cataract surgery, *PF* pranoprofen, *PF127* pluronic® F127, *PI* polydispersity index, *PLGA NP* poly(lactide-coglycolic acid nanoparticles, *PVA* polyvinyl alcohol, *PVP* polyvinyl pyrrolidone, *S100* Softisan® 100, *SDC* sodium deoxycholate, *TA* triamcinolone acetonide, *TW80* Tween® 80 (*Figures were adopted with permissions)*

potential, and encapsulation efficiency properties of paclitaxel nanoparticles. These factors are critical formulation and process parameters that must be carefully considered. The Plackett-Burman experimental design successfully identified critical factors, leading to the utilization of the Box-Behnken experimental design to delve further into the formulation and process parameters.

8.5 Conclusion

QbD is receiving greater attention and emphasis from pharmaceutical manufacturers than in previous years. The concepts and nomenclature in question hold immense significance, however, it is disheartening that a considerable number of individuals possess a restricted understanding of them. This limited comprehension can potentially deter individuals from implementing these principles in product development. Pharmaceutical manufacturers must implement QbD principles in their practices by gaining a thorough knowledge and understanding of the drawbacks of QbT. A thorough comprehension of CMAs (critical material attributes) and CPPs (critical process parameters) is essential for producing top-notch topical formulations that closely resemble commercial products. Paying attention to detail is absolutely crucial as the formulation process consists of numerous complex components. Laboratory scale testing is a crucial factor when developing a topical product. The information gained from this stage is essential for scaling up to pilot or pivotal development. Quality by Design, commonly known as QbD is an indispensable tool for recognizing and grasping CMAs and CPPs in pharmaceutical development. Understanding how these factors interact and contribute is crucial to achieving a high-quality end product. Incorporating QbD methods when creating a topical product is an absolute necessity. Implementing QbD not only expedites the product commercialization process, but also curtails expenses at every stage of development from a commercial perspective.

References

1. Yu, L. X., Amidon, G., Khan, M. A., Hoag, S. W., Polli, J., Raju, G. K., et al. (2014). Understanding pharmaceutical quality by design. *The AAPS Journal, 16*(4), 771–783. https://doi.org/10.1208/s12248-014-9598-3
2. Uner, B. D. P., & Baranauskaite, J. (2022). Ocular delivery systems and applications developed by QbD perspective: From the past to the future. In H. Kafadar (Ed.), *Multidisciplinary approach in medical science II* (pp. 49–63). IKSAD Publishing House.
3. Administration USFaD. (1995). *Guidance for industry: immediate release solid oral dosage forms scale-up and postapproval changes: chemistry, manufacturing, and controls, in vitro dissolution testing, and in vivo bioequivalence documentation.* Retrived June 21, 2023
4. CDER USFaDA. (2004). *PAT—a framework for innovative pharmaceutical development, manufacturing, and quality assurance. Guidance for industry.* CDER USFaDA.
5. ICH. (2005). *Guideline: validation of analytical procedures: Textand methodology Q2 (R1).* Harmonization ICo E.
6. Aksu, N. B., Yozgatlı, V., Okur, M. E., Ayla, Ş., Yoltaş, A., & Üstündağ, O. N. (2019). Preparation and evaluation of QbD based fusidic acid loaded in situ gel formulations for burn wound treatment. *Journal of Drug Delivery Science and Technology, 52*, 110–121. https://doi.org/10.1016/j.jddst.2019.04.015
7. Rosas, J. G., Blanco, M., González, J. M., & Alcalá, M. (2011). Quality by design approach of a pharmaceutical gel manufacturing process, part 1: Determination of the design space. *Journal of Pharmaceutical Sciences, 100*(10), 4432–4441. https://doi.org/10.1002/jps.22611
8. Chang, R. K., Raw, A., Lionberger, R., & Yu, L. (2013). Generic development of topical dermatologic products: Formulation development, process development, and testing of topical dermatologic products. *The AAPS Journal, 15*(1), 41–52. https://doi.org/10.1208/s12248-012-9411-0
9. Sivaraman, A. B. A. (2015). Quality by design approaches for topical dermatological dosage forms. *Research and Reports in Transdermal Drug Delivery, 4*(9), 9–21.
10. Uner, B., Ozdemir, S., Yıldırım, E., Yaba Ucar A, Tas C, Uner M, et al. (2022). Solid lipid nanoparticles, nanostructured lipid carriers and nanoemulsion of loteprednol etabonate for corneal delivery. Part II: Physical stability and, corneal retention and permeation. *Expert Opinion on Drug Delivery*. Article in Press.
11. Herkenne, C., Alberti, I., Naik, A., Kalia, Y. N., Mathy, F. X., Préat, V., et al. (2008). In vivo methods for the assessment of topical drug bioavailability. *Pharmaceutical Research, 25*(1), 87–103. https://doi.org/10.1007/s11095-007-9429-7
12. Raw, AS. (2015). *Quality by Design (QbD) for topical dermatologic products*
13. Arora, D., & Nanda, S. (2019). Quality by design driven development of resveratrol loaded ethosomal hydrogel for improved dermatological benefits via enhanced skin permeation and retention. *International Journal of Pharmaceutics, 567*, 118448.

14. Beg, S., Saini, S., Bandopadhyay, S., Katare, O., & Singh, B. (2018). QbD-driven development and evaluation of nanostructured lipid carriers (NLCs) of Olmesartan medoxomil employing multivariate statistical techniques. *Drug development and industrial pharmacy, 44*(3), 407–420.
15. Khurana, B., Arora, D., & Narang, R. K. (2020). QbD based exploration of resveratrol loaded polymeric micelles based carbomer gel for topical treatment of plaque psoriasis: In vitro, ex vivo and in vivo studies. *Journal of Drug Delivery Science and Technology, 59*, 101901. https://doi.org/10.1016/j.jddst.2020.101901
16. Sandhu, P. S., Kumar, R., Katare, O., & Singh, B. (2017). Surface-tailored nanomixed micelles containing quercetin–salicylic acid physical complex for enhanced cellular and in vivo activities: A quality by design perspective. *Nanomedicine, 12*(11), 1281–1303.
17. Rapalli, V. K., Banerjee, S., Khan, S., Jha, P. N., Gupta, G., Dua, K., et al. (2021). QbD-driven formulation development and evaluation of topical hydrogel containing ketoconazole loaded cubosomes. *Materials Science and Engineering: C, 119*, 111548. https://doi.org/10.1016/j.msec.2020.111548
18. Raina, H., & Kaur, S. (2017). Development of efavirenz loaded solid lipid nanoparticles: Risk assessment, quality-by-design (QbD) based optimisation and physicochemical characterisation. *Journal of Drug Delivery Science and Technology, 39*, 39–191. https://doi.org/10.1016/j.jddst.2017.02.013
19. Sangshetti, J. N., Deshpande, M., Zaheer, Z., Shinde, D. B., & Arote, R. (2017). Quality by design approach: Regulatory need. *Arabian Journal of Chemistry, 10*, S3412–S3S25. https://doi.org/10.1016/j.arabjc.2014.01.025
20. Zhang, L., & Mao, S. (2017). Application of quality by design in the current drug development. *Asian Journal of Pharmaceutical Sciences, 12*(1), 1–8. https://doi.org/10.1016/j.ajps.2016.07.006
21. Yu, L. X. (2008). Pharmaceutical quality by design: Product and process development, understanding, and control. *Pharmaceutical Research, 25*(4), 781–791. https://doi.org/10.1007/s11095-007-9511-1
22. Administration UFaD. (2006). *Guidance for industry: Q9 quality risk management*. ICH.
23. Roy, S. (2012). Quality by design: A holistic concept of building quality in pharmaceuticals. *International Journal of Pharmaceutical and Biomedical Research, 3*(2), 100–108.
24. Lionberger, R. A. (2008). FDA critical path initiatives: Opportunities for generic drug development. *The AAPS Journal, 10*(1), 103–109. https://doi.org/10.1208/s12248-008-9010-2
25. Üner, B., Özdemir, S., Taş, Ç., Özsoy, Y., & Üner, M. (2022). Development of lipid nanoparticles for transdermal loteprednol etabonate delivery. *Journal of Microencapsulation, 1–27*, 327–340.
26. Bagde, A., Dev, S., Madhavi, K., Sriram, L., Spencer, S. D., Kalvala, A., Nathani, A., et al. (2023). Biphasic burst and sustained transdermal delivery in vivo using an AI-optimized 3D-printed MN patch. *International Journal of Pharmaceutics, 636*, 122647. https://doi.org/10.1016/j.ijpharm.2023.122647
27. Pothuvan, U., Dubey, R., Bhamare, P., Upmanyu, N., Gupta, B. K., & Pandey, A. K. Development and characterization of sildenafil citrate loaded transdermal patches for treatment of pediatrics pulmonary arterial hypertension using Qbd approach. *A Journal for New Zealand Herpetology, 12*, 2230.
28. Gupta, T., Kenjale, P., & Pokharkar, V. (2022). QbD-based optimization of raloxifene-loaded cubosomal formulation for transdemal delivery: Ex vivo permeability and in vivo pharmacokinetic studies. *Drug Delivery and Translational Research, 12*(12), 2979–2992. https://doi.org/10.1007/s13346-022-01162-1
29. Alves, T., Arranca, D., Martins, A., Ribeiro, H., Raposo, S., & Marto, J. (2021). Complying with the guideline for quality and equivalence for topical semisolid products: The case of Clotrimazole cream. *Pharmaceutics, 13*, 555. https://doi.org/10.3390/pharmaceutics13040555
30. Adin, S. N., Gupta, I., Aqil, M., & Mujeeb, M. (2023). Baicalin loaded transethosomes for rheumatoid arthritis: Development, characterization, pharmacokinetic and pharmacodynamic

evaluation. *Journal of Drug Delivery Science and Technology., 81*, 104209. https://doi.org/10.1016/j.jddst.2023.104209
31. Kesharwani, D., Das Paul, S., Paliwal, R., & Satapathy, T. (2023). Development, QbD based optimization and in vitro characterization of Diacerein loaded nanostructured lipid carriers for topical applications. *Journal of Radiation Research and Applied Sciences, 16*(2), 100565. https://doi.org/10.1016/j.jrras.2023.100565
32. Borghi, A., Corazza, M., Minghetti, S., Toni, G., & Virgili, A. (2015). Continuous vs. tapering application of the potent topical corticosteroid mometasone furoate in the treatment of vulvar lichen sclerosus: Results of a randomized trial. *British Journal of Dermatology, 173*(6), 1381–1386.
33. Simões, A., Veiga, F., Vitorino, C., & Figueiras, A. (2018). A tutorial for developing a topical cream formulation based on the quality by design approach. *Journal of Pharmaceutical Sciences, 107*(10), 2653–2662.
34. Tas, C., Joyce, J. C., Nguyen, H. X., Eangoor, P., Knaack, J. S., Banga, A. K., et al. (2017). Dihydroergotamine mesylate-loaded dissolving microneedle patch made of polyvinylpyrrolidone for management of acute migraine therapy. *Journal of Controlled Release, 268*, 159–165.
35. JUB, B.-O., & Marksiene, J. (2021). Intranasal in situ gel as a promising approach for enhancing bioavailability and brain delivery. In *Updates in internal sciences for 2021* (pp. 87–108). IKSAD Publishing House.
36. Shingaki, T., Hidalgo, I. J., Furubayashi, T., Sakane, T., Katsumi, H., Yamamoto, A., et al. (2011). Nasal delivery of P-gp substrates to the brain through the nose–brain pathway. *Drug Metabolism and Pharmacokinetics, 26*(3), 248–255. https://doi.org/10.2133/dmpk.DMPK-10-RG-108
37. Gunes Ozunal, Z., Donmez Cakil, Y., Tekin, S., Saglam, E., & Gulhan, A. R. (2021). Metoclopramide increased cell proliferation in HepG2 cell line and sorafenib attenuated the effect. *Annals of Medical Research, 27*(4), 1188–1192.
38. Alsarra, I. A., Hamed, A. Y., Alanazi, F. K., & Neau, S. H. (2011). Rheological and mucoadhesive characterization of poly(vinylpyrrolidone) hydrogels designed for nasal mucosal drug delivery. *Archives of Pharmacal Research, 34*(4), 573–582. https://doi.org/10.1007/s12272-011-0407-6
39. Kaneishi, K., Imai, K., Nishimura, K., Sakurai, N., Kohara, H., Ishiki, H., et al. (2020). Olanzapine versus metoclopramide for treatment of nausea and vomiting in advanced cancer patients with incomplete malignant bowel obstruction. *Journal of Palliative Medicine, 23*(7), 880–881.
40. Scholar, G. (2023). *Studies on CNS diseases were conducted during the period of 2022–2023*
41. Michels, L. R., Fachel, F. N. S., Schuh, R. S., Azambuja, J. H., de Souza, P. O., Gelsleichter, N. E., et al. (2023). Nasal administration of a temozolomide-loaded thermoresponsive nanoemulsion reduces tumor growth in a preclinical glioblastoma model. *Journal of Controlled Release, 355*, 343–357. https://doi.org/10.1016/j.jconrel.2023.01.070
42. Manish, U., Vijay, G., & Soniwala, M. M. (2022). Design and optimization of midazolam loaded microemulsion using quality by design (Qbd) assisted statistical modelling. *Journal of Pharmaceutical Negative Results*, 765–776. https://doi.org/10.47750/pnr.2022.13.S09.088
43. Arora, D., Bhatt, S., Kumar, M., Verma, R., Taneja, Y., Kaushal, N., et al. (2022). QbD-based rivastigmine tartrate-loaded solid lipid nanoparticles for enhanced intranasal delivery to the brain for Alzheimer's therapeutics. *Frontiers in Aging Neuroscience, 14*, 960246. https://doi.org/10.3389/fnagi.2022.960246
44. Perkušić, M., Nižić Nodilo, L., Ugrina, I., Špoljarić, D., Jakobušić Brala, C., Pepić, I., et al. (2022). Tailoring functional spray-dried powder platform for efficient donepezil nose-to-brain delivery. *International Journal of Pharmaceutics, 624*, 122038. https://doi.org/10.1016/j.ijpharm.2022.122038
45. Rathod, M., Suthar, D., Patel, H., Shelat, P., & Parejiya, P. (2019). Microemulsion based nasal spray: A systemic approach for non-CNS drug, its optimization, characterization and statistical

modelling using QbD principles. *Journal of Drug Delivery Science and Technology, 49*, 286–300. https://doi.org/10.1016/j.jddst.2018.11.017
46. Eissa, E. M., Elkomy, M. H., Eid, H. M., Ali, A. A., Abourehab, M. A. S., Alsubaiyel, A. M., et al. (2022). Intranasal delivery of Granisetron to the brain via nanostructured cubosomes-based in situ gel for improved management of chemotherapy-induced emesis. *Pharmaceutics, 14*, 1374. https://doi.org/10.3390/pharmaceutics14071374
47. Ahuja, S. K., Ferreira, G. M., & Moreira, A. R. (2004). Application of Plackett-Burman design and response surface methodology to achieve exponential growth for aggregated shipworm bacterium. *Biotechnology and Bioengineering, 85*(6), 666–675. https://doi.org/10.1002/bit.10880
48. Chiang, Y.-D., Lian, H.-Y., Leo, S.-Y., Wang, S.-G., Yamauchi, Y., & Wu, K. C. W. (2011). Controlling particle size and structural properties of mesoporous silica nanoparticles using the Taguchi method. *The Journal of Physical Chemistry C., 115*(27), 13158–13165. https://doi.org/10.1021/jp201017e
49. Fangueiro, J. F., Andreani, T., Egea, M. A., Garcia, M. L., Souto, S. B., Silva, A. M., et al. (2014). Design of cationic lipid nanoparticles for ocular delivery: Development, characterization and cytotoxicity. *International Journal of Pharmaceutics, 461*(1–2), 64–73.
50. Yerlikaya, F., Ozgen, A., Vural, I., Guven, O., Karaagaoglu, E., Khan, M. A., et al. (2013). Development and evaluation of paclitaxel nanoparticles using a quality-by-design approach. *Journal of Pharmaceutical Sciences, 102*(10), 3748–3761.
51. Rincón, M., Espinoza, L. C., Silva-Abreu, M., Sosa, L., Pesantez-Narvaez, J., Abrego, G., et al. (2022). Quality by design of pranoprofen loaded nanostructured lipid carriers and their ex vivo evaluation in different mucosae and ocular tissues. *Pharmaceuticals, 15*(10), 1185.
52. Yousry, C., Zikry, P. M., Salem, H. M., Basalious, E. B., & El-Gazayerly, O. N. (2020). Integrated nanovesicular/self-nanoemulsifying system (INV/SNES) for enhanced dual ocular drug delivery: Statistical optimization, in vitro and in vivo evaluation. *Drug Delivery and Translational Research, 10*(3), 801–814. https://doi.org/10.1007/s13346-020-00716-5
53. Wu, Y., Vora, L. K., Mishra, D., Adrianto, M. F., Gade, S., Paredes, A. J., et al. (2022). Nanosuspension-loaded dissolving bilayer microneedles for hydrophobic drug delivery to the posterior segment of the eye. *Biomaterials Advances, 137*, 212767. https://doi.org/10.1016/j.bioadv.2022.212767

Chapter 9
Quality by Design for Parenteral Formulations

Ahmet Dogan Ergin and Burcu Uner

Abstract Quality by Design (QbD) is a systematic approach that focuses on understanding and controlling product quality through the integration of scientific principles, risk management, and regulatory requirements. This abstract provides a comprehensive overview of the application of QbD principles in the development of parenteral drug formulations, which are administered via injections, infusions, or implants.

Employing Quality by Design (QbD) during the development of parenteral drug formulations yields substantial advantages as it facilitates a thorough comprehension of the interplay amongst formulation variables, process parameters, and critical quality attributes (CQAs). By employing a proactive and science-based approach, manufacturers can optimize product quality, reduce batch failures, and enhance patient safety and efficacy.

The QbD approach in parenteral drug formulations involves several key elements. The first step is to identify and define the Critical Quality Attributes (CQAs) that are crucial to ensure the safety and effectiveness of the drug product. These attributes include parameters such as potency, purity, stability, and drug release profiles. Secondly, QbD encourages the establishment of a design space, which defines the acceptable ranges of formulation and process parameters that ensure product quality within the desired CQA specifications.

Furthermore, QbD promotes a lifecycle approach, encompassing the entire product lifecycle from development to post-approval changes. This involves continuous process verification, ongoing monitoring of product performance, and the implementation of quality control strategies to ensure consistency and reliability.

A. D. Ergin (✉)
Department of Pharmaceutical Technology, Faculty of Pharmacy, Trakya University, Edirne, Turkey

Department of Neuroscience, University of Turin, Turin, Italy
e-mail: adoganergin@trakya.edu.tr

B. Uner
Department of Pharmaceutical and Administrative Sciences, University of Health Science and Pharmacy in St. Louis, Saint Louis, MO, USA
e-mail: burcu.uner@uhsp.edu

N. K. Jain, N. Bajwa (eds.), *Introduction to Quality by Design (QbD)*,
https://doi.org/10.1007/978-981-99-8034-5_9

In conclusion, the application of QbD principles in parenteral drug formulations offers a robust and science-based approach to ensure product quality, safety, and efficacy. By leveraging these principles, manufacturers can optimize formulations, control critical parameters, and enhance process.

Keywords QTPP · Parenteral formulations · Design expert · QbD

9.1 Introduction

The word "quality" is used in a relative sense. The ISO defines this as "the totality of a product's or service's features and characteristics that affect its capacity to satisfy stated or implied needs." To put it another way, a product is said to be of high quality if it "complies with the requirements specified by the client" [1].

Numerous initiatives aimed at preserving the production of quality are covered in the conversations that follow. In theory, there are three distinct levels at which these activities are organized. These levels, starting at the top, are:

1. Quality Management (QM).
2. Quality Assurance (QA).
3. Quality Control (QC).

Organizations that aim to consistently exceed customer expectations adopt quality management as a highly structured and effective method. It involves the coordination of activities and processes within an organization to ensure that quality standards are defined, implemented, and maintained throughout all stages of a product or service's lifecycle. Furthermore, the organizational structure and roles for the effective functioning of the quality system are identified in this statement [2].

QA: QA is a QM component that aims to give assurance that quality requirements will be met; This refers to a set of organized and systematic actions that demonstrate the quality level achieved by the output of a process [2].

QC: QC is a QM component that focuses on meeting quality criteria; it encompasses activities that impose specified quality on a process and involves assessing a device or system's actual operating performance characteristics, comparing it to desired goals, and taking appropriate action in response to any discrepancies; QC examines a process's input to make sure that crucial components or parameters related to the process are accurate [2].

Quality assurance focuses on the systematic implementation of processes and activities to ensure that the desired quality is achieved. It involves establishing quality management systems, conducting audits, and providing training to ensure adherence to quality standards.

Quality control (QC) is a part of quality management that focuses on the activities and techniques used to assess, monitor, and control the quality of products, services, or processes. Its primary goal is to identify and correct any deviations or defects in order to ensure that the desired quality standards are Error prevention is the main goal of quality control. But mistakes will inevitably still be made despite all efforts. Consequently, the control system needs checks to find them [1].

Quality by design (QbD) was introduced first time as a quality should be built into a product from the start, and the majority of quality crises and issues come from poor product design by Dr. Joseph M. Juran who is prime mover of quality [3]. A high-quality drug product is one that is free of contamination and consistently provides the consumer with the therapeutic benefit stated on the label, according to Woodcock. The US Food and Drug Administration (FDA) promotes the use of risk-based strategies and QbD principles in the development, production, and regulation of pharmaceutical products [4, 5]. The key principle of QbD is that quality cannot be tested into products; quality should be made by design. Incorporated into products from the beginning, not tested into finished products [6]. By evaluating the factors that could affect quality, QbD seeks to ensure the intended level of product quality.

Quality components of pharmaceutical quality consist of Good Manufacturing Practices (GMP), Active Pharmaceutical Ingredient (API) Quality, Formulation and Dosage Form Quality, Analytical Testing, Stability Testing, Quality Control and Quality Assurance, Regulatory Compliance, Pharmacovigilance. Maintaining pharmaceutical quality is a collaborative effort involving pharmaceutical manufacturers, regulatory agencies, healthcare professionals, and other stakeholders. It is essential for ensuring that drugs are safe, effective, and reliable for patients who rely on them for their health and well-being.

Good manufacturing procedures (GMPs) and regulatory assessments served as the foundation for pharmaceutical product inspections prior to the twenty-first century.. Quality by design (QbD) approach first found its place in ICH guidelines in 2005, albeit in draft form, and was finally finalized and published in 2009 (ICH Q8, 2009) [4, 6]. The first steps in the adaptation process of ICH to new approaches were taken with Pharmaceutical Development ICH Q8, Quality Risk Assessment ICH Q9 (ICH Q9, 2005) [7], and then continued with the Quality Management System ICH Q10 (ICH Q10, 2008) and Q11 (ICH Q11, 2012) [8]. ICH Q8 provides principles and recommendations for pharmaceutical development, specifically focusing on the development of a quality target product profile (QTPP) and the use of quality risk management (QRM) in pharmaceutical development. ICH Q8 promotes a more flexible and science-based approach to pharmaceutical development, allowing for innovation and continuous improvement while ensuring the quality, safety, and efficacy of drug products. It encourages the application of quality risk management principles and the use of established scientific knowledge throughout the development process to make informed decisions and achieve the desired product quality. ICH Q8 describes pharmaceutical development with emphasis [6]. ICH Q9 provides principles and recommendations for quality risk management (QRM) in the pharmaceutical industry. It includes Quality Risk Management (QRM), Risk Assessment, Risk Control. ICH Q9 provides a systematic approach to managing risks in the pharmaceutical industry and supports the development of robust and science-based risk management practices. By implementing QRM principles, organizations can proactively identify and address potential risks, optimize resources, enhance decision-making, and ultimately ensure the quality and safety of pharmaceutical products [7]. ICH Q10 provides principles and recommendations for pharmaceutical quality systems, focusing on the implementation of an effective quality management system (QMS) throughout the pharmaceutical lifecycle. ICH Q10 provides guidance on the establishment and maintenance of a pharmaceutical quality system that facilitates consistent product quality, effective risk management, and continuous improvement. By implementing the principles outlined in ICH Q10, organizations can enhance their quality systems, meet regulatory expectations, and ultimately deliver safe and effective pharmaceutical products to patients [9].

In 2005, FDA (The Office of Generic Drugs' (OGD) issued a question-based review (QbR) for the Chemistry, Manufacturing, and Controls (CMC) evaluation, drug registration application By implementing and putting into practice the ideas and principles of the FDA's Pharmaceutical cGMPs for the twenty-first Century: A Risk-Based Approach and Process Analytical Technology initiatives, the Abbreviated New Drug Application (ANDA) or The QbR will transform the CMC review into a contemporary, scientific, and risk-based pharmaceutical quality assessment. ICH Q12, which was released in 2017, builds on the ideas presented in ICH Q8, Q9, and Q10 and offers detailed instructions on how to manage post-approval adjustments in a methodical and scientific way.To maintain product quality, patient safety, and regulatory compliance, it includes the principles of pharmaceutical development, quality risk management, and the pharmaceutical quality system. ICH Q12 encourages the adoption of pharmaceutical Quality System principles and the use of established tools and concepts, such as Quality Risk Management (QRM), to support effective change management and continual improvement. It promotes a systematic and harmonized approach to managing post-approval changes, including the implementation of established regulatory pathways, such as Variations and Post-Approval Change Management Protocols (PACMPs) [10].

The main purpose of the QbD approach conveyed in ICH Q8 is patient safety and product effectiveness.Quality by Design (QbD) is a systematic approach to pharmaceutical development that focuses on designing and controlling processes to ensure product quality. It is a concept promoted by regulatory agencies, such as the U.S. Food and Drug Administration (FDA) and the International Council for Harmonization of Technical Requirements for Pharmaceuticals for Human Use (ICH). QbD aims to build quality into the product by understanding the critical quality attributes (CQAs) and implementing effective control strategies based on scientific knowledge and risk management principles. The goal of Quality by Design is to shift the focus from traditional, post-approval testing to a more proactive approach where quality is built into the product and process from the beginning. By understanding the relationship between critical process parameters, critical material attributes, and critical quality attributes, manufacturers can design robust processes that consistently produce high-quality products. QbD aims to enhance patient safety, reduce variability, and increase manufacturing efficiency while complying with regulatory requirements [11].

There is different key components that should be known (Fig. 9.1):

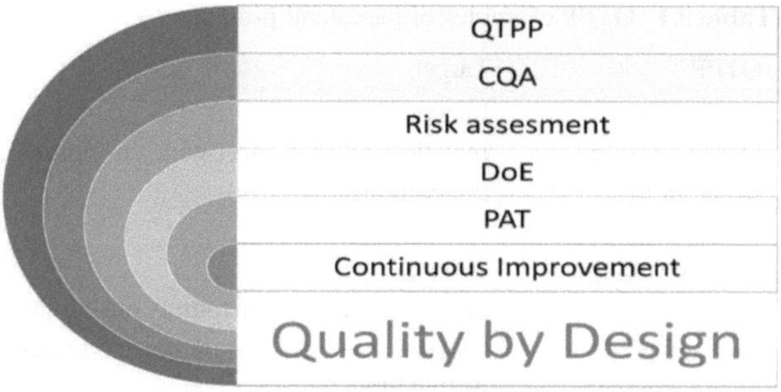

Fig. 9.1 Elements of QbD

9.1.1 Target Product Profile (TPP)

The TPP is a strategic document that defines the desired characteristics and attributes of a drug product. It outlines the target use, dosage form, route of administration, therapeutic indications, and other critical aspects that define the target product. The TPP captures both the clinical and quality aspects of the product. It describes the efficacy, safety, and performance requirements, as well as the desired quality attributes, such as the physical, chemical, and microbiological properties.

The patient and the prescription are the fundamental principles of TPP. The TPP need to be identical to both the reference original product and the generic product. To guarantee that the product to be developed offers the same TPP as the original product, various formulas and designs might be used. Clinical research as well as new safety and efficacy data are required if the TPP of the newly developed product differs from the original product. TPP also takes into account details like the administration route, the dosage amount and form, the maximum and lowest doses, the pharmaceutical appearance, and the intended patient group [12, 13].

9.1.2 Quality Target Product Profile (QTPP)

The Quality Target Product Profile (QTPP) is a key component of the Quality by Design (QbD) approach in pharmaceutical development. QTPP is a strategic planning for drug product development with the *Begin with the end in mind* [14]. It is a strategic document that defines the quality characteristics and attributes of a drug product that are critical for its safety, efficacy, and performance. The QTPP defines the desired quality characteristics of a drug product in terms of safety, efficacy, and performance. It includes attributes such as dosage form, strength, route of administration, release profile, stability, and packaging. QTPP can be specified according to dosage forms such as resuspension of an oral suspension, adhesion of a transdermal system, viscosity of a topical cream. Also, the comparative bioequivalence of the finished product with the reference product is within the scope of the TPQP.

Table 9.1 QTPP examples of parenteral products

QTPP	Target
Dosage form	Injection
Dosage design	Injection solution meet with the claimed criteria
Route of administration	Intravenous
Dosage strentgh	200 mg/2 mL
Stability	Should protect spesifications during the claimed time
Packaging	Closure container system needed to ensure the requirements of parenteral formulation

The QTPP is typically developed early in the drug development process, in conjunction with the Target Product Profile (TPP), which defines the clinical and non-clinical attributes of the product. The QTPP and TPP are interconnected, with the QTPP focusing specifically on the quality aspects of the product that need to be achieved. The QTPP serves as a reference point throughout the development process and guides decision-making related to formulation development, process optimization, and analytical methods. It provides a clear target for the product's quality characteristics and serves as a basis for the development of the control strategy, including the establishment of Critical Quality Attributes (CQAs) and critical process parameters (CPPs) during the QbD implementation.

To summarize in the QTPP, it is patient-related product performance. For example; When particle size affects the dissolution of the solid dosage form, the QTPP should include the dissolution profile of the finished product, not the particle size. Particle size is included here as CQA and is covered by the process definition and control strategy (Table 9.1) [12].

9.1.3 Critical Quality Attributes (CQAs)

CQAs are the key physical (Appearance, color, size, shape, clarity, viscosity, dissolution rate, particle size, moisture content, or density etc), chemical (active pharmaceutical ingredient (API) identity, purity, potency, impurity levels, degradation products, or stability), biological (biological activity, receptor binding, immunogenicity, or bioavailability), or microbiological (microbial contamination, sterility, or endotoxin levels) characteristics of a drug product that need to be controlled within predefined limits to ensure its safety, efficacy, and quality [12, 14].

CQAs are critical because they have a direct impact on the performance of the product and, if not controlled, may pose a risk to patient health or the desired therapeutic effect. CQAs are determined based on a thorough understanding of the product and its manufacturing process, as well as the requirements of the Target Product Profile (TPP) and the Quality Target Product Profile (QTPP). They are

Table 9.2 Examples of critical quality attribute parameters [15]

Operation	Critical Process Parameters	Critical Quality Attributes
Formulation development	Tonicity adjustment, solubility, aseptic environment, sterilization	The clear solution, drug release particle size, pH
Filling and sealing	Filling speed, needle and LAF air-flow speed, Elastic stopper, closing pumps	Appearance, Sterility assay, pH, particle siz, drug content and uniformity
Sterilization	Pressure (PSI), vapor initial temperature, internal air pressure, process duration time	Sterility, endotoxin limit, assay, pH, particle size

identified through scientific knowledge, data analysis, and risk assessment methodologies such as Quality Risk Management (QRM).

CQAs are typically established during the product development stage, and they serve as a basis for designing control strategies to ensure that the product consistently meets its quality requirements. These strategies may include setting appropriate specifications, selecting suitable analytical methods for testing, defining process parameters, and establishing acceptance criteria. Selection Criticial material attributes and critical process parameters (CPPs) has effect on the CQAs of products (Table 9.2). The risk matrix outlines the impact of different MAs and PPs on CQAs in both qualitative and quantitative ways, which aids in the identification of key material attributes (CMAs) and important process parameters. [16].

9.1.4 Critical Process Parameters (CPPs)

CPPs are the key variables or operating conditions that have a significant impact on the manufacturing process and directly influence Critical Quality Attributes (CQAs) of a pharmaceutical product. CPPs are identified and controlled within predefined limits to ensure that the product consistently meets its quality requirements.

The identification of CPPs is an important part of the Quality by Design (QbD) approach. By understanding and controlling these critical process variables, manufacturers can optimize the process and ensure consistent product quality. CPPs are typically established through a combination of prior knowledge, experimental design, and statistical analysis to determine their impact on CQAs. They are monitored, controlled, and documented throughout the manufacturing process to maintain process robustness and product quality. They are specific to each manufacturing process and can vary depending on the type of product, formulation, and manufacturing technology.

9.1.5 Risk Assessment and Management

Risk assessment plays a crucial role in the Quality by Design (QbD) approach. It is a systematic process of identifying, analyzing, evaluating, and controlling or mitigating risks associated with pharmaceutical development and manufacturing processes. Risk assessment helps in understanding the potential impact of variability and uncertainties on critical quality attributes (CQAs) and critical process parameters (CPPs) and enables informed decision-making throughout the product lifecycle.

The mosts common used risk analysis methods in QbD are [7, 17]:

- Failure Mode and Effect and CriticallyAnalysis (FMEA)
- Fault Tree Analysis (FTA)
- Hazard Analysis and Critical Control Points
- Preliminary Hazard Analysis
- Risk Priority Number (RPN)
- Ishikawa Diagram (Cause and Effect Diagram)

Mainly Ishikawa Diagrams and Failure Mode Effects Analysis are used in risk assesments. These risk analysis methods are not mutually exclusive, and they can be used in combination to conduct a comprehensive risk assessment in QbD. The selection of the appropriate method depends on the complexity of the process, available data, resources, and the specific requirements of the project (Table 9.3).

9.1.6 Design of Experiments (DOE)

DOE is a statistical method used in QbD to systematically study the effects of process parameters and formulation variables on product quality. It helps identify the critical process parameters (CPPs) and critical material attributes (CMAs) that have the most significant impact on CQAs.

Design of Experiments (DoE) is a statistical methodology used in the Quality by Design (QbD) approach to systematically evaluate and optimize critical process parameters (CPPs) and their impact on critical quality attributes (CQAs) of a pharmaceutical product. DoE enables efficient experimentation and data analysis to gain a deeper understanding of the relationship between process variables and product quality, leading to improved process control and robustness. DoE enables screening, Response Surface Methodology (RSM) to model and optimize the

Table 9.3 Example of initial risk assesment for injection products [15, 18]

Formulation CQAs	Drug-excipient compatibility	Drug content	Clearity	pH adjustment
Drug release	High	Medium	High	Medium
Buffer in formulation	High	Medium	Medium	Medium
Residual solvents	Low	Low	Medium	Medium

Table 9.4 Process optimization of different nanoparticle production using DoE

Drug	Nanoparticle Type	Design	CPPs	References
S-Adenosylmethionine	Polymeric nano-particle (chitosan, pectin)	2^3 full factorial design	Polymer concentration, drug concentration	[19, 20]
Coenzyme Q10	Liposome	3^2 full factorial design	Polymer concentration, drug concentration	[21]
Loteprednol etabonate	SLN, NLC, and NE	D-optimal mixture	Surfactant concentration, solid lipid/liquid lipid ratio	[22]
Vaccinium extract	Liposome loaded film	Factorial design	Phospholipid concentration, Polymer concentration	[23]
Timolol maleate	Ethosome	Box-Behnken	Solvent concentration, Surfactant concentration	[24]

relationship between CPPs and CQAs, full Factorial Design, design space determination, model validation, robustness. In the DoE studies, mostly Design Expert®, Modde®, Unscrambler®, JMP®, Statstica® and minitab® are used in drug development (Table 9.4) [15].

9.1.7 *Process Analytical Technology (PAT)*

Process Analytical Technology (PAT) is a key component of the Quality by Design (QbD) approach in pharmaceutical development and manufacturing. PAT involves the use of real-time monitoring, measurement, and control of critical process parameters during manufacturing. It enables a deeper understanding of the process, facilitates timely process adjustments, and ensures consistent product quality. The goal of PAT is to understand, control, and optimize the manufacturing process in order to consistently produce high-quality products. By integrating analytical tools and techniques into the manufacturing process, PAT enables real-time monitoring and control, facilitating a more proactive and science-based approach to quality assurance. In the parenteral injectable drug production, mainly (1) acoustic-resonance spectroscopy (ARS), (2) photoacoustic spectroscopy (PAS), (3) laser-induced breakdown spectroscopy (LIBS), and (4) dielectric spectroscopy were used [16].

9.1.8 Control Strategy

The control strategy outlines the measures and actions taken to ensure the process operates within defined limits and consistently produces a product of the desired quality. It includes specifications, control points, in-process controls, and the use of appropriate analytical methods.

9.1.9 Continuous Improvement

QbD promotes a culture of continuous improvement, where the knowledge gained from development, manufacturing, and post-approval activities is used to enhance processes, optimize control strategies, and improve product quality. Continuous improvement is an integral part of the Quality by Design (QbD) approach in pharmaceutical development and manufacturing. In ICH Q10 it is mentioned ‘’ A planned set of controls developed from current product technique understanding that secures Method performance and product quality” It is a systematic and ongoing effort to enhance processes, systems, and products with the aim of improving quality, efficiency, and overall performance. Procedure controls, process controls, lot release testing, process monitoring, characterisation testing, comparability testing, and stability testing are all examples of control strategies.A manufacturer can adopt a single point of control for that CQA or several points of control, depending on the risk associated with that CQA and the effectiveness of individual controls to identify a potential problem [25]. By embracing continuous improvement in QbD, pharmaceutical companies can drive ongoing advancements in product quality, process efficiency, and overall performance. It ensures that quality is continuously evaluated, monitored, and improved, resulting in enhanced patient safety and satisfaction, regulatory compliance, and business success.

The Quality by Design (QbD) approach offers several advantages for pharmaceutical development and manufacturing processes. Some key advantages of implementing the QbD approach include:

- Enhanced Product and Process Understanding: QbD promotes a systematic and science-based understanding of the product and its manufacturing process. It encourages the identification and evaluation of critical quality attributes (CQAs) and critical process parameters (CPPs). This deeper understanding allows for more effective control of product quality and process performance.
- Improved Risk Management: QbD incorporates risk assessment and management throughout the product lifecycle. By identifying and evaluating potential risks early in the development process, appropriate risk control strategies can be implemented to mitigate or eliminate those risks. This proactive approach enhances patient safety and reduces the likelihood of quality-related issues.
- Increased Process Robustness and Control: QbD focuses on designing processes that are robust and capable of consistently producing products with the desired

quality attributes. By identifying and controlling the critical process parameters and critical material attributes, process performance can be optimized, reducing variability and improving process control.

- Reduction in Post-Approval Changes and Regulatory Burden: With a thorough understanding of the product and process, QbD helps reduce the need for post-approval changes. By designing processes with built-in flexibility and understanding the impact of changes on product quality, manufacturers can minimize the number of regulatory submissions required for modifications.
- Enhanced Continuous Improvement: QbD promotes a culture of continuous improvement by encouraging the use of data-driven decision-making and process optimization. The knowledge gained from development, manufacturing, and post-approval activities can be utilized to make informed decisions, improve processes, and enhance product quality.
- Increased Efficiency and Cost Savings: QbD can lead to improved process efficiency, reduced batch failures, and increased productivity. By optimizing processes and controlling critical process parameters, manufacturers can minimize waste, reduce rejections, and achieve higher production yields, resulting in cost savings.
- Regulatory Support and Flexibility: Regulatory agencies, such as the FDA and ICH, support and encourage the implementation of QbD. Regulatory frameworks, such as the FDA's Process Validation Guidance and ICH Q8, Q9, and Q10, provide guidelines and tools to facilitate the application of QbD principles. Implementing QbD can help foster a collaborative relationship with regulatory authorities and streamline regulatory submissions.

While the Quality by Design (QbD) approach offers numerous advantages, there are also some potential disadvantages and challenges associated with its implementation.

- Increased Time and Resource Requirements: Implementing QbD involves conducting comprehensive studies, experiments, and data analysis to understand the critical quality attributes and optimize the process. This can require additional time, resources, and expertise, especially during the initial stages of development.
- Complexity: QbD can involve complex statistical analyses, experimental designs, and process understanding. This complexity may require specialized knowledge and expertise, particularly in areas such as design of experiments (DOE) and statistical modeling, which may pose challenges for smaller companies with limited resources.
- Regulatory Expectations and Documentation: The implementation of QbD requires extensive documentation to demonstrate the scientific understanding of the product and process. This includes the development of various QbD-related documents, such as the Quality Target Product Profile (QTPP), risk assessments, control strategies, and design space. Meeting regulatory expectations and ensuring compliance with documentation requirements can be demanding.
- Resistance to Change: Implementing QbD may require changes to existing processes, workflows, and mindsets within organizations. Resistance to change

from stakeholders, including employees and management, can pose challenges and hinder the successful adoption of the QbD approach.

- Data Requirements: QbD relies heavily on data collection, analysis, and interpretation to establish the relationship between critical process parameters, critical quality attributes, and product performance. Ensuring the availability and quality of relevant data can be a challenge, particularly if historical data is limited or if new testing methods or technologies are required.
- Cost Implications: The implementation of QbD may involve additional costs, such as investments in equipment, analytical methods, and training. Smaller companies or those with limited resources may find it challenging to allocate the necessary funds for these activities.
- Complexity of Scale-Up and Technology Transfer: Scaling up a process or transferring it to a different manufacturing site can present challenges in maintaining process understanding and ensuring consistent product quality. The complexity of QbD implementation can add an additional layer of complexity to these activities.

9.2 QbD Approaches for Parenteral Formulations

Parenteral formulations are pharmaceutical formulations that are administered by injection into the body, bypassing the gastrointestinal tract. These formulations are typically intended for rapid and direct delivery of drugs into the bloodstream, providing systemic distribution and avoiding the first-pass metabolism that occurs when drugs are taken orally. Parenteral routes of administration include intravenous (IV), intramuscular (IM), subcutaneous (SC), intradermal (ID), and intrathecal (IT) routes.

Parenteral formulations can take various forms depending on the specific route of administration and the characteristics of the drug being delivered. Some common types of parenteral formulations including solutions, suspensions, emulsions, lyophilized powders, liposomes, nanoparticles etc. Parenterals also classified as small volume and large volume parenterals.

By applying QbD principles, development of parenteral formulations can be more reliable, consistent, and safe. The proactive and systematic approach of QbD enables a better understanding of critical quality attributes, improved control over manufacturing processes, and a focus on continuous improvement. Ultimately, this leads to higher product quality, reduced variability, and enhanced patient safety in parenteral formulation development.

Using the literature, how to perform QbD is given below in items and in Fig. 9.2:

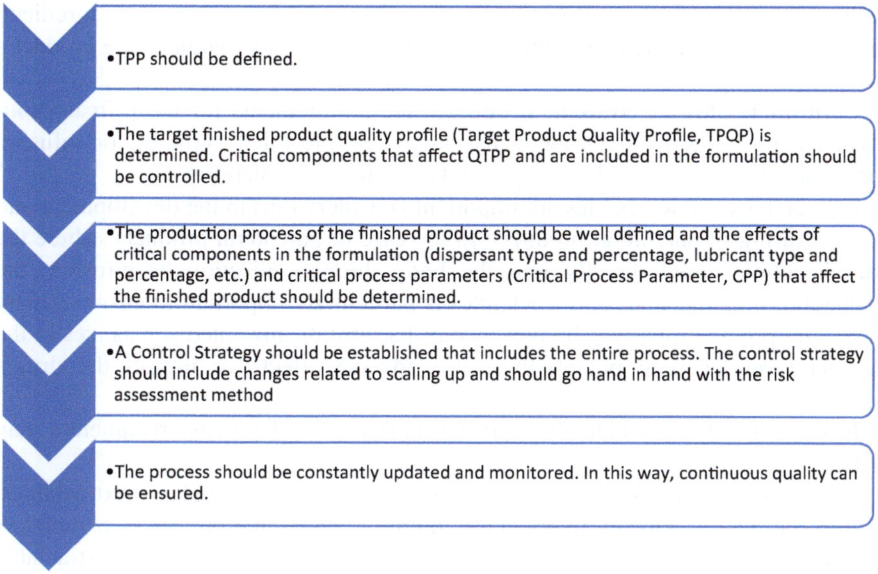

Fig. 9.2 QbD stages

9.2.1 Identification of QTPP of Parenteral Products

The Quality Target Product Profile (QTPP) for parenteral products outlines the desired quality characteristics of a specific parenteral formulation. It serves as a reference for formulation development, process design, and quality control. The QTPP takes into account various factors and attributes of the parenteral product that are critical for its safety, efficacy, and performance. Here are some common elements included in the QTPP of parenteral products:

Dosage Form: The QTPP specifies the desired dosage form of the parenteral product, such as solution, suspension, emulsion, or lyophilized powder [26].

Route of Administration: It identifies the intended route of administration, such as intravenous (IV), intramuscular (IM), subcutaneous (SC), intradermal (ID), or intrathecal (IT).

Strength: The QTPP defines the desired concentration or dose of the active pharmaceutical ingredient (API) in the parenteral product. There are several factors that can influence the active drug's assay value. It can be diminished by adsorption to the filter, deterioration brought on by a change in pH related to the major packaging elements, and insufficient light protection. The majority of the impacts are obvious.

Dose frequency: The dose frequency and QTPP of parenteral products are determined through a comprehensive development and regulatory process, considering the specific drug, therapeutic indication, patient needs, and regulatory requirements. The specific dose frequency should be defined during the drug development process and may be subject to further refinement and optimization based on clinical studies, patient feedback, and regulatory considerations. Potency, active ingredient, and functionality are the CQAs that must be met in order to guarantee the product's efficacy.

Container Closure System: It outlines the requirements for the primary container closure system, including vials, ampoules, pre-filled syringes, or infusion bags, considering factors like compatibility, sterility, and stability.

Extractables and leachables are important considerations in the development and manufacturing of parenteral products. In the case of parenteral products, where the drug is administered directly into the body through injection or infusion, the evaluation of extractables and leachables is particularly important to ensure patient safety and product quality. The presence of harmful or unwanted substances in the parenteral product can potentially cause adverse effects or impact the drug's efficacy.

In the context of extractables and leachables, critical process parameters may include:

Extraction Conditions: The extraction conditions used to evaluate extractables, such as the type of solvent, temperature, duration, and agitation. These parameters can influence the extraction efficiency and the types and amounts of extractable compounds that are released from the drug product or packaging materials.

Manufacturing Processes: Various manufacturing processes involved in the production of the drug product, such as mixing, filtration, sterilization, filling, and packaging. Each of these processes can potentially introduce or affect the levels of extractables and leachables.

The extractables and leachables are the main issues with any packaging. There shouldn't be any poisonous or harmful ingredients in the formulation provided by the primary or secondary packing material. Plasticizers, heavy metals, phthalates, and polyaromatic hydrocarbons are a few of the harmful substances that are frequently found. The "Container Closure Systems for Packaging of Human Drugs and Biologics" Guidance for Industry offers advice on the information of packaging materials required on pharmaceutical items [27].

The guidance's Attachment C contains details on numerous extraction studies. To be employed as packaging components, plastics and elastomers require both qualitative and quantitative profiles. Leachables may also enter the product by indirect contact, such as imprinting on the bottle or label adhesives, inks, or varnish, or through ambient air [26].

In the study conducted by Jenke, design space to the leachables were determined. The QbD approaches were used for 12 products and leachable profile was shown [28] (Table 9.5).

In sterile and parenteral drug product, mainly glass are used. Structured, scratched or broken glass containers cause the product to deteriorate, glass pieces to be mixed with the product and extra cleaning works. To reduce this, the CMAs of the product should be better defined and the mechanical resistance and exposure of vials to mechanical stress should be examined to prevent potential problems. Quantifying the frequencies that lead to damage in drug product containers due to glass exposure in industrial filling lines is crucial for preventing degradation and problems. It is

Table 9.5 Classification of Jenke

Variable	Design space boundaries
A	Aqueus product (pH 2–9), no polar agents
B	Same packaging system accepted material spesifications
C	Fill volume (50–1000 mL)
D	Underwent terminal sterilization and kept at RT for up to 24 months

imperative to identify specific forces and make process improvements based on the collected data. Eberle et al. conducted various measurements, including contact force, dynamic force, static force, and exposure measurements, to identify and define critical parameters that lead to glass scratches. Their aim was to improve the process [29].

Fill volume: Fill volume is another important spesification of parenterals. The specific fill volume of a parenteral product can vary depending on several factors such as.

Dosage (Higher dosages or higher API concentrations may require larger fill volumes), *Container Size:* This effect also packaging properties.

Overfill; In some cases, an overfill may be added to the fill volume to ensure that the intended dose can be withdrawn from the container without any loss of drug product during preparation or administration. The overfill volume is typically specified in the product's labeling or regulatory requirements.

Compatibility and Stability: The fill volume may also be influenced by considerations of drug compatibility with the container material, stability over time, and the need for headspace within the container to accommodate any gas or pressure changes.

Appearance: The QTPP describes the desired appearance of the parenteral product, including color, clarity, and absence of particulate matter.

Sterility: The sterility of parenteral products is a critical aspect of their manufacturing, storage, and administration. Sterility refers to the absence of viable microorganisms, including bacteria, fungi, and viruses, in the drug product. Parenteral products needs terminal sterilization process (which will provide a sterility assurance level (SAL) of 10–6 or greater) is preferred for a sterile drug product. Maintaining sterility is crucial because parenteral products are administered directly into the bloodstream, bypassing the body's natural defense mechanisms. Any contamination or presence of microorganisms in these products can pose serious risks to patient safety, including the potential for infections or adverse reactions.

In the sterilization of parenterals, selection of an appropriate sterilization method depends on the nature of the product and the materials involved. Common sterilization methods include steam sterilization (autoclaving), dry heat sterilization, ethylene oxide (ETO) sterilization, gamma irradiation, and electron beam sterilization. The choice of method depends on factors such as the product's heat sensitivity, compatibility with radiation, or the ability to withstand certain gases.

Temperature, exposure period, vacuum levels, pressure (for steam sterilization), dose (for radiation sterilization), and exposure (for x-ray sterilization) are crucial characteristics. This parameters rely on drug charactetistics. For a sterile drug product, a terminal sterilization procedure is preferred (which will yield a SAL of 106 or above). The stability of the pharmaceutical product should be evaluated during product development studies with a conventional "overkill" terminal sterilization procedure, such as an autoclave for 15 min at 121 °C. A less stringent sterilization cycle can be taken into consideration if the drug product cannot withstand this "overkill" sterilization cycle (for instance, because of undesirable impurity levels). By integrating it with a strict bioburden control technique, for instance, a less-heated autoclave cycle might be verified for supplying a 10^{-6} SAL. Combining an aseptic production process with a decreased terminal sterilization cycle (e.g., shortened exposure duration and/or lower temperature) is an extreme (but effective) example of a bioburden control method. When combined with the low (almost zero) product bioburden following the aseptic process, the decreased heat input can still deliver a validated sterility assurance level of 10^{-6}. For authorized sterile medicine products, this method of combining aseptic processing with shorter terminal sterilization cycles is used. However, an aseptic manufacturing procedure will need to be created if the drug product cannot sustain even a simple terminal sterilization procedure.

Pressure and Vacuum Levels (for Steam Sterilization): In steam sterilization, pressure and vacuum levels are critical parameters. Proper pressure and vacuum levels are maintained to ensure efficient steam penetration into the product and proper removal of air and non-condensable gases. Terminal sterilization process was generally conducting autoclave based heat sterilization process for 15 min at 121 °C. If drug is can't stand temperature, increase impurity, parameters should be re-evaluated. For this evaluation, frequent bioburden control strategy can be used. Aseptic condition with modified parameters (e.g., shortened exposure time and/or lower temperature) can be change.

For instance, a medicinal product for injection is designed as a microsuspension, with phospholipids stabilizing the particle size. One of the most significant CQAs of the therapeutic product is known to be particle size. For this drug product, sterilizing filtration is not an option, hence moist heat sterilization is used. To reduce the heat degradation of the phospholipid during autoclaving, trehalose is used as a thermoprotectant. To assure the physical/chemical properties as well as sterility of the drug product, the design space for the concentration of phospholipid, the concentration of trehalose, pH of the formulation, and the level of heat input may be developed in this scenario [30, 31].

Controlling the sterilization process parameters within a range that was identified through validation studies as providing a SAL of at least 106 and validating container closure integrity following exposure to the maximum sterilization procedure are both necessary to ensure sterility. After that, the sterile drug product can be distributed without having to subject it to a sterility test [32].

Controlling these critical process parameters is essential to achieving the desired sterility assurance level and ensuring the safety and efficacy of sterilized pharmaceutical products. Validation, monitoring, and documentation of the sterilization process are critical to ensuring consistency and compliance with regulatory requirements.

Preservative content: The presence of preservatives/antimicrobial content in parenterals helps maintain the sterility and stability of the product over its shelf life. The most common antimicrobials in the parenterals are benzyl alcohol, chlorobutanol, m-cresol, methylparaben,

phenol, phenoxyethanol, propylparaben, and thimerosal. The use of antimicrobial agents in parenteral products is regulated by various pharmacopeial standards and regulatory guidelines. These guidelines set limits on the allowable concentrations of antimicrobial agents and define the specific antimicrobial effectiveness testing requirements [33]. Product developers must demonstrate the reason for the antioxidant/antimicrobial uses and that this substance does not interact with other excipients. In addition, it should be shown that the effect is maintained throughout the shelf-life [34, 35].

Bacterial endotoxin: The bacterial endotoxin test, also known as the Limulus amebocyte lysate (LAL) test, is a widely used method to detect and quantify the presence of bacterial endotoxins in parenteral products. Bacterial endotoxins are lipopolysaccharides (LPS) found in the outer membrane of Gram-negative bacteria, and their presence in parenteral products can lead to adverse reactions and potential patient harm. The bacterial endotoxin test is performed to assess the level of bacterial endotoxins in parenteral products, such as injectable drugs, intravenous fluids, and medical devices. The presence of endotoxins can be harmful, triggering inflammatory responses, fever, and other serious reactions in patientsMainly there is three endotoxin test is available such as rabbit, LAL test and monocyte-activation testing (MAT) test. Among them LAL is the most used test. The LAL test is based on the clotting reaction of horseshoe crab blood (Limulus amebocyte lysate) in the presence of endotoxins. The LAL reagent contains a coagulation factor that forms a gel-like clot when it comes into contact with endotoxins. The degree of gel formation is proportional to the amount of endotoxins present in the sample.The LAL test is highly sensitive and can detect endotoxin levels as low as picograms per milliliter (pg/mL). It is capable of detecting even trace amounts of endotoxins that may be present in parenteral products.

The bacterial endotoxin test is a mandatory test for parenteral products, and its performance is governed by regulatory authorities such as the United States Pharmacopeia (USP), European Pharmacopoeia (Ph. Eur.), and other regional regulatory agencies (USP <71> Sterility tests and USP <85> Bacterial endotoxins). These guidelines provide detailed information on the test methodology, acceptance criteria, and validation requirements [30, 36–38].

pH: The QTPP specifies the desired pH range and osmolality of the parenteral product to ensure compatibility with the body's physiological conditions. pH testing is an important parameter to assess in parenteral products. The pH value of a solution indicates its level of acidity or alkalinity and can have implications for the stability, compatibility, and potential irritation of the product.

The pH of parenteral products can vary depending on the specific formulation, intended use, and route of administration. Different parenteral products may have different pH ranges based on factors such as the chemical nature of the active pharmaceutical ingredient (API), stability considerations, compatibility with body tissues, and regulatory requirements. By adjusting the pH, an ionizable substance can be solubilized to the appropriate concentration. Due to potential irritation issues, the permissible range for intravenous and intramuscular injection is pH 2–11, and pH 4–9 for subcutaneous injection. A buffer, such as acetate, phosphate, citrate, histidine, TRIS, and others, can be used to adjust the pH of the solution to a desirable range of pH 5–8. In order to immediately bring the formulation's pH to physiological pH, the buffer capacity should be kept to a minimum; otherwise, an irritation at the injection site may exist. This pH values alsa can change according to subtances physicochemical properties, large/small volumes and etc. [39, 40]

Osmolality: Parenteral formulations should be ideally isotonic whenever possible. Osmolality is expressed in terms of osmoles of solute particles per kilogram of solvent (osmol/kg). It provides a more accurate measure of the concentration of solutes compared to osmolarity, which is expressed in terms of osmoles per liter of solution. The osmolality of parenterals is an important parameter as it affects the compatibility with body fluids and can impact patient safety and comfort.

Tonicity: The osmolality of a parenteral product determines its tonicity, which refers to the effect of the solution on osmotic pressure across cell membranes. Tonicity is important for ensuring that the parenteral product is compatible with body fluids and does not cause adverse effects on cells or tissues.

The acceptable osmolality range for parenteral products can vary depending on the specific formulation, intended use, and route of administration. Regulatory guidelines, such as those provided by the United States Pharmacopeia (USP) or European Pharmacopoeia (Ph. Eur.), may define osmolality limits or provide general recommendations for different types of parenterals.

Parenterals should be isotonic. According to the USP (<785>), the osmolality of blood is between 285 and 310 mOsmol/kg [41]. Isotonic solutions have the same osmolality as the body's fluids (around 280–310 mOsm/kg). Isotonicity helps to minimize the osmotic pressure across cell membranes, ensuring that the parenteral product does not cause cell shrinkage or swelling. Hypertonic solutions have a higher osmolality than the body fluids, while hypotonic solutions have a lower osmolality. Hypertonic solutions can draw water out of cells, leading to cell shrinkage, while hypotonic solutions can cause water to enter cells, resulting in cell swelling. These conditions can have adverse effects on cells and tissues. Dextrose, glycerol, mannitol,, and sodium chloride are the mostly used agents to adjust the tonicity. It's important for manufacturers to accurately measure and control the osmolality of parenteral products to ensure their safety, compatibility, and efficacy when administered to patients [40].

Particulate matter: For particulate parenteral products, the QTPP defines the acceptable particle size range to prevent potential safety concerns and ensure administration without clogging or embolism risks. Particulate matter refers to any visible or subvisible particles present in parenteral products. These particles can originate from various sources, including the formulation components, manufacturing processes, packaging materials, or external contamination [42]. Controlling particulate matter is crucial in parenteral products as these particles can have potential safety risks and affect product quality. This problem can be result recalls of products. US two lots of Dianeal Low Calcium Peritoneal Dialysis Solution 2014, Baxter), bendamustine HCL for Injection (2012, Cephalon), carboplatin injection (2012, Hospira) were recalled due to particulate matter [43–45]. Regulatory authorities, such as the United States Pharmacopeia (USP) or European Pharmacopoeia (Ph. Eur.), provide guidelines and limits for particulate matter in parenteral products. These guidelines specify acceptance criteria and methods for particulate matter analysis, including both visible and subvisible particles. QbD approach is implementing to minimize particle risk. These approaches determine critical process that can cause this issue. Beside pharmaceutical production process, container manufacturers -especially glass producers- used QbD approach to understand fwhich material and process inputs can cause glass particles, which interactions can cause particles [46].

Stability and shelf life: It outlines the desired stability characteristics of the parenteral product, including chemical, physical, and microbiological stability, to ensure its shelf-life and efficacy throughout storage. Ensuring the stability of parenterals is crucial to ensure their safety, efficacy, and quality throughout their shelf life. Shelf life refers to the period during which a parenteral product is expected to remain stable and maintain its desired quality attributes within acceptable limits. Shelf life is determined based on stability data obtained from accelerated and long-term stability studies. The expiration date is assigned based on the results of these studies. As a satbility test, Physical (such as pH, clarity, color, particulate matter), chemical (chemical degradation of the active pharmaceutical ingredient (API), microbiological stability studies, container-closure system compatibility and stress tests should be done. While doing stability test, autorities such as the United States Pharmacopeia (USP), European Pharmacopoeia (Ph. Eur.), and other national regulatory agencies, guidelines can be follow.

Drug Release Profile: For controlled-release or extended-release parenteral products, the QTPP may specify the desired drug release profile over time.

Compatibility: The QTPP may include compatibility requirements with commonly used diluents, additives, or co-administered drugs to ensure compatibility during preparation and administration. Parenteral compatibility refers to the ability of different components in a parenteral formulation, including active pharmaceutical ingredients (APIs), excipients, and packaging materials, to coexist without causing undesirable interactions or compromising the quality, stability, or safety of the product. API and excipient compatibility: It is essential to evaluate the compatibility between the API and excipients used in the formulation of a parenteral product. Compatibility studies assess potential interactions, such as chemical reactions, degradation, or incompatibility, between the API and excipients. These studies help ensure that the selected excipients do not adversely affect the stability, efficacy, or safety of the API. For compatibility studies, excipient, container-closure system and drug-device (if applicable) compatibility should be study.

Miranda et al. has studies on systematization fo compatibility of injectables. They worked on 55 injectable parenteral drugs that is two type 2-in-1 and 3-in-1. The compatibility of drugs with the parenteral nutrition with or without lipids were assesed and drugs were classified as compatible, incompatible and untested. After analysis, they are summarized informations This study can be used in risk assesment and Qbd studies for formulators [47].

The QTPP serves as a reference for formulation scientists, regulatory authorities, and other stakeholders involved in the development, manufacturing, and quality control of parenteral products. It helps guide decisions and ensures that the final product meets the defined quality standards and the needs of patients and healthcare providers. These are some general quality targets for parenteral solutions. However, it's important to note that specific quality attributes and targets may vary depending on the formulation, drug product, and regulatory requirements. Pharmaceutical manufacturers follow rigorous quality control and assurance processes to meet these targets and ensure the safety and effectiveness of parenteral solutions. Beside these main QTPPS, there are additional QTPPs should be added according to different dosage forms.

Examples of delivery system specific QTPPs are as follows:

9.2.1.1 Lyophilized Products

In the lyophilized products, lyophilization process is most important unit operation. To enhance properties of product, process modelling was used by Mockus et al. In this model, they used bayseian model for the prediction of primary phase duration.

One aspect of the improved quality by design approach to product development is the prediction of crucial process parameters based on process modeling. On the lab scale lyophilizer, the expected primary drying time was experimentally confirmed. To decrease trial-and-error and cut expenses associated with pricey large-scale trials, the model is advised to be used during scale-up efforts [48].

Besides the required parenteral spesification, these target profiles should be investigate:

- Uniformity of dosage
- Properties of lyophilized cake
- Process for reconstitution: The lyophilized drug should reconstitute quickly and easily when mixed with the appropriate diluent. The goal is to have a short reconstitution time to facilitate ease of use and administration. The lyophilized drug should require a reasonable volume of diluent for reconstitution. The ideal target is to minimize the diluent volume needed while still achieving an appropriate concentration for administration. The reconstituted solution should have a clear, visually appealing appearance, free from any particulate matter or turbidity. A clear solution ensures that the drug is dissolved completely and ready for administration.
- Description of reconstituted solution: The lyophilized cake should have a uniform and consistent appearance, indicating proper lyophilization and absence of any defects or abnormalities.

- pH of reconstituted solution: The reconstituted solution should have a pH within an acceptable range that is compatible with the intended route of administration and the patient's body.
- Water content: The lyophilized drug should have an appropriate shelf life in its dry state. Lyophilized drug should have no or limited water content.
- Container closure integrity: The primary packaging, such as vials or ampoules, should maintain the integrity of the lyophilized product, protecting it from moisture, oxygen, and other environmental factors that could degrade its quality.

9.2.1.2 Emulsions/Nano-Emulsions

Development of parenteral emulsion formulation is very challenge fort he quality of drug.. There are some risks associated with comes from the manufacturing processes, including the potential for unstable emulsions and the formation of big droplets that can cause embolisms. To prevent this QbD approached were applying to minimize risks. Therefore critical processes and risk factors should be determined [25, 49, 50].

- Emulsion properties: Parenteral emulsions should have a visually appealing appearance, typically described as a homogeneous, milky white or translucent fluid. The absence of phase separation, sedimentation, or visible particulates is crucial for maintaining quality and ensuring patient acceptance.
- Emulsifier concentration: It's important to note that specific emulsifiers have recommended usage levels and concentration ranges provided by regulatory authorities, such as pharmacopeial monographs or guidelines. These recommendations help ensure the safety, efficacy, and quality of the parenteral emulsion products. During formulation development, emulsifier concentrations are typically adjusted to achieve the desired emulsion stability, droplet size, and other critical quality attributes. Stability studies and characterization tests, including visual inspection, droplet size analysis, and stability testing, are conducted to evaluate the effect of emulsifier concentration on the emulsion's stability and performance. Therefore, the appropriate concentration of emulsifiers in parenteral emulsions is determined through formulation development, optimization studies, and adherence to regulatory guidelines to achieve a stable and efficacious product.
- Globule size distribution: Parenteral emulsions should have a defined droplet size distribution, typically specified in terms of the mean droplet size and the range of droplet sizes. Smaller droplet sizes generally provide better stability, increased bioavailability, and reduced risk of adverse effects.
- Rheology
- Emulsion Stability: Parenteral emulsions should be stable over the intended shelf life, maintaining their physical and chemical properties. Stability studies are conducted to assess factors such as phase separation, creaming, coalescence, or

Ostwald ripening, which can impact the emulsion's long-term stability. Fort he stability, charge of globules should be measure (zeta potential).

- Drug release rate: The release rate of the active pharmaceutical ingredient (API) from the emulsion should be controlled and predictable, providing the desired pharmacokinetic profile. The rate of drug release can be influenced by the emulsion formulation, including the choice of emulsifiers and the API's solubility.

9.2.1.3 Suspensions/Nano-Suspensions

Suspensions are not applied often like the other dosage forms such as solution and emulsion. Because its main restriction is unsoluble particles that can mak embolism. By reducing particle size, parenteral suspension can be applied. Therefore main risk for parenteral suspension is big particle size that can be cause of the manufacturing process [49, 51, 52]. These are critical parameters that should be search:

- Suspension properties
- Surfactant concentration
- Particle size distribution: Nanosuspensions should have a defined and narrow particle size distribution, typically targeting a specific range suitable for the intended therapeutic application. The desired particle size may vary depending on factors such as drug properties, target site, and administration route.
- Particle morphology: The particles in the nanosuspension should exhibit a defined morphology, such as spherical or rod-like, depending on the manufacturing method and drug characteristics. Consistent particle morphology contributes to stability and reproducibility.
- Surface charge: Nanosuspensions may possess a surface charge, which can influence their stability, interactions with biological components, and pharmacokinetic behavior. The desired surface charge, whether positive, negative, or neutral, depends on the specific drug and intended therapeutic application.
- Rheology
- Stability: Nanosuspensions should exhibit good physical and chemical stability over the intended shelf life. Stability studies are conducted to assess factors such as particle aggregation, sedimentation, or Ostwald ripening, which can impact the long-term stability of the nanosuspension.
- Redispersibility: The nanosuspension should demonstrate good redispersibility upon shaking or gentle agitation, ensuring uniform distribution of the nanoparticles in the dispersion medium. Ease of redispersion is crucial for convenient and reliable administration.
- Viscosity: Viscosity is a measure of a suspension's resistance to flow. It determines the ease of administration, injection, or infusion. The viscosity of a parenteral suspension should be carefully controlled to ensure it can be delivered through the intended administration route, such as syringe or IV infusion, while maintaining stability.

- Yield stress: Yield stress refers to the minimum stress required to initiate flow in a suspension. Parenteral suspensions with a yield stress exhibit solid-like behavior when at rest and require a certain force to initiate flow. Yield stress can help prevent sedimentation or caking of solid particles during storage and facilitate uniform suspension re-dispersion upon shaking or gentle agitation.
- Thixotropy: Thixotropy is a reversible shear thinning behavior exhibited by some suspensions. It refers to the reduction in viscosity upon the application of shear stress, followed by a gradual recovery of viscosity when the stress is removed. Thixotropic behavior allows for easier administration and re-dispersion while maintaining stability when the suspension is at rest.
- Particle settling: Particle settling or sedimentation is a common phenomenon in suspensions, where solid particles settle over time due to gravity. The rate of particle settling can be influenced by the size, density, and shape of the suspended particles, as well as the formulation and rheological properties of the liquid medium. It is essential to minimize particle settling to maintain uniform suspension and prevent dose variability during storage.
- Redispersibility: Parenteral suspensions should exhibit good redispersibility, ensuring that the solid particles are readily dispersed throughout the liquid medium upon shaking or gentle agitation. Redispersibility is critical to achieving a homogeneous suspension and ensuring consistent dosing.
- Shear thinning behavior: Shear thinning refers to the decrease in viscosity with increasing shear rate or stress. Many parenteral suspensions exhibit shear thinning behavior, which facilitates administration by reducing resistance to flow during injection or infusion.

Besides the conventional dosage forms every nanoparticle systems has own unique characteristics. For example liposomes needs extra lipid and nonlipid component, surface characterization, size and size distribution, net charge, encapsulation efficiency, liposome phase transition temperature, leakage rate of drug from liposome, purity [53, 54], lipid nanoparticles needs, size and size distribution, net charge, encapsulation efficiency, Assay of lipid components, Degradation products related to lipid components, Internal volume, Lipid bilayer phase transition temperature, lamellarity [50, 55, 56].

While developing parenteral formulation, formulators face many technical problem which needs effective process. For optimization of formulation and process, different QbD tools such as DoE, risk assesment, statistical analysis is used by many researchers.. During formulation development, the researchers face severe technical obstacles, which necessitates the employment of an effective formulation development process. For the optimization of many formulation and process factors, DOE and statistical analysis are viable tools. The significant benefit of using DOE to design pharmaceutical product formulations is that it allows all critical factors to be assessed systematically and precisely. Once the crucial factors have been determined, the best formulations may be finished by employing precise DOE to optimise the values of all critical variables. For.

9.3 Conclusion

QbD is crucial for pharmaceuticals as it improves product quality, reduces risks, ensures regulatory compliance, supports cost savings, promotes process understanding, and ultimately enhances patient safety and satisfaction. It is an integral part of modern pharmaceutical development and manufacturing practices. By implementing QbD principles, pharmaceutical manufacturers can enhance product quality, reduce variability, improve process understanding, and optimize manufacturing processes for parenteral products. This systematic approach helps to ensure that quality is built into the product right from the design stage, leading to safer and more effective parenteral formulations.

References

1. van Reeuwijk, L. P., & Houba, V. J. G. (1998). *Guidelines for quality management in soil and plant laboratories. (FAO soils bulletin - 74).* Food and Agriculture Organization of the United Nations.
2. International Organization for Standardization. *ISO 9000:2015—Quality management systems—Fundamentals and vocabulary.* Retrived June 21, 2023, from https://www.iso.org/standard/45481.html
3. Juran, J. M. (1992). *Juran on quality by design: The new steps for planning quality into goods and services.* Free Press.
4. Yu, L. X., et al. (2014). Understanding pharmaceutical quality by design. *The AAPS Journal, 16*(4), 771–783.
5. Woodcock, J., & Woosley, R. (2008). The FDA critical path initiative and its influence on new drug development. *Annual Review of Medicine, 59*(1), 1–12.
6. Use, I.C.o.H.o.T.R.f.R.o.P.f.H., ICH Q8 (R2). (2009). *Pharmaceutical development.* ICH Harmonized Tripartite Guidelines.
7. (2006). International Conference on Harmonization of Technical Requirements for Registration of Pharmaceuticals for Human Use. *ICH Q9 Quality Risk Management.* ICH Harmonized Tripartite Guidelines.
8. (2012). International Conference on Harmonization of Technical Requirements for Registration of Pharmaceuticals for Human Use, *ICH guideline Q11 on development and manufacture of drug substances (chemical entities and biotechnological/biological entities).*
9. (2007). International Conference on Harmonization of Technical Requirements for Registration of Pharmaceuticals for Human Use. *ICH guideline Q10 on pharmaceutical quality system.* ICH Tripartite Guidelines.
10. (2020). International Conference on Harmonization of Technical Requirements for Registration of Pharmaceuticals for Human Use. *ICH guideline Q12 on technical and regulatory considerations for pharmaceutical product lifecycle management.* ICH Harmonized Tripartite Guidelines.
11. Grangeia, H. B., et al. (2020). Quality by design in pharmaceutical manufacturing: A systematic review of current status, challenges and future perspectives. *European Journal of Pharmaceutics and Biopharmaceutics, 147*, 19–37.
12. Yüce, M., & Çapan, Y. (2010). Pharmaceutical quality by design: Quality by design approach and its elements. *Journal of Faculty of Pharmacy of Ankara University, 39*(4), 369–390.
13. Pielenhofer, J., et al. (2023). Quality by design (QbD) approach for a Nanoparticulate Imiquimod formulation as an investigational medicinal product. *Pharmaceutics, 15*(2), 514.

14. Zurdo, J., et al. (2015). Early implementation of QbD in biopharmaceutical development: A practical example. *BioMed Research International, 2015*, 605427.
15. Mirajkar, R., Patil, N., & Basate, S. (2021). Quality by design in parenteral formulation: A review. *International Journal of All Research Education and Scientific Methods (IJARESM), 9*, 8.
16. Henriques, J., Cardoso, C., & Vitorino, C. (2019). On demand for new process analytical technologies applied to injectable drug products. *European Journal of Pharmaceutical Sciences, 137*, 104975.
17. Yu, L. X., & Kopcha, M. (2017). The future of pharmaceutical quality and the path to get there. *International Journal of Pharmaceutics, 528*(1–2), 354–359.
18. Troiano, G., et al. (2016). A quality by design approach to developing and manufacturing polymeric nanoparticle drug products. *The AAPS Journal, 18*(6), 1354–1365.
19. Ergin, A. D., et al. (2021). A novel delivery system for enhancing bioavailability of S-adenosyl-l-methionine: Pectin nanoparticles-in-microparticles and their in vitro—In vivo evaluation. *Journal of Drug Delivery Science and Technology, 61*, 102096.
20. Ergin, A. D., Sezgin-Bayındır, Z., & Yüksel, N. (2019). Characterization and optimization of colon targeted S-adenosyl-L-methionine loaded chitosan nanoparticles. *Journel Research Pharmacy, 23*(5), 914–926. https://doi.org/10.35333/jrp.2019.38
21. Ergin, A. D., & Uner, B. (2023). Characterization, optimization, and in vitro evaluation of cholesterol-free liposomes. *Journal of Drug Delivery Science and Technology, 84*, 104468.
22. Uner, B., et al. (2023). Loteprednol loaded nanoformulations for corneal delivery: Ex-vivo permeation study, ocular safety assessment and stability studies. *Journal of Drug Delivery Science and Technology, 81*, 104252.
23. Ockun, M. A., et al. (2022). Preparation, characterization and evaluation of liposomal-freeze dried anthocyanin-enriched Vaccinium arctostaphylos L. fruit extract incorporated into fast dissolving oral films. *Journal of Drug Delivery Science and Technology, 72*, 103428.
24. Uner, B., et al. (2023). Timolol-loaded ethosomes for ophthalmic delivery: Reduction of high intraocular pressure in vivo. *International Journal of Pharmaceutics, 640*, 123021.
25. Waghule, T., et al. (2021). Quality by design (QbD) in the formulation and optimization of liquid crystalline nanoparticles (LCNPs): A risk based industrial approach. *Biomedicine & Pharmacotherapy, 141*, 111940.
26. Joshi, H. N. (2012). *Quality by design (QbD) of sterile dosage form packaging.*
27. Food and Drug Administration. (1999). *Guidance for Industry: Container Closure System for Packaging Human Drugs and Biologics*. Food and Drug Administration.
28. Jenke, D. (2010). Application of quality by design (QbD) principles to Extractables/Leachables assessment. Establishing a design space for terminally sterilized aqueous drug products stored in a plastic packaging system. *PDA Journal of Pharmaceutical Science and Technology, 64*(6), 527–535.
29. Eberle, L., et al. (2017). Innovative approach for identifying root causes of glass defects in sterile drug product manufacturing. *European Journal of Pharmaceutical Sciences, 104*, 162–170.
30. Riley, B.S., & Li X. (2011). *Quality by design and process analytical technology for sterile products—where are we now? AAPS PharmSciTech, 12*(1), 114–118.
31. Witchey-Lakshmanan L., et al. (2006). Particulate-stabilized injectable pharmaceutical compositions of posaconazole.
32. Administration F.a.D. (2010). *Guidance for Industry-Submission of Documentation in Applications for Parametric Release of Human and Veterinary Drug Products Terminally Sterilized by Moist Heat Processes.*
33. Meyer, B. K., et al. (2007). Antimicrobial preservative use in parenteral products: Past and present. *Journal of Pharmaceutical Sciences, 96*(12), 3155–3167.
34. Committee for Proprietary Medicinal Products(cpmp) & Committee for Veterinary Medicinal Product(cvmp). (1997). *Note for guidance on inclusion of antioxidants and antimicrobial preservatives in medicinal products.*

35. Committee for Proprietary Medicinal Products(cpmp) & Committee for Veterinary Medicinal Product(cvmp). (2003). *Note for guidance on excipients, antioxidants and antimicrobial preservatives in the dossier for application for marketing authorisation of a medicinal product.*
36. Food and Drug Administration (FDA). (2017). *Drug products, including biological products, that contain nanomaterials—guidance for industry*. Food and Drug Administration (FDA).
37. United States Pharmacopeia (USP) and the National Formulary (NF). (2013). Sterility tests.
38. United States Pharmacopeia (USP) and the National Formulary (NF). (2018). *Bacterial endotoxins.*
39. Chaturvedi, S., Anand, C., & Saini, L. (2022). A detailed concepts on parenteral preparation. *International Journal of Pharmaceutical Research and Applications, 7*(4), 1.
40. Röthlisberger, D., et al. (2016). If Euhydric and isotonic do not work, what are acceptable pH and osmolality for parenteral drug dosage forms? *Journal of Pharmaceutical Sciences, 106*, 446–456.
41. United States Pharmacopeia (USP). (2016). *USP<785>Osmolality and osmolarity.*
42. Stephen, E. L. (2013). Particulate matter in injectable drug products. *PDA Journal of Pharmaceutical Science and Technology, 67*(3), 186–200.
43. Baxter. (13 Aug 2014). *"Baxter Voluntarily Initiates U.S. Recall of Two Lots of Peritoneal Dialysis Solution Due to Presence of Particulate Matter".*
44. Cephalon. (27 Jan 2012) *Cephalon, Inc. issues a voluntary nationwide recall of Treanda (bendamustine HCL) for Injection 25mg/Vial Due to Particulate Matter.*
45. Hospira. (14 Dec 2012). *Hospira Issues Voluntary Nationwide Recall of Three Lots of Carboplatin Injection Due to Visible Particulate Matter.*
46. Challener, C.A., Parenterals, particulates, and quality by design. Pharmaceutical Technology, 2014. 38(11).
47. Miranda, T. M., & Ferraresi Ade, A. (2016). Compatibility: Drugs and parenteral nutrition. *Einstein (Sao Paulo), 14*(1), 52–55.
48. Mockus, L., et al. (2011). A QbD case study: Bayesian prediction of lyophilization cycle parameters. *AAPS PharmSciTech, 12*(1), 442–448.
49. Chobisa, D. (2019). QbD-based development of pharmaceutical parenteral drug products: An overview. *Pharmaceutical Quality by Design*, 151–172.
50. Deng, Y., et al. (2019). Quality by design approach for the preparation of fat-soluble vitamins lipid injectable emulsion. *International Journal of Pharmaceutics, 571*, 118717.
51. Ghosh, I., et al. (2012). Identification of critical process parameters and its interplay with nanosuspension formulation prepared by top down media milling technology—A QbD perspective. *Pharmaceutical Development and Technology, 18*, 729.
52. Soni, G., et al. (2020). Quality by design (QbD) approach in processing polymeric nanoparticles loading anticancer drugs by high pressure homogenizer. *Heliyon, 6*(4), e03846.
53. Alina, P., et al. (2019). Pharmaceutical development of liposomes using the QbD approach. In C. Angel (Ed.), *Liposomes*. Rijeka.
54. Barros, C., et al. (2021). Quality by design approach for the development of liposome carrying ghrelin for intranasal administration. *Pharmaceutics, 13*(5), 686.
55. Zagalo, D. M., et al. (2022). A quality by design (QbD) approach in pharmaceutical development of lipid-based nanosystems: A systematic review. *Journal of Drug Delivery Science and Technology, 70*, 103207.
56. Amasya, G., et al. (2019). QbD guided early pharmaceutical development study: Production of lipid nanoparticles by high pressure homogenization for skin cancer treatment. *International Journal of Pharmaceutics, 563*, 110–121.

Chapter 10
Quality By Design Approach for Novel Drug Delivery Systems

Mazen Al-Mohaya and Burcu Mesut

Abstract QbD offers several unique benefits, including deep material and process understanding, flexibility change within the design space without post-approval, application of effective control strategies, and a more robust and reproducible product. The purpose of this chapter is to identify and highlight the fabrication of novel drug delivery systems under the implementation of the QbD approach. It provides a map of QbD, and its elements (material attributes, process parameters, quality attributes, risk assessment tools, and DoE). It also includes the most common novel drug delivery systems formulated using the QbD approach with examples. The knowledge mentioned in this chapter will facilitate understanding novel drug delivery systems using QbD and encourage increasing these novel systems to reach the market in the near future.

Keywords Quality by design · Quality attributes · Process parameters · Design space · Design of experiments

10.1 Introduction

According to updated data, 70% or more of the novel drugs are low-solubility drugs; this is the limiting factor in absorption, which can affect bioavailability. Low stability, low metabolism, low active influx, and high first-pass metabolism can also influence active compounds' bioavailability and are factors in the failure of clinical trials [1]. In order to solve problems, scientists have been developing novel drug delivery systems (NDDS). A new approach to drug delivery, known as novel drug delivery systems, focuses on formulating pharmaceutical forms with optimized

M. Al-Mohaya
Institute of Health Sciences, Istanbul University, Istanbul, Turkey

B. Mesut (✉)
Pharmaceutical Technology Department, Faculty of Pharmacy, Istanbul University, Istanbul, Turkey
e-mail: bmesut@istanbul.edu.tr

N. K. Jain, N. Bajwa (eds.), *Introduction to Quality by Design (QbD)*,
https://doi.org/10.1007/978-981-99-8034-5_10

characteristics, including smaller particle size, enhanced permeability parameters, enhanced solubility and stability, and targeted delivery to specific sites [2]. NDDSs have the potential to improve the performance of new compounds compared to conventional dosage forms. Innovative fabrication methods and nanotechnology have transformed the field of pharmaceutical science, paving the way for new therapeutic strategies. The advantages of nanotechnology-based formulations are vast and diverse, thanks to their minuscule particle size and unique properties. These benefits include the encapsulation of small molecules and biologics, protection of drugs in biological environments, enhancement of drug biodistribution, controlled or stimuli-responsive drug release, targeted and intracellular delivery of therapeutics, and integration of diagnostic and imaging agents for theranostic purposes. Utilizing nanotechnology-based formulations is an exceptional option for numerous applications and has the potential to enhance the efficacy of therapeutic treatments. The comprehensive exploration of drug delivery enhancement strategies through formulation systems with a focus on therapeutic applications and development approaches was exhaustively covered in the special issue titled "Novel Drug Delivery Systems, Devices, and Fabrication Methods". Experts in the field discussed a wide range of topics, including microparticles, microbubbles, nanoparticles, microneedles, organogels, micelles, lipid-based systems, polymer-based systems, inorganic nanoparticles, laser-generated micropores, and nanoemulsion. Despite the significant amount of effort invested by researchers in developing innovative drug delivery systems for various medical uses, the transformation of these products into commercial products poses significant challenges [3].

Some commercial-approved products, such as Doxil (liposome systems) and Estrasorb (nanomicelle systems), have been approved to be used clinically [4]. However, there are some obstructions to applying novel drug delivery systems in industrial productions and clinical transformations. For example, factors that affect the physicochemical characteristics are still not completely identified and illustrated. Particle size consistency, high encapsulation, suitable surface charge, ligand coated, and homogeneous shapes are challenges for industrial production because they must meet the quality criterion of the Currently Good Manufacturing Practice (CGMP) guidelines. Moreover, novel drug delivery system productions are away from simple, scalable, and cost-effective. To get over these obtacles, formulation and process designs need to be standardized based on a scientific and systematic approach [5].

10.2 Quality By Design (QbD)

In 2005, the FDA approved the implementation of quality by design (QbD) as a new and improved method for quality testing. This decision was made in compliance with the ICH Q8 guidelines for pharmaceutical development. As per the FDA, QbD entails a methodical approach to development that commences with predetermined goals and places paramount importance on product and process control, backed by reliable scientific methods and quality risk management. Prioritizing quality through design is a must. Merely relying on quality testing won't suffice to improve the product. The production process should involve meticulous product design and extensive control throughout manufacturing to ensure top-notch quality. QbD is a clear and systematic process that guarantees a high-quality end product. This approach is exceedingly practical and encompasses various key elements, including product identification and comprehension, production methodology, statistical data analytics, and the evaluation of parameter risks. The aim of QbD is to instill a planning mindset that prioritizes quality and maximizes the efficiency of the drug development process [6]. Incorporating Quality by Design (QbD) practices is crucial in boosting the safety and effectiveness of medications. Although its initial goal was not to enhance the probability of regulatory approval, it undoubtedly enhances the dependability and uniformity of the product and the process as a whole. Maximizing the probability of obtaining FDA approval [7].

10.3 Quality By Design (QbD) Objectives

To achieve objectives in Quality by Design (QbD), it is crucial to connect product quality with the desired clinical performance Developing a robust formulation and manufacturing process is essential to ensure the consistent delivery of a superior product. For pharmaceutical QbD objectives, it is imperative to establish clear and concise statements outlining the desired outcomes [8]:

- Establishing product quality standards that are founded on clinical performance and bear substantial significance.
- The primary objective is to optimize product and process design, comprehending and regulating it to enhance process capability, curtail product variation, and minimize defects.
- The aim to improve the efficiency of product development and manufacturing processes.
- Our root cause analysis and post-approval change management processes must be improved.

10.4 Elements of Quality By Design

Furthermore, the QbD process for pharmaceutical product development encompasses several essential steps, including [9]:

1. Defining the quality target product profile (QTPP) by considering aspects such as quality, safety, and efficacy, taking into account factors like administration route, dosage form, bioavailability, strength, and stability.
2. It is imperative that we establish a clear and concise strategy for the development and production of the medication.

3. It is absolutely vital to identify possible critical quality attributes (CQAs) associated with the drug substance, raw materials, and drug product to comprehend and regulate characteristics that have a substantial influence on product quality.
4. Identifying potential critical process parameters.
5. The establishment of functional relationships between critical material attributes (CMAs), critical process parameters (CPPs) for unit operations, and the drug product's critical quality attributes (CQAs) is based on the utilization of risk assessment tools and experimental methods.
6. Rigorously enhance both the formulation and manufacturing processes to fulfill the Quality Target Product Profile (QTPP) that was established in the initial stage.
7. Establishing the design space and control strategy. First and foremost, it is imperative to establish and specify the design parameters and control tactics.

10.4.1 Quality Target Product Profile (QTPP)

In order to guarantee the safety and efficiency of a drug product, Quality by Design (QbD) necessitates the creation of a Quality Target Product Profile (QTPP). This document precisely outlines the quality characteristics that the drug must possess. The QTPP functions as a summary of the ideal quality of the drug, while incorporating considerations for its safety and efficacy [7]. When determining the Quality Target Product Profile (QTPP), numerous factors must be considered. FDA regulations, patient needs, and manufacturing considerations all play a critical role in establishing acceptance criteria. The Quality Target Product Profile (QTPP) takes into account important features of the medication, including its purity, stability, ability to release the drug, sterility, and criteria for release that affect how the medication is absorbed and processed by the body. Understanding the QTPP is absolutely crucial for determining the success or failure of a product. It's imperative that all aspects, including the form, method, strength, and container used for the dosage, are precisely defined. Failure to do so could have disastrous consequences. Ensuring bioequivalence is crucial when creating generic drug products. However, if the QTPP definition is inaccurate or incomplete, or if there is a lack of a specific target product profile, there may be significant disparities in drug efficacy between the generic drug and the reference listed drug. To gain a comprehensive understanding, it is imperative that one refers to perspective papers that stress the significance of QTPP and underscore Identifying the key possessions of a drug product at the beginning of the development process is absolutely crucial, as it offers a host of advantages [8, 10].

10.4.2 Critical Quality Attributes (CQAs)

Critical Quality Attributes (CQAs) are non-negotiable when it comes to ensuring a product meets the necessary standards for quality. By adhering to these standards, the end product will unequivocally meet all FDA safety, efficacy, and performance regulations. It is crucial to comprehend that CQAs are applicable not only to the drug substance or product but also to intermediates and excipients. Identifying CQAs demands a meticulous grasp of the product design and an understanding of the characterization techniques required to evaluate these attributes within their scope and limitations. When working with nano-sized formulations, it is imperative to evaluate CQAs on a case-by-case basis. Nanoparticle size is significant, but it is crucial to identify the specific aspects of size determination that are vital for the final product. Sometimes, a narrow polydispersity index (PI) may not be necessary, but it is crucial to establish a maximum particle size limit. The decision to determine whether a characteristic is a critical attribute requires evaluating both its impact and associated uncertainty [11, 12].

10.4.3 Critical Material Attributes (CMAs)

The ICH guidelines may not offer a crystal-clear definition of critical material attributes (CMAs), but these are absolutely crucial components of input materials such as drug substances, excipients, and in-process materials. They must be maintained within specific ranges to ensure the quality of the end product. CMAs are discerned during the quality risk management process by analyzing the impact of both material attributes and process parameters on critical quality attributes (CQAs). It is imperative to use various tools, such as Ishikawa diagrams and failure mode and effects analysis (FMEA), to ensure high-quality attributes. These tools evaluate the impact of input variables, while theoretical modeling and design of experiments (DoE) can test high-risk variables. It is important to note that CQAs relate to output materials, whereas CMAs pertain to input materials. Additionally, CQAs of an intermediate step can become CMAs if it serves as input material in a subsequent step. Material attributes can also interact with process parameters and affect the CQAs of granulation. In nanoparticle-based formulations, CMAs are crucial because different components of the nanoparticles interact and collectively influence the properties of the API and final product. Even a small difference in the quality of the input material can have a big effect on the overall quality of the final product. An example of this is the production of Abraxane® (paclitaxel bound to albumin protein), where the source and quality of albumin can vary greatly and impact the physical properties of the final product [13]. The interaction between materials is a crucial factor that can significantly impact the properties of nanoparticle drugs. Although albumin may be considered safe by itself, when combined with a drug like Abraxane®, it triggers an immune response in pigs. This highlights the importance of thoroughly examining the interaction between materials to ensure the safety and efficacy of nanoparticle drugs [13, 14].

10.4.4 Critical Process Parameters (CPPs)

Process parameters in pharmaceutical manufacturing are crucial input variables that significantly affect the physical, chemical, or biological changes in a product during unit operations such as melting, mixing, filtering, and drying. Critical process parameters (CPPs) are specific processes or unit operations that have a profound impact on the quality and performance of the pharmaceutical product. They are considered as process parameters that can affect the critical quality attributes of the product [7]. The identification of Critical Process Parameters (CPPs) according to the ICH Q9 guidelines requires strict adherence to four main steps that follow a risk-based approach:

1. to assist you in identifying the significant parameters and unit operations that are pertinent and responsive to your requirements.
2. Assessing the repercussions of the tools and machinery employed in the manufacturing or formation of an item.
3. The execution of this task requires the implementation of experiments that manipulate process parameters to observe the impact on critical quality attributes.
4. Ensuring that the crucial quality features fall within the acceptable range by establishing processing specifications.

Establishing the link between critical material attributes and critical quality attributes for a product solely depends on the identification of critical process parameters (CPPs) derived from the starting materials. The identification of these CPPs is of utmost importance as it aids in comprehending and controlling the manufacturing process, which inevitably affects the quality of the final product [8].

10.4.5 Risk Assessment (RA)

Risk assessment aims to evaluate the influence of specific variables or critical attributes associated with raw materials (active pharmaceutical ingredients and excipients) and packaging materials. It also identifies the CPPs of the drug product. Each attribute of a CPP is categorized as high, medium, or low risk based on its impact on the final product quality. High-risk attributes are subjected to further investigation to mitigate potential risks. Risk assessment employs different methods, including failure mode effects analysis (FMEA) and relative risk-based matrix assessment (RRMA), to assess and analyze the risks involved [15, 16].

10.4.6 Design of Experiments (DoE)

DoE is a methodology employed to conduct multivariate experiments to study the impact of multiple factors at the same time. It allows for investigating and understanding the relationship between material attributes, process parameters (all inputs), and the resulting critical quality attributes (CQAs). By employing mathematical models, computer-aided process design, and process simulation, DoE aids in defining the design space. The design space represents the range of process input settings that lead to desired CQAs. Various models can be used in DoE, including factorial design, Box-Behnken design, Plackett-Burman design, and Taguchi design, among others [17].

10.4.7 Design Space (DS)

According to ICH Q8(R2) guidelines, the design space is defined as "the multidimensional combination and interaction of input variables and process parameters that have been demonstrated to provide assurance of quality" [7]. To establish a design space, it is necessary to first determine CQAs, CMAs, and CPPs. The design space encompasses the variables related to CQAs, CMAs, and CPPs, ensuring that the product meets the desired specifications outlined in QTPP. During the When determining the design space for a drug product, various factors are analyzed simultaneously to determine their impact. The design space is a subset of the knowledge base where the output meets the predetermined acceptance criteria or specifications. It is essential to establish a design space to guarantee that a product meets safety, identity, purity, and strength standards. Any modifications made to specific factors that affect quality attributes within the design space are not deemed significant changes. Engineers must define a design space to ensure a reliable and sturdy process with a consistent output that meets the internal target range. It is imperative to note that any deviation from the design space post-FDA approval will require following a post-approval change process. Adhering to the design space is non-negotiable for maintaining the safety and reliability of the process [18].

10.4.8 Control Strategy

A vital aspect of consistent development of high-quality products is the implementation of a rigorous control strategy. This entails controlling input materials such as active pharmaceutical ingredients, excipients, and packaging material, maintaining predetermined product specifications, monitoring CPPs, and implementing real-time release testing (RTRT) alongside an overall monitoring program. Through the implementation of these processes, we must systematically attain the desired quality standards throughout the manufacturing process [17].

10.5 Novel Drug Delivery Systems

Novel drug delivery made a revolution in pharmaceutical technology and increased at an enormous rate; thus, the next section highlights the most common NDDSs using QbD.

10.5.1 Liposomal

A liposome is a carrier composed of a lipid bilayer that encloses a watery center. The carriers can be classified into two main types based on their particle size, number of bilayers, and preparation methods: unilamellar vesicles (ULVs) and multilamellar vesicles (MLVs). ULVs are composed of a single phospholipid bilayer, while MLVs consist of multiple concentric phospholipid bilayers like an onion. Within ULVs, there are two subcategories: small unilamellar vesicles (SUVs) that are less than 100 nm in size and large unilamellar vesicles (LUVs) that are over 100 nm [19–21].

Developed liposomes containing the anti-proliferative drug erlotinib HCl using the QbD approach and the thin-film hydration technique. We conducted a thorough risk assessment analysis using an Ishikawa fishbone diagram and identified three variables for optimization: molar ratio of drug to lipid (X1; CMA), sonication time (X2; min; CPP), and hydration time (X3; min; CPP). The optimization study was carried out using the response surface method, which involved testing each variable at two different levels and including four center points. We utilized a Box Behnken design to investigate the impact of various variables on liposome preparation. The collected data was meticulously analyzed using Design Expert 7.0.2 software, and we carefully assessed two response variables: particle size (measured in nm) and entrapment efficiency (measured in percentage). It is worth noting that, during our drug release study, the liposomes we formulated demonstrated sustained release of the drug erlotinib HCl over 24 h in simulated lung fluid with a pH of 7.4. The liposomal formulation's ability to inhibit blood vessel growth was put to the test through a zebrafish tail fin regeneration assay. The outcome clearly demonstrated that the formulation effectively suppressed tail fin regeneration for 14 days. This solidifies its potential as a powerful anti-angiogenic treatment [22].

The research aimed to create a topical insulin that could effectively and safely heal wounds while maintaining its stability and release rate. To achieve this, scientists thoroughly studied the formulation of insulin-filled vesicles in bio-adhesive hydrogels, which would deliver the insulin to wounds in a controlled manner. They used QbD principles to investigate how CPPs affected the CQAs, and the Ishikawa diagram was utilized to identify the highest-risk factors. Additionally, a combination of fractional factorial design and Box-Behnken design was employed for screening and optimization. During the screening phase, factors were assessed such as solvent volumetric ratio (chloroform: methanol) (X1), hydration time (X2), sonication time (X3), and temperature of liposomes and film formation (X4) in relation to entrapment efficiency. During the optimization phase, the researchers strictly focused on the three critical factors (XA, XB, and XC) that had the potential to significantly affect particle size (Y1), zeta potential (Y2), entrapment efficiency (Y3), and release rate (Y4). These factors were the lipid content, phosphatidylcholine: cholesterol ratio, and insulin amount. Once the optimization was completed, the formula was swiftly incorporated into a mucoadhesive gel and underwent rigorous clinical and stability trials. The optimized formula yielded a remarkable particle size of 257.751 nm, a zeta potential of −20.548 mV, an entrapment efficiency of 87.379%, and a release rate of 91.521 μg/cm^2/h. Furthermore, stability tests have confirmed that the liposomal insulin remains stable for a period of 6 months when stored in an aqueous dispersion at 4 °C, while also providing a sustained release for 24 h. The clinical study conducted clearly demonstrates that the use of this liposomal chitosan gel leads to a significant 16 times increase in wound healing rate when compared to the control group. Moreover, there is a reduction in ulcer redness and no occurrences of hypoglycemia. In conclusion, the insulin-loaded liposomal chitosan gel demonstrates highly promising characteristics as a drug delivery system, offering exceptional stability and sustained release properties [23].

10.5.2 Solid Lipid Nanoparticles (SLNs)

SLNs emerged in the early 1990s as NDDS that bridged the gap between available lipid systems, such as emulsions and liposomes, and PNPs. SLNs offer a unique combination of properties derived from these established systems. Their production involves exchanging the liquid lipid component of an emulsion (o/w) with a solid lipid, resulting in submicron-sized particles with an average size ranging from approximately 40 to 1000 nm. A surfactant is typically used to maintain system stability. SLNs provide several advantages, including biocompatibility, protection of the drug from degradation, and controlled release. It is important to note that SLNs differ from nanostructured lipid carriers (NLCs) based on the structure of the solid particle matrix [24].

The focal point of previous research has been exploring the factors that impact the properties and quality of SLNs containing carvedilol. The methodology utilized a combination of homogenization and ultra-sonication. To determine the impact of different variables, a Plackett-Burman design was employed through Statgraphics Centurion XV, version 15.2.05. The variables that were scrutinized consisted of lipid quantity (X1), surfactant quantity (X2), drug loading (X3), lecithin quantity (X4), charge-inducing agent quantity (X5), homogenization time (X6), and sonication time (X7). The experimentation on nanoparticles encompassed various properties such as size, zeta potential, entrapment efficiency, amount released after 1 h, and time for 85% of the drug to be released. We have gained significant insights into the variables that have a considerable impact on the formulation's quality. The nanoparticles underwent thorough evaluation in terms of particle size, zeta potential, entrapment efficiency, and in vitro drug release. The combination of independent variables resulted in a range of variability for particle size, zeta potential, and entrapment efficiency. The in vitro drug release showed an initial burst release followed by sustained release for up to 12 h, with the time for 85% drug release varying from 4.25 to 11.25 h. We conclude that the Plackett-Burman design played a vital role in determining and understanding the process and formulation variables that affect the quality of carvedilol-loaded SLNs [25].

A study recently conducted delved into the development and characteristics of surface-engineered SLNs of rosuvastatin calcium, with a particular focus on targeting low-density lipoprotein-receptors. The study utilized a QbD approach and an Ishikawa fishbone diagram to establish a cause-effect relationship between CMAs/CPPs and CQAs of the SLNs. To identify the most critical product and process variables affecting the SLNs, the Taguchi design with Minitab 17 software was implemented. The study analyzed factors such as sonication time, sonication amplitude, stirring time, stirring speed, type of manufacturing process, lipid concentration, and surfactant concentration on CQAs including particle size, zeta potential, entrapment efficiency, and the time required for 80% drug release. The study yielded promising results for the surface-engineered SLNs of rosuvastatin in managing conditions like hyperlipidemia with improved biopharmaceutical performance [26].

10.5.3 Nanostructured Lipid Carriers (NLCs)

NLCs were developed in the late 1990s by Muller et al. as an improvement over SLNs to overcome their limitations. They typically range in size from 100 to 500 nm and consist of a combination of solid and liquid lipids in a ratio of 70:30 (or up to a maximum of 99.9:0.1). NLCs have a higher drug-loading capacity and lower water content than SLNs, minimizing the potential release of drugs during storage. NLCs can have a solid content increased by up to 95%. Both SLNs and NLCs have hydrophobic cores, making them ideal for encapsulating hydrophobic drugs. NLCs consist of physiological and/or biodegradable lipids, making them highly biocompatible and well-tolerated [24].

The development and analysis of aceclofenac-loaded NLCs were accomplished through the use of QbD. A Taguchi orthogonal array design was employed, which rigorously tested various CMAs, including the type of liquid lipid, solid lipid, surfactant, percentage of surfactant, and the ratio of lipid mixture to liquid lipid, as well as CPPs such as stirring speed and temperature. We rigorously tested 8 different formulations based on our design and analyzed them using Design Expert1 version 10.0. Our evaluation included specific qualities such as particle trapping efficiency and size. After careful observation, we confidently determined the critical material attributes (CMAs) and critical process parameters (CPPs). From this analysis, we firmly selected solid lipid (X1), lipid phase and surfactant phase (Smix) (X2), and liquid lipid (X3) as the CMAs necessary for producing NLCs using the microemulsion technique. We optimized and evaluated various quality attributes of NLCs using a 33 Box-Behnken design. Our systematic assessment of CMAs, such as lipids, oil-to-lipid ratio, and surfactant concentration, allowed us to optimize the NLCs and enhance their CQAs like particle size and entrapment efficiency. The optimized ACE-NLCs boasted a spherical shape, nano size, higher drug loading, and entrapment efficiency. Additionally, the drug release from the NLCs followed a biphasic pattern, with an early burst release and sustained release for up to 48 h. Our optimized NLCs-based gel was exceptional, with excellent texture, rheological profile, and enhanced cell uptake efficiency on hyperkeratinocytic cells (HaCaT cell lines), along with improved ex vivo skin permeability compared to the marketed formulation. Lastly, we demonstrated that aceclofenac-loaded NLCs hydrogel was a novel delivery alternative for targeting various skin layers by using dermatokinetic modeling and a pharmacodynamic study on carrageenan-induced edema mice [27].

The QbD method was utilized by the author in a separate research study to optimize two NLC formulations containing diazepam for nose-to-brain delivery. One formulation had a negative surface charge, while the other had a positive surface charge. Prior to conducting the experiments, suitable diazepam-loaded NLC formulation for nose-to-brain delivery was created by identifying the QTPP and CQAs. The CQAs consisted of particle size, PDI, zeta potential, entrapment efficiency, and biocompatibility. The NLC size, PDI, zeta potential, encapsulation efficiency, and cell biocompatibility were defined as the CQAs. The desired limit values were set as particle size <200 nm, PDI < 0.3, zeta potential $> \pm 20$ mV, entrapment efficiency $>80\%$, and biocompatibility with human neuronal cells. To optimize diazepam-loaded NLC, the process was divided into three important parts:

1. The CMAs related to the ratio of solid and liquid lipids and the ratio of emulsifiers were thoroughly evaluated.
2. CPPs were extensively examined by comparing two different NLC production methods: ultrasound technique and high-pressure homogenization.
3. An anionic emulsifier in the diazepam-loaded NLC formulation was forcefully replaced with a cationic lipid to change the surface charge from negative to positive, as nasal formulations must have a positive charge to interact with the negative charge of the nasal mucus.

The central composite design (CCD) was employed to optimize the parameters and assess the impact of the independent variables on the dependent variables. The optimization results of the CMAs indicated that the most suitable ratios of lipids and emulsifiers were 6.7:2.9 and 4.2:0.3 (% w/w), respectively. High-pressure homogenization was forcefully determined to be the most applicable production technique for the optimized formulation (F1C15) with a negative surface charge. The optimized diazepam-loaded NLC formulation (F2A8) was asserted to have a positive charge. The formulations exhibited different particle sizes, PDI, zeta potential, and entrapment efficiency. Formulation F1C15 was forcefully deemed more suitable for targeting the brain compared to formulation F2A8. Diazepam integration in the NLC led to sustained drug release. Formulation F1C15 showed no significant changes in CQAs after 3 months of storage, while formulation F2A8 forcefully exhibited a significant increase in particle size. Biocompatibility studies forcefully revealed that formulation F2A8 was more cytotoxic than formulation F1C15 [28].

10.5.4 Self-Nano-Emulsified Drug Delivery Systems (SNEDDs)

SNEDDS are mixtures composed of oil, surfactant, cosurfactant, and drug that can form fine emulsions (o/w) with gentle agitation when added to water. There has been increasing concern in recent years regarding the use of lipid-based formulations for delivering low-solubility drugs. Normally, self-emulsified delivery systems are made by dissolving active ingredients in fixed oils or triglycerides, which are then mixed with solubilizing agents. However, this method tends to have low drug loading and the active ingredient or excipient may precipitate over time due to limited solubility in the oils. Therefore, finding an essential oil that allows for solubilization without the risk of precipitation during handling and storage became important. Pseudo-ternary phase diagrams must be utilized when illustrating the formation of emulsions. However, there is limited published information on the process of self-emulsification and how the characteristics of the final product relate to the variations within the prepared systems [29].

Panigrahi et al. carried out a study utilizing Quality by Design principles to create a solid self-nano-emulsifying drug delivery system (S-SNEDDS) with cinacalcet HCl (CH). They identified potential risk factors that could affect the CQAs of S-SNEDDS using an Ishikawa fish-bone diagram, followed by an FMEA to analyze and screen the influential factors. Using a 2-level Taguchi design with 7 factors, they performed a screening of factors to determine the ones significantly impacting the CQAs, including mean globule size, PDI, emulsification time, and cumulative % drug release at 30 min. The CH-loaded S-SNEDDS underwent a systematic optimization process using a Box Behnken design (BBD) with the aid of Design Expert ver. 11.1.01 software. The purpose was to thoroughly investigate the impact of surfactant concentration, co-surfactant concentration, and adsorbent concentration on the CQAs. Each formulation underwent a comprehensive analysis. The S-SNEDDS formulation was optimized and displayed an impressive average drug release of 97.6%, emulsification time of 23.3 min, mean globule size of 89.5 nm, and PDI of 0.211. In vivo pharmacokinetic studies demonstrated improved pharmacokinetic parameters with the optimized S-SNEDDS. Therefore, it can be concluded that the optimized S-SNEDDS formulation effectively achieved the desired objectives [30].

The QbD approach was utilized in a research study to enhance the dissolution and oral absorption of gemfibrozil through SNEDDS. An initial screening was conducted to determine the optimal combination of components. The formulation variables, namely surfactant (X1), co-surfactant (X2), and oily phase (X3), were optimized using a Box-Behnken experimental design as a statistical tool. Various parameters, including emulsification efficacy (Y1), turbidity (Y2), mean droplet size (Y3), PDI (Y4), and drug release, were analyzed in the SNEDDS formulations. The quadratic model formulations were analyzed using Design-Expert® version 7.1.6 software, resulting in fifteen different mixtures. Through the optimization process, the formulations were tested in various pH levels to determine the ideal values for the components (X1, X2, and X3) which resulted in 32.43%, 29.73%, and 21.62% respectively, with a gemfibrozil content of 16.22%. The spherical droplet morphology of the SNEDDS was confirmed by transmission electron microscopy. A study was conducted comparing the SNEDDS formulation to commercial tablets and revealed a significantly higher dissolution rate for the optimized SNEDDS formulation of gemfibrozil. Both formulations met the Weibull mathematical model for drug release, however, the SNEDDS formulation had significant variations in parameters. The SNEDDS formulation had a significantly higher dissolution efficiency, proving its superiority over the commercial formulation in terms of in vitro dissolution profile. This study highlights the potential of SNEDDS for gemfibrozil as a promising alternative to enhance oral drug absorption [31].

10.5.5 Polymeric Nanoparticles (PNPs)

In recent years, there has been a lot of attention on PNPs, which are small colloidal particles. These particles have unique properties and behavior due to their small size. PNPs have been developed using a variety of synthetic and natural polymers, including proteins, sugars, and other macromolecules that are biodegradable or nonbiodegradable. PNPs are available in two main forms: nanospheres and nanocapsules. Nanospheres have a rigid core made from an insoluble polymeric matrix where the drug molecule can be directly attached to the polymer backbone, either on the surface or embedded within the matrix. Nanocapsules, on the other hand, consist of a liquid core surrounded by a cross-linked polymeric membrane that provides rigidity. These nanocapsules are especially useful for carrying lipophilic drugs due to their hydrophobic core. In addition, the outer hydrophilic part of nanocapsules can act as a stimuli-responsive block, enabling controlled drug release under specific conditions such as pH, temperature, redox, or external factors [24, 32].

To optimize the development of Poly (Lactic-Co-Glycolic Acid) (PLGA) NPs encapsulating Liraglutide, a DoE approach was used as part of the QbD method. The focus of this study was on particle size and entrapment efficiency of the NPs, along with PDI and surface charge, which were considered crucial parameters. The DoE study was conducted to screen numerous variables and identify the most influential ones with minimal experimental runs while maintaining a high level of accuracy. The Plackett-Burman Design, which consisted of 8 runs and 7 independent variables, was used with the STATISTICA 13 program, and an analysis of variance was performed. Independent variables for formulation and process were chosen based on prior risk assessment and included concentration of PLGA (X1), concentration of Liraglutide (X2), second sonication time (X3), polyvinyl alcohol concentration (X4), type of lyoprotectant (X5), lyoprotectant concentration (X6), and the ratio of external aqueous phase to organic phase (w2/o) (X7). The dependent variables chosen were particle size (Y1), PDI (Y2), entrapment efficiency (Y3), and zeta potential (Y4). The goal of optimizing the NP formulation was to reduce particle size and increase encapsulation efficiency. Although PDI and zeta potential were not considered critical quality attributes, particle size and entrapment efficiency were the primary areas of focus. The study produced spherical-shaped NPs with a uniform distribution, measuring 188.95 nm in size and achieving 51.81% encapsulation efficiency. Furthermore, Liraglutide was successfully entrapped in the PNPs without compromising its natural amorphous state or structural integrity [33].

Another example adopted a highly systematic QbD strategy to tailor formulations that meet specific requirements. The formulation design focused on complex lipid-polymer hybrid nanoparticles (LPNs), which present a significant challenge due to the intricate interplay of various factors. The objective of the study was to increase the loading of a luciferase splice-correcting phosphorothioate RNA ASO (Luc-ASO) with 2′-O-methyl modifications into LPNs as well as optimize the formulations for Luc-ASO delivery and safety in HeLa 705 cells, which harbor an abnormal luciferase gene. The assessment included splice-correction efficiency and cell viability. The DoE encompassed two independent variables: the content of lipidoid 5 and the ratio of lipidoid 5 to antisense oligonucleotides (ASO). The lipidoid 5 content ranged from 10% to 20% (w/w), while the weight ratio of lipidoid 5 to ASO ranged from 10: 1 to 30:1. A quadratic model and an I-optimal design were used during the study. The dependent response variables were z-average, PDI, zeta potential, entrapment efficiency, drug loading (physicochemical properties) as well as biological responses. For the statistical analysis, we utilized Design Expert software. We thoroughly assessed the data both numerically and graphically, creating desirability plots and overlay plots. We made the decision to exclude the z-average and PDI from the modeling, as they had no impact on the critical formulation parameters and could not be incorporated into statistically valid models. To validate the model, we chose five formulations within the DS for preparation, all of which met the predetermined QTPP for CQAs. This confirms the validity of the constructed DS [34].

10.5.6 Polymeric Micelles (PMs)

Polymeric micelles (PMs) have been extensively researched for their ability to enhance the delivery of hydrophobic drugs. These self-assembled units form from amphiphilic copolymers in aqueous solutions when they reach the critical micelle concentration (CMC). The special PMs with a core-shell structure are created due to the difference in solubility between the hydrophilic and hydrophobic parts of a copolymer. These PMs can vary in size from a few tens to several hundred nanometers. PMs have many advantages over surfactant micelles, such as improved stability, low CMC, slower break down, controlled drug release, and longer retention times. The size of PMs is critical for their ability to permeate and be retained by the body. Their inner core can hold hydrophobic drugs, while the outer shell protects them from surrounding water. Shell-forming polymers enable PMs to stay in the bloodstream longer, evading the immune system and delivering drugs to specific areas. This makes PMs an essential tool in drug delivery. It's worth noting that PMs have several advantages over liposomes. They can be easily prepared, without requiring any specific drug modifications for effective loading, and can achieve precise control of drug release [35].

A study was conducted by Gupta and colleagues to investigate the ongoing processing of polymeric micelles and evaluate CMAs and CPPs. They used a block copolymer of mPEG (5kD)-PCL (2kD) as a model drug carrier. They utilized a new co-axial turbulent jet with the co-flow continuous method to precisely control the physicochemical properties. In order to attain the optimal particle size and PDI for polymeric micelles, a rigorous study was implemented utilizing a full factorial design ($3 \times 3 \times 4$). This study scrutinized the effects of independent variables, including polymer concentration in ethanol, aqueous phase temperatures, and flow rates, on the dependent variables (CQAs). To demonstrate the efficacy of the approach, the researchers utilized curcumin, a water-insoluble drug. Polymeric micelles are commonly employed to enhance the solubility and stability of hydrophobic drugs. A further study was conducted to maximize the amount of drug that could be incorporated into the micelles. The micelles produced using this method were smaller, more uniform in size, and contained a greater concentration of the drug in comparison to those generated using an alternative technique. This conclusively proves that the continuous method is highly proficient in ensuring the production of top-quality micelles [36].

10.5.7 Nanosuspensions (NSs)

Nanosuspensions (NSs) are a new and hopeful way of delivering drugs. They involve submicron colloidal dispersion of drug particles, which are stabilized by polymers, surfactants, or a combination of the two. The size of the particles in nanosuspensions generally falls in the range of 200 to 600 nm [37]. Nanosuspensions possess small particle sizes that provide a significantly greater surface area for drugs. This translates into poorly soluble drugs dissolving at an accelerated rate. Pharmaceutical companies have extensively adopted nanosuspensions due to their highly advantageous form of drug formulation, which offers a plethora of benefits. These benefits include enhanced bioavailability, rapid onset of action, and minimal food-related effects. Nanosuspensions exhibit maximum efficacy for drugs classified as BCS Class II and IV. Moreover, this technology is produced without the use of harsh chemicals or co-solvents, making it an extremely safe and easily scalable option. It's no surprise that nanosuspensions have gained popularity within the industry [38]. Wet milling in water with a surfactant is the typical method for producing nanosuspensions. Their usage is increasingly widespread in improving drug absorption for ailments that are hard to dissolve in water, such as malaria, cancer, and heart disease [37].

Significant progress has been made in the creation of nanosuspensions through the implementation of QbD principles. A recent study was conducted to improve the critical process parameters (CPPs) for a low-solubility drug using a top-down media milling technique combined with various polymer systems. The QbD approach was utilized to thoroughly analyze specific process parameters including drug content, size of grinding media, and agitation rate (RPM). Three distinct types of polymers and a surfactant were put to the test. To generate reliable data, a DOE was carried out using a 3-factor, 2-level design, with two replicates for each formulation. Based on statistical analysis, it is evident that the RPM of the media milling process has the most significant impact on particle size. Comparatively, the drug content and the size of grinding media have less significant effects than RPM. Moreover, the selection of polymeric stabilizer plays a crucial role in the production of nanoparticles, with HPMC-based systems showing superior performance in consistently producing effective nanoparticles and preventing crystal growth during storage among different polymers. This study established the relationships between formulation parameters, enabling identification and prioritization of critical/noncritical variables related to the CQAs of nanosuspension formulations in the early stages of product development [39].

10.6 Conclusion

In recent years, there has been a growing interest in using Quality by Design (QbD) to enhance the safety, effectiveness, and stability of new drugs delivered through innovative methods. While QbD has already been implemented in traditional drug delivery systems, novel drug delivery systems are also showing great potential and some have already been approved for use. This chapter aims to help you understand how to implement the QbD approach in new drug delivery systems. It's important information for anyone interested in this field. Implementing QbD methodology can effectively address the obstacles that arise during the manufacturing process and guarantee the consistency and reliability of the process. To achieve the desired outcome in manufacturing, it's important to set clear objectives and have a deep understanding of the product and manufacturing process. Expert knowledge, experience, quality risk management, and artificial intelligence can then be used to control the process effectively. To ensure a dependable and efficient new drug delivery system, it is crucial to have an in-depth comprehension of the product, materials, processes, risk assessment, and experimental design methodologies.The product's performance is greatly influenced by factors such as stability, solubility, absorption, and bioavailability. These factors ultimately impact the clinical study.

References

1. Rahman, H. S., Othman, H. H., Hammadi, N. I., Yeap, S. K., Amin, K. M., Samad, N. A., et al. (2020). Novel drug delivery systems for loading of natural plant extracts and their biomedical applications. *International Journal of Nanomedicine, 15*, 2439–2483.
2. Faheem, A. M., & Abdelkader, D. H. (2020). Novel drug delivery systems. In *Engineering drug delivery systems* (pp. 1–16). Elsevier.
3. Agrahari, V. (2018). Novel drug delivery systems, devices, and fabrication methods. *Drug Delivery and Translational Research, 8*(2), 303–306.
4. Bobo, D., Robinson, K. J., Islam, J., Thurecht, K. J., & Corrie, S. R. (2016). Nanoparticle-based medicines: A review of FDA-approved materials and clinical trials to date. *Pharmaceutical Research, 33*(10), 2373–2387. https://doi.org/10.1007/s11095-016-1958-5
5. Li, J., Qiao, Y., & Wu, Z. (2017). Nanosystem trends in drug delivery using quality-by-design concept. *Journal of Controlled Release, 256*, 9–18. https://doi.org/10.1016/j.jconrel.2017.04.019
6. Mesut, B., Özsoy, Y., & Aksu, B. (2015). The place of drug product critical quality parameters in quality by design (QBD). *Turkish Journal of Pharmaceutical Sciences, 12*(1), 75–92. https://www.researchgate.net/publication/281925128_The_place_of_drug_product_critical_quality_parameters_in_quality_by_design_QBD%0A. http://www.embase.com/search/results?subaction=viewrecord&from=export&id=L604295641
7. ICH Q8(R2). (2009). ICH Q8(R2) pharmaceutical development. *Work Qual by Des Pharm, 8*, 28.
8. Yu, L. X., Amidon, G., Khan, M. A., Hoag, S. W., Polli, J., Raju, G. K., et al. (2014). Understanding pharmaceutical quality by design. *The AAPS Journal, 16*(4), 771–783.
9. Gibson, M., Carmody, A., & Weaver, R. (2018). Development and manufacture of drug product. *Pharmaceutical Quality by Design: A Practical Approach, 1*, 117–156. https://doi.org/10.1002/9781118895238.ch6
10. Raw, A. S., Lionberger, R., & Yu, L. X. (2011). Pharmaceutical equivalence by design for generic drugs: Modified-release products. *Pharmaceutical Research, 28*(7), 1445–1453.
11. Troiano, G., Nolan, J., Parsons, D., Van Geen, H. C., & Zale, S. (2016). A quality by design approach to developing and manufacturing polymeric nanoparticle drug products. *The AAPS Journal, 18*(6), 1354–1365. https://doi.org/10.1208/s12248-016-9969-z
12. Bansal, A. K., Balwani, G., & Sheokand, S. (2018). Critical material attributes in wet granulation. In *Handbook of pharmaceutical wet granulation* (p. 421).
13. Desai, N. (2012). Challenges in development of nanoparticle-based therapeutics. *The AAPS Journal, 14*(2), 282–295.
14. Trynda-Lemiesz, L. (2004). Paclitaxel-HSA interaction. Binding sites on HSA molecule. *Bioorganic Medicinal Chemistry, 12*(12), 3269–3275.
15. ICH Q9. (2005). Quality risk management. *Q9, Current Step, 4*, 408.
16. Beg, S., Hasnain, M. S., Rahman, M., & Swain, S. (2019). Introduction to quality by design (QbD): Fundamentals, principles, and applications. In *Pharmaceutical quality by design: Principles and applications* (pp. 1–17). Elsevier Inc. https://doi.org/10.1016/B978-0-12-815799-2.00001-0
17. Waghule, T., Dabholkar, N., Gorantla, S., Rapalli, V. K., Saha, R. N., & Singhvi, G. (2021). Quality by design (QbD) in the formulation and optimization of liquid crystalline nanoparticles (LCNPs): A risk based industrial approach. *Biomedicine & Pharmacotherapy, 141*, 111940. https://doi.org/10.1016/j.biopha.2021.111940
18. Garcia, T., Cook, G., & Nosal, R. (2008). PQLI key topics—Criticality, design space, and control strategy. *Journal of Pharmaceutical Innovation, 3*(2), 60–68.
19. Vemuri, S., & Rhodes, C. T. (1995). Preparation and characterization of liposomes as therapeutic delivery systems: A review. *Pharmaceutica Acta Helvetiae, 70*(2), 95–111.
20. Malam, Y., Loizidou, M., & Seifalian, A. M. (2009). Liposomes and nanoparticles: Nanosized vehicles for drug delivery in cancer. *Trends in Pharmacological Sciences, 30*(11), 592–599.

21. Akbarzadeh, A., Rezaei-sadabady, R., Davaran, S., Joo, S. W., & Zarghami, N. (2013). Liposome: classification, preparation, and applications. *Nanoscale Research Letters, 8*, 1–9.
22. Dhoble, S., & Patravale, V. (2019). Development of anti-angiogenic erlotinib liposomal formulation for pulmonary hypertension: A QbD approach. *Drug Delivery and Translational Research, 9*(5), 980–996.
23. Dawoud, M. H. S., Yassin, G. E., Ghorab, D. M., & Morsi, N. M. (2019). Insulin Mucoadhesive liposomal gel for wound healing: A formulation with sustained release and extended stability using quality by design approach. *AAPS PharmSciTech, 20*(4), 158.
24. Rehman, A. U., Akram, S., Seralin, A., Vandamme, T., & Anton, N. (2019). Lipid nanocarriers: Formulation, properties, and applications. In *Smart nanocontainers: Micro and nano technologies* (pp. 355–382). Elsevier Inc. https://doi.org/10.1016/B978-0-12-816770-0.00021-6
25. Hosny, K. M., El-Say, K. M., & Alkhalidi, H. M. (2018). Quality by design approach to screen the formulation and process variables influencing the characteristics of carvedilol solid lipid nanoparticles. *Journal of Drug Delivery Science and Technology, 45*, 168–176. https://doi.org/10.1016/j.jddst.2018.03.010
26. Beg, S., Jain, S., Kushwah, V., Bhatti, G. K., Sandhu, P. S., Katare, O., et al. (2017). Novel surface-engineered solid lipid nanoparticles of rosuvastatin calcium for low-density lipoprotein-receptor targeting: A quality by design-driven perspective. *Nanomedicine, 12*(4), 333–356.
27. Garg, N. K., Sharma, G., Singh, B., Nirbhavane, P., Tyagi, R. K., Shukla, R., et al. (2017). Quality by design (QbD)-enabled development of aceclofenac loaded-nano structured lipid carriers (NLCs): An improved dermatokinetic profile for inflammatory disorder(s). *International Journal of Pharmaceutics, 517*(1–2), 413–431. https://doi.org/10.1016/j.ijpharm.2016.12.010
28. Costa, C. P., Cunha, S., Moreira, J. N., Silva, R., Gil-Martins, E., Silva, V., et al. (2021). Quality by design (QbD) optimization of diazepam-loaded nanostructured lipid carriers (NLC) for nose-to-brain delivery: Toxicological effect of surface charge on human neuronal cells. *International Journal of Pharmaceutics, 607*, 120933. https://doi.org/10.1016/j.ijpharm.2021.120933
29. Zidan, A. S., Sammour, O. A., Hammad, M. A., Megrab, N. A., Habib, M. J., & Khan, M. A. (2007). Quality by design: Understanding the product variability of a self-nanoemulsified drug delivery system of cyclosporine A. *Journal of Pharmaceutical Sciences, 96*(9), 2409–2423.
30. Panigrahi, K. C., Patra, C. N., & Rao, M. E. B. (2019). Quality by design enabled development of oral self-nanoemulsifying drug delivery system of a novel Calcimimetic Cinacalcet HCl using a porous carrier: In vitro and in vivo characterisation. *AAPS PharmSciTech, 20*(5), 1–17.
31. Villar, A. M. S., Naveros, B. C., Campmany, A. C. C., Trenchs, M. A., Rocabert, C. B., & Bellowa, L. H. (2012). Design and optimization of self-nanoemulsifying drug delivery systems (SNEDDS) for enhanced dissolution of gemfibrozil. *International Journal of Pharmaceutics, 431*(1–2), 161–175. https://doi.org/10.1016/j.ijpharm.2012.04.001
32. Borandeh, S., van Bochove, B., Teotia, A., & Seppälä, J. (2021). Polymeric drug delivery systems by additive manufacturing. *Advanced Drug Delivery Reviews, 173*, 349–373. https://doi.org/10.1016/j.addr.2021.03.022
33. Ismail, R., Sovány, T., Gácsi, A., Ambrus, R., Katona, G., Imre, N., et al. (2019). Synthesis and statistical optimization of poly (lactic-co-glycolic acid) nanoparticles encapsulating GLP1 analog designed for oral delivery. *Pharmaceutical Research, 36*(7), 99.
34. Thanki, K., Papai, S., Lokras, A., Rose, F., Falkenberg, E., Franzyk, H., et al. (2019). Application of a quality-by-design approach to optimise Lipid-polymer hybrid nanoparticles loaded with a splice-correction antisense oligonucleotide: Maximising loading and intracellular delivery. *Pharmaceutical Research, 36*(3), 37.
35. Amin, M. C. I. M., Butt, A. M., Amjad, M. W., & Kesharwani, P. (2017). Polymeric micelles for drug targeting and delivery. In *Nanotechnology-based approaches for targeting and delivery of drugs and genes* (pp. 167–202). Elsevier Inc. https://doi.org/10.1016/B978-0-12-809717-5.00006-3
36. Gupta, A., Costa, A. P., Xu, X., Lee, S. L., Cruz, C. N., Bao, Q., et al. (2020). Formulation and characterization of curcumin loaded polymeric micelles produced via continuous processing.

International Journal of Pharmaceutics, 583, 119340. https://doi.org/10.1016/j.ijpharm.2020.119340

37. Singh, B., Saini, S., Lohan, S., & Beg, S. (2017). Systematic development of nanocarriers employing quality by design paradigms. In *Nanotechnology-based approaches for targeting and delivery of drugs and genes* (pp. 110–148). Elsevier Inc. https://doi.org/10.1016/B978-0-12-809717-5.00003-8
38. Verma, S., Lan, Y., Gokhale, R., & Burgess, D. J. (2009). Quality by design approach to understand the process of nanosuspension preparation. *International Journal of Pharmaceutics, 377*(1–2), 185–198.
39. Ghosh, I., Schenck, D., Bose, S., Liu, F., & Motto, M. (2013). Identification of critical process parameters and its interplay with nanosuspension formulation prepared by top down media milling technology—A QbD perspective. *Pharmaceutical Development and Technology, 18*(3), 719–729.

Chapter 11
Quality-By-Design Strategy for Developing Novel Herbal Products

Pawan Devangan, Gopal Bajad, Soham Loharkar, Nitin Wadate, Spandana Gollapalli, Tathagata Dutta, Ashish Baldi, Snigdha Singh, Anju Katyal, Ramesh Chandra, Preet Amol Singh, and Jitender Madan

Abstract A quality control strategy for herbal materials, taking into account the alignment between process characteristics and material quality attributes. The application of this approach holds significant potential for ornamental the building of quality herbal products. The concept of Quality by Design (QbD) has emerged as a dominant framework in the pharmaceutical industry to ensure the quality, safety, and efficacy of herbal products, as well as optimization of product development, manufacturing, and process control to elevate overall quality. In recent years, QbD principles have found application in the development and manufacturing of herbal formulations due to its emerging popularity and demand for natural remedies. This chapter explores the implementation of QbD for herbal products, highlighting its

P. Devangan · G. Bajad · S. Loharkar · N. Wadate · S. Gollapalli · J. Madan (✉)
Department of Pharmaceutics, National Institute of Pharmaceutical Education and Research, Hyderabad, Telangana, India

T. Dutta
JodasExpoim Private Limited, Hyderabad, Telangana, India

A. Baldi
Department of Pharmaceutical Science and Technology, Maharaja Ranjit Singh, Punjab Technical University, Bathinda, Punjab, India

S. Singh
Department of Chemistry, University of Delhi, Delhi, India

A. Katyal
Dr. B.R Ambedkar Centre for Biomedical Research, University of Delhi, Delhi, India

R. Chandra
Department of Chemistry, University of Delhi, Delhi, India

Dr. B.R Ambedkar Centre for Biomedical Research, University of Delhi, Delhi, India

Institute of Nano Medical Science, University of Delhi, Delhi, India

Maharaja Surajmal Brij University, Bharatpur, Rajasthan, India

P. A. Singh
University Centre of Excellence in Research, Baba Farid University of Health Sciences, Faridkot, Punjab, India

N. K. Jain, N. Bajwa (eds.), *Introduction to Quality by Design (QbD)*,
https://doi.org/10.1007/978-981-99-8034-5_11

importance in surmounting the challenges posed by the inherent variability of natural ingredients. A profound comprehension of the identification of critical quality attributes (CQAs), which represent measurable characteristics influencing its performance, and the critical process parameters (CPPs) specific to herbal products, encompassing the variables affecting the manufacturing process, is indispensable for the effective implementation of QbD principles. Numerous examples showcase the successful application of QbD in optimizing the manufacturing process of herbal formulations. Furthermore, this chapter presents regulatory aspects and guidelines related to QbD for the herbal formulation, advancements, and future directions in QbD for herbal products, shedding light on potential areas for further research and improvement. Overall, this chapter serves as a comprehensive guide for researchers, formulators, and industry professionals interested in leveraging QbD principles to enhance the quality and reliability of herbal formulations.

Keywords Herbal formulation · Quality by design · Regulatory aspects · Critical quality attributes · Critical process parameters

Abbreviations

API	Active pharmaceutical ingredients
AYUSH	Ayurveda, yoga & naturopathy, unani, siddha, and homoeopathy
BBB	Blood brain barrier
BBD	Box behnken design
CCD	Central composite design
CCS	Croscarmellose sodium
CMA	Critical material attributes
CPP	Critical process parameter
CQA	Critical quality attributes
CSD	Colloidal silicon dioxide
DBP	Di-butyl phthalate
DoE	Design of experiment
DT	Disintegration time
EC	Ethyl cellulose
EMA	European medicines agency
FMEA	Failure mode and effect analysis
FTIR	Fourier transform infrared
GMP	Good manufacturing practices
HE	Hypoglycaemic effect
HPLC	High-performance liquid chromatography
HPMC	Hydroxy propyl methyl cellulose
ICH	International council for harmonisation
PAT	Process analytical technologies
PCL	Poly ε-caprolactone
PDI	Polydispersity index

PVA	Polyvinyl alcohol
PVPK-30	Polyvinyl pyrrolidone K30
QbD	Quality by design
QRM	Quality risk management
QTPP	Quality target product profile
RSM	Response surface methodology
UV	Ultra-violate
WHO	World Health Organization
WV	Weight variation

11.1 Introduction

Herbal formulations refer to a pharmaceutical dosage form that incorporates one or more phytoconstituents or processed herbs in precise quantities, designed to confer targeted nutritional or cosmetic advantages. These formulations are intended for employment in the diagnosis, treatment, or mitigation of diseases in human beings or animals, as well as to induce modifications in the structure or physiology of human beings or animals [1]. In addition, herbal formulation encompasses the systematic progression of developing a product or composition by harnessing diverse herbs and botanical constituents. This intricate process entails the deliberate selection, harmonious amalgamation, and meticulous processing of specific herbs, yielding a medicinal or therapeutic formulation. These formulations manifest in diverse modalities, such as capsules, tablets, powders, tinctures, syrups, creams, ointments, and lozenges intended to administer the desired therapeutic effects of the herbal constituents in a manner that is both convenient and efficacious [2]. The fundamental objective of these formulations frequently revolves around harnessing the inherent potential of the natural compounds present in herbs to promote health and well-being [3, 4].

The expanding usage of herbal formulations is recognized to various reasons, involving; inadequate accessibility of conventional formulations, particularly in the rural area, increased reports of failure and adverse events allied with conventional formulations; perceived herbal medication affordability, prevalent availability, inexpensive, and perceived safety and efficacy of herbal formulation. Subsequently, there has been a notable upsurge in the societal inclination towards accessible and superior herbal compositions in response to public demand [5].

Quality by Design (QbD) is scientifically defined as a systematic methodology for product development that initiates with pre-established objectives and prioritizes comprehension of the product and process, along with process control, founded on robust scientific principles and effective management of quality risks [6]. To establish effective synchronization among the quality attributes of raw materials, in-process materials, and final products, development of dependable, precise, and sensitive quality control methods has become imperative. This necessitates utilization of a combination of traditional and advanced instrumental approaches of analysis, enabling robust and accurate assessment of product quality throughout the manufacturing process.

The experimental design encompassed the selection of research objects, reagents, and extractants, as well as the determination of extraction process parameters, chromatography parameters, and initial pharmacological assay. To streamline the experimentation process, a final screening design was employed to examine the impact of these parameters on the manufacturing process. This approach effectively reduced the number of required experiments, enabling the study of numerous process parameters with minimal trials [7].

This chapter intends to evaluate the significance of QbD in enhancing the understanding of the development process for herbal drug products or formulations. In addition, discussion on identification of critical quality attributes (CQA) and critical process parameters (CPP) in case of herbal product development, has also been incorporated. Moreover, the usage of different types of QbD models such as Box Behnken design (BBD) and Central Composite Design (CCD) with usage of surface response methodology for herbal formulation with specific examples and case studies. Furthermore, risk assessment, design space, process understanding, optimization, control strategy and real-time quality assurance for herbal products are elaborated with specific point. Apart from that lighting on regulatory aspects, guidelines related to QbD for herbal formulation, challenges and opportunities in implementing QbD for herbal formulations.

11.1.1 Significance of QbD in the Development of Herbal Products

QbD principles are employed during the development of herbal products to secure consistent quality, enhance scientific comprehension, mitigate risks, optimize processes, and meet regulatory requisites [8]. Due to the intricate nature of their natural constituents, herbal products often demonstrate variability in composition and therapeutic effects. By implementing QbD, researchers can systematically design and regulate critical parameters that impact product quality, safety, and efficacy. QbD facilitates the identification of CQAs and their association with formulation characteristics and process parameters, thereby deepening the scientific understanding of herbal products as shown in Fig. 11.1. This enhanced understanding aids informed decision-making during formulation development and enables the optimization of manufacturing processes to achieve consistent quality and performance. QbD also emphasizes the assessment and management of risks throughout the product lifecycle, thereby mitigating potential issues linked to the variability of herbal ingredients and manufacturing processes. By adhering to regulatory expectations and guidelines, QbD supports compliance with complex regulatory frameworks. In conclusion, QbD fosters a systematic and efficient approach to the development of

Fig. 11.1 Different QbD approaches for quality assurance in herbal products

herbal products, resulting in high-quality products that meet regulatory standards and deliver dependable therapeutic outcomes.

11.2 Critical Quality Attributes (CQAs) in Herbal Products

CQAs are physical-chemical or microbiological specifications of the formulation or products which have a significant impact on the product's quality along with its safety and efficacy. CQAs are crucial variables that must be regulated and tracked in order to ensure that the finished product continually satisfies the necessary quality standards. ICH Q8 (R1) defines CQAs as "physical, chemical, biological or microbiological properties or characteristics that should be within an appropriate limit, range, or distribution to ensure the desired product quality". During the herbal product development, many excipients are added to modify desired properties or to inhibit the undesired properties of phyto-active constituents. The risks related to CQAs posed by the process and the formulation ingredients (phyto-bioactive and excipients) are discovered, examined, and then evaluated. Finally, by monitoring and controlling these critical attributes safety efficacy and desired quality levels can be ensured and hence CQAs play an important role in herbal pharmaceuticals.

CQAs of herbal products are generally those aspects that affect the purity of product, drug release, its strength and stability. CQAs generally involve lots of physicochemical quality attributes and product-specific aspects. In herbal dosage forms hardness, weight variation, disintegration time, friability and content uniformity in herbal tablets [9]; particle size distribution, bulk density, moisture content, flowability, dispersibility, impurity level, phyto-active contents in herbal powder; fermentation time, alcohol content in asava; consistency and texture of lepa; specific gravity, viscosity in taila; are considered as CQAs related to conventional herbal formulations. In novel herbal formulation entrapment efficiency, particle size and particle size distribution, zeta potential, percentage drug release, drug loading, yield are considered as CQAs [10, 11]. To achieve these CQAs, CPP and critical material attributes (CMA) and must be studied and optimized.

For instance, *Rosmarinus officinalis* L. is one of the most promising plants of the Lamiaceae family having antioxidant and anti-inflammatory activity as well as synergic activity to inhibit the acetylcholinesterase and butyrylcholinesterase; but due to low water solubility and bioavailability, its clinical use is inadequate. *Shalabalija* et al., formulated PEGylated liposomes of rosemary extract which were optimized using a CCD for the treatment of Alzheimer's disease. The main aim of the study was to obtain optimized formulation in terms of the biopharmaceutical and physicochemical properties like particle size, maximum drug encapsulation, efficient BBB transport and controlled release of active constituents in the brain. In optimization, 20 experiments were performed in which three independent variables such as the amount of rosemary extract, soya lecithin and polyethyleneglycol (PEG) considered as critical formulation parameters and particle size, absolute amount of encapsulated rosemary extract and absolute amount of rosemary extract released in 1, 4, 6, 24 h as a response. The optimized liposomal formulation was found to have a D50 value in a range from 120 $\pm$ 1.65 to 133 $\pm$ 2.60 nm with narrow unimodal size distribution and high entrapment efficiency (~90%) [12].

In another example, *Savic* et al., developed and optimized anti-wrinkle emulsion serum of *Acmella oleracea* extract. In which, CCD was employed to study simultaneous influence of two independent factors emollient and emulsifier concentration on four different response variables i.e., apparent viscosity, loss factor, yield point and structural changes. The QbD design study revealed the significant effect of emollient and emulsifier concentration on critical rheological attributes of emulsion. Different emollient concentrations from 10 to 20%, emulsifier concentration from 1 to 3% was evaluated and formulation containing emollients 15% and emulsifier 1% was chosen as optimized formulation. Finally, the authors concluded that the results of *A. oleracea* extract containing emulsion serum showed promising potential effects as well as it is a safe, effective and non-invasive herbal anti-wrinkle product [13].

11.2.1 Identification and Definition of CQAs for Herbal Products

CQAs are determined on the basis of intended use of the herbal product as well as regulatory requirements. These critical attributes are identified by a combining historical data, scientific knowledge, risk assessments, and experimental studies. Once these CQAs identified, serve as critical checkpoints during the formulation development, manufacturing process, and quality control processes. The CPPs and

CMAs are the process-related and materials-related variables whose variability has an effect on the CQA and should be tracked or regulated to make sure the process output has the necessary quality. According to ICH Q8 guideline, CPP are the specific parameters within the manufacturing process that directly influence a CQA and, as a result, require monitoring or control to ensure the desired product quality. CMA refers to a physical, chemical, biological, or microbiological property or characteristic of an input material that must fall within a suitable limit range or distribution to guarantee the desired quality of the output material. CMA is not defined in the ICH guidelines but these material-related variables or attributes create a very huge impact on the final product.

Shariare et al., formulated the liposomes for the delivery of *Aphanamixis polystachya* leaf extract and this liposomal formulation was optimized by Design of experiment (DoE) with four factors at two levels [14]. In optimization, researchers monitored the injection rate, stirring speed, cholesterol: phospholipid ratio and drug in solvent were as CPP and their impact on particle size and polydispersity index was considered as preferable quality attributes. The liposomal delivery system of *Aphanamixis polystachya* leaf extract was effectively optimized by desirable quality attributes like average particle size and poly dispersity index. DoE study depicted that quality attributes were dependent on interactions between process parameters and material attributes compared to their individual effects. The stirring speed was most common CPP that significantly affected particle size and drug in the solvent was a CMA that significantly affected the polydispersity index of liposomes. Similarly, *Shariare* et al., have also developed the liposomes of *Corchorus olitorius* leaf extract by employing full factorial design [15]. Process parameters like injection rate, stirring speed, stirring time and temperature as well as material attributes like buffer pH, phospholipid and cholesterol ratio, drug content, drug phospholipid and cholesterol ratio were assessed at two levels for their effect on particle size and polydispersity index of liposomes. It was found that, drug: lipid, cholesterol ratio and buffer pH have major impact on particle size of liposomes. Liposomal prepared by using high pH buffer i.e., 7.4 and a high lipid and cholesterol ratio showed low particle size of the liposome. In addition, isolation processing temperature and buffer phospholipid: cholesterol and drug: lipid and cholesterol ratio found to have a significant impact on PDI of liposomes.

Nandhini et al., developed and optimized the phytosomes of vasaka by using BBD. In optimization, independent variables or CPPs/CMAs were phosphatidylcholine concentration, stirring temperature and stirring time and particle size (nm), percent yield (%) and entrapment efficiency (%) of formulation were evaluated as quality attributes. The results showed optimal quality attributes at 1:1 extract and phosphatidylcholine ratio, stirring for 120 min at 146 °C. The optimized vasaka phytosomes showed 95.89% yield, 231.0 nm particle size, and 95.48% entrapment efficiency [16].

11.2.2 Importance of CQAs for Ensuring Product Quality and Efficacy

QbD is a multipurpose tool that helps in many situations like screening of variables, optimization, transfer function identification, design for comparison, etc. [17]. One such fruitful use of QbD is to monitor the interaction of different chemicals, ingredients and phytopharmaceuticals to monitor product quality and efficacy. Three-dimensional response surface methodology (RSM) is an applicable tool for optimizing various ratios and also provides a complete description about combination effect. *Monton et al.,* studied the interaction of herbal ingredients of Trikatuk recipe by using simplex lattice experimental design [18]. Trikatuk is Thai traditional medicine used as anti-flatulent, mucolytic and in the management of runny nose. Trikatuk recipe is composed of three medicinal plants, i.e., rhizomes of *Zingiber ligulatum*Roxb., fruits of *Piper nigrum* L., and Infructescences of *Piper retrofractum* Vahl. The authors monitored the weight ratio of *Z. offcinale, P. nigrum, and P. retrofractum* as independent variables and extraction yield, 6-gingerol content and piperine content as three response variables. The objective of this study was to predict the interaction of herbal ingredients in Trikatuk recipe. The 12 herbal formulas obtained by simplex lattice experimental design was extracted by maceration technique and analyzed by high-performance liquid chromatography (HPLC) to quantify piperine and 6-gingerol content. Authors concluded that the level of interaction in an equal weight ratios of *Z. offcinale, P. nigrum, and P. retrofractum* was 200% higher which suggests that the use of all these three plants in equal ratios are more efficacious than other ratios.

Uncaria tomentosa also identifiedas 'cat's claw' is widely used for the treatment of various health problems like inflammation, fever, viral infection and cancer. *Ribero et al,* significantly improved the anti-tumorigenic activity of *Uncaria tomentosa* extract by formulating optimized poly ε-caprolactone (PCL) nanoparticles by using three factors, three-level BBD [19]. The effect of three independent variables i.e., concentration in the organic phase, polyvinyl alcohol (PVA) concentration in aqueous phase and aqueous phase to organic phase ratio on four different variable responses i.e., entrapment efficiency (%), nanoparticles diameter (nm), polydispersity index, zeta potential (mV) was determined. The BBD was successfully employed for the optimisation of nanoparticle formulation. The optimum formulation condition ethyl acetate: acetone (3:2) mixture (5 mL), 150 mg polycaprolactone with 8.5 mL of 1.0% w/v PVA solution as aqueous phase showed 81.6% entrapment efficiency, 247.3 nm mean particle size and 0.062 polydispersity index. The optimized desired quality attributes of nanoparticles would be able to ensure the tumor targeting of *U. tomentosa* alkaloid.

11.2.3 Response Surface Design for Herbal Formulation

Response surface design can be elaborated as the collective collaboration of the mathematical and statistical techniques which further assist in the optimization of the process parameters as well as the formulation variables and restricts to the determination of quadratic effects. The RSM framework provides access to BBD, CCD, and three-level designs within the Design Expert software [20].

The CCD design remains the most frequent design used for the optimization of pharmaceutical formulations as it is well accommodated to the current state of art. The design remains of a choice as they provide several variants such as face-centered, cubic and hybrid designs. The design considers a sequential approach that remains unaffected by the missing data, also previously conducted factorial screening can be well integrated into a CCD design [21]. Literature demonstrates several screening designs including full factorial and fractional factorial which report CCD as the final optimization design [22, 23]. BBD is another design that is widely used for the purpose of optimization of a design. This design allows for estimation of the parameters in a quadratic fashion. Furthermore, BBD design remains a little advantageous over CCD, as CCD requires experimental study of factors at five different levels while the former requires only three levels [24].

RSM designs including BBD and CCD have been widely used in the optimization of pharmaceutical formulations including conventional as well as nanotechnology-based formulations and have given rise to fruitful outcomes in determining the best formulation. Considering the number of critical factors present in the herbal formulations the RSM designs are of great advantage as they offer the study of factors at varying number of levels. Owing to which literature has witnessed RSM-optimized herbal formulations including herbal extract gel [25], oral herbal disintegrating tablets [26, 27], herbal extract effervescent tablets [28], polyherbal phytosomal gel [29] to name a few.

The literature describes the formulation of a transdermal film containing curcumin, an herbal drug used for managing osteoarthritis, using the BBD. The design focused on the formulation variables as the type of polymers hydroxy propyl methyl cellulose (HPMC), di-butyl phthalate (DBP) and ethyl cellulose (EC) at three different levels (low, medium and high), effect of studied on the responses film elongation (R1) and drug release (R2). The polymers were selected based on their significantly influence the two responses. The design suggested twenty runs in which the concentration of polymers varied. The effects of the factors on the R1 revealed that both HPMC as well as DBP had a significant positive impact on R1 where increase in HPMC led to an increase in the tensile strength of the film which is also supported by the response surface graphs and contour graphs. Overall output justifies the film with higher tensile strength is obtained as the proportions of HPMC and DBP increased. On considering the R2 it can be observed that both HPMC as well as DBP significantly influence the response R2 positively. The increase in the concentration of HPMC as well as DBP causes an increase in the drug release as supported by the response surface graphs, on the contrary it is seen that the rise in the concentration of EC causes a decrease in the drug release. The results revealed that the increase in concentration of DBP causes increased film elongation values. Furthermore, increase in the concentration of ethyl cellulose was found to decrease the drug release moreover increased HPLC concentration led to increase in the drug release. The optimized batch suggested by the design showed an exceptional in-vitro drug release of 80.03% and 94.35% of % film elongation [30].

In another study, authors report the formulation of an oral herbal tablet containing *Merremia tridentata* extract for the management of diabetes. The design which was optimized using CCD focused on the variation in the concentration of the formulation variables including polyvinyl pyrrolidone K30 (PVPK-30) X_1, aerosil-colloidal silicon dioxide (CSD) X_2 and croscarmellose sodium (CSD) X_3 at five different levels. The effect of these factors was studied on the responses which included in vivo hypoglycaemic effect (HE), disintegration time (DT), average percent dissolution (D), friability (F), hardness (H), weight variation (WV), and thickness (T). The design of experiment suggested twenty different runs with varying factors. The HE was seen to be higher when the levels of CCS and CSD increased as they positively influence the solubility and absorption of active constituents leading to increased, HE. In case of the DT and dissolution, it shown that an increase in the CCS causes a decrease in DT and an increase in dissolution on the other hand an increase in amount of CSD leads to decrease in the DT as well as an increase in dissolution. In addition to this reduced concentration of PVPK-30 is depicted to decrease DT and substantially increase dissolution. In the case for the response friability, it can be observed that an increase in concentration of both CCS and CSD cause an increase in the friability while an increased concentration of PVPK30 leads to decreased friability. The design outlined the optimum formulation with 4.00% wt/wt PVP K30,13.82% wt/wt CCS, and 2.56% wt/wt CSD which led to HE of ↓69.58% at day 15, dissolution of 98.94%, and DT of 501 s [31] Table 11.1.

11.2.4 Factors Influencing CQAs in Herbal Products

Herbal products have CQAs that are distinct characteristics significantly affecting the final product's safety, efficacy, and quality. The influence of various factors on CQAs in herbal products is crucial for successful formulation development and manufacturing processes. Variability in phytoactives, herbal products have their efficacy from natural phyto-active ingredients and factors like geographical origin, soil conditions, climate, harvesting practices, and post-harvest handling contribute to the variability of phyto-actives. The variability in these botanicals can affect the CQAs of herbal products. Moreover, botanical sources and species, many phyto-active have different botanical sources which may contain a different concentration of active components. This variability may affect the potency and therapeutic effects of herbal products hence, confirming correct species is crucial for consistent product quality. Additionally, the extraction method also significantly affects the composition and concentration of phyto-actives in the herbal product. Different extraction

Table 11.1 QbD-based herbal formulation development

S. No.	Formulation	QbD	Factorial design	Independent variable	Response	References
1	Nanostructured lipidic carriers of *Smilax China* and *Salix alba*	Box-Behnken design	Three factors at three levels	(1) Sonication time (2) Concentration of lipid (3) Surfactant	(1) Particle size (2) Polydispersity index (3) Entrapment efficiency	Qadir et al. [32]
2	Nanoemulsion of curcumin	Full-factorial design	Four factors at two levels	Concentration of (1) Oil (2) Surfactant and co-surfactant (3) Chitosan	(1) Globules size (2) Zeta potential	Sood et al. [33]
3	Self-micro emulsifying drug delivery systems of Rotundine	Central composite face design	Two factors at three levels	Concentration of (1) Capryol 90 (2) Cremophor RH 40 and Transcutol HP mixture	(1) Droplet size (2) PDI (3) Dissolution efficiency	Tung et al. [34]
4	Sustained release tablet containing ivy leaf extract	Response surface design	Two factors at three levels	(1) Concentration of polymer (2) Types of polymers	Dissolution of Hederacosides-C at (1) 6 h (2) 12 h (3) 24 h	Na et al. [35]
5	Lecithin/chitosan Nanoparticle of curcumin	Box-Behnken design	Three factors at three levels	(1) Stirring speed Concentration of (2) Chitosan (3) Lecithin	(1) Average particle size (2) Optimal zeta potential (3) Maximum entrapment efficiency	Walbi et al. [36]
6	Lipid-based oral formulation of curcumin (LBOF)	Box-Behnken design	Three factors at three levels	Concentration of (1) Gelucire 44/14	(1) Solubility of CRM in *LBOFs*	Pawar et al. [37]

(continued)

Table 11.1 (continued)

S. No.	Formulation	QbD	Factorial design	Independent variable	Response	References
				(2) Labrasol (3) TPGS	(2) Droplet size of diluted *LBOFs*	
7	Herbal transdermal of *Lysiphyllum strychnifolium*	Response surface design	Four factors at three levels	(1) Concentration of ethanol (2) Power (3) Extraction time (4) Cycle	(1) Extraction yield (%) (2) Astilbin content (%)	Monton et al. [38]
8	Vaginal suppository of curcumin	Mixture design	Three factors at two levels	Concentration of (1) PEG-1500 (2) PEG-6000 (3) Poloxamer-407	(1) Hardness (2) Deformation time (3) Drug release (%)	Ghazwani et al. [39]
9	Nanoemulgel containing-quercetin	Box-Behnken design	Three factors at three levels	Concentration of (1) Oil (2) Surfactant (3) Co-surfactant	(1) Mean globule size (2) PDI (3) Entrapment efficiency	Chitkara et al. [40]
10	Herbal tablets containing *Merremia tridentata*	Central composite face design	Three-factor at five level	Concentration of (1) PVPK-30 (2) Croscarmellose sodium (3) Colloidal silicon dioxide	(1) Hypoglycemic effect (2) Disintegration time (s) (3) Average dissolution (%) (4) Friability (%) (5) Hardness (N) (6) Weight variation (%) (7) Thickness (mm)	Canh Pham et al. [31]
11	Tablet and capsule containing-*Satureja kitaibelii*	Box-Behnken design	Three factors at three levels	(1) Ethanol concentration (2) Extraction temperature (3) Extraction time	(1) Extraction yield (2) Total phenolic content	Mudrić et al. [41]

					(3) Rosmarinic acid content (4) Clinopodic acid content	
12	Emulsion based Nanotailored gel of Silymarin	D-optimal mixture design	Three factors at two levels	Concentration of (1) Oil (2) Surfactant and co-surfactant (3) Water	(1) Globule size (2) Drug release (%)	Dhawan and Nanda [42]

methods like supercritical fluid extraction, maceration and percolation result in different CQAs [43]. Furthermore, formulation design aspects like selection of excipients and processing methods have greater impact on the CQAs of herbal products. Compatibility between phyto-active and excipients as well as manufacturing process affects the bioavailability and stability of herbal products [44]. Similarly, process parameters during manufacturing such as mixing speed, temperature, drying time, and pressure affect CQAs. Therefore, these parameters should be monitored or controlled to ensure the desired qualities related to safety and efficacy. In addition, storage conditions, such as temperature, light exposure and humidity, can affect the shelf life and stability of herbal products hence, appropriate storage is necessary to maintain CQAs over time. Identity, potency and purity of herbal ingredients are very crucial for reliable product quality hence, quality of raw materials used in herbal products have direct influence on CQAs. Contamination of raw materials with foreign substances, such as pesticides, microorganisms and heavy metals, can negatively affect CQAs. Rigorous testing and adherence to quality standards are essential to ensure product purity [2].

11.3 Risk Assessment and Design Space

Risk assessment is a crucial step for the herbal product development by QbD as well as a precursor for suitable design space studies. Risk assessment links material attributes like moisture content, particle size, density, etc. and process-related parameters like rate of drying, temperature, mixing speed, etc., to the specified quality attributes of the products. Risk assessment is a scientific procedure that is part of quality risk management; it helps in determining which CMAs and CPPs could have an impact on product CQAs. When these parameters are identified, higher-level process understanding is done by using different mathematical tools [45].

Risk assessment and management are commonly done by simple method like check sheets, flow charts, process mapping, cause and effect diagrams, etc. Process mapping technique relates CMA, CPP and CQA to response surface resultant from experimental data. For example, product specification development to assure bioequivalence with the help of acceptable specifications related to dissolution of the product [46, 47]. The Ishikawa diagram also known as a fishbone diagram or cause-effect diagram represents all the aspects that influence product quality. This diagram has horizontal lines pointing toward the product's quality and diagonal lines representing major influencing factors. The influence of CPP and CMA is represented on sublines of the diagonal lines [48]. Ishikawa diagram allows the visualization and concurrently shows parameters that have the potential to affect the product quality, however, this method is not able to determine the association of parameters [49]. Failure mode effect analysis is another tool that identifies inadequacies in product development. It is a proactive systemic and qualitative risk analysis that detects, ranks, and rates all possible issues, as well as the consequences of those issues. This method thoroughly identifies, prioritizes and estimates the risk as well as prevents failure [50] (Fig. 11.2).

Gaonkar et al., developed herbal antioxidant-rich supplement by QbD approach [51]. Researchers have used the extract of *Curcuma longa* i.e., turmeric, *Camelia sinensis* i.e., green tea, *Piper nigrum* i.e., black pepper and *Emblica officinalis* i.e., amla as potent antioxidants and compressed them into a tablet by wet granulation method by employing CCD. Before starting experimentation, *Gaonkar* et al., performed a risk assessment of herbal extract and excipient variables among which amla extract was found as critical input material. The risk assessment of material attributes related to amla extract (moisture content, flow property, hygroscopicity, residual solvent, impurity) and formulation variable (starch level, croscarmellose sodium level, talc level, lactose level) over CQAs (friability and disintegration) were performed. Attributes related to amla extract showed that hygroscopicity of amla extract has a high risk on the disintegration time of the tablet and rest all attributes have low risk of disintegration and friability of tablet. Attributes related to excipient showed that lactose and talc level have low risk while starch which is used as a binder has a high risk on both CQAs and the level of CSS have a high risk of disintegration and medium risk on friability. The attributes having low risk are acceptable and do not require further investigation in contrast, high-risk attributes are unacceptable and need further investigation to lower the risk. After risk assessment study formulation parameters having a greater impact on CQAs studied in detail and selection of these formulation parameters within the proposed design space, higher risk can be converted to lower risk.

Currently, pharmaceutical manufacturers are under a regulatory burden to quantify the robustness and reliability of their product development or manufacturing process. However, QbD has given prospects to improve the reliability and robustness of the product development or manufacturing process [52]. ICH Q8 guidelines have introduced the design space concept in which changes in the manufacturing process and formulation can be implemented without additional regulatory approval. ICH Q8 defines design space as "the multidimensional combination and interaction of input variables (e.g., material attributes) and process parameters that have been demonstrated to provide assurance of quality" (ICH Q8). Design space is proposed by applicant who sets the limits, within these limits process parameters and input variables can be changed to improve product quality and process capability without additional regulatory approval. Changes outside the specified design space parameter values require post-approval changes [53]. *Benjasiri mongkol* et al., developed the redispersible dry emulsion of resveratrol by applying design space to an optimized spray-drying process and formulation [54]. The BBD employing RSM was used for optimization of formulation and relationship between independent attributes and responses demonstrated using design space which determines independent attributes ranges in which reliable responses will be achieved. The optimized formulation prepared with a concentration of 2.75% low methoxy pectin and 7% caprylic glyceride, spray dried at 10.1 mL/min pump speed in design space limit achieved all requirements of good flow, small redispersed size, and high spray drying efficiency.

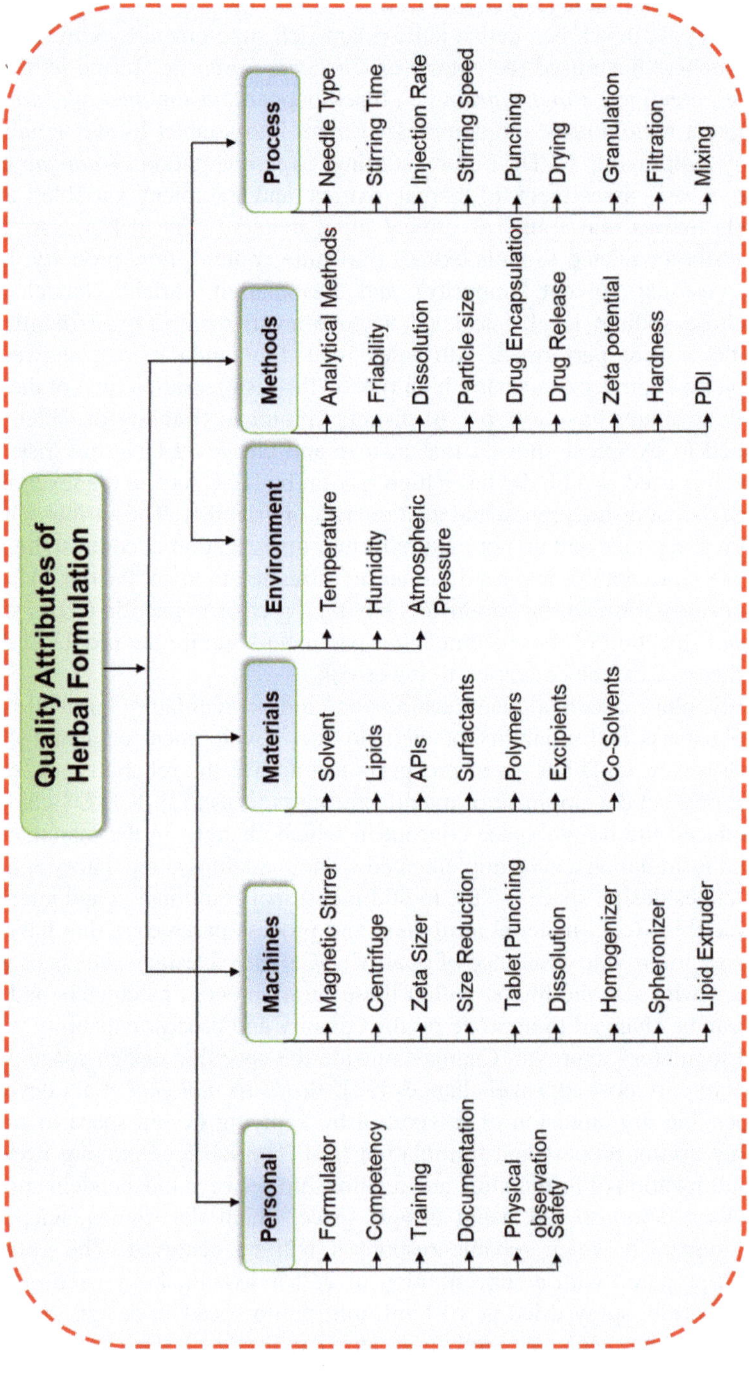

Fig. 11.2 Different quality attributes related to herbal formulations

11.4 Process Understanding and Optimization

Herbal product development is a complex process involving many multivariate unit operations. Owing to complex chemical constituents of plant-based products, it is very difficult to establish a relationship between CMAs, CPPs and CQAs. Traditionally, herbal product process development or manufacturing process is majorly experienced based on which process variables runs without controlled strategies and effect of these process variables on CQAs is not well understood. Hence, an enhanced process understanding of relationship between CMAs, CPPs and CQAs is very important to improve herbal product development and manufacturing performance to achieve desired product quality [26, 27]. Recently, QbD is playing a vital in the development of herbal products with robust processes and aims to bypass trial and error approaches as well as direct the process to a controlled and predictable environment. QbD helps to ensure the quality of herbal products through understanding and controlling the manufacturing or formulation variables [55, 56].

Zhang et al., demonstrated the use of QbD to improve the process understanding of herbal product manufacturing by taking the example of danshen i.e., *Salvia miltiorrhiza Bunge* [26, 27] in which, researchers employed RSM of BBD having four center points. Ethanol precipitation which is a widely used unit operation for herbal product manufacturing employed to demonstrate the use of QbD and recovery of four phytoactive (danshensu, rosmarinic acid, protocatechuic aldehyde and salvianolic acid B) and removal of saccharides used to determine ethanol precipitation performance. Ethanol consumption, density of concentrate and settling temperature were evaluated as critical variables through failure analysis and risk assessment. After performing 16 runs given by QbD researchers observed that operation conditions promoting removal of saccharides cause more loss of phytoactives. More dense concentrate significantly affected recovery of each phytoactive while ethanol consumption showed significance on recovery of danshensu, protocatechuic aldehyde and salvianolic acid B but not on rosmarinic acid and increased settling temperature showed higher recovery of danshensu and salvianolic acid B. Similarly, ethanol consumption mainly affected the saccharides recovery, followed by concentrate density. The results facilitated a better understanding of corelation between multiple factors like CMAs and CPPs and ethanol precipitation as well as demonstrated QbD as a powerful tool for manufacturing and development of herbal products.

11.5 Control Strategy for Herbal Products

11.5.1 Setting Appropriate Specifications for Herbal Product Quality Attributes

Setting appropriate specifications for herbal formulation quality attributes is vital to confirm the safety, efficacy, and consistency of herbal formulations. Specifications include the criteria and limits that must be met throughout herbal product development and evaluation. Numerous key considerations are necessary for this process. Firstly, aligning specifications with regulatory guidelines specific to herbal products is essential for market approval and compliance. Identifying and prioritizing CQAs that significantly impact product safety and effectiveness, such as the concentration of active compounds, microbiological limits, physical characteristics, and dissolution rates, are imperative [57]. Defining the QTPP provides a reference point for setting specifications based on desired attributes and performance characteristics. In addition, utilizing recognized pharmacopeial standards for usually utilized herbal ingredients ensures consistency and quality. Acknowledging the inherent variability of herbal ingredients owing to aspects like geographic origin and growing conditions allows for appropriate acceptance criteria. Considering data from safety and toxicological parameters aids to establish permissible levels of potentially harmful compounds like heavy metals or pesticides. Moreover, assessing the capability of the development process to meet desired quality attributes ensures realistic and achievable specifications. Conducting stability studies to establish shelf life and storage conditions informs the establishment of shelf-life specifications. Validating analytical methods used for quality testing ensures its accuracy, precision, specificity, and robustness, supporting the reliability of specified limits [58]. Furthermore, periodically reviewing and updating specifications based on scientific findings, process improvements, or regulatory changes confirms continued relevance. Collaboration among formulation scientists, analytical chemists, regulatory experts, and stakeholders is crucial in this dynamic process, aiming to establish scientifically sound criteria that build trust with consumers and regulatory authorities, guaranteeing the safety, efficacy, and consistency of herbal products.

11.5.2 Implementation of Process Controls and Monitoring Techniques

Implementation of process controls and monitoring techniques is important to ensure the consistent quality and efficacy of herbal product manufacturing. As integral components of the QbD approach, process controls and monitoring aim to optimize and maintain the production process within predefined quality parameters. Real-time monitoring methods, such as process analytical technologies like spectroscopy, chromatography, and sensors integrated into production lines, provide immediate feedback on CPP [59]. In addition, statistical control charts assist to monitor process variability and identify trends or deviations, distinguishing between common and special cause variation. In-process evaluation at various manufacturing stages assesses CQAs to ensure the product meets desired specifications. Moreover, data collection, analysis of process performance, consistency, identifying opportunities for improvement and supporting informed decision-making. Adjusting CPP based on real-time data leads to improved product quality and diminished variability through process optimization. Integrating risk management strategies proactively identifies and mitigates potential process risks. Process validation demonstrates the capability of consistently producing herbal products meeting quality attributes. Apart from that personnel training ensures competent implementation of process controls and monitoring techniques, contributing to accurate data collection and prompt intervention in case of deviations. A culture of continuous improvement fosters regular process data review and corrective actions to address variations or inefficiencies. Effective implementation of process controls and monitoring ensures herbal product quality, safety, and efficacy, enhancing process efficiency, reducing waste, and building consumer trust [60].

11.5.3 Quality Control Methods for Assessing the Quality of Herbal Products/Formulations

The quality assessment of herbal formulations necessitates the utilization of diverse quality control methods to ensure its safety, efficacy, and uniformity. For the identification of herbal ingredients, various techniques such as macroscopic and microscopic analysis, thin-layer chromatography and high-performance liquid chromatography are employed to authenticate specific herbal constituents. Chemical composition analysis involves HPLC, gas chromatography, mass spectrometry, and nuclear magnetic resonance spectroscopy to quantify active constituents and markers [61]. Moreover, heavy metal and toxicological analysis using atomic absorption spectroscopy and inductively coupled plasma-mass spectrometry assess heavy metal content and potential contaminants. In addition, microbiological testing involves the evaluation of total bacterial count, yeast and mold count, and specific pathogen tests to verify the absence of harmful microorganisms. Apart from that, physical characterization methods like particle size analysis, density measurements, and rheological properties offer insights into the physical attributes of herbal formulations. Furthermore, dissolution testing evaluates the release of active compounds, ensuring appropriate bioavailability and effectiveness of solid oral dosage. Stability studies are conducted to govern shelf life and storage conditions, while residual solvent analysis detects any remaining solvents from the manufacturing process [62]. Pesticide residue analysis ensures compliance with acceptable limits for harmful pesticides. Additionally, moisture content analysis guarantees the stability and quality of herbal formulations. Total ash and acid-insoluble ash tests evaluate the purity and authenticity of herbal ingredients, while foreign matter detection identifies and quantifies any foreign substances present. The selection of suitable quality control methods depends on the specific herbal formulation, its intended application, and regulatory requirements. Implementation of robust quality control procedures ensures that herbal products consistently meet predefined quality standards, thus providing consumers with safe and effective formulations [63].

11.6 Real-Time Quality Assurance for Herbal Products

11.6.1 Importance of Real-Time Monitoring in QbD for Herbal Products

Real-time monitoring is a crucial step in QbD for the herbal formulation, offering esteemed benefits throughout development and manufacturing. Complex compositions and inherent variability in natural ingredients, herbal formulation requires precise control to ensure consistent quality and efficacy [61]. Real-time monitoring allows for early detection of deviations, enabling immediate corrective actions to prevent substandard products and reduce waste. It supports continuous process optimization, ensuring enhanced product consistency and reliable therapeutic outcomes. Data-driven decision-making is facilitated by real-time data analysis, helping manufacturers adhere to quality targets and regulatory requirements. Proactive risk management identifies and mitigates potential issues, ensuring product integrity. Real-time monitoring accelerates new product development by providing prompt feedback for formulation and process adjustments. It fosters continuous improvement and empowers the herbal product industry to achieve high-quality products, regulatory compliance, and customer satisfaction [60].

11.6.2 In-Process Testing and Process Verification Techniques

In-process testing and process verification techniques are essential for ensuring the quality, uniformity, and efficacy of herbal formulations during the manufacturing process. These methodologies entail the continuous monitoring and assessment of diverse parameters throughout production to ascertain conformity with predetermined quality standards. Widely employed approaches encompass the physical characterization of raw materials, chemical analysis employing inductive coupled plasma-mass spectroscopy, supercritical fluid chromatography, high-performance thin-layer chromatography, high-performance liquid chromatography and gas chromatography, spectroscopic analysis (Fourier Transform Infrared and UV-Visual spectrophotometer), microbiological testing, dissolution testing, stability testing, process verification, and in-line monitoring [64]. These practices serve to guarantee that herbal products consistently fulfil quality criteria, adhere to regulatory mandates, and furnish consumers with safe and effective formulations [61].

11.6.3 Handling Deviations and Variations During Herbal Product Manufacturing

Effectively managing deviations and variations during herbal product manufacturing is paramount for upholding product quality, safety, and efficacy. The inherent variability of natural ingredients necessitates a comprehensive approach to address potential discrepancies arising from raw materials, environmental factors, or processing methodologies. To tackle these challenges adeptly, several key strategies can be employed throughout the manufacturing process. Firstly, the implementation of robust process control and monitoring systems assumes critical importance. Real-time monitoring enables continuous tracking of CPP, facilitating timely detection of deviations and enabling immediate corrective actions [64]. Conducting a thorough root cause analysis emerges as another vital step in managing deviations. Understanding the underlying reasons for variations empowers manufacturers to devise targeted solutions and prevent reoccurrence. Moreover, detailed documentation and investigations become imperative to assess the impact of deviations on product quality. This information forms the basis for developing appropriate corrective and preventive actions, reinforcing process improvement endeavours. Additionally, prompt implementation of corrective actions becomes essential to address identified deviations effectively. Manufacturers may need to adjust process parameters, modify formulation methods, or optimize raw material selection to restore product consistency [65].

Furthermore, conducting risk assessments aids in evaluating the potential impact of deviations on product quality and patient safety. Prioritizing actions based on risk levels ensures efficient resource allocation. Apart from that, the validation of manufacturing processes also assumes criticality in ensuring consistent product quality. Process validation confirms that the manufacturing process reliably produces herbal products that meet predefined quality standards. The establishment of a robust change control system assumes significance in managing planned changes in the manufacturing process. This prevents unintended variations and ensures proper evaluation and approval of changes. Rigorous quality control testing at various production stages aids in detecting deviations and verifying adherence to quality specifications. Additionally, comprehensive training and skill development for personnel involved in the manufacturing process enhances the identification and resolution of deviations. Moreover, ensuring that suppliers of herbal raw materials meet specific quality criteria and adhere to GMP mitigates variations originating from raw materials. Besides that, by diligently implementing these strategies, manufacturers can effectively handle deviations and variations during herbal product manufacturing, resulting in the consistent fulfilment of desired quality attributes, ensuring consumer safety and satisfaction.

11.6.4 Ensuring Product Quality Throughout the Production Life-Cycle

To deliver safe and effective herbal formulations that cater to consumers' needs, it is crucial to ensure product quality throughout the entire production lifecycle. The intricate and variable nature of herbal ingredients necessitates a comprehensive approach spanning from sourcing to distribution. Key measures include meticulous selection of high-quality raw materials, strict adherence to GMP, and application of QbD principles during formulation development [66]. In-process monitoring, comprehensive quality control testing and thorough process validation play pivotal roles in upholding product consistency and safety. Additionally, stability testing is vital for determining product shelf life and appropriate storage conditions, while change control procedures effectively manage planned adjustments to prevent undesirable variations [67]. Quality assurance oversees every aspect of production, and continuous improvement efforts foster heightened quality and efficiency. Ensuring compliance with regulatory standards is paramount to ensuring that herbal products meet all required specifications. Integrating these strategies results in a harmonized approach to maintaining unwavering product quality, instilling consumer confidence, and effectively meeting the stringent demands of the herbal product industry [68].

11.7 Regulatory Aspects and Guidelines Related to QbD for Herbal Formulation

The herbal formulation must adhere to rigorous criteria of quality, safety, and efficacy akin to those imposed on synthetic pharmaceutical products. The integration of the International Council for Harmonisation ICH Q8, Q9, and Q10 guidelines has amalgamated the phases of pharmaceutical product development, wherein the assurance of quality for the herbal product is established throughout each production stage, aligning with the principles of GMP. In this context, the World Health Organization (WHO) has established numerous regulations pertaining to the standardization protocols employed during the various stages of herbal product development [69].The concept of QbD is an integral component of the ICH Q8 (R2) "Pharmaceutical Development" guideline, aiming to assure product quality through the modeling and regulation of the manufacturing process itself. As delineated in the ICH Q8 Pharmaceutical Development guidelines, it is described as a methodical and systematic undertaking that commences with predetermined objectives, with a prime emphasis on comprehending the product and process, while employing sound scientific principles and implementing robust quality risk management strategies for process control and oversight. This paradigm emphasizes the need to go beyond solely relying on the final product as the sole determinant of quality at the end of the process. By adopting this approach, it becomes possible to establish dependable and controlled processes and operations, thereby yielding a product of superior quality [70].

The worldwide escalation in the utilization of traditional medicines, spanning developed as well as developing countries, has engendered a surge in the global demand for herbal products within the realm of healthcare. The amplified global requisition for herbal-based products has placed a dual burden. On one hand, it entails the herbal industry to generate products of exceptional quality. On the other hand, it necessitates the federal regulatory authorities to devise stringent regulations that ensure the provision of uniformly high-quality herbal products [71]. This collective effort addresses the burgeoning market demand. Therefore, European Medicines Agency (EMA) tailored the QbD approach to geta better understanding of the herbal formulations manufacturing process [72]. Regulatory guidelines related to QbD principles for herbal formulations may vary depending on the country or region. However, some general regulatory frameworks and guidelines that can be considered include:

1. **ICH Guidelines**: The International Council for Harmonisation of Technical Requirements for Pharmaceuticals for Human Use (ICH) has established numerous standards and guidelines that can be appropriate to QbD implementation for herbal formulations. These include:

   ICH Q8 (R2) Pharmaceutical Development: Provides guidance on pharmaceutical development and emphasizes the importance of a systematic approach to formulation development, process understanding, and risk management [82].

   ICH Q9 Quality Risk Management: Offers guidance on implementing quality risk management principles, including risk assessment, risk control, and risk communication, which can be applied to herbal formulations.

   ICH Q10 Pharmaceutical Quality System: Focuses on the establishment of a pharmaceutical quality system that ensures consistent product quality, including the application of QbD principles [73].
2. **National Pharmacopoeias**: National pharmacopeia often contains monographs or guidelines specific to herbal products and formulations. These monographs may provide specific requirements and recommendations for quality attributes, testing methods, and manufacturing considerations. Such as Ayurvedic Pharmacopoeia of India establishes quality standards for Ayurvedic products. Published by the Department of Ayurveda, Yoga & Naturopathy, Unani, Siddha, and Homoeopathy (AYUSH) under the Ministry of Health and Family Welfare, Government of India, the API serves as a comprehensive reference for ensuring the quality of Ayurvedic formulations [72].
3. **Herbal medicine regulations**: Some countries have specific regulations and guidelines for herbal medicines or traditional medicine products. These

regulations may outline requirements for safety, efficacy, quality control, and manufacturing practices, which may align with QbD principles. Such as the United States Dietary Supplement Health and Education Act and Current Good Manufacturing Practices for the United States, European Medicines Agency and Traditional Herbal Medicinal Products Directive for European, Chinese Pharmacopoeia for China, and Australian Regulatory Guidelines for Complementary Medicine for Australia.

4. **Good manufacturing practices**: GMP regulations and guidelines applicable to pharmaceutical products offer valuable insights for incorporating QbD principles into herbal formulations. By emphasizing the consistent quality, safety, and efficacy of medicinal products throughout their manufacturing processes, GMP guidelines provide pertinent considerations for implementing QbD in herbal formulation practices.

Regulatory bodies across various nations may release dedicated guidance documents or policies concerning the implementation of QbD principles in the context of herbal formulations. These documents serve to delineate anticipated standards, optimal methodologies, and obligatory criteria governing the development, manufacturing, and quality control of herbal products. It is imperative to consult the respective regulatory authorities within the intended market or jurisdiction to access the latest and region-specific guidelines and requirements pertaining to QbD for herbal formulations.

11.8 Challenges and Opportunities in Implementing QbD for Herbal Formulations

11.8.1 Challenges of Implementing QbD for Herbal Formulations

Employing QbD for herbal formulations brings about numerous challenges that are requisite to be tackled. The application of QbD for herbal medications still has some major problems that need to be resolved. The first problem is the quality fluctuation of herbal raw materials, and the other problem is the challenge of establishing acceptable ranges of CQAs [57]. Outlined below are several prominent challenges encountered when implementing QbD for herbal formulations:

1. **Complex and variable composition**: Herbal formulations frequently contain multiple active constituents with complex compositions, making it challenging to determine the precise impact of each ingredient on the formulation's quality attributes. The inherent variability of herbal materials adds an additional layer of complexity [74, 75].
2. **Limited scientific knowledge**: There may be a lack of scientific understanding of the interactions between herbal ingredients, formulation characteristics, and

process variables. The formulation and manufacturing processes may be difficult to optimize and cause and effect links may be challenging to determine [76].

3. **Lack of comprehensive characterization methods**: Herbal formulations contain numerous active ingredients, and it is difficult to establish trustworthy analytical procedures to precisely quantify and identify these constituents. The application of QbD may be hampered by the absence of standardized procedures for characterization and quality assessment [77].
4. **Availability of reference standards**: Obtaining appropriate reference standards for herbal components can be challenging attributable to the variability and complexity of plant-based substances. It is so challenging to create precise quality criteria and control procedures.
5. **Limited process understanding**: It might be challenging to comprehend how manufacturing processes affect the quality attributes of herbal formulations, especially when taking into account extraction procedures, drying methods, and other processing requirements particular to herbal materials [26, 27].
6. **Data availability and quality**: Acquiring dependable and comprehensive data pertaining to herbal formulations can pose a challenge. Incomplete or inconsistent information within the literature can impede the effective application of QbD principles [78].
7. **Cultural and traditional knowledge integration**: The integration of traditional knowledge and cultural practices into the QbD framework for herbal formulations presents both challenges and significance. Balancing scientific principles with traditional knowledge is vital to uphold the authenticity and cultural relevance of herbal formulations, although it can be a complex endeavor. Overcoming these challenges will enable the effective application of QbD principles and promote the quality, safety, and efficacy of herbal formulations [21].

11.8.2 Opportunities of Implementing QbD for Herbal Formulations

Numerous opportunities can be realized by applying QbD principles to herbal compositions. The following are some important benefits of applying QbD to herbal formulations:

1. **Enhanced product understanding**: QbD enables a more profound comprehension of the interconnections among raw materials, formulation, process parameters, and the resultant product attributes. This enhanced understanding facilitates the optimization of formulations and processes, culminating in enhanced product quality and performance.
2. **Risk mitigation**: A key component of QbD is risk assessment and control, which enables the proactive identification and mitigation of hazards pertaining to the quality, safety, and effectiveness of products. This increases consumer trust in product performance and lowers the possibility of quality problems.

3. **Process optimization**: The application of QbD principles has the capacity to stimulate process optimization, resulting in heightened efficiency, diminished manufacturing costs, and amplified productivity [79, 80].
4. **Consistent quality and efficacy**: QbD encourages the development of CQAs and trustworthy control methods to guarantee the consistency of the quality and effectiveness of herbal products/formulations across various batches and manufacturing facilities [57].
5. **Regulatory compliance**: Regulatory compliance entails adhering to specific regulations, and guidelines set forth by governing bodies. When applying QbD principles, organizations showcase their commitment to ensuring product quality and safety. By aligning with regulatory expectations and guidelines, the implementation of QbD can aid in achieving and maintaining regulatory compliance. This approach enhances the organization's ability to meet regulatory requirements, promotes transparency, and fosters a culture of quality throughout the product development and manufacturing processes [78].
6. **Continuous improvement**: QbD promotes a culture of continuous improvement by advocating for ongoing monitoring, rigorous data analysis, and proactive process refinement. This approach allows organizations to consistently enhance product quality, safety, and efficacy. By continuously assessing and optimizing processes based on real-time data and feedback, organizations can identify areas for improvement, implement corrective actions, and drive positive change throughout the product lifecycle. Through the application of QbD, organizations can achieve higher levels of quality, meet evolving customer expectations, and remain competitive in the market [8].
7. **Faster development and time to market:** The adoption of QbD expedites the process of herbal product development by enabling a systematic and efficient approach. By emphasizing risk-based decision-making and in-depth process understanding, organizations may streamline development activities, minimize development time, and accelerate the introduction of herbal products to the market [21].
8. **Scientific advancements and innovation**: QbD stimulate research and development endeavors within the herbal product industry, fostering a deeper understanding of herbal components, formulation design, and manufacturing processes. This emphasis on QbD cultivates scientific advancements, encourages innovation, and facilitates the exploration of novel approaches to herbal product development.

Manufacturers of herbal formulations can improve product quality, streamline products [81] on procedures, satisfy regulatory standards, and ultimately provide consumers with safe and efficient herbal products by taking advantage of these opportunities.

11.9 Conclusion and Future Perspectives

QbD in herbal products has witnessed great advancements over time which is supported by continuous research and development for the improvement of product quality, manufacturing process and regulatory compliance. The development in analytical techniques and molecular modelling has provided access to a detailed understanding of herbal components and their interactions during formulation development leading to more promising CQAs and CPPs. In addition to this, the evolution of process analytical technologies (PAT) has desirably integrated PAT tools such as spectroscopy, chromatography, and in-line sensors with the real-time monitoring of the CPPs leading to enhanced process control. Furthermore, the improvement in quality risk management has fostered risk assessment, hence avoiding potential sources of risk and variability. The combination of the DoE methods with herbal formulation development has given rise to optimum herbal formulation with inbuilt quality in fewer experimental runs. QbD has also helped to greatly optimize the extraction process for natural extracts with excellent potency and composition. Moreover, processes such as continuous manufacturing, in silico modeling and process validation strategies have led to enhanced process efficiency, accelerated formulation development and production of high-quality herbal products. To be brief QbD for herbal products has revolutionized the herbal product industry accounting for formulations with built-in quality, safety and therapeutic efficacy.

The future prospects in QbD aim to focus on formulation optimization in order to keep track of the evolving regulatory aspects as well as consumer demands. The integration of advanced analytical techniques including metabolomics, genomics and proteomics provides a deep understanding of the herbal product behavior as well as an understanding of additional CQAs hence improving process control in formulation development. QbD may also be applied to personalized medicine and precision medicine to develop formulations that satisfy individual patient requirements. Artificial intelligence and machine learning hand in hand with QbD also support the real-time analysis of CPPs to generate an optimized formulation process. Furthermore, integrating QbD with herbal nanomedicines provides an enhanced delivery platform for herbal compounds, thereby offering a ground for enhanced efficacy for herbal medicines.

Implementing QbD in the herbal product industry comes with many benefits in the manufacturing of herbal products. QbD sets the seal on consistent product quality with a diminished chance of variability. The execution of quality risk management (QRM) allows for the beforehand identification of risks and mitigates. QbD helps in the development of a product with enhanced regulatory compliance, better safety, and excellent quality. The data obtained from QbD promotes decision making facilitating cost-effective development; QbD focuses on continuous improvement which repeatedly improves product quality and efficiency. Eventually, QbD has gained consumer confidence due to the manufacture of herbal products with consistent quality based on a critical understanding of CQAs.

References

1. Desai, N. P., & Shah, C. N. (2016). *A review of recent research on herbal formulation: An ayurvedic medicine- the nature' s boon* (Vol. 6, pp. 699–705).
2. Chaudhary, R., & Kumari, P. (2022). Stability aspects of herbal formulation. *World Journal of Pharmaceutical and Life Sciences, 8*, 103–110.
3. Kumadoh, D., & Ofori-kwakye, K. (2017). Dosage forms of herbal medicinal products and their stability considerations—An overview considerations-an overview. *Journal of Critical Reviews, 4*(4), 1–8. https://doi.org/10.22159/jcr.2017v4i4.16077
4. Musthaba, M., Baboota, S., Athar, T. M. D., Thajudeen, K. Y., Ahmed, S., & Ali, J. (2010). Patented herbal formulations and their therapeutic applications. *Recent Patents on Drug Delivery & Formulation, 4*, 231–244. https://doi.org/10.2174/187221110793237538
5. Id, E. L. P., & Sesaazi, C. D. (2022). *D-optimal mixture design optimized solid formulation containing fruits extracts of Momordica charantia and Abelmoschus esculentus* (pp. 1–18). https://doi.org/10.1371/journal.pone.0270547
6. Bastogne, T. (2017). Quality-by-design of nanopharmaceuticals—A state of the art. *Nanomedicine: Nanotechnology, Biology and Medicine, 13*, 2151–2157. https://doi.org/10.1016/j.nano.2017.05.014
7. Pan, J., He, S., Zheng, J., Shao, J., Li, N., Gong, Y., & Gong, X. (2019). The development of an herbal material quality control strategy considering the effects of manufacturing processes. *Chinese Medicine, 14*, 1–12. https://doi.org/10.1186/s13020-019-0262-9
8. Calixto, J. B. (2000). Efficacy, safety, quality control, marketing and regulatory guidelines for herbal medicines (phytotherapeutic agents). *Brazilian Journal of Medical and Biological Research, 33*, 179–189. https://doi.org/10.1590/s0100-879x2000000200004
9. Oh, E., Kim, U., Lee, B. J., & Moon, C. (2019). Multivariate statistical optimization of tablet formulations incorporating high doses of a dry herbal extract. *Pharmaceutics, 11*, 1. https://doi.org/10.3390/pharmaceutics11020079
10. Güngör-Ak, A., Küpeli-Akkol, E., Aksu, B., & Karataş, A. (2022). Preparation and optimization of berberine phospholipid complexes using QbD approach and in vivo evaluation for anti-inflammatory, analgesic and antipyretic activity. *Journal of Research in Pharmacy, 26*, 370–382. https://doi.org/10.29228/jrp.135
11. Jain, P., Taleuzzaman, M., Kala, C., Kumar Gupta, D., Ali, A., & Aslam, M. (2021). Quality by design (Qbd) assisted development of phytosomal gel of aloe vera extract for topical delivery. *Journal of Liposome Research, 31*, 381–388. https://doi.org/10.1080/08982104.2020.1849279
12. Shalabalija, D., Mihailova, L., Crcarevska, M. S., Karanfilova, I. C., Ivanovski, V., Nestorovska, A. K., Novotni, G., & Dodov, M. G. (2021). Formulation and optimization of bioinspired rosemary extract loaded PEGylated nanoliposomes for potential treatment of Alzheimer's disease using design of experiments. *Journal of Drug Delivery Science and Technology, 63*, 102434. https://doi.org/10.1016/j.jddst.2021.102434
13. Savic, S. M., Cekic, N. D., Savic, S. R., Ilic, T. M., & Savic, S. D. (2021). 'All-natural' anti-wrinkle emulsion serum with Acmella oleracea extract: A design of experiments (DoE) formulation approach, rheology and in vivo skin performance/efficacy evaluation. *International Journal of Cosmetic Science, 43*, 530–546. https://doi.org/10.1111/ics.12726
14. Shariare, M. H., Rahman, M., Lubna, S. R., Roy, R. S., Abedin, J., Marzan, A. L., Altamimi, M. A., Ahamad, S. R., Ahmad, A., Alanazi, F. K., & Kazi, M. (2020). Liposomal drug delivery of Aphanamixis polystachya leaf extracts and its neurobehavioral activity in mice model. *Scientific Reports, 10*, 1–16. https://doi.org/10.1038/s41598-020-63894-9
15. Shariare, M. H., Noor, H. B., Khan, J. H., Uddin, J., Ahamad, S. R., Altamimi, M. A., Alanazi, F. K., & Kazi, M. (2021). Liposomal drug delivery of Corchorus olitorius leaf extract containing phytol using design of experiment (DoE): In-vitro anticancer and in-vivo anti-inflammatory studies. *Colloids and Surfaces B: Biointerfaces, 199*, 111543. https://doi.org/10.1016/j.colsurfb.2020.111543

16. Sundaresan, N., & Kaliappan, I. (2021). Development and characterization of a nano-drug delivery system containing vasaka phospholipid complex to improve bioavailability using quality by design approach. *Research in Pharmaceutical Sciences, 16*, 103–117. https://doi.org/10.4103/1735-5362.305193
17. Yu, L. X., Amidon, G., Khan, M. A., Hoag, S. W., Polli, J., Raju, G. K., & Woodcock, J. (2014). Understanding pharmaceutical quality by design. *The AAPS Journal, 16*, 771–783. https://doi.org/10.1208/s12248-014-9598-3
18. Monton, C., & Suksaeree, J. (2021). Interaction of herbal ingredients contained in the Trikatuk recipe: Design of experiment (DOE) and chemical analysis point of view. *Advances in Traditional Medicine, 21*, 443–452. https://doi.org/10.1007/s13596-021-00569-4
19. Ribeiro, A. F., de Oliveira Rezende, R. L., Cabral, L. M., & de Sousa, V. P. (2013). Poly ε-caprolactone nanoparticles loaded with Uncaria tomentosa extract: Preparation, characterization, and optimization using the Box-Behnken design. *International Journal of Nanomedicine, 8*, 431–442. https://doi.org/10.2147/IJN.S38491
20. Behera, S. K., Meena, H., Chakraborty, S., & Meikap, B. C. (2018). Application of response surface methodology (RSM) for optimization of leaching parameters for ash reduction from low-grade coal. *International Journal of Mining Science and Technology, 28*, 621–629. https://doi.org/10.1016/j.ijmst.2018.04.014
21. Grangeia, H. B., Silva, C., Simões, S. P., & Reis, M. S. (2020). Quality by design in pharmaceutical manufacturing: A systematic review of current status, challenges and future perspectives. *European Journal of Pharmaceutics and Biopharmaceutics, 147*, 19–37. https://doi.org/10.1016/j.ejpb.2019.12.007
22. Faulhammer, E., Llusa, M., Wahl, P. R., Paudel, A., Lawrence, S., Biserni, S., Calzolari, V., & Khinast, J. G. (2016). Development of a design space and predictive statistical model for capsule filling of low-fill-weight inhalation products. *Drug Development and Industrial Pharmacy, 42*, 221–230. https://doi.org/10.3109/03639045.2015.1040416
23. Lourenço, V., Lochmann, D., Reich, G., Menezes, J. C., Herdling, T., & Schewitz, J. (2012). A quality by design study applied to an industrial pharmaceutical fluid bed granulation. *European Journal of Pharmaceutics and Biopharmaceutics, 81*, 438–447. https://doi.org/10.1016/j.ejpb.2012.03.003
24. Narayan, R., Pednekar, A., Bhuyan, D., Gowda, C., Koteshwara, K. B., & Nayak, U. Y. (2017). A top-down technique to improve the solubility and bioavailability of aceclofenac: In vitro and in vivo studies. *International Journal of Nanomedicine, 12*, 4921–4935. https://doi.org/10.2147/IJN.S141504
25. Taleuzzaman, M., Sartaj, A., Kumar Gupta, D., Gilani, S. J., & Mirza, M. A. (2023). Phytosomal gel of Manjistha extract (MJE) formulated and optimized with central composite design of quality by design (QbD). *Journal of Dispersion Science and Technology, 44*, 236–244. https://doi.org/10.1080/01932691.2021.1942036
26. Zhang, L., Yan, B., Gong, X., Yu, L. X., & Qu, H. (2013a). Application of quality by design to the process development of botanical drug products: A case study. *Aaps Pharmscitech, 14*, 277–286. https://doi.org/10.1208/s12249-012-9919-8
27. Zhang, W., Wang, Y., Gao, X.-m., Gao, X., Peng, S., Zheng, Y., & Okeke, C. I. (2013b). Optimization of Jiawei Qing'e oral fast disintegrating tablets based on response surface-central composite design. *Chinese Herbal Medicines, 5*, 138–144. https://doi.org/10.3969/j.issn.1674-6348.2013.02.011
28. Suksaeree, J., Monton, C., Charoenchai, L., Chankana, N., & Wunnakup, T. (2023). Optimization of process and formulation variables for Semha-Pinas extract effervescent tablets using the Box-Behnken design. *AAPS PharmSciTech, 24*, 52. https://doi.org/10.1208/s12249-023-02514-x
29. Khan, A. D., Singh, M. K., & Lavhale, P. M. (2023). Polyherbal phytosomal gel for enhanced topical delivery: Design, optimization by central composite design, in vitro and ex-vivo evaluation. *Journal of Dispersion Science and Technology, 1–14*, 1–14. https://doi.org/10.1080/01932691.2023.2195927

30. Kriplani, P., Guarve, K., & Singh Baghel, U. (2021). Formulation optimization and characterization of transdermal film of curcumin by response surface methodology. *Chinese Herbal Medicines, 13*, 274–285. https://doi.org/10.1016/j.chmed.2020.12.001
31. Canh Pham, E., Vo Van, L., Viet Nguyen, C., Nguyen Duong, N. T., & Le Thi, T. V. (2023). Formulation development, optimization, in vivo antidiabetic effect and acute toxicity of directly compressible herbal tablets containing Merremia tridentata (L.) extract. *Journal of Drug Delivery Science and Technology, 84*, 104445. https://doi.org/10.1016/j.jddst.2023.104445
32. Qadir, A., Aqil, M., Ali, A., Warsi, M. H., Mujeeb, M., Ahmad, F. J., Ahmad, S., & Beg, S. (2020). Nanostructured lipidic carriers for dual drug delivery in the management of psoriasis: Systematic optimization, dermatokinetic and preclinical evaluation. *Journal of Drug Delivery Science and Technology, 57*, 101775. https://doi.org/10.1016/j.jddst.2020.101775
33. Sood, S., Jain, K., & Gowthamarajan, K. (2014). Optimization of curcumin nanoemulsion for intranasal delivery using design of experiment and its toxicity assessment. *Colloids and Surfaces B: Biointerfaces, 113*, 330–337. https://doi.org/10.1016/j.colsurfb.2013.09.030
34. Tung, N.-T., Tran, C.-S., Pham, T.-M.-H., Nguyen, H.-A., Nguyen, T.-L., Chi, S.-C., Nguyen, D.-D., & Bui, T.-B.-H. (2018). Development of solidified self-microemulsifying drug delivery systems containing l-tetrahydropalmatine: Design of experiment approach and bioavailability comparison. *International Journal of Pharmaceutics, 537*, 9–21. https://doi.org/10.1016/j.ijpharm.2017.12.027
35. Na, Y.-G., Jeon, S.-H., Byeon, J.-J., Kim, M.-K., Lee, H.-K., & Cho, C.-W. (2019). Application of statistical design on the early development of sustained-release tablet containing ivy leaf extract. *Journal of Drug Delivery Science and Technology, 54*, 101319. https://doi.org/10.1016/j.jddst.2019.101319
36. Walbi, I. A., Ahmad, M. Z., Ahmad, J., Algahtani, M. S., Alali, A. S., Alsudir, S. A., Aodah, A. H., & Albarqi, H. A. (2022). Development of a curcumin-loaded lecithin/chitosan nanoparticle utilizing a Box-Behnken design of experiment: Formulation design and influence of process parameters. *Polymers (Basel), 14*, 3758. https://doi.org/10.3390/polym14183758
37. Pawar, Y. B., Purohit, H., Valicherla, G. R., Munjal, B., Lale, S. V., Patel, S. B., & Bansal, A. K. (2012). Novel lipid based oral formulation of curcumin: Development and optimization by design of experiments approach. *International Journal of Pharmaceutics, 436*, 617–623. https://doi.org/10.1016/j.ijpharm.2012.07.031
38. Monton, C., Sampaopan, Y., Pichayakorn, W., Panrat, K., & Suksaeree, J. (2022). Herbal transdermal patches made from optimized polyvinyl alcohol blended film: Herbal extraction process, film properties, and in vitro study. *Journal of Drug Delivery Science and Technology, 69*, 103170. https://doi.org/10.1016/j.jddst.2022.103170
39. Ghazwani, M., Hani, U., Osmani, R. A. M., Rahamathulla, M., Begum, M. Y., Wahab, S., Gowda, D. V., Ravikumar, A. A., Kumar, H. Y., Ather, H., & Moin, A. (2021). An efficient herbal approach for treating fungal infection in cervical cancer patients by developing and optimizing a vaginal suppository. *International Journal of Polymeric Science, 2021*, 1–11. https://doi.org/10.1155/2021/9198387
40. Chitkara, A., Mangla, B., Kumar, P., Javed, S., Ahsan, W., & Popli, H. (2022). Design-of-experiments (DoE)-assisted fabrication of quercetin-loaded nanoemulgel and its evaluation against human skin cancer cell lines. *Pharmaceutics, 14*, 2517. https://doi.org/10.3390/pharmaceutics14112517
41. Mudrić, J., Arsenijević, J., Maksimović, Z., Ibrić, S., Gopčević, K., & Đuriš, J. (2021). Tablet and capsule formulations incorporating high doses of a dry optimized herbal extract: The case of Satureja kitaibelii. *Journal of Drug Delivery Science and Technology, 66*, 102776. https://doi.org/10.1016/j.jddst.2021.102776
42. Dhawan, S., & Nanda, S. (2023). Implementation of quality by design (QbD) concept for the development of emulsion based nanotailored gel for improved antiphotoageing potential of Silymarin. *Journal of Drug Delivery Science and Technology, 81*, 104201. https://doi.org/10.1016/j.jddst.2023.104201

43. Sridhar, A., Ponnuchamy, M., Kumar, P. S., Kapoor, A., Vo, D.-V. N., & Prabhakar, S. (2021). Techniques and modeling of polyphenol extraction from food: A review. *Environmental Chemistry Letters, 19*, 3409–3443. https://doi.org/10.1007/s10311-021-01217-8
44. Thakur, L., Ghodasra, U., Patel, N., & Dabhi, M. (2011). Novel approaches for stability improvement in natural medicines. *Pharmacognosy Reviews, 5*, 48–54. https://doi.org/10.4103/0973-7847.79099
45. Kalyan, S., & Vihar, P. (2019). Quality by design: Changing outlook of pharmaceutical development. *International Journal of Pharmaceutical Sciences and Research, 10*, 4100–4108. https://doi.org/10.13040/IJPSR.0975-8232.10(9).4100-08
46. Mishra, V., Thakur, S., Patil, A., & Shukla, A. (2018). Quality by design (QbD) approaches in current pharmaceutical set-up. *Expert Opinion on Drug Delivery, 15*, 737–758. https://doi.org/10.1080/17425247.2018.1504768
47. Pramod, K., Tahir, M., Charoo, N., Ansari, S., & Ali, J. (2016). Pharmaceutical product development: A quality by design approach. *International Journal of Pharmaceutical Investigation, 6*, 129–138. https://doi.org/10.4103/2230-973x.187350
48. Kolipaka, T., Sen, S., Mane, S. S., Bajad, G. D., Dengale, S. J., & Ranjan, O. P. (2023). Development of posaconazole nanosuspension for bioavailability enhancement: Formulation optimization, in vitro characterization, and pharmacokinetic study. *Journal of Drug Delivery Science and Technology, 83*, 104434. https://doi.org/10.1016/j.jddst.2023.104434
49. Park, G., Kim, M. K., Go, S. H., Choi, M., & Jang, Y. P. (2022). Analytical quality by design (AQbD) approach to the development of analytical procedures for medicinal plants. *Plants, 11*, 1. https://doi.org/10.3390/plants11212960
50. van Leeuwen, J. F., Nauta, M. J., de Kaste, D., Odekerken-Rombouts, Y. M. C. F., Oldenhof, M. T., Vredenbregt, M. J., & Barends, D. M. (2009). Risk analysis by FMEA as an element of analytical validation. *Journal of Pharmaceutical and Biomedical Analysis, 50*, 1085–1087. https://doi.org/10.1016/j.jpba.2009.06.049
51. Parab Gaonkar, V. M., Mannur, V. S., Mastiholimath, V. S., & Hullatti, K. K. (2020). Development and evaluation of herbal supplement: A quality by design approach. *Indian Journal of Pharmaceutical Sciences, 82*, 640–649. https://doi.org/10.36468/pharmaceutical-sciences.690
52. Peterson, J. J. (2008). A Bayesian approach to the ICH Q8 definition of design space. *Journal of Biopharmaceutical Statistics, 18*, 959–975. https://doi.org/10.1080/10543400802278197
53. Chatzizacharia, K. A., & Hatziavramidis, D. T. (2014). Design space approach for pharmaceutical tablet development. *Industrial and Engineering Chemistry Research, 53*, 12003–12009. https://doi.org/10.1021/ie5005652
54. Benjasirimongkol, P., Piriyaprasarth, S., & Sriamornsak, P. (2019). Improving dissolution and photostability of resveratrol using redispersible dry emulsion: Application of design space for optimizing formulation and spray-drying process. *Journal of Drug Delivery Science and Technology, 51*, 411–418. https://doi.org/10.1016/j.jddst.2019.03.005
55. Jariwala, N., Putta, C. L., Gatade, K., Umarji, M., Ruhina Rahman, S. N., Pawde, D. M., Sree, A., Kamble, A. S., Goswami, A., Chakraborty, P., & Shunmugaperumal, T. (2023). Intriguing of pharmaceutical product development processes with the help of artificial intelligence and deep/machine learning or artificial neural network. *Journal of Drug Delivery Science and Technology, 87*, 104751. https://doi.org/10.1016/j.jddst.2023.104751
56. Yu, L. X. (2008). Pharmaceutical quality by design: Product and process development, understanding, and control. *Pharmaceutical Research, 25*, 781–791. https://doi.org/10.1007/s11095-007-9511-1
57. Yan, B., Li, Y., Guo, Z., & Qu, H. (2014). Quality by design for herbal drugs: A feedforward control strategy and an approach to define the acceptable ranges of critical quality attributes. *Phytochemical Analysis, 25*, 59–65. https://doi.org/10.1002/pca.2463
58. Indrayanto, G. (2018). Recent development of quality control methods for herbal derived drug preparations. *Natural Product Communications, 13*, 1599–1606. https://doi.org/10.1177/1934578x1801301208

59. Neethu Kannan, B., Devi, V. G., John, L. G. A., Natarajan, M., & Kanagarajan, A. (2019). Pharmacognostic evaluation and HPTLC fingerprinting of Pirantai Vatakam, a Siddha formulation. *Journal of Phytopharmacol, 8*, 33–37. https://doi.org/10.31254/phyto.2019.8201
60. Muyumba, N. W., Mutombo, S. C., Sheridan, H., Nachtergael, A., & Duez, P. (2021). Quality control of herbal drugs and preparations: The methods of analysis, their relevance and applications. *Talanta Open, 4*, 100070. https://doi.org/10.1016/j.talo.2021.100070
61. Balekundri, A., & Mannur, V. (2020). Quality control of the traditional herbs and herbal products: A review. *Future Journal of Pharmaceutical Sciences, 6*, 67. https://doi.org/10.1186/s43094-020-00091-5
62. Alamgir, A. N. M. (2017). Herbal drugs: Their collection, preservation, and preparation; evaluation, quality control, and standardization of herbal drugs. In *Therapeutic use of medicinal plants and their extract: Volume 1: Pharmacognosys* (pp. 453–495). Springer International Publishing. https://doi.org/10.1007/978-3-319-63862-1_10
63. Karbwang, J., Crawley, F. P., Na-Bangchang, K., & Maramba-Lazarte, C. (2019). Herbal medicine development: Methodologies, challenges, and issues. *Evidence-Based Complementary and Alternative Medicine, 2019*, 4935786. https://doi.org/10.1155/2019/4935786
64. Singh, S. K., Jha, S. K., Chaudhary, A., Yadava, R. D. S., & Rai, S. B. (2010). Quality control of herbal medicines by using spectroscopic techniques and multivariate statistical analysis. *Pharmaceutical Biology, 48*, 134–141. https://doi.org/10.3109/13880200903059388
65. Mollah, S., Abu, M., & Nyeem, B. (2019). *Quality control and evaluation of herbal drugs*. https://doi.org/10.1016/c2016-0-04232-8
66. Orman, E., Bekoe, S. O., Jato, J., Spiegler, V., Asare-Nkansah, S., Agyare, C., Hensel, A., & Bekoe, E. O. (2022). Quality assessment of African herbal medicine: A systematic review and the way forward. *Fitoterapia, 162*, 105287. https://doi.org/10.1016/j.fitote.2022.105287
67. Mahady, G. B. (2015). Chapter 11—Good quality and clinical practices for the future development of herbal medicines. In P. K. Mukherjee (Ed.), *Evidence-based validation of herbal medicine* (pp. 247–258). Elsevier. https://doi.org/10.1016/B978-0-12-800874-4.00011-8
68. Pan, S., Neeraj, A., Srivastava, K. S., Kishore, P., Danquah, M. K., & Sarethy, I. P. (2013). A proposal for a quality system for herbal products. *Journal of Pharmaceutical Sciences, 102*, 4230–4241. https://doi.org/10.1002/jps.23732
69. Fang, L., Zhu, R., Sang, Z., Xu, X., Zheng, L., Xiang, Y., Guo, C., Li, Z., Espinosa, S. A., Zhang, Q., & Shan, C. (2023). World Health Organization survey on the level of integration of traditional Chinese medicine in Chinese health system rehabilitation services. *Integrative Medicine Research, 12*, 100945. https://doi.org/10.1016/j.imr.2023.100945
70. Ramya Sri, K., Bhavani, K., Teja, T. S., Srinivasa Babu, P., & Kurra, P. (2021). An outline of quality by design in pharmaceutical development. *Journal of Pharmaceutical Sciences and Research, 13*, 200–205.
71. Choudhury, A., Singh, P. A., Bajwa, N., Dash, S., & Bisht, P. (2023). Pharmacovigilance of herbal medicines: Concerns and future prospects. *Journal of Ethnopharmacology, 309*, 116383. https://doi.org/10.1016/j.jep.2023.116383
72. Kumar, S., Walia, R., Saxena, S., Dey, P., & Madaan, R. (2022). Chapter 16—Regulatory considerations of herbal bioactive–based formulations. In I. S. Bakshi, R. Bala, R. Madaan, & R. K. Sindhu (Eds.), *Herbal bioactive-based drug delivery systems* (pp. 419–436). Academic. https://doi.org/10.1016/B978-0-12-824385-5.00009-1
73. Das, A. K., Mandal, S. C., Mandal, V., Beg, S., & Singh, B. (2014). QbD as an emerging paradigm in extraction technology for developing optimized bioactives. *Pharma Times, 46*, 50–56.
74. Karole, S., Shrivastava, S., Thomas, S., Soni, B., Khan, S., Dubey, J., & Dubey, S. P. (2019). Polyherbal formulation concept for synergic action: A Review. *Journal of Drug Delivery and Therapeutics, 9*, 453–466. https://doi.org/10.22270/jddt.v9i1-s.2339
75. Kiyohara, H., Matsumoto, T., & Yamada, H. (2004). Combination effects of herbs in a multi-herbal formula: Expression of Juzen-taiho-to's Immuno-modulatory activity on the intestinal

immune system. *Evidence-based Complementary and Alternative Medicine, 1*, 83–91. https://doi.org/10.1093/ecam/neh004

76. Parveen, A., Parveen, B., Parveen, R., & Ahmad, S. (2015). Challenges and guidelines for clinical trial of herbal drugs. *Journal of Pharmacy & Bioallied Sciences, 7*, 329–333. https://doi.org/10.4103/0975-7406.168035
77. Parveen, B., Parveen, A., Parveen, R., & Ahmad, S. (2020). Challenges and opportunities for traditional herbal medicine today with special reference to its status in India. *Annals of Phytomedicine, 9*(2), 97–112. https://doi.org/10.21276/ap.2020.9.2.8
78. Patwardhan, B. K. (2014). Botanicals: Quality and regulatory issues botanicals: Quality and regulatory issues. *Journal of Scientific & Industrial Research, 64*, 83–92.
79. BalaYadav, R., Pathak, D. P., Varshney, R., & Arora, R. (2022). Design and optimization of a novel herbosomal-loaded PEG–poloxamer topical formulation for the treatment of cold injuries: A quality-by-design approach. *Drug Delivery and Translational Research, 12*, 2793–2823. https://doi.org/10.1007/s13346-022-01140-7
80. Mykhailenko, O., Ivanauskas, L., Bezruk, I., & Petrikait, V. (2022). Application of quality by design approach to the pharmaceutical development of anticancer crude extracts of Crocus sativus Perianth. *Scientia Pharmaceutica, 90*(1), 19. https://doi.org/10.3390/scipharm90010019
81. Marziyeh, S., Moghddam, M., Ahad, A., Aqil, M., Sarim, S., & Sultana, Y. (2017). Optimization of nanostructured lipid carriers for topical delivery of nimesulide using Box—Behnken design approach. *Artificial Cells, Nanomedicine, and Biotechnology, 45*, 1401. https://doi.org/10.3109/21691401.2016.1167699
82. ICH Q8 (R2). (2009). *Pharmaceutical development*. . Retrieved July 31, 2023, from https://database.ich.org/sites/default/files/Q8_R2_Guideline.pdf

Chapter 12
Quality By Design in Quality Assurance

Suchitra Nishal, Parmita Phaugat, Monika Gulia, and Vikas Jhawat

Abstract Juran introduced the concept of quality by design (QbD) initially, who held view that quality ought to be scheduled and believed that the majority of quality crises and issues stem form how quality was initially planned. New processes can be explored by learning about the current processes, and improve the performance of these processes to be globally competitive, experimental design is a potent technique and instrument for QbD. A crucial component of contemporary approach to quality is quality by design. When it comes to the proper vocabulary and element of quality by design, professional experts from industries are often confused. The goal of quality assurance (QA), a part of quality management, is to give customers the certainty that their needs will be addressed. This chapter briefly examines the procedures and cycles of QA in many industries, along with all the tools, benefits, and regulatory aspects of QbD. The relationship between QbD and QA as well as the difficulties in getting industry to embrace QbD are also covered. It may be concluded from the chapter that QA and QbD are essential parts of the quality management system (QMS), QbD has been discovered to be a modern method of achieving quality with predetermined objectives.

Keywords Quality by design · Experimental design · Quality assurance · Regulatory · Relationship · Quality management system

12.1 Historical Development of the Quality Paradigm

In the 1950s, Philip Crosby introduced notion of "zero defects," which was recognized to be significant advancement in commercial quality control. Quality is neither intangible nor unquantifiable. It is a crucial strategic decision that can be measured

S. Nishal
College of Pharmacy, Pt. B.D. Sharma University of Health Sciences, Rohtak, Haryana, India

P. Phaugat · M. Gulia · V. Jhawat (✉)
Department of Pharmacy, School of Medical and Allied Sciences, GD Goenka University, Gurugram, Haryana, India

N. K. Jain, N. Bajwa (eds.), *Introduction to Quality by Design (QbD)*,
https://doi.org/10.1007/978-981-99-8034-5_12

and put to use to raise profits. Acceptable levels of defect or quality, as well as conventional quality control procedures, are indicators of failure rather than proof of success. According to him, the terms "good," "bad," "high," and "low quality" have no real significance because conformity to specifications defines quality. Items that management neglected to specify or regulate are referred to as non-conforming items. Non-conformance costs the same as doing something incorrectly the first time and failing to identify and correct flaws [1]. Deming, an American statistician, also asserted in 1982 that businesses might increase the quality of their products while lowering costs by adopting the concepts of quality management [2]. The idea of Six Sigma (6σ) was also projected in 1986 through Bill Smith, a leading engineer in Motorola's Communications Division, as a result of growing concerns from the field sales strength over claims for warranty. Newest approach to establishing a uniform system for counting flaws, with 6σ approaching very near to excellence [3, 4]. By steadily lowering the organizational faults, the six sigma method is a project-driven attempt to management that intends to improve an organization's goods, services, and operations. It is a business approach that emphasizes enhancing the comprehension of client requirements, organizational structures, output, and financial outcomes.

The major goal of six-sigma is to lower prospective irregularity from practices and outcomes by either employing a constant improvement approach or a design/redesign methodology known as design for 6σ i.e. DFSS, that works on the segments of define, measure, analyze, improve in addition to control. Hence, the approach, also stated as DMAIC, is exercised in present-day processes as well as products. In the course of designing or redesigning goods or processes, design for six sigma exercises IDOV technique i.e. identify, design, optimize, along with verify. Unlike the improvement six sigma technique, design/redesign for six sigma allows for the replacement of existing processes with new ones. To assess the use of DFSS over six-sigma, it is necessary to discover when one approach becomes more important than another and to test the applicability of multicriterion decision techniques such as the AHP (analytic hierarchy process) [5]. Engineer and management consultant Dr. Joseph M. Juran, an American-Roma, is credited with creating and coining the term "Quality by Design" (QbD). In his opinion, final testing was unnecessary because quality possibly will be scheduled from the beginning of building process. The automobile business was where the idea initially executed [6, 7]. FDA indicated that complex controls were considered necessary for drug manufacturing practices with a view to fabricate valuable drugs, and certainly, to get stronger regulatory decision-making. It headed to a stringent regulatory background being created. The FDA changed procedures in 2002 under Pharmaceutical Current Good Manufacturing Practices (cGMP) for twenty-first century to direct this problem. A practice for developing, assessing, as well as managing constructing practices grounded on facts of science along with elements that impact quality of final item was discussed in Process Analytical Technology (PAT). The moment to implement QbD for a more systematic method arrived in 2005, and the USFDA requested that some businesses submit their CMC in QbD [8]. Figure 12.1 is used to portray order of events occurred in the saga of quality.

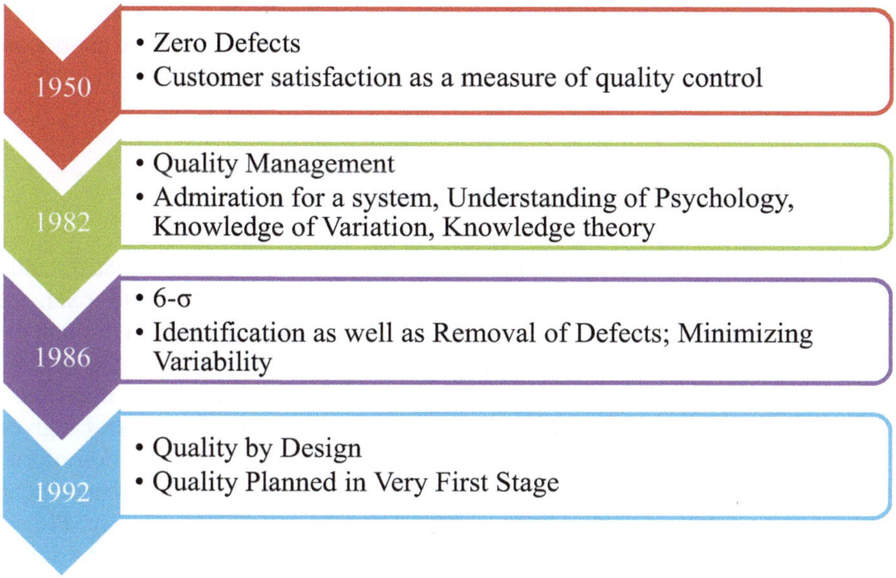

Fig. 12.1 Depicting history of quality concept

12.2 QbD Evolution and Regulatory Reforms

Joseph M. Juran, a well-known quality professional, presented quality by design (QbD) idea in book "The New Steps for Planning Quality into Goods and Services" from 1992. In his opinion, final testing was unnecessary because quality needs to be designed at initial stage. The automobiles business was where the idea initially appeared. In an article titled "Elucidation: Lessons from the Auto Industry" published in June 2007, it was claimed that Toyota Automobiles was the first business to use numerous Quality by Design concepts to enhance its automobiles in the 1970s. Because of this, we may say that the QbD concept is novel for sectors that are under FDA regulation, not for sectors like telecommunication, aviation, technology, etc. [2]. Quality by Design was used by many medical device manufacturers in the 1990s, which decreased risk and production costs while simultaneously strengthening both product effectiveness and patient safety. The FDA representatives recommended that since QbD was so successful in the production of medical devices, this idea needed to be applied to the production of pharmaceuticals and biologics as well. Therefore, internal debate within FDA commenced during 1990s' closing years, and in 2002, they ultimately issued a concept paper on cGMP in the twenty-first century. Pilot programmes to teach other companies about the use and methodology of Quality by Design were launched with the enormous assistance of some pharmaceutical corporations. According to FDA, QbD is "Developing a product to meet predefined product quality, safety and efficacy; and designing a manufacturing process to meet predefined product quality, safety and efficacy." One of Cerulean Associations LLC's consultants, John Avellanet, claims that "when Quality by Design is planned and implemented properly, the benefits are enormous. But if Quality by Design is tackled haphazardly, the benefits fizzle" [9–11].

12.3 QbD in Pharmaceutical Industry

Dr. Janet Woodcock, Deputy Commissioner & CMO, US FDA, said that "Quality by Design is going to be an evolution, but eventually, years down the road, the regulations will reflect what successful companies have done- those first through the quality by design gates [12, 13]." With minimal dependence on logical processes along with scientific opinions, pharmaceutical trade had exercised conventional tools to develop, for instance, transforming one element or one distinct parameter at given moment, which led to partial acquaintance besides behavior identification of end product as well as process. Goal of cGMPs was to revolutionize and enhance FDA oversight of pharmaceutical industry, bringing it into twenty-first century. Project that is extensively covered in this final report, aimed to update FDA's supervision of quality for certain human and veterinary medications including vaccines [10, 14, 15]. The assessment of the regulatory programmes for both pharmaceutical and chemistry, manufacture and controls took into account the objectives described in Fig. 12.2.

The FDA released PAT- A Framework for Innovative Pharmaceutical Development, Manufacturing, Quality systems Approach to Pharmaceutical Current Good Manufacturing Practice Regulations in same year to establish a thorough quality system for pharmaceutical development. The FDA and ICH released an array

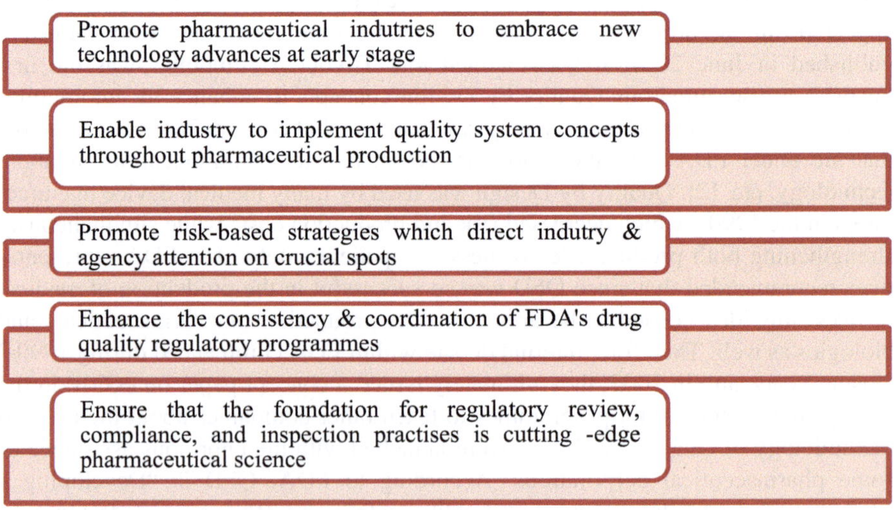

Fig. 12.2 Depicting the objectives for evaluation of regulatory programmes [8, 16]

regarding quality guidelines in an effort to harmonize quality standards around the word: ICH Q8—Pharmaceutical Development in 2004, ICH Q9- Quality Risk Management in 2005, and ICH Q10- Pharmaceutical Quality System in 2008. In accordance to these recommendations, FDA updated its Pharmaceutical Process Validation: General Principles and Practices guidelines in 2011, and included a section on using risk-based experimental design to build and improve processes.

Number of manufacturing supplements to proposals for New Drug Applications (NDAs), Abbreviated New Drug Applications (ANDAs), and Biological License Applications (BLAs) actually increased significantly in 2007, with FDA receiving over 5000 supplements. The FDA noted an increase in total firms acquiescing ANDAs and NDAs after closing date together with substantial supplementary applications were presented for each production improvement. Principal focus of data in both original application supplements was on chemistry, whereas additional crucial production processes like product development and engineering received the least attention [8, 17, 18]. Through a number of seminars and workshops, feedback from the industry was solicited in order to better understand key QbD standards along with their practical appliance for advancement of pharmaceuticals. In order to address ongoing concerns on the subject, FDA released Q&A on Q8, Q9, and Q10 guidelines in 2011. All ICH guidelines (Q8-Q10) were also acknowledged by the European Medicine Agency (EMA), in addition to the FDA. Programme of FDA and EMA trials for the concurrent evaluation of functions covering QbD features started in March, 2011 [17, 19–24].

12.4 Foundations of QbD and Advantages

ICH standards Q8, Q9, and Q10 serve as the fundamental principles of QbD, with Q8 being particularly important. The Fig. 12.3 describes the outline of foundations of QbD. Quality is described as "The suitability of either a drug substance or drug product for its intended use" in ICH Q8. Identity, purity, and strength are only a few elements of the term quality. In accordance with ICH Q8, QbD is a systematized scheme for advancement that starts through established goals and accentuates product in addition to understanding and control of process, based on strong

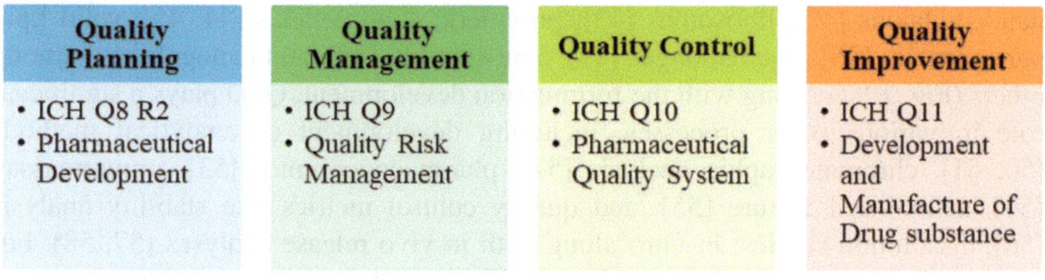

Fig. 12.3 Framework of foundations of QbD [25, 26]

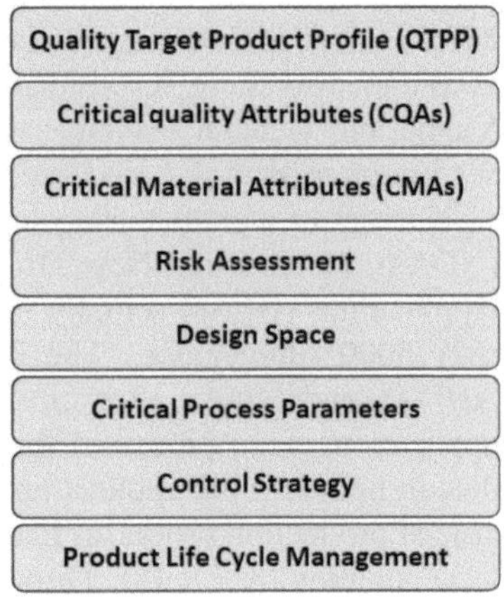

Fig. 12.4 Overview of components of QbD

scientific learning alongside quality risk management. Product testing by itself cannot ensure that a process regularly yields a product that meets predefined standards. Adequate process design, knowledge of and control over the variables that affect the production of processes, successful validation studies, and product testing [27–29].

Evaluation is mandatory for progress of final formulations; QbD establishes structures that guarantee quality of ultimate product. The elements described in Fig. 12.4 are included in QbD according to ICH Q8 [30, 31]. The tools and benefits of QbD are depicted in Figs. 12.5 and 12.6, respectively [34, 35].

12.5 Fields Implementing QbD

The QbD technique has currently been used successfully in a number of pharmaceutical research domains, including the development of conventional drug delivery systems, for instance, tablets [36, 37] and capsules [38]; novel drug delivery systems [39], such as controlled release formulations [40]; and nanocarriers [41], such as nanoemulsions [42], liposomes [43]; polymeric nanoparticles [44, 45]; solid lipid nanoparticles [46]; microsponges [47]; nanosponges [48]; and nanogels [49] among others (Fig. 12.7) Along with the formulation development, QbD plays a significant role in various other processes, including development of analytical methods [50, 51], chromatographic analysis [52], pharmacogenomics [53], immunoassay [54], animal cell culture [55], and quality control metrics like stability analysis [56], dissolution studies, in vitro along with in vivo release analyses [57, 58]. For industrialized production of vaccine, QbD principles can be used to accelerate process development. Using the current production platform and integrated

Fig. 12.5 Outline of tools of QbD

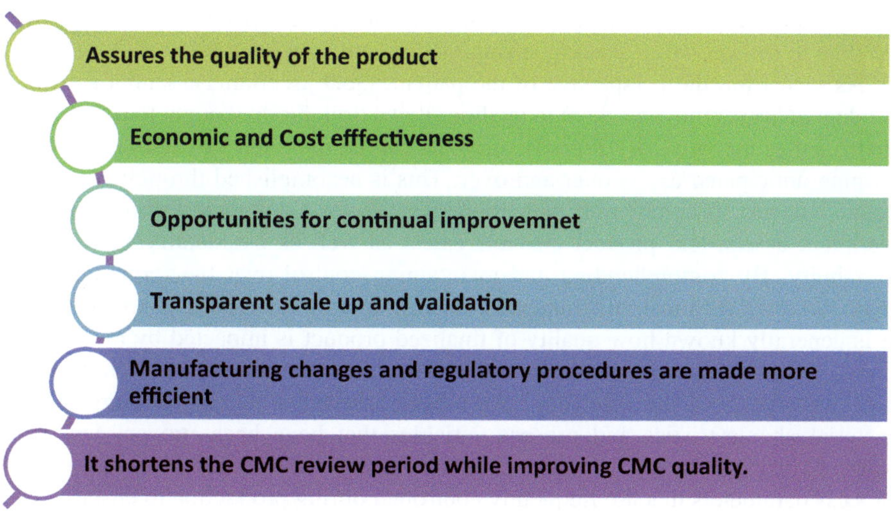

Fig. 12.6 A description of QbD's benefits [32, 33]

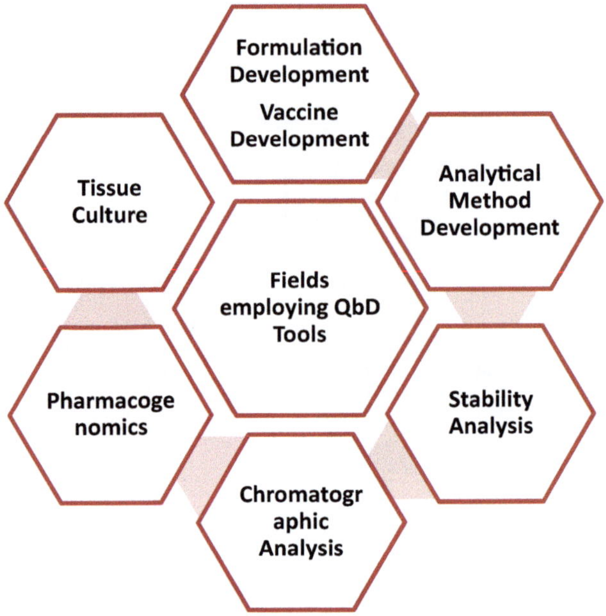

Fig. 12.7 Representation of fields employing QbD tools

knowledge encompassing historical in addition to scientific manufacturing, risk review was used to isolate process factors, and these constraints accompanied by their interactions were computed *via* DoE runs [59, 60]. Effectiveness of processes can be increased by the use of QbD, which will minimize operating costs and speed up time to market while also enhancing quality and consistency [59, 61].

As seen from the perspective of the patient, QbD ascertains essential assets for product quality, renovates them to qualities that a drug product ought to possess, and then verifies the way, key process variables can be modified to yield drug product owning anticipated assets over and over. This is accomplished through creation of acquaintances of formulation with processing factors, embracing drug substance, excipient assets, and product characteristics, as well as by identifying causes of variability. By comprehending and maintaining control over the formulation and manufacturing factors, QbD framework ensures quality for generic pharmaceuticals. It is generally known how quality of finalized product is impacted by raw components, including drug substance along with excipients. Accordingly, manufacturers are aware of both the operating range accompanied with range of crucial raw material characteristics and process variables that have been proven. Operating range is distinct as range of values for the qualities of the raw materials and the process parameters that are frequently controlled during production to ensure reproducibility [18, 24, 62].

12.6 Regulatory Aspects for QbD

According to ICHQ8 (R1), QbD is a systematized way of developing pharmaceuticals, starts with predetermined goals and places stress on both product, process control and understanding. To establish product and process control, however, it is crucial to develop suitable analytical techniques. The analytic target profile is an appliance that may be exercised to construct any analytical practice, regardless of whether it is intended to be a part of a QbD approach or in the development methodologies that have been utilized traditionally. The ATP can be used in any scenario to explain the technique specifications required to accurately measure the drug product's defined CQAs. For every method used in the overall product control strategy, an ATP would be created; it would specify what the approach must measure and the appropriate measurement is necessary namely precision, accuracy, working range, and sensitivity [63–65]. Process analytical technology integration with bioproduction practice permits transition from a rigid QbT to a more adaptable QbD methodology. By combining cutting-edge sensor technology and mathematical modeling methods, it is possible to forecast key quality characteristics online, which enables real-time product quality control. Roche [66], gained first ever QbD authorization for Gazyva, a monoclonal antibody, indicating that the QbD's positive effects had finally reached the biopharmaceutical sector [67–69]. The gaps in our capacity for supervising numerous constraints related to production cycle will progressively close as the food production sector keeps embracing the concept of in-line, online, or at-line sensors and real-time identification for procedural control and surveillance. Investments in creation of methods and tools that promote high output analytical and process development, PAT, experiment design, risk assessment, information organization, and improvement of process/product interpretation would set advanced operational as well as financial advantages in commercialization process along with additional product pipelines. Recent regulatory acceptance of the first QbD application by the United States FDA indicates the effectiveness of this initiative. Recent developments have demonstrated near infrared spectroscopy's strong potential for modeling and real-time monitoring of several food processes [18, 70, 71].

12.7 Quality Assurance: A Valued Feature in Any Industry

Quality assurance is expressed as "the collections of policies, procedures, systems, and practices internal or external to the organization designed to achieve, maintain, and enhance quality" in the Analytic Quality Glossary [72]. Assuring quality and integrity of data, goods, and devices while building trust between all parties involved would be made possible by the application of quality management standards in medical research plants. But because it can be used to any system, ISO9001:2000 is preferred by the majority of centers because it would be impossible to operate a

Fig. 12.8 Outlines the eight quality systems that help ensure the finished pharmaceutical product is of a high caliber [75, 76]

separate quality assurance programme for each system [73]. The strategy for quality assurance includes three primary actions: outlining quality, quantifying quality, and expanding quality for boosting health services and individual performance [74]. The quality systems of QA are expressed in Fig. 12.8 along with the cycle of QA in Fig. 12.9. Together, these activities cover the full spectrum of QA procedures and techniques that are mutually beneficial and operate synergistically to assure high-quality care as a system outcome. None of the core activity can boost or sustain quality on its own; rather, it necessitates the cooperation and synergy of all three to provide long-lasting gains. As part of defining quality, systems for quality must be designed as well as expectations or criteria for quality. Standards may be clinical or administrative and may be developed for inputs, processes, or outputs. The processes of quantifying current performance or conformity with expected requirements, such as customer satisfaction, are known as quality measurement. This entails defining indicators, creating and adapting information systems, and analyzing and interpreting the outcomes. Improving quality entails the execution of quality enhancement schemes and tools to close the gap between present and expected levels of quality by assessing and addressing system flaws in order to enhance, or in certain instances re-design health services systems. There are numerous methods for enhancing the quality of products and services, including individual

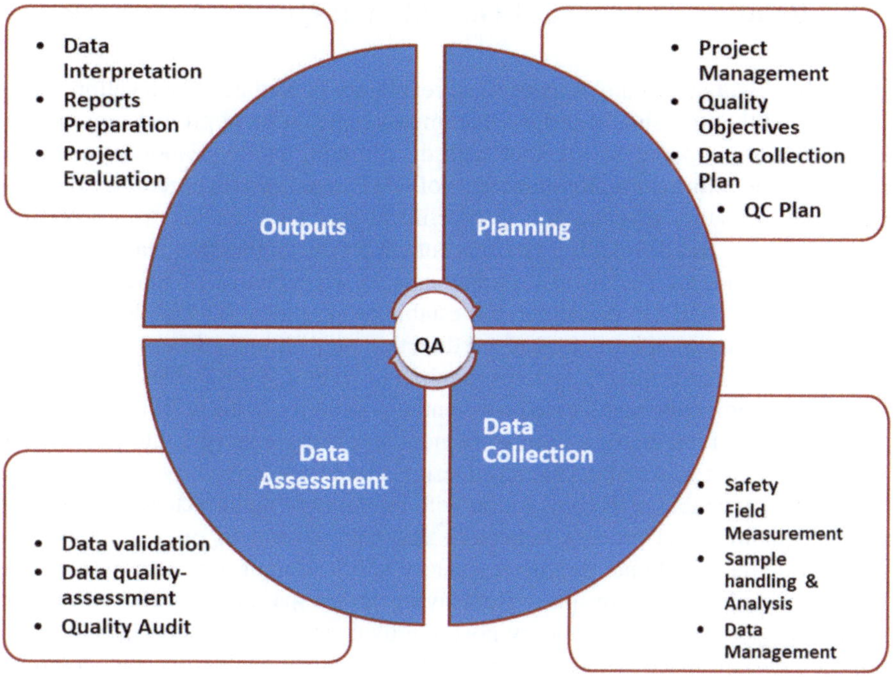

Fig. 12.9 Framework of cycle of quality assurance

problem-solving, systems and process redesign, and organizational restructuring [74, 77]. In addition to the pharmaceutical industry, higher education institutes such as the EUA (European University Association) have developed recommendations for quality assurance implementation to raise the standard of the educational system [78]. European Association for Quality Assurance in Higher Education (ENQA) oversees coordination of European quality assurance policies. A collection of European quality assurance standards has been created by the ENQA. Internal standards accompanied by guidelines, external standards with guidelines, and requirement for outside quality assurance organizations make up these standards. There are numerous institutions and organizations that are in charge of auditing higher education in each European nation. Our strategy for helping HEIs enhance the quality of education is though quality assurance and auditing. In Finland, the national higher education policy, national evaluations of HEIs, and quality assurance of HEIs make up three components of the national quality assurance of higher education [79, 80].

The basic cycle of QA can be explained with the help of figure.

12.8 Relation Among ICH Q8 (R2) and Quality Assurance

The quality assurance and pharmaceutical regulatory procedures that a nation's drug regulatory authorities should address encompass registration of products, guidelines for production, importation together with distribution, and regulation of medicine promotion and information. Requirements of GMP together with other related codes, like good laboratory practice and good clinical practices, are taken into concern while designing and developing pharmaceutical products. International Conference on Harmonization of Technical Requirements for Registration of Pharmaceuticals for Human Use (ICH), in particular, has established pertinent ICH Q8 R2, Q9, Q10, & Q11 guidelines to offer a generalized direction on path towards a pharmaceutical product, including the quality systems required to favor it [81–83]. Design space and design of experiments have gained substantial popularity in the scientific literature, and ICH Q8(R2) integrates a number of new regulatory concepts into pharmaceutical development operations [84, 85]. Design space is described by ICHQ8(R2) as multifaceted assemblage in conjunction with interplay of input factors, for instance, material qualities and process indicators that were shown to offer assurance for quality. Regulations do not require regulatory notification of parameter movements that take place inside design space. But leaving design space is regarded as variation and would typically start regulatory post-approval amendment practices. Applicant proposes design space, which is subject to regulatory evaluation and approval. Design space is thus a direct result of DoE data analysis or verified models, for instance, first-principle models [24, 86]. Equipment and scale dependencies may exist in the design space. In order to employ the design space established at laboratory scale at commercial scale, it may be necessary to justify its use. Dimensionless numbers, heat and mass transport, kinematics, geometrical considerations, and continuous verification in commercial manufacturing are some of the justification techniques. The design space can be scaled across when mechanical insight or trustworthy empirical models are present. This is essential since mechanistic insight of unit operations may be limited and scale-up is mostly reliant on general guidelines accompanying trial-and-error methods. Process understanding could be applied to each individual unit operation or to a collection of unit operations in order to establish CMAs, CPPs, CQAs [87, 88]. A control strategy is established as a result of the knowledge obtained *via* properly planned development studies. Three tiers of controls could be used in a control approach (Fig. 12.10). Real-time CQA monitoring of the output materials is done at level 1 exercising automatic engineering control. It is most compliant to degree of control. To assure that CQAs continually adhere to defined acceptance criteria, input material properties are scrutinized, and process restrictions are automatically modified. Comparing level 1 control to traditional end-product testing, the former can provide real-time release testing and latter offers advanced degree of quality assurance. Incorporating real-time release testing does not necessitate use of process analytical technology (PAT), it should be stated that using PAT is not sole attempt to do real-time release testing [89, 91, 92].

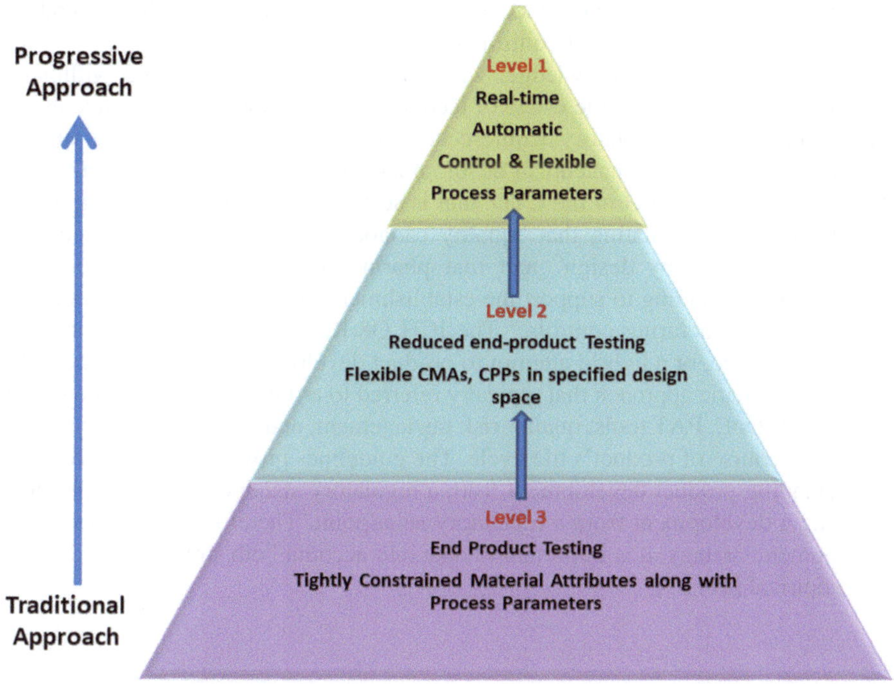

Fig. 12.10 Transitioning from traditional to progressive approach [89, 90]

Pharmaceutical control at level 2 includes adjustable material attributes together with process parameters inside defined design are, as well as fewer end-product tests. QbD promotes an insight of product and process, further makes it easier to find differentiating factors affecting product quality. It is possible to move restraints upstream and lessen dependency over end-product testing by comprehending how variability affects in-process ingredients, downstream maneuvers, and quality of drug product. The level of control that has always been applied in the pharmaceutical sector is level 3. This control technique is based on rigorous testing of the final product and strictly regulated material properties and process parameters. Any major change in these necessitates regulatory control because the causes of variability are not well-characterized, and impacts of CMAs & CPPs on the CQAs of medicinal products are not sufficiently understood. Industry and regulatory resources are heavily invested in discussions on acceptable variation, the need for added controls, and creation of acceptance criteria. In practice, a hybrid strategy integrating levels 1 and 2 may be employed. Control strategy is described as an anticipated set of controls, generated by means of current product along with process knowledge which guarantees process execution and product quality in ICH Q8(R2). Parameters together with characteristics of drug substance along with drug product ingredients and components, operating conditions of resource and devices, the in-process controls, descriptions of end product, and correlated techniques in addition to rate of monitoring are all examples of controls [84, 93–95].

The use of previous knowledge, which is defined as documented within the organization, and the possible usage of technology platforms, however have gotten far less attention. In regulatory acquiescence, such as NDAs or marketing authorization applications, pharmaceutical development section's contents are outlined in ICHQ8(R2). In module 3 of the CTD, that is sect. 3.2P.2. The guideline introduces the idea of QbD by stating that "quality cannot be tested into products; quality should be built in by design" and that pharmaceutical development "provides scientific understanding to support the establishment of the design space, specifications, and manufacturing controls." The ICH Q8(R2) guideline provides an pragmatic, systematic, or a hybrid attempt to product development. The guideline states that the systematic approach that typically referred to as QbD, may incorporate prior knowledge, DoE, PAT tools, quality risk management, and knowledge management during the course of product's life cycle. The guidelines provide a fresh approach to understanding product development from a regulatory standpoint. Therefore, when choosing a development from a regulatory standpoint. Therefore, when choosing a development strategy, it is important to take into account both strategies, systematic and empirical [96, 97].

12.9 Quality By Design: A Contemporary Approach to Quality Assurance

Adopting systematic QbD principles has had a positive impact on drug quality assurance, encouraging pharmaceutical production through advancements in drug quality, safety, and efficacy. There are four major stages in the quality assurance processes such as designing a quality assurance strategy, project quality audit, analyze, and reevaluation of the project. QA examines the procedures and makes improvements to the ones that produced the final product. QA does not actually test the product; instead it concentrates on methods and procedures. It mainly involves the process-focused activities. 'Fit for purpose' and 'Right first time' are the two basic principles of quality assurance. The key norm of QbD is that 'quality can't be assessed into products, but it should be assembled in *via* design'. Hence, QbD strives to guarantee the required product quality by evaluating the factors that could affect the quality. QbD is a meticulous, proactive, risk-based approach to pharmaceutical development commencing with programmed goals and put emphasis on product and process understanding as well as process control grounded in consistent research. According to the aforementioned remarks, QA and QbD concentrate on the procedures that are taken to ensure product's quality [24, 30, 98, 99]. Figure 12.11, describes the basic actions of QA and QbD which ensures the dependence of QbD in QA. Both identify the steps involved in achieving the goal for the target product [100].

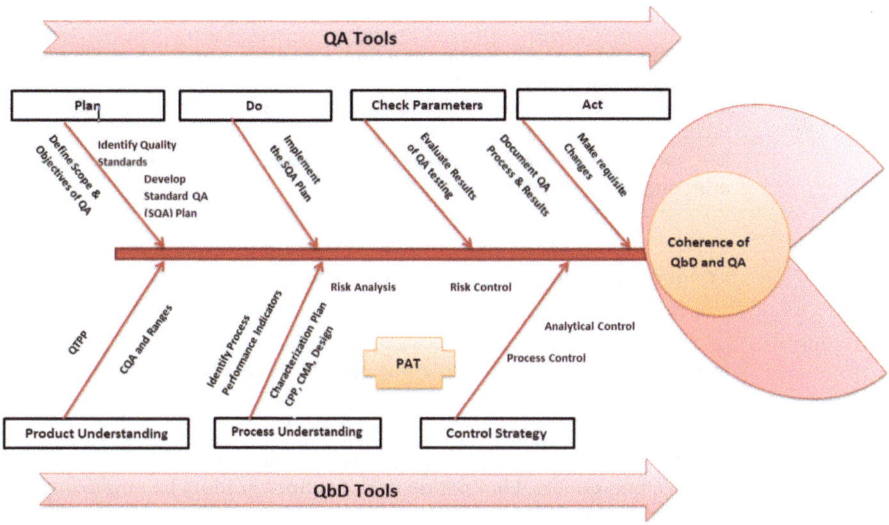

Fig. 12.11 Outlining tools of QA and QbD symbolizing their coherence [35, 100, 101]

PAT- A Framework for Innovative Pharmaceutical Development, Manufacturing, and Quality Assurance was a set of guidelines for industry released by the USFDA in September 2004. To encourage novel and risk-based assessments by industry and the Agency, the framework is built on an understanding of how processes work. The framework is made up of two parts: a strategy for implementing regulations that will take innovation into account and a collection of scientific philosophies and tools that assist innovation. The PAT team approach to chemistry manufacturing and control (CMC) review and current good manufacturing practices (cGMP) inspections, as well as combined coaching along with certification of PAT assessment and inspection professionals, are all parts of the regulatory implementation plan [102, 103]. Our new approach is meant to allay manufacturer's fears with the suggestions in the guidance. The Agency encourages manufacturers to create and put into practice valuable and competent innovative techniques in pharmaceutical development, manufacturing, and quality assurance by using PAT framework discussed here [104, 105]. The production of pharmaceuticals is constantly changing, with a growing focus on scientific and engineering concepts. Right through the life cycle of a product, efficient application of the most recent pharmaceutical engineering as well as science principles and information can increase both the production and regulatory procedures 'effectiveness. By regulating the quality of pharmaceutical products using an integrated systems approach, this FDA programme aims to achieve just that. The strategy for identifying and reducing risks associated with subpar product and process quality is based on scientific and engineering principles. Concern among manufactures that improvements in manufacturing and quality control would lead to a regulatory deadlock can be used to describe the ideal condition of pharmaceutical manufacturing and regulation

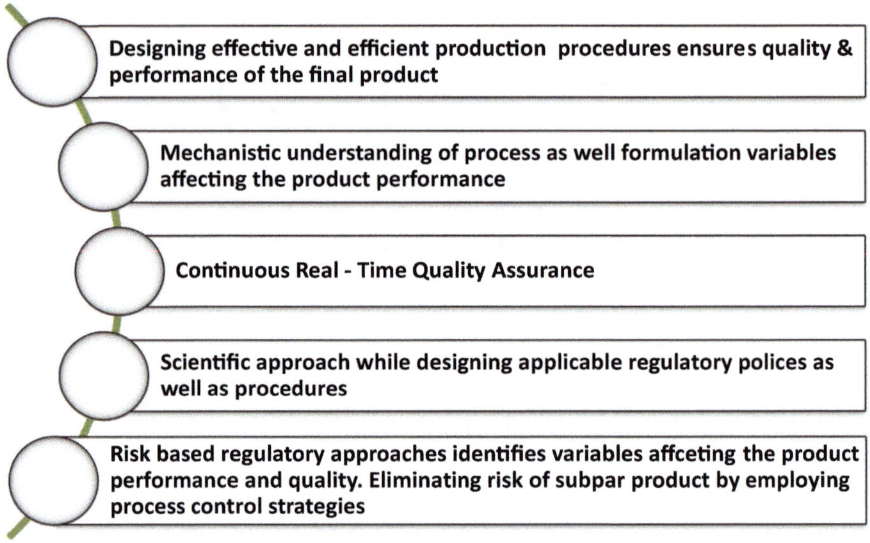

Fig. 12.12 A description of optimal environment for production and control of pharmaceuticals

in this regard. Following the guidelines referenced by FDA, Fig. 12.12 provides a description of the ideal situation for pharmaceutical manufacturing and regulation [106–109].

In an ideal QbD framework, formulation and manufacturing variables are understood and controlled to ensure pharmaceutical quality, and end-product testing is only carried out to verify the product's quality. As a result, specifications should be described in terms of how they relate to in vivo performance, such as dissolution, and how they vary within "design space" for the formulation and procedure that has been established and approved. Theoretically, any modifications done within this design space should not entail regulatory approval and should result in a persistent product quality. Nevertheless, a design space is not always the result of a QbD filing because the effort to establish process understanding should be proportionate to complexity and risk of the process and product [97]. Table 12.1 aids in explaining differences between QA and QbD [110, 111].

Requirement analysis is the first stage of quality assurance and it is known as the process of finding, comprehending, capturing, validating, and managing a component's needs. Requirement analysis aims to create comprehensive, consistent, and pertinent requirements for a process (Fig. 12.13). The target specifications are used to build up each prerequisite and constraints in this step.

In QbD, the first step is to define the quality target product profiles (QTPP). QTTP is a hypothetical overview of quality attributes of a drug product, which should be attained to assure intended quality, taking into consideration safety as well as effectiveness of medicinal product. It serves as framework for the product's design [112, 113]. It may also comprise factors like therapeutic effect, dosage form, route of

Table 12.1 Summary of distinctions among QA and QbD

S. No.	Features	Traditional QA Approach	QbD Approach
1.	**Regulation**	21 CFR parts 210 and 211 for CGMP	ICH Q8, Q9, & Q10
2.	**Pharmaceutical development**	Experimental, random, focus optimization	Design space, multivariate systematic approach, proactive in incorporating quality into products, emphasis on control strategy and robustness
3.	**Production method**	Rigid	Variable in design space
4.	**Process control**	Product quality testing and in-process inspection	PAT controlled and process operations tracking
5.	**Product qualifications**	Mainly quality control method based on batch information	A component of overall quality control approach. Dependent on product performance
6.	**Control strategy**	Testing and inspection	Risk based surveillance and PAT real time release testing
7.	**Lifecycle management**	Corrective measures and post-approval changes desired	On-going improvement within design space contingent on feedback approach

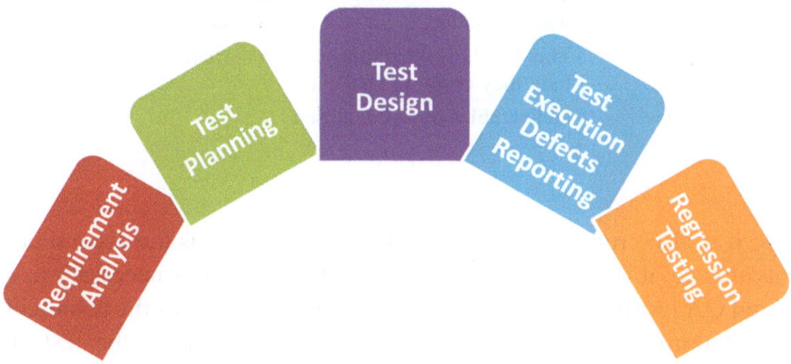

Fig. 12.13 Outline of QA process and stages

administration, dose strength, target site, physical attributes, pH, homogeneity, in vitro drug release, pharmacokinetics, safety, sterility, and stability. Within the QbD process, these can be managed and optimized to create the desired end result. This will encourage in initial stages of new product development by providing clear direction for the process. As well as reducing the overall development time and costs, it will help identify medical device development failures in the early stages. To best satisfy future product expectations, it is crucial to define the QTTP prior to beginning the creation of a new product. In addition, consumer's demands related to product sales and market success should be considered when defining QTTP. The formation of QTTP is an active process that gets response from all QbD organized

Table 12.2 Describes the QA test plan [117]

Steps	Name	Description
Step 1	Analyze the product	Describe every feature of the product Number of test cases required for each feature, in each circumstance The entire required test should be listed
Step 2	Analyze target audience	Target audience should be taken into account while developing a QA test plan. Making sure to prioritize the needs of the consumer is essential. Additionally, this will guarantee that the user is considered throughout the testing process
Step 3	Design a strategy	Record all the test cases to determine what needs to be tested along with when it should be tested. Scope of the test is defined. Recognition of different testing types and designing of test strategy
Step 4	Define the objectives	Finding out what the QA testing goals are is the next stage. It also entails deciding who us in charge of testing, what will be tested, when it should be finished, and how the outcomes will be evaluated
Step 5	Determine criteria of test	The requirements that must be met for the test are categorized as suspension criteria (are the circumstances that would call for a temporary halt to the testing) and exit criteria Exit criteria must be met for a test to be deemed successful. The test can advance to advance stage when an exit requirement is satisfied
Step 6	Plan resources	After development of a strategy and test plan, resources are to be decided for completion of the task. These resources may include number of person, time duration, management tools and budget
Step 7	Plan test environment	This covers everything, from the location of the exams to the best way to conduct them and who will conduct them
Step 8	Schedule and estimation	This involves scheduling tests for when they must be run and how long they ought to take
Step 9	Determine test deliverables	Deliverables must be chosen

processes, it is also crucial to note [9, 114]. Test planning is the second process for quality assurance; it is a written description of the procedures necessary to carry out the required QA testing. It is a thorough document that outlines the resources needed to carry out testing and produce the best results, as well as the test strategies, objectives, itinerary, forecasting, and success standards. It also specifies who in your organization will be in charge of each assignment, the subjects being tested, and the deadline for completion. It is necessary to concentrate on establishing processes in order to develop the ideal test plan [115, 116]. The structure for developing a QA test plan is presented in Table 12.2.

The test design in QA addresses the project objectives. For knowing the functionality of a suitable feature, the requirements and the procedures must be determined in test design. CQAs is the second step for QbD process, a biological, chemical, physical, microbiological quality or attribute that falls in required range, or distributed in order to maintain the intended level of quality of product [115, 118, 119]. Process design stage begins here. A drug product's quality traits may incorporate identity, drug content uniformity, assay, products of degradation, residuals, dissolution, percent moisture content, microbiological limitations, and physical aspects like odor, color, and configuration. The degree of criticality of any attribute is defined by its impact on consumer's limits. Critical Process parameter (CPP) is factor that could affect the CQA. In order to facilitate and precise identification of variations beyond approved limits that impacts product quality, CPPs must be monitored. CMA is another phrase frequently used when analyzing CPPs and how they affect CQAs. The acronym for this is Critical Material Attributes, it is concerned with the input materials' chemical, physical, biological or microbiological characteristics. In order to establish CMAs, CPPs, and CQAs, process insight could be applied to each individual unit operation or to a group of unit activities [24, 120, 121]. By comparing test plan and test design of QA with QbD processes mentioned above, it may be concluded that the test plan for QA incorporates QbD practices such as CQAs, CPPs, and CMA of QbD. Next step for quality assurance is to identify and correct the defects in quality. Cause and effect analysis and Pareto analysis are two crucial industrial engineering approaches for addressing quality issues. A Pareto diagram, often known as a Pareto chart, is a straightforward statistical diagram. It helps with quality control. This was created by renowned Italian economist Vilfredo Pareto at the beginning of the nineteenth century. Today, it is common practice in quality management to distinguish between the significant few and unimportant many. It resembles a cumulative bar graph. With the longest bars on the left and smallest bars to the right, the lengths of the bars signify frequency or expense. The critical few are represented by the longest bar. It applies the Pareto principle, which holds that only a small number of significant factors (20%) account for the vast majority of issues (80%). Important phases in production of pharmaceutical drugs include processing and compression. But a lot of unsatisfactory products are made here. Cause-effect analysis is employed to notify the causes linked to the specific problem. After conducting a brainstorming exercise, the causes were divided into groups, including man, machine, technique, measurement, material, and environment [122, 123]. QbD in pharmaceuticals focuses on improvement of process efficiency and decrease product variability, which frequently results in defects in product, rejections, and recalls. Risk assessment through QbD facilitates in prior identifications risks associated with the process and hence a control approach can be decided within the optimized design space. Understanding and predicting the sources of variability in the manufacturing process is a major goal of risk assessment for pharmaceutical development. This makes it possible to out into practice an applicable control strategy to validate that the CQAs adhere to the desired standards. CPPs and CQAs are identified iteratively throughout the development process. Scarcity of enough process/ product expertise of product being developed at early stages of development, past knowledge is the main basis for the designation. The risks that are initially recognized are therefore apparent risks, and when process/ product expertise is increased, true risk become more obvious and a management strategy can be more clearly outlined. Therefore, risk-assessment approaches employed in initial stages of development happen to be primarily qualitative and act as a way to prioritize testing [124–127]. The QbD-QTPP, CQA, CMA, and CPP aspects strengthen product approach to development, resulting in a complete improvement. Finally, during product lifecycle management, product and process capability are continuously evaluated and enhanced after approval.

12.10 Critical Challenges of Introducing QbD in Traditional QA Process

QA and QC are the traditional approaches followed by the industries to achieve quality and these have now turned into an integral measure of quality management system of every kind of industry. In contrast to QA, QbD involves various modern tools and techniques in order to introduce quality in the product from the very beginning stage. The industries can face many challenges during implementation of the modern QM approach *i.e.* QbD, these challenges may be simple or complex depending upon the infrastructure of the industries. Table 12.3 lists the main issues that industries are facing along with potential remedies from personnel of FDA's Center for Drug Evaluation and Research (CDER) [128, 129].

12.11 Discussion

The FDA and other regulatory authorities have established highly strict standards and guidelines for the pharmaceutical industries. As the potential risks are quite high, it is essential for the pharmaceutical corporations to follow these standards. To ensure a high-quality final product, pharmaceutical businesses must adhere to GMP protocols and regulations set forth by regulatory organizations. These standards' main objective is to ensure that patients' and consumers' health is protected by creating high-quality pharmaceuticals. Since GMP compliance is required for medications, all pharma companies place strong emphasis on quality. Quality Control (QC), Quality Assurance, and Quality Management are the three basic traditional methods for ensuring quality. An inspection-based technique is used in QC. It is used to guarantee a certain degree of quality in a good or service. The process of quality control involves comparing the product's quality to the appropriate standard. When the product is ready, it checks the non-conformance and flaws. It is clear that the flaws are unacceptable. Examining the finished product or service is what quality control is all about. Analyzing the process that leads to the final product is a key component of QA. In order to reduce or completely eliminate potential issues with quality of final product, industries employ QA to make sure that a product is made correctly. Pharma corporations have historically placed a strong emphasis on a strategy that combines QA and QC. Few pharma behemoths have adopted the QA strategy *via* Six Sigma deployment, but the great majority has not yet switched from QC approach. Even those who moved to QA had their process excellence efforts restricted to manufacturing. But quality management (QM), the modern strategy that many cutting-edge pharmaceutical businesses have adopted, is an end-to-end strategy. Beginning with consideration of consumer needs, it moves on to product design and concludes with post-purchase feedback. Although it has long been a staple of manufacturing jargon, it has recently begun to spread to other sectors as well. In pharma industry, constant improvements are considered at all

Table 12.3 Challenges of QA to QbD Quality Paradigm Shift

S. No.	Challenges	Remedial Approach from FDA's CDER Personnel
1.	Because there is requirement of specialized instrumentation and additional testing, controlling variability	Lack of equipment and data management techniques are required for tracking, controlling along with knowledge execution in existing pharmaceutical industries. The business and regulatory advantages will be easier to see when more industries adopt QbD and PAT strategies
2.	Scale-up and legacy product issues with QbD	Commercial data, along with further development work, may result in something from stronger process regulators to realizing demand for updating application with updated product qualifications. This may or may not entail the creation of a design space. During all stages of product, embracing scaling up and technology transfer, the ICH Q10 PQS offers guidance for adopting these methods. The FDA never stops educating its compliance and investigational employees about cutting-edge technology, quality systems, risk-management. Technical CDER compliance specialists oversee as well as take part in preapproval along with post-approval assessments. Coordination of joint inspections with international counterparts has been greatly aided by CDER's Office of Compliance
3.	Regulatory viewpoint on dimensions of design space including proximity of a company's process results to the design space's edges	Importance of a design space is independent of size or process location inside it, but rather in the process understanding it symbolizes, which is based on sound research together with knowledge gathered all through commercialized process during product's life cycle after design space is altered. ICH Q10 provides efficient control strategy. FDA's new process validation regulation makes it clearer that validation is not a firm activity but should be ongoing approach instead which can be used regardless of design space in addition modern tools that continuously monitor process
4.	A quality system is used by ICH Q10 to focus on a life-cycle approach. Which approaches can the industry use to implement this strategy most effectively?	By using data gathered in a commercial process to spur ongoing development, the challenges can be overcome. The process-performance along with product-quality monitoring system turns out as crucial PQS component. If forecasts made throughout development have given a

(continued)

Table 12.3 (continued)

S. No.	Challenges	Remedial Approach from FDA's CDER Personnel
		good grasp of the process, that system will confirm it
5.	Why is it suggested that an industry start considering QbD ideas near ending of phase II? Does employing a quality based approach ever reach a conclusion?	At the conclusion of phase II, the rigor of some development studies might rise. Real-time release testing and the types of process controls that will be implemented will also are taken into account. The division of manufacturing and product quality of the OC has examined propositions at diverse stages, particularly after new resources or manufacturing tools are being projected. Each stage of product life cycle is highlighted in ICHQ10. They can be reached through the website's subject-matter specialists: www.fda.gov/Drugs/DevelopmentApprovalProcess/Manufacturing/default.html. Preclinical, clinical, and marketing aspects of a company's operations should be covered by a PQS
6.	Will FDA change the requirements for submitting QbD-based applications for new pharmaceuticals now that the chemistry, manufacturing, and controls pilot Programme are coming to an end?	Applications are submitted in the same way for both traditional and QbD methods. It is advantageous to us administratively if the applicant highlights the QbD components or other distinctive parts of the application in the cover letter
7.	Which of the CMC pilot's key lessons did FDA learn most effectively?	Learned how to establish regulatory choices on suggested manufacturing methods and QbD data in a science-and risk-based manner. Accumulated knowledge of several flexible regulatory approaches, including RTRT, in-process testing, and design spaces. The effectiveness of QbD for pharmaceuticals depends on both quality risk-management and pharmaceutical quality techniques, according to one of primary conclusions from the pilot
8.	As of May 2009, FDA had received more submissions based on QbD than it had during the CMC Pilot Programme (12 NDAs, 18 INDs, and 3 supplementary NDAs). Does the industry appear to understand what the FDA is seeking in QbD-based submissions?	Before CMC pilot published, early submissions had most of their development finished. The applications resembled hybrids more since they combined conventional and QbD methodologies. Some of the most recent applications seeing were created entirely using QbD techniques

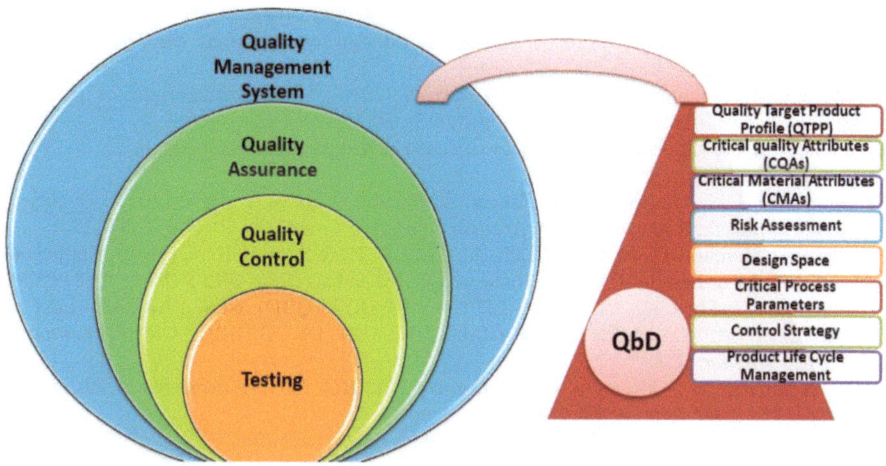

Fig. 12.14 A rationalization how QbD and QA are related

stages in addition to the manufacturing of drugs, with the goal of lowering the industry-wide high molecule-to-market time. Customer focus, people involvement, leadership, process approach, recurrent improvement, and other ideologies are adopted by QM. QM approach, which has so far been used by the world's largest automakers, incorporates QbD through Design for Six Sigma strategy, which makes sure that quality is fostered into product through product and process design. QbD is a concept that has recently gained a lot of attention in the pharmaceutical sector. Veteran chemists are actively discussing QbD. QbD is a special DFSS adaptation for pharmaceutical industries. It addresses the 'Quality in Design' component of pharmaceutical products, which encompasses process design and implementation. To ensure a set level of product quality, QbD belongs to planning and developing formulations and manufacturing procedures. Although QbD is a modern technique that considers every aspect of the process, product, and manufacturing resources, it is also reasonable to say that QbD and QA are components of QM. The figure does a better job of expressing the relationship between QbD and QA.

The above figure describes that the QMS is a composition of conventional testing, QC, and QA. On the other hand QbD with different tools competes with the conventional approach facilitates the production of quality product with assurance of quality. QbD can be marked as improvised version of QA (Fig. 12.14).

References

1. Crosby, P. (2001, 22 August). Developer of the zero-defects concept. *The New York Times*.
2. Deming, W. E. (1993). *The new economics for industry* (Vol. 1, p. 235). Government, Education, Massachusetts Institute of Technology.
3. Barney, M. (2002). Motorola's second generation. *Six Sigma Forum Magazine, 1*(3), 13–16.

4. Pyzdek, T. (1997). Motorola's six sigma program. *Quality Digest.*
5. Banuelas, R., & Antony, J. (2004). Six sigma or design for six sigma? *The TQM Magazine, 16*(4), 250–263.
6. McConnell, J., Nunnally, B. K., & Mcgarvey, B. (2010). The forgotten origins of quality by design. *Journal of Validation Technology, 16*(3), 30.
7. Juran, J. M. (1992). *Juran on quality by design: The new steps for planning quality into goods and services.* Simon and Schuster.
8. Sangshetti, J. N., Deshpande, M., Zaheer, Z., Shinde, D. B., & Arote, R. (2017). Quality by design approach: Regulatory need. *Arabian Journal of Chemistry, 10*, S3412–S3425.
9. Martinez-Marquez, D., Mirnajafizadeh, A., Carty, C. P., & Stewart, R. A. (2018). Application of quality by design for 3D printed bone prostheses and scaffolds. *PLoS One, 13*, e0195291.
10. Purohit, P., & Shah, K. (2013). QUALITY BY DESIGN (QBD): NEW PARAMETER FOR QUALITY IMPROVEMENT & PHARMACEUTICAL DRUG DEVELOPMENT. *Pharma Science Monitor, 3*, 3.
11. Weckenmann, A., Akkasoglu, G., & Werner, T. (2015). Quality management–history and trends. *The TQM Journal, 27*, 281–293.
12. Avellanet, J. (2007). *Is quality by design right for my organization…?*
13. Walker, T. (2006). FDA's' critical Path'initiative could lay groundwork for expedited drug development, quality improvement. *Formulary, 41*, 255–255.
14. FDA, U. (2004). *Pharmaceutical CGMPs for the 21st century—A risk-based approach.* US Food & Drug Administration.
15. Immel, B. K. (2001). A brief history of the GMPs for pharmaceuticals. *Pharmaceutical Technology, 25*, 44–53.
16. Bhutani, H., Kurmi, M., Singh, S., Beg, S., & Singh, B. (2004). Quality by design (QbD) in analytical sciences: An overview. *Quality Assurance, 3*, 39–45.
17. Beg, S., Hasnain, M. S., Rahman, M., & Swain, S. (2019a). Introduction to quality by design (QbD): Fundamentals, principles, and applications. In *Pharmaceutical quality by design.* Elsevier.
18. Jain, S. (2014). Quality by design (QBD): A comprehensive understanding of implementation and challenges in pharmaceuticals development. *International Journal of Pharmacy and Pharmaceutical Sciences, 6*, 29–35.
19. Bendale, H. P., Bakliwal, A. A., Talele, S. G., Deshmukh, S. A., & Chaudhari, G. N. (2015). Quality by design: A modern approach in pharmaceutical development of formulation. *World Journal of Pharmaceutical Research, 4*, 402–422.
20. Kumar, R., Mittal, A., & Kulkarni, M. P. (2022). Quality by design in pharmaceutical development. In *Computer aided pharmaceutics and drug delivery: An application guide for students and researchers of pharmaceutical sciences.* Springer.
21. Moore, C. M. (2008). *FDA update on implementation of quality by design.* The 2008 Annual Meeting.
22. Saharan, V. A. (2022). *Computer aided pharmaceutics and drug delivery: An application guide for students and researchers of pharmaceutical sciences.* Springer Nature.
23. Shirohiwala, R., Shah, C., & Upadhyay, U. (n.d.). *Implementation of Qbd & AQbD approach in pharmaceutical formulation and analytical method development: A comprehensive research-review.*
24. Yu, L. X., Amidon, G., Khan, M. A., Hoag, S. W., Polli, J., Raju, G., & Woodcock, J. (2014). Understanding pharmaceutical quality by design. *The AAPS Journal, 16*, 771–783.
25. Beg, S., & Hasnain, M. S. (2019). *Pharmaceutical quality by design: Principles and applications.* Academic.
26. Beg, S., Rahman, M., & Kohli, K. (2019b). Quality-by-design approach as a systematic tool for the development of nanopharmaceutical products. *Drug Discovery Today, 24*, 717–725.
27. Chavan, S. D., Pimpodkar, N. V., Kadam, A. S., & Gaikwad, P. S. (2016). Quality by design. *Asian Journal of Research in Pharmaceutical Science, 6*, 45–50.

28. Kadam, V. R., Patil, M., Pawar, V. V., & Kshirsagar, S. (2017). A review on: Quality by design (QbD). *Asian Journal of Research in Pharmaceutical Sciences, 7*, 197–204.
29. Mishal, A., & Rathod, S. (2014). Quality by design: A new era of development of quality in pharmaceuticals. *Research Journal of Pharmacy and Technology, 7*, 581–591.
30. Nadpara, N. P., Thumar, R. V., Kalola, V. N., & Patel, P. B. (2012). Quality by design (QBD): A complete review. *International Journal of Pharmaceutical Sciences Review and Research, 17*, 20–28.
31. Taticek, R., & Liu, J. (2015). Definitions and scope of key elements of QbD. In *Quality by design for biopharmaceutical drug product development* (pp. 31–46).
32. Aksu, B., & Mesut, B. (2015). Quality by design (QbD) for pharmaceutical area. *Journal of Faculty of Pharmacy of Istanbul University, 45*, 233–251.
33. Zacché, M. & Andersson, M. (2019). *The advantages of a quality by design approach in pharma drug development.*
34. Betterman, S. M., Levy, S. E., & Brown, B. A. (2012). A tale of two drugs: How using QbD tools can enhance the development process. *Journal of GXP Compliance, 16*, 34.
35. Gandhi, A., & Roy, C. (2016). Quality by design (QbD) in pharmaceutical industry: Tools, perspectives and challenges. *Pharma Tutor, 4*, 12–20.
36. Simão, J., Chaudhary, S., & Ribeiro, A. (2023). Implementation of quality by design (QbD) for development of bilayer tablets. *European Journal of Pharmaceutical Sciences, 1*, 106412.
37. Singh, R., Muzzio, F. J., Ierapetritou, M., & Ramachandran, R. (2015). A combined feed-forward/feed-back control system for a QbD-based continuous tablet manufacturing process. *Processes, 3*, 339–356.
38. Stegemann, S., Connolly, P., Matthews, W., Barnett, R., Aylott, M., Schrooten, K., Cadé, D., Taylor, A., & Bresciani, M. (2014). Application of QbD principles for the evaluation of empty hard capsules as an input parameter in formulation development and manufacturing. *AAPS PharmSciTech, 15*, 542–549.
39. Šahinović, M., Hassan, A., Kristó, K., Regdon, G., Vranić, E., & Sovány, T. (2023). Quality by design-based development of solid self-emulsifying drug delivery system (SEDDS) as a potential carrier for oral delivery of lysozyme. *Pharmaceutics, 15*, 995.
40. Nagy, B., Szabados-Nacsa, Á., Fülöp, G., Nagyné, A. T., Galata, D. L., Farkas, A., Mészáros, L. A., Nagy, Z. K., & Marosi, G. (2023). Interpretable artificial neural networks for retrospective QbD of pharmaceutical tablet manufacturing based on a pilot-scale developmental dataset. *International Journal of Pharmaceutics, 633*, 122620.
41. Mehrotra, S., & Kumar, L. (2023). Implementation of quality by design in the formulation and development of nanocarrier-based drug delivery systems. *Critical Reviews™ in Therapeutic Drug Carrier Systems, 40*, 1–46.
42. Patra, C. N., Mishra, A., Jena, G. K., Panigrahi, K. C., Sruti, J., Ghose, D., & Sahoo, L. (2023). QbD enabled formulation development of nanoemulsion of nimodipine for improved biopharmaceutical performance. *Journal of Pharmaceutical Innovation, 1*, 1–19.
43. Alshaer, W., Nsairat, H., Lafi, Z., Hourani, O. M., AL-Kadash, A., Esawi, E., & Alkilany, A. M. (2022). Quality by design approach in liposomal formulations: Robust product development. *Molecules, 28*, 10.
44. Ghose, D., Patra, C. N., & Swain, S. (2023). QbD-based formulation development and characterization of Isradipine loaded freeze-dried nanoparticles. *Journal of Pharmaceutical Negative Results, 1*, 451–465.
45. Soundarya, R., Halagali, P., Preethi, S., Vikram, H. P., Mehdi, S., & Singadi, R. (2023). Quality by design (QBD) approach in processing of nanoparticles loading antifungal drugs. *Journal of Coastal Life Medicine, 11*, 279–290.
46. Harish, V., Almalki, W. H., Alshehri, A., Alzahrani, A., Gupta, M. M., Alzarea, S. I., Kazmi, I., Gulati, M., Tewari, D., & Gupta, G. (2023). Quality by design (QbD) based method for estimation of Xanthohumol in bulk and solid lipid nanoparticles and validation. *Molecules, 28*, 472.

47. Dev, A., Dwivedi, J., & Momin, M. (2019). Quality by design based formulation and evaluation of acyclovir microsponges. *Journal of Drug Delivery and Therapeutics, 9*, 54–60.
48. Rodrigues, K., Nadaf, S., Rarokar, N., Gurav, N., Jagtap, P., Mali, P., Ayyanar, M., Kalaskar, M., & Gurav, S. (2022). QBD approach for the development of hesperetin loaded colloidal nanosponges for sustained delivery: In-vitro, ex-vivo, and in-vivo assessment. *OpenNano, 7*, 100045.
49. Kesharwani, D., Paul, S. D., Paliwal, R., & Satapathy, T. (2023). Exploring potential of diacerin nanogel for topical application in arthritis: Formulation development, QbD based optimization and pre-clinical evaluation. *Colloids and Surfaces B: Biointerfaces, 223*, 113160.
50. Pant, A., Sharma, G., Saini, S., Jain, A., Barnwal, R. P., Singh, G., & Singh, B. (2023). Quality by design (QbD)-steered development and validation of analytical and bioanalytical methods for Raloxifene: Application of Monte Carlo simulations and variance inflation factor. *Biomedical Chromatography, 37*, e5641.
51. Xiao, C., Zhang, X., Wang, W., Yang, X., Sun, Y., Zhang, W., He, S., Yang, J., Lv, Z., & Hu, C. (2023). Developing an improved UHPLC method for impurity profile analysis of ceftriaxone using analytical quality by design. *Analytical Methods: Advancing Methods and Applications, 15*, 639–647.
52. Kumbhar, P., Waghmare, P., Nadaf, S., Manjappa, A., Shah, R., & Disouza, J. (2023). QbD and six sigma quality approach for chromatographic estimation of repurposed simvastatin from nanostructured lipid carriers. *Microchemical Journal, 185*, 108310.
53. Jhawat, V., Gulia, M., Gupta, S., Maddiboyina, B., & Dutt, R. (2020). Integration of pharmacogenomics and theranostics with nanotechnology as quality by design (QbD) approach for formulation development of novel dosage forms for effective drug therapy. *Journal of Controlled Release, 327*, 500–511.
54. Verch, T. (2014). Application of quality by design and statistical quality control concepts in immunoassays. *Bioanalysis, 6*, 3251–3260.
55. Lee, H. W., Bhatia, H., Park, S. Y., Kamga, M. H., Reimonn, T., Sha, S., Huang, Z., Galbraith, S., Liu, H., & Yoon, S. (2019). Process analytical technology and quality by design for animal cell culture. In *Cell culture engineering: Recombinant protein production* (pp. 365–390).
56. Waterman, K. C. (2011). The application of the accelerated stability assessment program (ASAP) to quality by design (QbD) for drug product stability. *AAPS PharmSciTech, 12*, 932–937.
57. Mahmood, A., Rapalli, V. K., Gorantla, S., Waghule, T., & Singhvi, G. (2022). Dermatokinetic assessment of luliconazole-loaded nanostructured lipid carriers (NLCs) for topical delivery: QbD-driven design, optimization, and in vitro and ex vivo evaluations. *Drug Delivery and Translational Research, 12*, 1118–1135.
58. Pawar, J., Tayade, A., Gangurde, A., Moravkar, K., & Amin, P. (2016). Solubility and dissolution enhancement of efavirenz hot melt extruded amorphous solid dispersions using combination of polymeric blends: A QbD approach. *European Journal of Pharmaceutical Sciences, 88*, 37–49.
59. Chang, L., Blue, J. T., Schaller, J., Phelan, L., Khandke, L., & Green, B. A. (2015). Applicability of QbD for vaccine drug product development. In *Quality by design for biopharmaceutical drug product development* (pp. 437–473).
60. Haas, J., Franklin, A., Houser, M., Maraldo, D., Mikola, M., Ortiz, R., Sullivan, E., & Otero, J. M. (2014). Implementation of QbD for the development of a vaccine candidate. *Vaccine, 32*, 2927–2930.
61. Yarovoi, H., Frey, T., Bouaraphan, S., Retzlaff, M., & Verch, T. (2013). Quality by design for a vaccine release immunoassay: A case study. *Bioanalysis, 5*, 2531–2545.
62. Raw, A. S., Lionberger, R., & Yu, L. X. (2011). Pharmaceutical equivalence by design for generic drugs: Modified-release products. *Pharmaceutical Research, 28*, 1445–1453.
63. Das, V., Bhairav, B., & Saudagar, R. (2017). Quality by design approaches to analytical method development. *Research Journal of Pharmacy and Technology, 10*, 3188–3194.

64. Rozet, E., Lebrun, P., Michiels, J.-F., Sondag, P., Scherder, T., & Boulanger, B. (2015). Analytical procedure validation and the quality by design paradigm. *Journal of Biopharmaceutical Statistics, 25*, 260–268.
65. Schweitzer, M., Pohl, M., Hanna-Brown, M., Nethercote, P., Borman, P., Hansen, G., Smith, K., & Larew, J. (2010). Implications and opportunities of applying QbD principles to analytical measurements. *Pharmaceutical Technology, 34*, 52–59.
66. FDA. (2013). *Food and Drug Administration BLA Market Approval Letter for Gazyva*. https://www.accessdata.fda.gov/drugsatfda_docs/nda/2013/125486Orig1s000Approv.pdf
67. Kelley, B. (2016). Quality by design risk assessments supporting approved antibody products. In *MAbs* (pp. 1435–1436). Taylor & Francis.
68. Sommeregger, W., Sissolak, B., Kandra, K., von Stosch, M., Mayer, M., & Striedner, G. (2017). Quality by control: Towards model predictive control of mammalian cell culture bioprocesses. *Biotechnology Journal, 12*, 1600546.
69. Staby, A., Ahuja, S., & Rathore, A. S. (2017). Model-based preparative chromatography process development in the QbD paradigm. In *Preparative chromatography for separation of proteins* (pp. 1–10).
70. Rathore, A. S., & Kapoor, G. (2016). Implementation of quality by design for processing of food products and biotherapeutics. *Food and Bioproducts Processing, 99*, 231–243.
71. Rathore, A. S., & Kapoor, G. (2017). Implementation of quality by design toward processing of food products. *Preparative Biochemistry and Biotechnology, 47*, 435–440.
72. Williams, J. (2016). *Quality assurance and quality enhancement: Is there a relationship?* (Vol. 22, pp. 97–102). Taylor & Francis.
73. Fabregas-Fernández, A., García-Montoya, E., Pérez-Lozano, P., Suné-Negre, J. M., Ticó, J. R., & Minarro, M. (2010). Quality assurance in research: Incorporating ISO 9001: 2000 into a GMP quality management system in a pharmaceutical R+ D+ I center. *Accreditation and Quality Assurance, 15*, 297–304.
74. Silimperi, D. R., Miller Franco, L., Veldhuyzen Van Zanten, T., & Macaulay, C. (2002). A framework for institutionalizing quality assurance. *International Journal for Quality in Health Care, 14*, 67–073.
75. Kidney, G., Cummings, L., & Boehm, A. (2007). Toward a quality assurance approach to e-learning courses. *International Journal on E-learning, 6*, 17–30.
76. Petropoulu, A. (2018). Quality assurance/pharmaceutical quality systems in manufacturing medicinal products. *European Pharmaceutical Review, 1*, 1.
77. Tzavaras Catsambas, T., Kelley, E., Legros, S., Massoud, R., & Bouchet, B. (2002). The evaluation of quality assurance: Developing and testing practical methods for managers. *International Journal for Quality in Health Care, 14*, 75–081.
78. Loukkola, T., & Zhang, T. (2010). *Examining quality culture: Part 1-quality assurance processes in higher education institutions*. European University Association Brussels.
79. Kontio, J. (2012). *Quality assurance at higher education institutes: The role of educational initiatives* (pp. 27–31). International Conference on Engineering Education.
80. Lindstrom, J. (1994). A system for quality improvement in higher education. *European Journal of Engineering Education, 19*, 255–261.
81. Granato, G. E., Bank, F. G., & Cazenas, P. A. (1998). *Data quality objectives and criteria for basic information, acceptable uncertainty, and quality-assurance and quality-control documentation*. US Dept. of the Interior, US Geological Survey; Branch of Information.
82. Jones, B. E. (2003). Principles and practices for quality assurance and quality control. In *The National highway runoff data and methodology synthesis* (p. 23).
83. Kulkarni, S. V., & Bertoni, M. J. (1996). *Environmental sampling quality assurance*. American Chemical Society.
84. Holm, P., Allesø, M., Bryder, M. C., & Holm, R. (2017). Q8 (R2) Pharmaceutical development. In *ICH quality guidelines: an implementation guide* (pp. 535–577).

85. Singh, J. (2015). International conference on harmonization of technical requirements for registration of pharmaceuticals for human use. *Journal of Pharmacology and Pharmacotherapeutics, 6*, 185–187.
86. Patil, H. D., Patil, C. B., Patil, V. V., & Patil, P. S. (2023). An overview on quality by design. *Asian Journal of Research in Pharmaceutical Science, 13*, 1.
87. Saxena, V., & Deshkara, S. (n.d.). *Quality by design: A scientific approach.*
88. Sousa, A., Serra, J., Estevens, C., Costa, R., & Ribeiro, A. (2023). A quality by design approach in oral extended release drug delivery systems: Where we are and where we are going? *Journal of Pharmaceutical Investigation, 53*, 269–306.
89. Rios, M., Herwig, C., Graham, L., & Satzer, P. (2020). Developing process control strategies for continuous bioprocesses. *Bioprocess International, 18*, 12–16.
90. Jelsch, M., Roggo, Y., Kleinebudde, P., & Krumme, M. (2021). Model predictive control in pharmaceutical continuous manufacturing: A review from a user's perspective. *European Journal of Pharmaceutics and Biopharmaceutics, 159*, 137–142.
91. Chen, Z., Lovett, D., & Morris, J. (2011). Process analytical technologies and real time process control a review of some spectroscopic issues and challenges. *Journal of Process Control, 21*, 1467–1482.
92. Guerra, A., von Stosch, M., & Glassey, J. (2019). Toward biotherapeutic product real-time quality monitoring. *Critical Reviews in Biotechnology, 39*, 289–305.
93. Hasnain, M. S., Ahmed, S. A., Khatoon, A., Afzal, M., Ansari, M. T., Khatoon, S., Tabish, M., AL-Marshad, F. M., & Nayak, A. K. (2021). Pharmaceutical product development: A quality by design (QbD) approach. In *Advances and challenges in pharmaceutical technology*. Elsevier.
94. ICH. (2009). *Harmonised tripartite guideline. Guidance for industry: Q8 (R2), pharmaceutical development, International Conference on Harmonisation of Technical Requirements for Registration of Pharmaceuticals for Human Use (ICH).* US Department of Health and Human Services, Food and Drug Administration.
95. Pramod, K., Tahir, M. A., Charoo, N. A., Ansari, S. H., & Ali, J. (2016). Pharmaceutical product development: A quality by design approach. *International Journal of Pharmaceutical Investigation, 6*, 129.
96. Rozet, E., Lebrun, P., Hubert, P., Debrus, B., & Boulanger, B. (2013). Design spaces for analytical methods. *TrAC Trends in Analytical Chemistry, 42*, 157–167.
97. Teasdale, A., Elder, D., & Nims, R. W. (2017). *ICH quality guidelines: An implementation guide*. John Wiley & Sons.
98. Patil, A. S., & Pethe, A. M. (2013). Quality by design (QbD): A new concept for development of quality pharmaceuticals. *International Journal of Pharmaceutical Quality Assurance, 4*, 13–19.
99. Vynnyk, T. (2015). Key elements of quality assurance of higher education in Ukraine. In *Збірник тез доповідей Міжнародної науково-технічної конференції «Фундаментальні та прикладні проблеми сучасних технологій» присвяченої 55-річчю заснування ТНТУ та 170-річчю з дня народження Івана Пулюя* (pp. 232–232).
100. Dietrich, J., Forrest, A., & Meeker-O'connell, A. (2018). QbD in clinical trials: A focused approach to quality assurance and risk management. *Guest Column, Clinical Leader.*
101. Choudhary, V. (2021). Design and conduct of clinical studies by integration of quality by design (QbD) into quality management system (QMS)-A pragmatic approach. *Clinical Research Open Access, 7*, 2.
102. Valþórsson, H. (2006). *PAT implementation in pharmaceutical manufacturing and its economical impact.* University_of_Basel.
103. Watts, C. (2004). *PAT–A framework for innovative pharmaceutical development manufacturing and quality assurance.* FDA/RPSGB Guidance Workshop.
104. Inhalation Technology Focus Group/International Pharmaceutical Aerosol Consortium on Regulation and Science Tests and Methods Technical Team, & Nagao, L. M. (2002). Recommendations to the food and drug administration: Metered dose inhaler tests and methods in the

chemistry, manufacturing, and controls draft guidances for metered dose inhalers and dry powder inhalers. *Drug Information Journal, 36*, 549–556.
105. Shah, D. N. (2004). Obtaining approval of new drug applications and abbreviated new drug applications from a chemistry, manufacturing, and controls perspective. In *The pharmaceutical regulatory process*. CRC Press.
106. Braithwaite, J. (1993). Transnational regulation of the pharmaceutical industry. *The Annals of the American Academy of Political and Social Science, 525*, 12–30.
107. Cullen, P., O'donnell, C. P., & Fagan, C. C. (2014). Benefits and challenges of adopting PAT for the food industry. In *Process analytical technology for the food industry* (pp. 1–5).
108. Hinz, D. C. (2006). Process analytical technologies in the pharmaceutical industry: The FDA's PAT initiative. *Analytical and Bioanalytical Chemistry, 384*, 1036–1042.
109. Lawrence, X. Y., & Kopcha, M. (2017). The future of pharmaceutical quality and the path to get there. *International Journal of Pharmaceutics, 528*, 354–359.
110. Barshikar, R. (2020). Quality by design (QbD) and its implementation in pharma industry. *Express Pharma*. https://www.expresspharma.in/business-strategies/quality-by-design-qbd-and-its-implementation-in-pharma-industry/Accessed, 31
111. Kengar, M. D., Tamboli, J. A., & Magdum, C. S. (2019). Quality by design-a review. *Pharma Tutor, 7*(4), 48–51.
112. Mohapatra, S., Mirza, M. A., Ahmad, S., Farooq, U., Ansari, M. J., Kohli, K., & Iqbal, Z. (2023). Quality by design assisted optimization and risk assessment of black cohosh loaded ethosomal gel for menopause: Investigating different formulation and process variables. *Pharmaceutics, 15*, 465.
113. Namjoshi, S., Dabbaghi, M., Roberts, M. S., Grice, J. E., & Mohammed, Y. (2020). Quality by design: Development of the quality target product profile (QTPP) for semisolid topical products. *Pharmaceutics, 12*, 287.
114. Martinez-Marquez, D., Mirnajafizadeh, A., Carty, C. P., & Stewart, R. A. (2019). Facilitating industry translation of custom 3d printed bone prostheses and scaffolds through quality by design. *Procedia Manufacturing, 30*, 284–291.
115. Feldman, S. (2005). Quality assurance: Much more than testing: Good QA is not only about technology, but also methods and approaches. *Queue, 3*, 26–29.
116. Miller, M. (1990). QA aims to get it right first timë. *Concrete, 24*, 2.
117. Anastasiia S., & Elena K. (2020). *Retrieved May 27, 2023, from https://rubygarage.org/blog/software-qa-test-planning-activities*
118. Berridge, J. C. (2004). ICH Q8: Pharmaceutical development. *Pharmaceutical Quality Forum, 1*, 1–22.
119. Peterson, J. J. (2008). A Bayesian approach to the ICH Q8 definition of design space. *Journal of Biopharmaceutical Statistics, 18*, 959–975.
120. Potter, C. (2009). PQLI application of science-and risk-based approaches (ICH Q8, Q9, and Q10) to existing products. *Journal of Pharmaceutical Innovation, 4*, 4–23.
121. Vvssn, R., Useni Reddy, M., Arunkanth Krishnakumar, N., Maheshwar Reddy, M., & Hanimi Reddy, B. (2015). Quality improvement with scientific approaches (QbD, aQbD and PAT) in generic drug substance development: Review. *International Journal of Research and Development in Pharmacy & Life Sciences, 4*(102), 2278–0238.
122. AL-Baldawi, Z., & Hussein, I. A. (2020). Integration Pareto distribution and Pareto analysis to analyse and diagnose defects and the root of causes for the air cooling motor. *Diyala Journal of Engineering Sciences, 13*, 49–57.
123. Paul, S., & Abdullahil, A. (2009). *Defect identification and analysis pharmaceutical product using Pareto and cause-effect analysis*. International Conference on Mechanical Engineering.
124. Bhattacharya, J., Pharm, M., & Phil, M. (2015). Quality risk management–Understanding and control the risk in pharmaceutical manufacturing industry. *International Journal of Pharmaceutical Science Invention, 4*, 29–41.
125. Mishra, V., Thakur, S., Patil, A., & Shukla, A. (2018). Quality by design (QbD) approaches in current pharmaceutical set-up. *Expert Opinion on Drug Delivery, 15*(8), 737–758.

126. Sawarkar, K. T., Asnani, A. J., Chaple, D. R., & Ikhar, J. S. (n.d.) *Quality by design (QbD): Building and enhancing quality of pharmaceuticals.*
127. Singh, R. K., Nayak, P., Bharath Raj, K., Khandige, P. S., Sivaa Arumugam, R., Sindhu, K., & Rajeshkumar, R. (2021). Compliance and assurance of complete manufacturing of tablets following Ich Q10. *NVEO-NATURAL VOLATILES & ESSENTIAL OILS Journal|NVEO, 1*, 6426–6436.
128. Drakulich, A. (2009). Critical challenges to implementing QbD: A Q & A with FDA. *Pharmaceutical Technology, 33*, 10.
129. Van Arnum, P. (2007). *A FDA perspective on quality by design.*

Chapter 13
Quality by Design in Pharmaceutical Packaging

Tanweer Haider, Wasim Akram, Ankita Kishore, Neeraj Mishra, and Prathap Madeswara Guptha

Abstract It is imperative that a safe and effective medication product be produced not only with high-quality ingredients but also with acceptable packaging components since those components are in direct contact with the medication. An interaction between major packaging components and the therapeutic product may result in failure to meet the quality target product profile (QTPP). It is possible to avoid this by implementing a Quality by Design (QbD) strategy. QbD is a systematic development method that starts with established objectives and focuses on product and process understanding and process control, all of which are founded on strong science and quality risk management. Critical Quality Attributes (CQAs) are physical, chemical, biological, or microbiological features or characteristics that must fall within a specific limit, range, or distribution to ensure the intended product quality. During the packaging component selection process, QbD can assist in establishing a link between the materials attributes of packaging components and the quality of the medicinal product. Monitoring specific quality parameters of the medicinal product can help assess the influence of each type of packaging component. QbD has been extensively discussed, primarily in the context of pharmaceutical formulation manufacture. However, QbD applications for pharmaceutical packaging are not generally sought. This chapter is only concerned with the application of QbD to the pharmaceutical packaging subset of dosage forms.

Keywords QbD · Quality by design · Packaging materials · Critical quality attributes · Quality target product profile · Primary packaging · Secondary packaging

13.1 Introduction

Quality by Design (QbD) is a systematic approach to pharmaceutical development and manufacturing that focuses on ensuring product quality by design rather than relying on post-production testing. QbD principles can be applied to various aspects

T. Haider · W. Akram · A. Kishore · N. Mishra · P. M. Guptha (✉)
Amity Institute of Pharmacy, Amity University, Gwalior, Madhya Pradesh, India

N. K. Jain, N. Bajwa (eds.), *Introduction to Quality by Design (QbD)*,
https://doi.org/10.1007/978-981-99-8034-5_13

of the pharmaceutical industry, including drug formulation, manufacturing processes, and packaging. In the context of pharmaceutical packaging, QbD emphasizes the integration of quality considerations throughout the entire packaging development lifecycle [1]. It involves a proactive and scientific approach to understanding and controlling the critical quality attributes (CQAs) of the packaging materials and processes. Doing so helps minimize the risks associated with packaging-related issues and ensures the delivery of safe and effective drugs to patients. QbD starts with a comprehensive understanding of the product and its packaging requirements. It involves identifying the drug product's critical quality attributes (CQAs) and the packaging components and understanding how the packaging process can impact these attributes.

QbD incorporates a risk-based approach to identify and manage potential risks throughout the packaging lifecycle. It includes evaluating and prioritizing risks associated with packaging materials, processes, and interactions with the drug product. By understanding and mitigating risks, the quality and safety of the packaged drug can be improved. Design of experiments (DOE) QbD utilizes statistical experimental design techniques to systematically study the effects of various packaging parameters on critical quality attributes. By applying DOE principles, packaging scientists can optimize packaging processes and identify the optimal packaging conditions that ensure product quality and stability. QbD promotes using advanced monitoring technologies and process controls to ensure consistency and reliability in packaging operations. Real-time monitoring helps detect and correct any variations or deviations in the packaging process, allowing for timely adjustments and maintaining product quality. Packaging processes and materials may evolve over time. It emphasizes the need for ongoing monitoring, evaluation, and continuous improvement throughout the product lifecycle. It includes periodic reassessment of the packaging design and process to accommodate changes in materials, technologies, regulations, and customer needs. Manufacturers can enhance their understanding of packaging risks by adopting a QbD approach to pharmaceutical packaging, optimizing processes, and ensuring consistent product quality. This proactive approach helps minimize the potential for packaging-related issues, such as contamination, stability problems, or inadequate protection, ultimately leading to safer and more effective drugs reaching the patients.

The key principles of Quality by Design in pharmaceutical packaging include:

(a) *Product and process understanding:* QbD starts with a comprehensive understanding of the product and its packaging requirements. It involves identifying the critical quality attributes (CQAs) of the drug product and the packaging components, as well as understanding how the packaging process can impact these attributes.
(b) *Risk assessment and management:* QbD incorporates a risk-based approach to identify and manage potential risks throughout the packaging lifecycle. It includes evaluating and prioritizing risks associated with packaging materials,

processes, and interactions with the drug product. By understanding and mitigating risks, the quality and safety of the packaged drug can be improved.

(c) *Design of experiments (DOE):* QbD utilizes statistical experimental design techniques to systematically study the effects of various packaging parameters on critical quality attributes. By applying DOE principles, packaging scientists can optimize packaging processes and identify the optimal packaging conditions that ensure product quality and stability.

(d) *Real-time monitoring and control:* QbD promotes using advanced monitoring technologies and process controls to ensure consistency and reliability in packaging operations. Real-time monitoring helps detect and correct any variations or deviations in the packaging process, allowing for timely adjustments and maintaining product quality.

(e) *Continuous improvement and lifecycle management:* QbD recognizes that packaging processes and materials may evolve over time. It emphasizes the need for ongoing monitoring, evaluation, and continuous improvement throughout the product lifecycle. It includes periodic reassessment of the packaging design and process to accommodate changes in materials, technologies, regulations, and customer needs.

(f) *Better understanding on packaging risks:* Manufacturers can enhance their understanding of packaging risks by adopting a QbD approach to pharmaceutical packaging, optimizing processes, and ensuring consistent product quality. This proactive approach helps minimize the potential for packaging-related issues, such as contamination, stability problems, or inadequate protection, ultimately leading to safer and more effective drugs reaching the patients.

13.2 QbD- In Selecting Packaging Components

For a drug product to be effective and safe, ingredients must be high quality, and packaging components must be carefully selected since they can directly connect with the drug. Failure to fulfill the quality target product profile (QTPP) may come from interactions between the major packaging elements and the therapeutic product [2]. Relying on a QbD strategy can help to prevent this. QbD is a systematic approach to development that starts with established goals and focuses on product/process understanding and process control. This strategy should be founded on strong research and quality risk management (QRM). As part of the packaging component selection process, QbD can contribute to establishing a correlation between the quality of a drug's packaging component and its material attributes. Monitoring specific quality parameters of a drug can help assess the influence of each type of packaging component [3].

The Critical Material Attributes (CMAs) of the packaging components must be determined after the identification of the CQAs. These characteristics include the building material, the container size to fill volume ratio, and any other processes like sterilization or depyrogenation that may affect the CQA of the medicinal product.

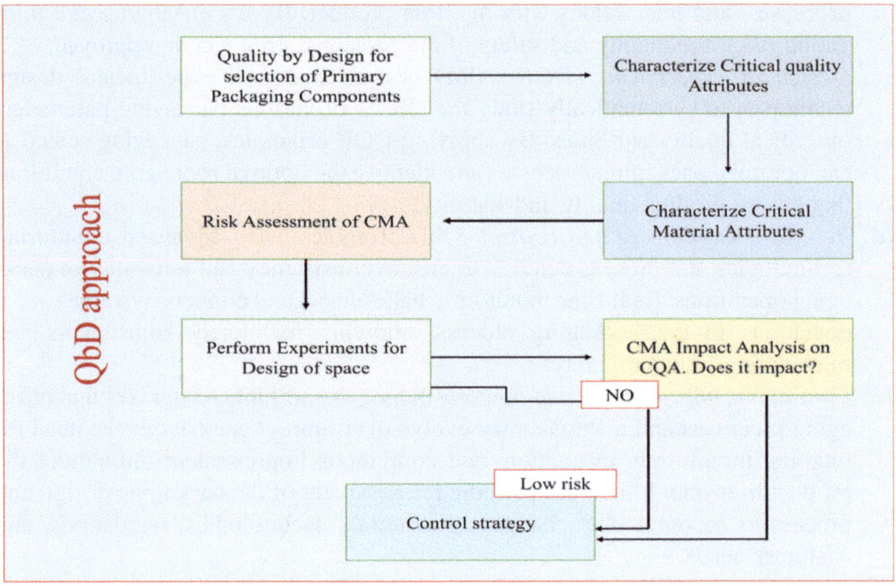

Fig. 13.1 QbD Approach Chart

CMAs for the packaging components that pose a high risk must be identified when the CMAs have been defined. Using a risk management approach, these CMAs are further assessed (Fig. 13.1). It can be classified depending on the severity and probability of risk [4]. For instance, if light and heat are present, a hypothetical packaging element may interact with a therapeutic product. The severity is rather remarkable in this instance. According to data collected during initial development, the component is incompatible with the medicinal product in the presence of light and heat. Therefore, there is a high likelihood that the drug product and the packaging component will interact. Because of the increased severity and high probability, employing this packaging component carries a higher risk. For instance, if light and heat are present, a hypothetical packaging element may interact with a therapeutic product [5]. The severity is rather significant in this instance. According to data collected during initial development, the component is incompatible with the medicinal product in the presence of light and heat. Therefore, there is a high likelihood that the drug product and the packaging component will interact. Because of the increased severity and high probability, employing this packaging component carries a higher risk. Sufficient experimentation or scientific evidence should be presented to mitigate the identified risks. To comprehend this process, let's revisit the hypothetical packaging component. Given that it has been recognized as a high-risk factor, it is necessary to either replace the component or implement additional measures to control the drug product's exposure to light and heat.

When selecting packaging components, it is crucial to consider their impact on the product's stability, compatibility, and protection. Here are some critical considerations in applying QbD to packaging component selection:

(a) *Risk assessment:* Conduct a thorough risk assessment to identify potential risks associated with the packaging components. This assessment should consider factors such as material compatibility, potential interactions with the drug product, permeation properties, and the potential for leachable and extractable.
(b) *Material selection:* Choose packaging materials with a proven track record of compatibility with the drug product. Consider factors such as chemical resistance, barrier properties, and stability under various storage conditions.
(c) *Performance requirements:* Define the critical quality attributes (CQAs) of the packaging components that are essential for ensuring the quality and integrity of the drug product. This may include factors such as moisture resistance, photoresistance, and tamper-evident features [1].
(d) *Design of experiments (DoE):* Utilize the DoE principle to systematically evaluate and optimize the packaging component attributes. It involves conducting controlled experiments to understand the impact of various factors (e.g., material composition, design parameters) on the performance of the packaging components [1].
(e) *Quality control and testing:* Implement robust quality control measures to ensure the consistency and reliability of the selected packaging components. It may include testing for dimensional accuracy, mechanical strength, and integrity.
(f) By applying QbD principles to packaging component selection, pharmaceutical manufacturers can make informed decisions that minimize risks, enhance product quality, and improve patient safety. It is crucial to collaborate closely with packaging suppliers, regulatory agencies, and other relevant stakeholders throughout the process to ensure compliance with applicable regulations and guidelines.

For example, if a particular packaging component is identified as posing a high risk, appropriate actions should be taken. It can involve replacing the component with a more suitable alternative or implementing additional controls to mitigate risks. These controls may include measures to protect the product from light and heat exposure, as these factors can potentially impact product quality. By employing QbD principles in selecting packaging components, manufacturers can make informed decisions based on scientific understanding and minimize the potential risks to product quality. This approach ensures the packaged product's safety, efficacy, and quality throughout its lifecycle [6].

13.2.1 Critical Material Attributes in Packaging Components

Critical Material Attributes (CMAs) are essential characteristics of packaging components that significantly impact the quality, safety, and effectiveness of the packaged product. When selecting packaging components, it is necessary to consider various CMAs. Here are some examples:

(a) *Compatibility:* The packaging components should be compatible with the product they are intended to contain. It involves assessing the potential interactions between the packaging material and the product, such as chemical reactions, leaching of substances, or adsorption, which could compromise the quality or stability of the product.
(b) *Container Closure Integrity:* The integrity of the container closure system is critical to prevent microbial contamination, moisture ingress, or loss of product integrity. The packaging components, such as vials, bottles, closures, and seals, must provide a reliable barrier and maintain their integrity throughout the product's shelf life.
(c) *Mechanical Strength and Durability:* Packaging components should possess adequate mechanical strength and durability to withstand handling, transportation, and storage conditions. They should resist physical stress, such as compression, impact, or vibration, to prevent damage or leakage that could compromise the product.
(d) *Protection from Environmental Factors:* Packaging components should protect the product from environmental factors, such as light, moisture, oxygen, or temperature variations, depending on the product's sensitivity. The components' ability to provide the required protection is crucial to maintain product stability and prevent degradation [7].
(e) *Sterility:* In the case of sterile products, packaging components must be sterile to prevent microbial contamination. The sterility assurance of components, such as vials, syringes, or ampoules, is critical to ensure the safety and efficacy of the product [8].

13.2.2 Quality Target Product Profile (QTPP) in Packaging Components

The Quality Target Product Profile (QTPP) is used in pharmaceutical development to define a final drug product's desired characteristics and attributes. In the context of pharmaceutical packaging, the QTPP for packaging can be described as the desired quality attributes, specifications, and performance criteria of the packaging system that contribute to the overall quality and effectiveness of the product [2]. Here are some key elements that can be included in the QTPP for pharmaceutical packaging:

(a) *Container Closure System:* The QTPP should define the desired characteristics of the container closure system, including the type of container (e.g., bottle, vial, blister pack), closure (e.g., cap, stopper, foil), and any specific requirements for the system (e.g., child-resistant closures, tamper-evident features) [9].
(b) *Compatibility:* The QTPP should specify the desired level of compatibility between the packaging materials and the drug product. It includes chemical

compatibility, stability, and prevention of interactions that could impact the product's efficacy, safety, or stability [10].

(c) *Container Closure Integrity:* The QTPP should outline the requirements for container closure integrity, ensuring that the packaging system provides an effective barrier to microbial contamination, moisture ingress, and product leakage throughout the product's shelf life.

(d) *Protection from Environmental Factors:* The QTPP should define the desired level of protection the packaging provides against environmental factors such as light, moisture, oxygen, and temperature variations. It should specify the necessary level of protection to maintain the stability and quality of the drug product.

(e) *Ease of Use:* The QTPP may include aspects related to the user-friendliness of the packaging system, such as ease of opening, resealing, or administration of the product. It is particularly important for patient compliance and medication adherence.

(f) *Regulatory Compliance:* The QTPP should consider any specific regulatory requirements related to the packaging, such as child-resistant packaging, labeling, serialization, or tamper-evident features.

(g) It is important to note that the specific elements and attributes included in the QTPP for pharmaceutical packaging may vary depending on the nature of the drug product, dosage form, and regulatory requirements. The QTPP guides ensure that the packaging system meets the desired quality standards and effectively preserves the drug product's integrity, safety, and efficacy.

13.3 Role of Packaging Design in Drug Development

Packaging design plays a crucial role in the overall drug development process. It encompasses the creation and development of the packaging materials and visual elements that are used to contain, protect, and present pharmaceutical products. In the pharmaceutical field, packaging refers to a wide range of elements that surround a pharmaceutical product from its production to its use. It is a comprehensive and diverse effort that serves the important purpose of providing life-saving medicines, medical devices and therapies, as well as innovative products such as medical nutrition, to people around the world. Packaging ensures that these pharmaceutical products, whether supplements, poultices, liquids, solids, powders, suspensions, or drops, reach their intended users safely and effectively. When done right, the packaging is invisible to the consumer, but any flaw in its execution or design can be criticized by all parties involved [11, 12].

The distribution of goods, including pharmaceuticals, is now more widespread than ever in the modern, globalized world. Due to this trend, packaging design needs to factor in mass customization to ensure it is suitable for different markets. Given the difficulties and opportunities facing pharmaceutical companies doing business in various sectors, the issue of optimization in packaging design deserves in-depth explanation and discussion. Sustainability and environmental concerns should always be taken into account when designing packaging. The subjective aspect of the environmental impact of packaging materials and waste management requires careful consideration and decision-making during the design process. A key objective of this sector is to strike a balance between the need for efficient packaging and environmentally friendly practices [12, 13].

Packaging design is a growing field of science and engineering, as well as an art. It plays a vital role in the success of the pharmaceutical industry. It consists of several corporate divisions including research and development, engineering, operations, purchasing, marketing, and general administration. Depending on the organizational structure and the priorities of each company, the duties and controls over the packaging may change. Packaging represents the largest total purchase of materials required for most pharmaceutical products. It is necessary to ensure that pharmaceuticals are distributed, sold and protected. The importance of packaging in maintaining product integrity, meeting regulatory requirements, and meeting consumer expectations cannot be overstated [14]. Following are some of the important elements of packaging design in drug development:

13.3.1 Product Protection and Preservation

Packaging design ensures the integrity, stability, and safety of pharmaceutical products throughout their shelf life. Product protection and preservation are fundamental considerations in pharmaceutical packaging design during the product development phase. Proper packaging design helps maintain the potency, purity, and sterility of drugs, preventing contamination and ensuring patient safety. It protects drugs from environmental factors such as light, moisture, oxygen, and temperature fluctuations that could degrade their quality or efficacy. The packaging design aims to safeguard the integrity, stability, and efficacy of pharmaceutical products throughout their lifecycle.

Pharmaceutical products are physically protected by packaging design from various external factors that may degrade their quality. This includes protection against physical damage during handling, transportation and storage as well as against light, moisture, oxygen, temperature changes and other environmental factors. To protect against potential contamination, spoilage or tampering, the packaging material acts as a barrier [15]. The stability and shelf life of pharmaceutical products is strongly influenced by packaging design. To minimize interactions between the drug and its environment, it is essential to choose the right packaging material and format. To protect the drug from degradation by exposure to air or moisture, packaging techniques such as blister packs, vials, or ampoules are used with materials that are resistant to moisture, light, or oxygen penetration [16].

Pharmaceutical products are not contaminated by packaging design features. This includes sterile packaging components that keep an injectable medication sterile, such as vials, syringes, or pre-filled cartridges. The packaging may also include a tamper-evident seal or closure to ensure that the product has remained clean and undamaged until it is delivered to the customer [16]. Pharmaceutical packaging often includes child-resistant ingredients, especially in products intended for home use. The child-resistant packaging is made to prevent children from accidentally swallowing it, reducing the risk of injury or poisoning. This usually involves mechanisms that adults can reach but are difficult for young children to open [17]. The goal of pharmaceutical packaging design is to protect the product from external influences that may compromise its quality or safety. This involves selecting materials with appropriate barrier properties to protect against contamination from microbes, moisture, oxygen or light. For example, blister packs often use aluminium foil as a moisture and oxygen barrier [18].

13.3.2 Regulatory Compliance

Pharmaceutical packaging design must comply with various regulatory guidelines and requirements set by health authorities such as the Food and Drug Administration (FDA) in the United States or the European Medicines Agency (EMA) in Europe. These regulations cover aspects like labelling, child-resistant packaging, tamper-evident features, and patient information leaflets. Packaging designers need to ensure that their designs meet all applicable regulations to ensure legal compliance [16]. Regulatory bodies have mandated specific labelling requirements to convey important information about a pharmaceutical product. This information includes details such as the product name, active ingredients, dosage form, dosage strength, method of administration, lot number, expiration date, storage requirements, and any warnings or protections. Relevant. Labelling on the package must be accurate, clear and legible so that patients and healthcare workers can quickly identify and understand the product [19]. Patient information leaflets (PILs) are an essential component of pharmaceutical packaging, providing crucial information about the medication to patients. Regulatory authorities mandate the inclusion of PILs with pharmaceutical products to enhance patient understanding of the drug's proper use, potential side effects, contraindications, and precautions. Packaging designers must ensure that there is sufficient space and clear formatting to include comprehensive and accurate information in the PIL, complying with the regulatory requirements [20].

Pharmaceutical products that pose a risk if accidentally swallowed by children require child-resistant packaging. According to regulatory agencies, some medications must be packaged in child-resistant containers, especially those for home use. These containers are designed to be difficult for children to access or open, reducing the chance of accidental swallowing. Packaging designers must incorporate appropriate child-resistant features into their design to meet the stringent requirements set by regulatory bodies [21].

To ensure legal compliance, packaging designers must keep up with the increasing regulations issued by health authorities. They should closely monitor updates from regulatory agencies like the FDA or EMA to ensure their designs align with the latest requirements. Collaboration with regulatory experts or consultants can also provide valuable guidance in ensuring compliance during the package design process. By following regulatory requirements and guidelines, packaging designers contribute to the safe and effective use of pharmaceutical products. Compliance with these regulations helps protect patients, minimize risk, and maintain the integrity and quality of pharmaceutical packaging throughout the product life cycle.

13.3.3 Drug Identification and Information

The packaging design helps in the identification and differentiation of different medicines. This includes things like product names, brands, logos, and unique identifiers like barcodes or serial numbers. Clear and accurate labelling provides essential information, such as dosing instructions, warnings, side effects, and expiration dates, so that healthcare professionals and patients can use medications correctly and safely [22].

13.3.4 User Experience and Convenience

Packaging design in the pharmaceutical industry places great emphasis on taking the user experience into account to ensure that medicines can be handled, opened and used effectively by patients. Pharmaceutical packaging prioritizes user-friendly design elements to enhance convenience and utility, and ultimately improve treatment outcomes. The easy-to-use packaging includes easy-to-open features, especially for those with poor dexterity or grip. Patients can easily access the drug with the help of simple opening mechanisms, such as blister packs with perforated tabs or peel-off seals. These features reduce frustration and make getting medication simple and fast. Another feature of packaging design is tamper-evident seals, which are used in packaging design to provide a visual indication that the package has been opened or tampered with. These features allow patients to feel confident about the quality and safety of the product. The tamper-evident seal makes it more likely that patients will receive medications as intended, which also increases patient confidence. Other than these, appropriate container sizes, clear instructions for administration and enhanced medication adherence should be followed for enhanced user experiences [23].

13.3.5 Branding and Marketing

In the pharmaceutical and cosmetic sector, it is crucial to generate brand recognition and differentiation. Helps develop the visual identity of the drug by reflecting the positioning and core values of the company. Pharmaceutical companies can influence consumer perception, build trust, and support their marketing initiatives through the use of eye-catching design, compelling graphics, and consistent brand elements [24].

13.3.6 Counterfeit Prevention

Packaging design incorporates security features and anti-counterfeiting measures to protect against the illicit distribution of counterfeit drugs. These may include holograms, security labels, tamper-evident packaging, or unique serial numbers that can be authenticated by consumers or regulatory authorities. Robust packaging design helps ensure the authenticity and traceability of pharmaceutical products, safeguarding public health. Holographic elements are commonly used in pharmaceutical packaging design as a visual security feature. Holograms are three-dimensional images or patterns that possess unique optical properties, making them difficult to replicate. They serve as a visual indicator of authenticity, allowing consumers and regulatory authorities to identify genuine products by verifying the presence of the holographic feature [25]. Security labels are another tamper-evident feature applied to packaging to detect any unauthorized opening or tampering. These labels are designed to break or leave a visible mark when tampered with, providing clear evidence of product integrity. Security labels serve as an important deterrent to counterfeiters and help consumers and regulatory authorities identify potentially compromised or tampered products [26]. Packaging designs incorporate unique serial numbers or codes that can be verified by customers or regulatory agencies. Usually, a barcode or QR code containing the serial number is printed on the packaging or included in the printed serial number. People can confirm the authenticity and traceability of a product by looking at the unique identifier, which allows them to differentiate between genuine and counterfeit products. Modern packaging incorporates track-and-trace technologies, allowing pharmaceutical products to be tracked and controlled throughout the supply chain. These innovations track the flow of goods from manufacturing to distribution using special identifiers such as serial numbers or RFID tags, ensuring their authenticity and preventing diversion or counterfeiting [27].

Overall, drug packaging design is a multidisciplinary process that considers user experience, safety, regulatory compliance, branding, and security. Throughout the life cycle of a pharmaceutical product, it is essential for the safety of the product, the dissemination of information and the improvement of patient safety and comfort.

13.4 Elements of QbD for Developing Container and Closure System

Quality by Design (QbD) is an organized strategy for designing and developing products and processes to ensure quality. It is commonly applied in the pharmaceutical industry to ensure that drug products meet the required safety, efficacy, and quality standards. One key aspect of drug product development is the container and closure system, which plays a crucial role in ensuring the integrity and stability of the drug product during its shelf life. This section will discuss the elements of QbD for developing a container and closure system for a pharmaceutical product [1, 28].

13.4.1 Element 1: Critical Quality Attributes (CQAs)

Identifying the drug product's critical quality attributes (CQAs) is the initial aspect of QbD for building a container and closure system. The physical, chemical, biological, or microbiological features or characteristics crucial to guaranteeing product quality are called CQAs. It is critical to determine the CQAs of drug products early in the development phase to guide the container and closure system selection. CQAs for the container and closure system may include features such as the ability to keep moisture and oxygen out of the container, the ability to prevent leachable and extractable materials from migrating into the drug product, and the capacity to keep the drug product sterile [1, 29].

13.4.2 Element 2: Risk Assessment

The second element of QbD for developing a container and closure system is the identification and assessment of risks associated with the container and closure system. Risk assessment includes identifying potential sources of risk, evaluating the likelihood and severity of those risks, and implementing approaches to mitigate them. For example, the risk of leachable and extractable migrating from the container and closure system into the pharmaceutical product may be mitigated by choosing materials that are compatible with the pharmaceutical product and have low levels of leachable and extractable [30].

13.4.3 Element 3: Design of Experiments (DoE)

The design of experiments (DoE) to optimize the system is the third element of QbD for designing a container and closure system. DoE is a statistical tool that systematically studies many aspects that may affect the container and closure system's performance. DoE can be used to determine the best mix of parameters that will ensure the system's desired performance. For example, DoE may be used to discover the optimal sealing parameters for a closure system to ensure that it provides the desired level of moisture and oxygen protection [31].

13.4.4 Element 4: Material Selection and Compatibility Studies

Choosing appropriate materials for the container and closure system is the fourth element of QbD for developing a container and closure system. Material selection is critical to assure drug product compatibility, stability, and quality. It entails assessing the physicochemical qualities of various materials, including glass, plastics, and elastomers, as well as their potential interactions with the drug product. Compatibility tests evaluate the possibility of leachable and extractable materials from the container and closure system migrating into the medicinal product. These studies involve subjecting the materials to various stress conditions and analyzing the resulting extract solutions for potential contaminants [32].

13.4.5 Element 5: Process Analytical Technology (PAT)

The fifth element of QbD for developing a container and closure system is using process analytical technology (PAT) to monitor and control the manufacturing process. PAT monitors real-time critical process parameters (CPPs) using sensors, analyzers, and other measurement instruments. PAT allows for the identification and control of variability in the manufacturing process, which can affect the quality of the container and closure system. PAT can monitor the sealing temperature of a closure system to ensure that it fulfils the required specifications [33].

13.4.6 Element 6: Control Strategy

The sixth element of QbD for developing a container and closure system is implementing a control strategy to ensure the system's quality. A control strategy is a planned set of controls implemented to assure product quality and process performance based on current product and process understanding. The control strategy for the container and closure systems may comprise in-process testing, release testing, and stability testing. These tests are intended to confirm that the container and closure system meet the necessary performance criteria while reducing the possibility of product failure.

By implementing a control strategy, pharmaceutical companies can ensure that the container and closure system meets appropriate quality standards. It enables continuous monitoring and optimization of the manufacturing process to preserve product quality and performance [34].

In summary, applying Quality by Design (QbD) principles in developing a container and closure system for a drug product is crucial for ensuring its quality, safety, and efficacy. The QbD elements discussed above, including the identification of critical quality attributes (CQAs), risk assessment, design of experiments (DoE), material selection and compatibility studies, process analytical technology (PAT), and control strategy, provide a systematic and science-based approach to developing and maintaining a high-quality container and closure system.

By following these elements, pharmaceutical companies can improve their understanding of the product and process, identify, and mitigate risks, optimize performance through statistical analysis, select suitable materials, monitor and control the manufacturing process, and implement effective quality control measures. This systematic approach contributes to the overall assurance of product quality and patient safety throughout the drug product's lifecycle. Regulatory agencies also recognize the value of QbD and increasingly emphasize its importance in assessing the quality of drug products during the approval process.

13.5 QbD- Extractable and Leachable Study for Packaging Components

Biopharmaceutical packaging is essential for the safety of drug products because it acts as a barrier, sterilizes the contents, and protects them from light damage, oxidation, and moisture absorption. It must specify its components in detail, be able to incorporate dose details and guarantee proper dosage. Testing is crucial for parts and materials such as metal springs, adhesives, plastic containers, elastomeric valves, gaskets, and coatings to ensure quality and stop dangerous contaminants from getting into the drug product [16].

Chemical substances may leak out of plastic container systems, increasing the toxicity of drug products. By analyzing the product, manufacturers must show that leaching has no detrimental effects on the safety or efficacy of dosage forms. It is not just a hypothetical situation; there is a long history of dangerous packaging additives leaking into stored goods. Up to the 1980s, rubber was boosted in pliability by using carbon black. Before it was discovered, elastomers created with carbon black leached cancer-causing poly-aromatic hydrocarbons, they were added to the rubber, ranging from asthma inhalers to infant bottle nipples.

The QbD principle states predictable test results replace further testing by ensuring the system operates within a defined design space [35]. We are discussing about the extractables and leachables in this section, so basically, extractables are organic and inorganic chemical entities released from pharmaceutical packaging under laboratory conditions, which can accelerate or exaggerate normal storage conditions (e.g., temperature, solvent systems, stoichiometry, etc.) depending on the study's purpose, leachables are organic and inorganic chemical components, usually a subset of extractable, which migrate into a pharmaceutical product's formulation from any product contact material due to direct contact under regular manufacturing/ storage environment or accelerated drug product stability investigations [36, 37]. A Container Closure System (CCS) consists of primary or secondary components made of polymers with various formulation ingredients. When container/closure system directly interacts with the drug product, these ingredients can be released through the package thickness, development, and package-product interface and are known as migrating substance [36]. A typical extractable and leachable program comprises exaggerated extraction tests to extract possible leachables from packaging or processing components. In these tests, components are shredded and placed in solvents, heated at elevated temperatures to remove potential impurities. Extracted compounds are identified using analytical techniques such as inductively coupled plasma mass spectrometry (ICP–MS), gas chromatography-mass spectrometry (GC-MS), and high-pressure liquid chromatography-mass spectrometry (HPLC-MS). Although not all leachables are extractable, techniques have been developed to find them in the product matrix [38].

In product testing, leachable impurity is quantified in the drug product and validated for accuracy, precision, specificity, linearity, range, and limit of quantification. These validated impurity tests are added to the product stability study suite. Extraction, identification, development, and validation of new methods for these components is a costly and labour-intensive endeavour, resembling a "quality-by-QC test" strategy. New materials must be chosen and qualified if harmful leachables are found during stability studies. It is uncertain which components could leach out during storage; hence techniques are being developed to identify these extractable in the final product matrix [39]. There are some regulations that are already in Place for Extractable and Leachable Materials. These are as follows:

(a) GMP CFR 211.94—Drug Product Containers and Closures
(b) Federal Food, Drug, and Cosmetic Act
(c) General Requirements, GMP CFR 211.160
(d) Schedule M of the Drug & Cosmetic Act of 1940's Rules: India
(e) EU regulations
(f) International recommendations from the EMEA (European Medicine Evaluation Agency) and Health Canada [36].

With the help of the above regulations a suitable protocol can be designed for the study of leachables and extractables.

The components of QbD employed in pharmaceutical development are described in Food & Drug Administration Guidelines for Industry, Q8 (R2) [40].

13.5.1 Elements of QBD in Packaging Material, Extractables, and Leachables

13.5.1.1 Quality Targeted Product Profile (QTPP)

The QbD element, QTPP, is the basis for designing drug products and packaging. It describes the intended use, administration route, dosage form, patient population, and chronic/short-term use. Any changes to QTPP must be evaluated for the impact on Critical Quality Attributes (CQA) [41].

13.5.1.2 Critical Quality Attribute (CQA)

A physical, biological, or microbiological characteristic called CQA must be within acceptable limits to guarantee the intended level of product quality. Although it can also apply to container closure systems, it is generally linked with drug ingredients, intermediates, excipients, and drug products. Baseline CQA mandates basic entry-level quality requirements for packaging and production components. For example, for plastic containers buffering capacity, heavy metals, residue on ignition, and gravimetric leachable parameters for non-volatile residues listed in United States Pharmacopoeia (USP) <671> [42]. Same as for the glass containers etching and hydrolytic resistance criteria that are mentioned in USP 37 <660> must be fulfilled [43]. For elastomeric and plastic components, gross toxicity screening needs to be confirmed, such as that offered by USP <88> [44].

After assessing the risk for extractable, CQAs are included to estimate the relative risk of design parameters to patient health. For instance, dosage form analysis considers drug-product interactions and the danger of leached contaminants in parenteral or inhalation products. Drug packaging for solvent-propelled inhalants or liquid parenteral suggests a wider variety of leachables at lower detection limits [45].

13.5.1.3 Risk Assessment

Risk assessment is a useful scientific procedure that is utilised in quality risk management to determine which characteristics of the materials and process variables have an impact on the CQAs of the products. This risk assessment is often carried out early in the pharmaceutical development process, and it may be beneficial to repeat it when more knowledge and facts become available.

An Ishikawa diagram (Fig. 13.2), called a fishbone diagram or a cause-and-effect diagram, is a graphic tool used to pinpoint and examine the various reasons for a specific issue or occurrence. Japanese quality control specialist Dr. Kaoru Ishikawa created it [46]. The diagram is shaped like a fishbone, with the "head" denoting the issue or effect under investigation and the "bones" denoting potential causes or

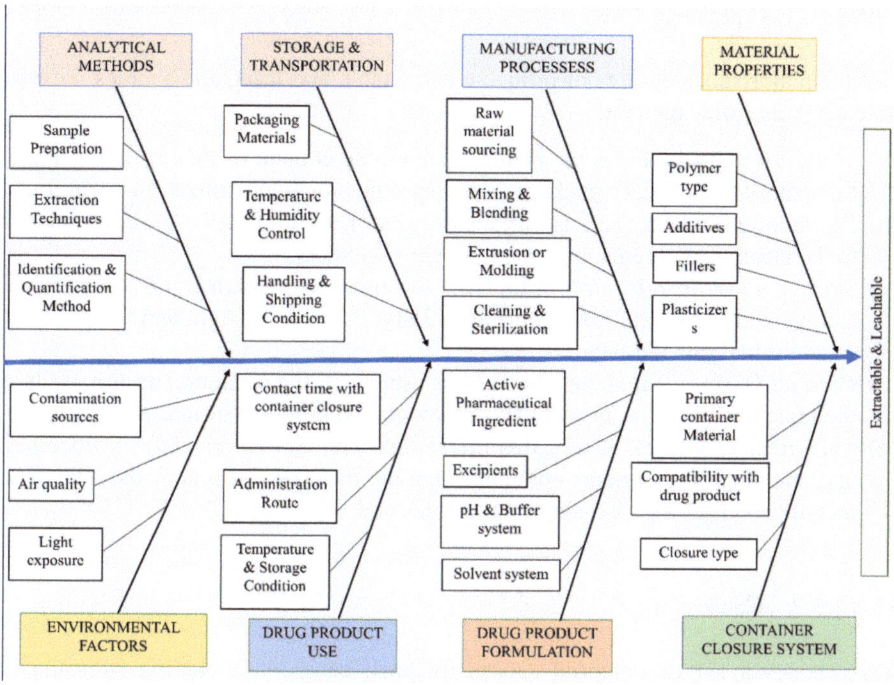

Fig. 13.2 Leachable risk assessment using the Ishikawa diagram to relate to design variables

groups of causes. Ishikawa diagrams are a valuable tool to promote a systematic approach to problem-solving and decision-making by visually organizing and structuring the analysis of possible causes. The design and quality team can create an Ishikawa diagram to evaluate potential risks and arrive at an accurate extractable and leachables assessment. Materials sourced, formulation, storage conditions, and sterilization methods should all be considered. In the context of drugs or other products, a diagram created by Ishikawa can be a valuable tool for evaluating extractable and leachables. Here is an example of how to use it:

13.5.1.3.1 Problems or Effects

Start by stating the problem or consequence of leachables and extractables in unambiguous terms. For instance, there can be unidentified chemicals in a pharmaceutical product.

13.5.1.3.2 Categories

List the major subcategories of probable extractable and leachable sources. Several potential categories include:

(a) *Materials:* This includes the elements about the content of the primary packaging material, secondary packaging, or any other elements utilized in the product.
(b) *Production Process:* This is various aspects of the production process, such as tools, cleaners, solvents, and working environments.
(c) *Storage and transportation:* It includes the elements that affect the conditions of storage and transportation, such as humidity, exposure to light and temperature, or additional outside factors.
(d) *Product Design:* It includes the various aspects of the product's design, such as the material selection, product arrangement, and component interaction.
(e) *Analytical processes:* Under this the sample preparation, extraction processes, and analytical instruments come and that are different aspects of the analytical techniques used for assessing extractables and leachables.

13.5.1.3.3 Causes

This creates a list of potential causes for each category, including substandard packaging, contaminants, and insufficient compatibility testing.

13.5.1.3.4 Analysis

Analysis includes examining and elaborating on identified causes, considering elements affecting extractables and leachables, using diagrams to connect categories.

13.5.1.3.5 Solutions

For a solution one can rank reasons in likelihood and significance and develop appropriate responses, such as testing, modifying production, or improving product design. By using an Ishikawa diagram, one can methodically locate and evaluate probable extractables and leachables causes. It aids in creating efficient tactics to lessen or eliminate them, ensuring the product's safety and quality [47]. Table 13.1 shows the relationship between QTPP, CQA, and FDA risk assessment, that have been discussed in above sections [35, 36, 48].

Table 13.1 Depicting the relationship among QTPP, CQA, and risk assessment

Route of administration and associated risk level	Probability of packaging component-drug interaction		
	High	Medium	Low
High	Injections and injectable suspensions; aerosols and solutions for inhalation;	Sterile powders, powders for injection, and powders for inhalation	
Medium	Nasal aerosols and sprays, transdermal ointments and patches, and ophthalmic solutions and suspensions		
Low	Topical and lingual aerosols, topical and oral solutions and suspensions	Powders for topical and oral use	Oral tablets and capsules (soft and hard gelatin both)

13.5.1.4 Design Space

The design space describes the relationship between materials and critical quality attributes in manufacturing. While working with packaging component suppliers is crucial to identify potential variability in each component, especially those with high product contact and risk. This information can assess leachable components and their impact on drug product quality [49].

13.5.1.5 Control Strategy

The next step in the QbD process is the control strategy, which focuses on predefined controls that utilize the design space to guarantee the quality of the finished product. Controlling extractable component specifications upstream helps reduce variability and ensure accurate material assessment [50].

13.5.1.6 Life Cycle Management

In QbD, partner with material suppliers to understand potential changes, establish a vendor quality agreement, and proactively source and assess alternative materials in case of business failures or unacceptable changes [51]. Various steps of the QbD approach are depicted in Fig. 13.3.

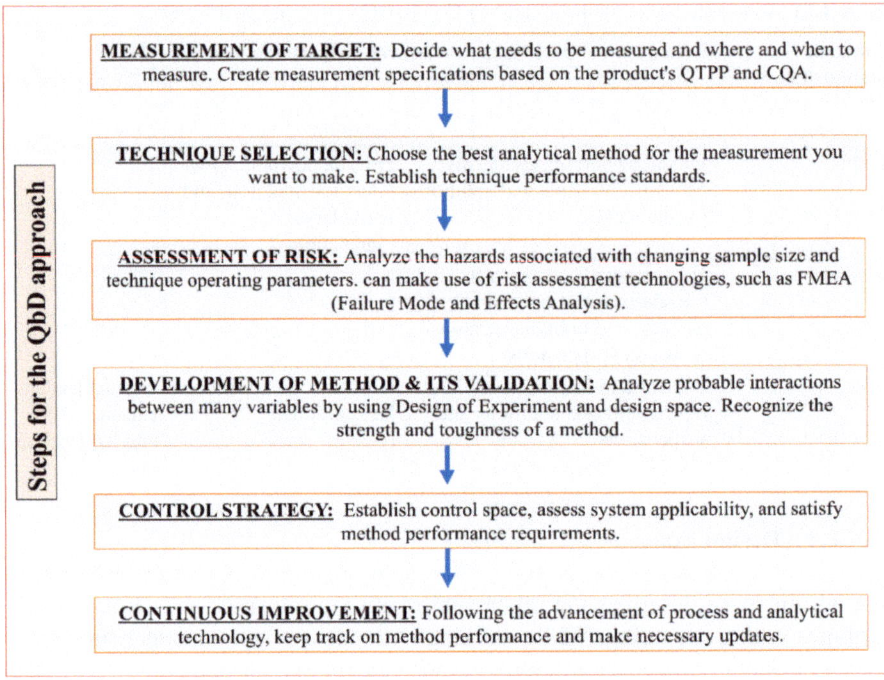

Fig. 13.3 Steps of QbD approach

13.5.2 QBD in Container/Closure System

A container/closure system with QbD principles is crucial for ensuring product quality, reducing risks, enhancing patient safety, complying with regulations, and optimizing cost-effectiveness. By considering critical quality attributes and critical process parameters, QbD helps minimize variability and ensure the system meets performance criteria. It also reduces risks by identifying and controlling potential sources of variability and leachables early in the development process. QbD also minimizes the risk of leachables migrating into the drug product, especially for sensitive patient populations. It also aligns with regulatory expectations, facilitating regulatory approval and compliance with Good Manufacturing Practices. Overall, QbD principles contribute to a systematic and proactive approach to delivering safe and effective drug products to patients. The following are important factors to take into account while designing a container/closure system with the help of QbD concepts to minimize the effects of leachable:

13.5.2.1 Selection of Material

Choose materials for the container and closures that are well-characterized, low-leachable, have a history of success, and have tested extractable and leachable traits.

13.5.2.2 Assessment of Risk

Conduct risk analysis, taking into account aspects like material qualities, to find leachable substances and their effects on pharmaceutical products.

13.5.2.3 Designs of Experiments (DOE)

Utilize DOE to analyze factors affecting leachables, identifying critical parameters and optimizing container/closure system design by varying and analyzing profiles.

13.5.2.4 Compatibility Studies

Compatibility studies evaluate drug product-container interaction, assessing potential leachable migration, using accelerated, forced, and real-time stability methods.

13.5.2.5 Control Strategy

Create a control strategy to reduce leachable risk, implementing material specifications, process controls, and testing protocols, and using analytical methods like HPLC and GC-MS.

13.5.2.6 Continuous Monitoring

The lifecycle of the container/closure system should be monitored for an efficient control approach using regular evaluations, process parameters, and ongoing monitoring after marketing.

These QbD concepts can be used to build a container/closure system offering to reduce the possible impact of leachables on the medication, hence ensuring the drug product's quality and safety. Food & Drug Administration Guidelines for Industry, Q8 (R2), highlights proactive measures to incorporate quality into the finished product. The construction of container and closure using QbD [3] depicts in Fig. 13.4.

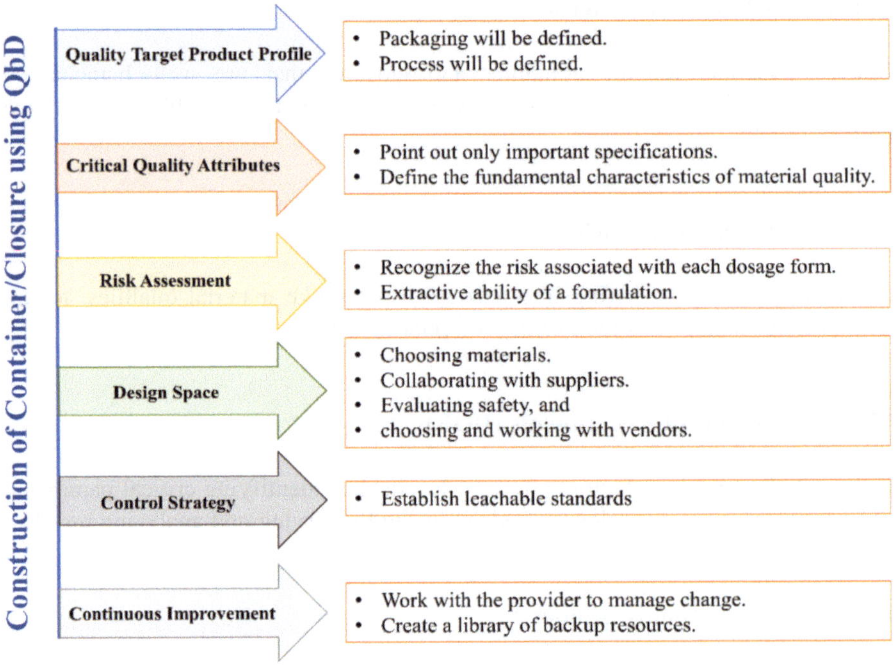

Fig. 13.4 Construction of a container/closure system with quality by design (QbD) aimed at reducing the effects of leachable

QbD principles enable a more effective extractable and leachable program, focusing on leachables with potential risks to product integrity and considering external factors like vendor qualification and supply chain disruptions.

13.6 Conclusion

QbD has rapidly grown into an essential and widely used technology in the research and development of pharmaceutical products. Though applying QbD at the product/process design level is where it is most successful, it ought to be done in the manufacturing and quality assessment contexts. Whenever a Quality by Design (QbD) approach is applied during product development, patients are going to get medications of outstanding quality, manufacturers will see enhancements in manufacturing with significantly fewer batch failures, and drug regulatory agencies will have greater trust in the robust quality of the products. This method enables the process to establish priorities and flexible limits.

When selecting primary packaging components, a QbD strategy must be considered in account. To assess its effect on the final product's quality, it is important to take into account the material characteristics of packaging components. Therefore,

QbD is a strategy for reducing the risk related to the main packaging and can be emphasized at an earlier stage of the development of a drug product. When generating SDFs, packaging concerns must be made. The quality of the finished product may be affected by packaging process characteristics. Assessing the impact of material qualities and process conditions on CQAs is essential when designing packaging for sterile items. Identifying and handling the sources of unpredictability is crucial. In addition, it is crucial to keep an eye on them during the product's lifespan.

Pharmaceutical QbD aims to enhance postapproval change management and product development and manufacturing efficiency by decreasing product variability and faults. It is achieved by developing therapeutically applicable requirements and an accurate formulation and manufacturing process. The QTPP, design of products and processes and knowledge, scale-up, control strategy, and constant improvement can all be considered major components of pharmaceutical QbD. The use of DoE, PAT, prior knowledge, and risk estimation as tools can make QbD deployment easier. After approval, during the product lifecycle management procedure, the ability of the product and the procedure is evaluated and continuously enhanced.

References

1. Yu, L. X., Amidon, G., Khan, M. A., Hoag, S. W., Polli, J., Raju, G. K., & Woodcock, J. (2014). Understanding pharmaceutical quality by design. *The AAPS Journal, 16*(4), 771–783. https://doi.org/10.1208/s12248-014-9598-3
2. Namjoshi, S., Dabbaghi, M., Roberts, M. S., Grice, J. E., & Mohammed, Y. (2020). Quality by design: Development of the quality target product profile (QTPP) for semisolid topical products. *Pharmaceutics, 12*(3). https://doi.org/10.3390/pharmaceutics12030287
3. Patel, S., & Patel, N. (2017). *How to use QbD to select packaging components*. PDA letter. https://www.pda.org/pda-letter-portal/home/full-article/how-to-use-qbd-to-select-packaging-components (Date of access 3 Nov 2023).
4. Sangshetti, J. N., Deshpande, M., Zaheer, Z., Shinde, D. B., & Arote, R. (2017). Quality by design approach: Regulatory need. *Arabian Journal of Chemistry, 10*, S3412–S3425. https://doi.org/10.1016/j.arabjc.2014.01.025
5. Parag, D., Animesh, M., & Shital, P. (2018). Quality by design (QbD): A new horizon for robust analytics in extractable and leachable study for packaging components. *International Journal of Pharmaceutics and Drug Analysis, 6*(2), 89–104.
6. Verch, T., Campa, C., Chéry, C. C., Frenkel, R., Graul, T., Jaya, N., Nakhle, B., Springall, J., Starkey, J., Wypych, J., & Ranheim, T. (2022). Analytical quality by design, life cycle management, and method control. *The AAPS Journal, 24*(1), 34. https://doi.org/10.1208/s12248-022-00685-2
7. Alamri, M. S., Qasem, A. A. A., Mohamed, A. A., Hussain, S., Ibraheem, M. A., Shamlan, G., Alqah, H. A., & Qasha, A. S. (2021). Food packaging's materials: A food safety perspective. *Saudi Journal of Biological Sciences, 28*(8), 4490–4499. https://doi.org/10.1016/j.sjbs.2021.04.047
8. USP. USP: Chemical/Physical Tests: <381> Elastomeric closures for injections. USP Revised Bulletin Official. Page 1–7. https://www.uspnf.com/sites/default/files/usp_pdf/EN/USPNF/generalChapter381.pdf (date of access 3 Nov 2023).

9. FDA. (1999). *Container closure systems for packaging human drugs and biologics, chemistry manufacturing and controls documentation*. In C.C.F.G.f. Industry (Ed.).
10. ICH. (2009). *ICH Q8 (R2): Pharmaceutical development*. Page no. 1–28. https://database.ich.org/sites/default/files/Q8_R2_Guideline.pdf. (date of access 3 Nov 2023).
11. Carter, S. (2005). Copper and Gunn's packaging in tutorial pharmacy, structural approaches to texture. *Proceedings of the IEEE, 67*(5), 786–804.
12. Zadbuke, N., Shahi, S., Gulecha, B., Padalkar, A., & Thube, M. (2013). Recent trends and future of pharmaceutical packaging technology. *Journal of Pharmacy & Bioallied Sciences, 5*(2), 98–110. https://doi.org/10.4103/0975-7406.111820
13. Jain, U., & Nayak, S. (2008). *Pharmaceutical packaging technology* (1st ed.).
14. Edward, J. (2009). *Pharmaceutical packaging handbook* (1st ed.). Informa Healthcare.
15. Han, J. H. (2014). A review of food packaging technologies and innovations. In J. H. Han (Ed.), *Innovations in food packaging* (2nd ed., pp. 3–12). Academic. https://doi.org/10.1016/B978-0-12-394601-0.00001-1
16. WHO. (2002). *Guidelines on packaging for pharmaceutical products*. https://cdn.who.int/media/docs/default-source/medicines/norms-and-standards/guidelines/regulatory-standards/trs902-annex9.pdf?sfvrsn=82b4c57d_2.
17. Petroski, H. (2005). Engineering painful design. *American Scientist, 93*(2), 113–117.
18. Singh, A., Sharma, P. K., & Malviya, R. (2011). Eco friendly pharmaceutical packaging material. *World Applied Sciences Journal, 14*(11), 1703–1716.
19. Holbein, M. E. (2009). Understanding FDA regulatory requirements for investigational new drug applications for sponsor-investigators. *Journal of Investigative Medicine, 57*(6), 688–694. https://doi.org/10.2310/JIM.0b013e3181afdb26
20. Herber, O. R., Gies, V., Schwappach, D., Thürmann, P., & Wilm, S. (2014). Patient information leaflets: Informing or frightening? A focus group study exploring patients' emotional reactions and subsequent behavior towards package leaflets of commonly prescribed medications in family practices. *BMC Family Practice, 15*, 163. https://doi.org/10.1186/1471-2296-15-163
21. White, N. D., & Kibalama, W. (2018). Prevention of pediatric pharmaceutical poisonings. *American Journal of Lifestyle Medicine, 12*(2), 117–119. https://doi.org/10.1177/1559827617745014
22. Manoj, S. (2023). *Product branding, packaging and labelling*. Retrieved July 9, 2023, from https://www.yourarticlelibrary.com/marketing/product/product-branding-packaging-and-labelling/49054..
23. Carli Lorenzini, G., & Olsson, A. (2022). Exploring how and why to develop patient-centered packaging: A multiple-case study with pharmaceutical companies. *Therapeutic Innovation & Regulatory Science, 56*(1), 117–129. https://doi.org/10.1007/s43441-021-00338-0
24. Srivastava, P., Ramakanth, D., Akhila, K., & Gaikwad, K. K. (2022). Package design as a branding tool in the cosmetic industry: consumers' perception vs. reality, SN. *Business Economics, 2*(6), 58. https://doi.org/10.1007/s43546-022-00222-5
25. Paul, D. (2022). Can holograms help in the frontline fight against pharmaceutical counterfeiting? https://www.europeanpharmaceuticalreview.com/article/176751/holograms-pharmaceutical-counterfeiting/.
26. Ziance, R. J. (2008). Roles for pharmacy in combatting counterfeit drugs. *Journal of the American Pharmacists Association, 48*(4), e71–e91.
27. Apoorva, H. (2023). *QR codes for smart packaging is the future of packaging*. Retrieved July 9, 2023, from https://blog.beaconstac.com/2020/03/qr-codes-smart-packaging/.
28. Nail, S., & Searles, J. (2008). Elements of quality by design in development and scale-up of freeze-dried parenterals. *BioPharm International, 21*(1), 40–45.
29. USFDA. (1999). *FDA guidance for industry: Container closure systems for packaging human drugs and biologics*. Retrieved July 9, 2023, from https://www.fda.gov/media/70788/download.
30. Norwood, D. L., Paskiet, D., Ruberto, M., Feinberg, T., Schroeder, A., Poochikian, G., Wang, Q., Deng, T. J., DeGrazio, F., & Munos, M. K. (2008). Best practices for extractables and

leachables in orally inhaled and nasal drug products: An overview of the PQRI recommendations. *Pharmaceutical Research, 25*, 727–739. https://doi.org/10.1007/s11095-007-9521-z
31. USFDA. (2009). *Q8(R2) pharmaceutical development*. Retrieved July 9, 2023, from https://www.fda.gov/regulatory-information/search-fda-guidance-documents/q8r2-pharmaceutical-development.
32. USP. (2017). *General Chapter <659> Packaging and storage requirements*. Retrieved July 7, 2023, from https://www.uspnf.com/sites/default/files/usp_pdf/EN/USPNF/revisions/659_rb_notice.pdf.
33. USFDA. (2004). *Guidance for industry: PAT—A framework for innovative pharmaceutical development, manufacturing, and quality assurance*. Retrieved July 7, 2023, from https://www.fda.gov/media/71021/download.
34. Li, L. (2013). *Advances in container closure integrity testing, sterile product development: Formulation, process, quality and regulatory considerations* (pp. 315–329). Springer. https://doi.org/10.1007/978-1-4614-7978-9_12
35. Grilli, A. (2015). Quality by design and extractable and leachable testing. *BioPharm International, 28*(2), 34–37.
36. Das, P., Maity, A., & Pathane, S. (2018). Quality by design (QbD): A new horizon for robust analytics in extractable and leachable study for packaging components. *International Journal of Pharmaceutics and Drug Analysis, 6*(2), 89–104.
37. Singh, G., Lu, D., Liu, C., & Hower, D. (2021). Analytical challenges and recent advances in the identification and quantitation of extractables and leachables in pharmaceutical and medical products. *TrAC Trends in Analytical Chemistry, 141*, 116286. https://doi.org/10.1016/j.trac.2021.116286
38. Jenke, D. (2022). *Extractables and leachables: characterization of drug products, packaging, manufacturing and delivery systems, and medical devices*. Wiley.
39. Cuadros-Rodríguez, L., Lazúen-Muros, M., Ruiz-Samblás, C., & Navas-Iglesias, N. (2020). Leachables from plastic materials in contact with drugs. State of the art and review of current analytical approaches. *International Journal of Pharmaceutics, 583*, 119332. https://doi.org/10.1016/j.ijpharm.2020.119332
40. Li, K., Rogers, G., Nashed-Samuel, Y., Lee, H., Mire-Sluis, A., Cherney, B., Forster, R., Yeh, P., & Markovic, I. (2015). Creating a holistic extractables and leachables (E&L) program for biotechnology products. *PDA Journal of Pharmaceutical Science and Technology, 69*(5), 590–619. https://doi.org/10.5731/pdajpst.2015.01073
41. Wakankar, A. A., Wang, Y. J., Canova-Davis, E., Ma, S., Schmalzing, D., Grieco, J., Milby, T., Reynolds, T., Mazzarella, K., Hoff, E., Gomez, S., & Martin-Moe, S. (2010). On developing a process for conducting extractable and leachable assessment of components used for storage of biopharmaceuticals. *Journal of Pharmaceutical Sciences, 99*(5), 2209–2218. https://doi.org/10.1002/jps.22012
42. Hunt, D. G. (2006). *Containers Permeation USP29-NF24* (p. 456).
43. Zaidi, K. (2014). Compendial approvals for USP37-NF32 1S inhalation and nasal drug products general information and product quality tests PF 37, USP37-NF32, online.
44. Trumalai, R. *Biological reactivity tests, in vivo, USP29-NF24* (p. 2526).
45. Gore, A., & Pujara, C. (2022). *Topical ophthalmic dosage form development: Key components and critical quality attributes, ophthalmic product development: From bench to bedside* (pp. 153–180). Springer. https://doi.org/10.1007/978-3-030-76367-1_7
46. Jenke, D. (2018). Identification, analysis and safety assessment of leachables and extractables. *TrAC Trends in Analytical Chemistry, 101*, 56–65. https://doi.org/10.1016/j.trac.2017.10.024
47. Lopes, A. G. (2016). *Practical guide to single-use technology: Design and implementation*. Smithers Rapra.
48. I.C.o.H.E.W. Group. (1997). International conference on harmonisation of technical requirements for registration of pharmaceuticals for human use. In *ICH harmonised tripartite guideline. Guideline for good clinical practice*, 1997 CFR & ICH Guidelines.

49. Yoon, S., & Regn, M. (2022). *Packaging development: Multi-dose container closure for preservative free products, extractable/leachables from packaging, new technologies, ophthalmic product development: From bench to bedside* (pp. 229–245). Springer.
50. Broschard, T. H., Glowienke, S., Bruen, U. S., Nagao, L. M., Teasdale, A., Stults, C. L. M., Li, K. L., Iciek, L. A., Erexson, G., Martin, E. A., & Ball, D. J. (2016). Assessing safety of extractables from materials and leachables in pharmaceuticals and biologics—Current challenges and approaches. *Regulatory Toxicology and Pharmacology, 81*, 201–211. https://doi.org/10.1016/j.yrtph.2016.08.011
51. Kushwaha, P., & Madan, A. (2008). Extractables and leachables: An overview of emerging challenges. *Pharmaceutical Technology, 32*(8), 1–14.

Chapter 14
Quality by Design in Relation to Clinical Trials

Alka Kumari, Neha Bajwa, Preet Amol Singh, Virender Sachdeva, Tamana, and Gaurav Joshi

Abstract Clinical trials play a crucial role in evaluating the safety and efficacy of new drugs and treatments. Ensuring the quality and integrity of clinical trial data is essential for regulatory approval and patient safety. The quality by design (QbD) approach encompasses a systematic and proactive strategy to identify and mitigate potential risks to quality throughout the clinical trial process. This chapter aims to provide an overview of the QbD approach in relation to clinical trials.The chapter begins by discussing the principles and key elements of QbD, including the identification of critical quality attributes (CQAs) and critical process parameters (CPPs). It highlights the importance of a comprehensive risk assessment to identify potential sources of variability and their impact on trial outcomes. The chapter also emphasizes the need for robust design principles, starting from the protocol development stage, to ensure quality considerations are incorporated into the study design.

Furthermore, the chapter explores the integration of QbD principles into the various stages of clinical trial execution. It discusses the implementation of risk-based monitoring strategies, which involve the use of risk assessment tools to prioritize monitoring activities based on potential risks to data quality. Additionally, the chapter highlights the importance of comprehensive data management systems and data integrity measures to ensure accurate and reliable data collection, analysis, and reporting. The chapter also addresses the role of QbD in the selection and qualification of vendors and investigational sites, emphasizing the need for rigorous quality assessments. It discusses the importance of effective communication and

A. Kumari · Tamana
University Institute of Pharma Sciences, Chandigarh University, Mohali, Punjab, India

N. Bajwa · P. A. Singh
University Centre of Excellence in Research, Baba Farid University of Health Sciences, Faridkot, Punjab, India

V. Sachdeva
College of Pharmacy, Pt. B.D. Sharma University of Health Sciences, Rohtak, Haryana, India

G. Joshi (✉)
Department of Pharmacy Practice, University Institute of Pharma Sciemnces (UIPS), Chandigarh University, Mohali, Punjab, India

N. K. Jain, N. Bajwa (eds.), *Introduction to Quality by Design (QbD)*,
https://doi.org/10.1007/978-981-99-8034-5_14

collaboration between stakeholders, including sponsors, investigators, and regulatory authorities, to ensure a shared understanding of quality expectations and adherence to quality standards. Lastly, the chapter discusses the regulatory landscape and initiatives promoting the adoption of the QbD approach in clinical trials. It highlights the benefits of QbD, including improved trial efficiency, enhanced data quality, and increased patient safety. In summary, the QbD approach offers a proactive methodology for ensuring the quality and reliability of clinical trial data. By integrating quality considerations throughout the trial process, from design to execution, the QbD approach can help mitigate risks, enhance data integrity, and ultimately contribute to the successful completion of clinical trials.

Keywords Quality by Design (QbD) · Clinical trials · Quality assurance · Critical quality attributes (CQAs) · Critical process parameters (CPPs) · Risk assessment · Study design · Protocol development

14.1 Introduction

ICH Q8(R2) defines QbD as "a systematic approach to development that begins with predefined objectives and emphasizes product and process understanding and process control [1], based on sound science and quality risk management." An organization that designs, analyses, and controls production by regularly measuring critical performance and quality characteristics of new and in-progress materials and processes to ensure the safety of the finished product. In QbD, the formulation and manufacturing factors are understood and controlled to ensure the quality of the final product [2]. As a result, the manufacturing procedure, formulation control, and design contribute to consistent product quality. Applying the QbD technique can result in solid formulations and a high rate of regulatory approvals for pharmaceutical products. Multivariate statistics and the design of experiments are used in QbD [3]. Quality by Design (QbD) is a systematic approach to pharmaceutical development that is becoming increasingly important in designing and conducting clinical trials. QbD is a proactive approach to quality that focuses on developing quality into the product rather than testing for it later. By doing so, it can help to ensure that clinical trials are conducted in a more efficient, effective, and consistent manner [4]. The QbD approach involves defining a drug product's critical quality attributes (CQAs) and designing a manufacturing process that consistently produces products that meet those CQAs. This approach is supported by several tools and techniques, including risk assessment, statistical design of experiments (DOE), and process analytical technology (PAT) [5]. In clinical trials, QbD can be used to design and optimize the clinical trial process itself, to improve the efficiency and effectiveness of the prosecution while minimizing risk to patients. For example, QbD can be used to identify the critical quality attributes of the study drug, the key variables that may affect the safety and efficacy of the drug, and the essential points of the trial where these variables need to be closely monitored [6]. Quality by Design (QbD) is a systematic approach to product and process development that emphasizes the importance of understanding the underlying science and using that knowledge to design a product and manufacturing process that meets predefined quality attributes. QbD originated in the pharmaceutical industry but has since been adopted by other sectors, such as food and beverage, cosmetics, and medical devices [7]. The goal of QbD is to ensure that the product is consistently manufactured to meet its intended use and that the process is robust, efficient, and capable of delivering the desired quality attributes. QbD involves several steps, including defining the quality target product profile (QTPP), identifying critical quality attributes (CQAs), determining the essential process parameters (CPPs) that affect the CQAs, and designing experiments to establish the relationship between the CPPs and CQAs. The data obtained from these experiments are used to develop a control strategy that ensures the process operates within a specified range and provides consistent quality. Overall, QbD is a proactive approach to quality management that can result in improved product quality, reduced manufacturing costs, and increased efficiency. It emphasizes the importance of designing quality into the product and process rather than relying on end-product testing to detect and correct quality issues [8] (Fig. 14.1).

14.2 Key Aspects of Quality by Design

14.2.1 TPP (Target Product Profile)

The TPP contains information about the drug at a particular stage of development as well as a description of the general goal of the drug development program. The TPP often links drug development activities to certain concepts intended for inclusion in drug labelling and is structured by the significant parts of drug labelling [9].

14.2.2 TPQP (Target Product Quality Profile)

The TPQP is the foundation for the overall pharmaceutical product development strategy. This characteristic determines the design and level of development. It is a performance-based quality program or a quality feature that a product must have to comply with TPP. TPQP has been defined as the "Safety and efficacy of drug product after ensuring the desired quality and quality characteristics of a drug product." It is a strategic tool for drug development, and the steps involved lead to the end goal with maximum results. The TPP can be crucial in every step of the drug development process, including efficient optimization of a drug candidate, organizational decision-making, clinical research plans, and effective interaction with regulatory agencies [5]. Tests that are important for the performance of dosage forms to achieve the therapeutic objective are included in TPQP. It is related to identity, assay, dosage form, purity, and stability. TPQP is the extension of the target product profile focussing on quality which gives guidance to formulation scientists to prepare efficient formulations by maintaining product quality.

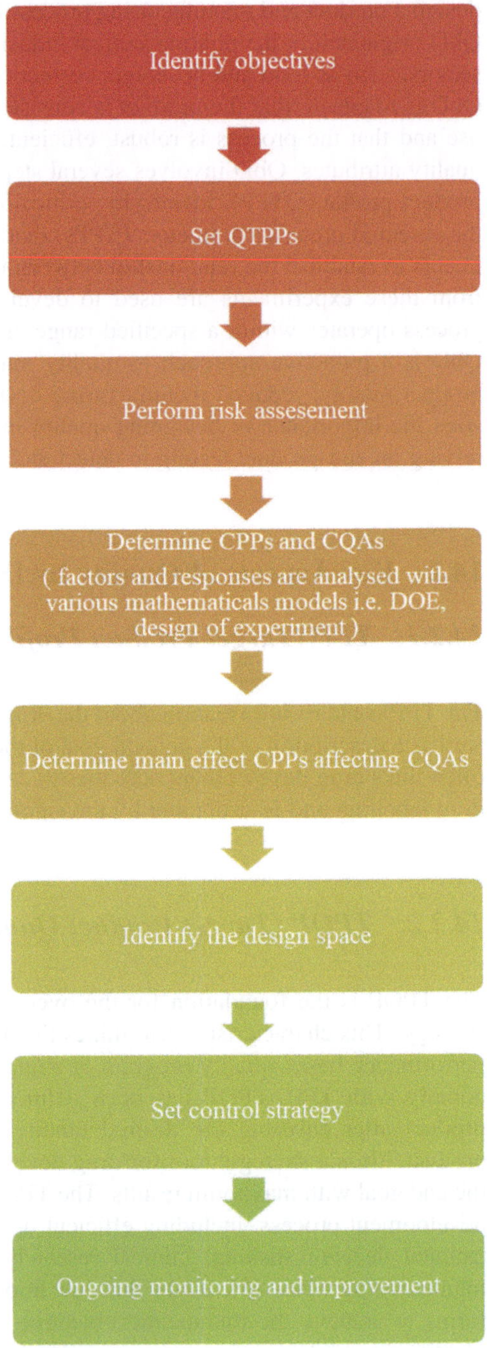

Fig. 14.1 Outlining the elements of QBD

14.2.3 Critical Quality Attributes

After identifying TPP and TPQP, the next step is identifying relevant CQAs. A CQA is "a physical, chemical, biological, or microbiological property or characteristic within an appropriate limit, range, or distribution to ensure the desired product quality." Making these risk evaluations requires using preliminary product information, such as the collected laboratory, nonclinical, and clinical experience with a given product-quality attribute. A detailed grasp of the physicochemical and biological characteristics of the API is the first step in formula creation. Risk analysis leads to the identification of CQAs following ICH Guideline Q9 [10]. For the identification of CQAs, various risk assessment methods are used. Generally speaking, CQAs of the solid oral dosage form impact the product's purity, strength, drug release, and stability. More product-specific CQAs are also included for additional distribution methods, such as aerodynamic qualities for inhaled medications, sterility for parenteral medications, and adhesion properties for transdermal patches. The drug product CQAs, which may include CQAs for the drug ingredients, raw materials, and intermediaries, are influenced by particle size distribution and bulk density [3] (Figs. 14.2 and 14.3).

14.3 Critical Process Parameters

Critical process parameters (CPPs) are "parameters whose variability impact a CQA and therefore should be monitored or controlled to ensure the process produces the desired quality." Critical process parameters (CPPs) directly and considerably impact essential quality attributes when changed within the normal operating range. Process robustness is the capacity of a process to exhibit acceptable quality and performance while simultaneously tolerating input unpredictability [11]. Process capability should be investigated to demonstrate a process's repeatability and

Fig. 14.2 Various parameters of TPQP

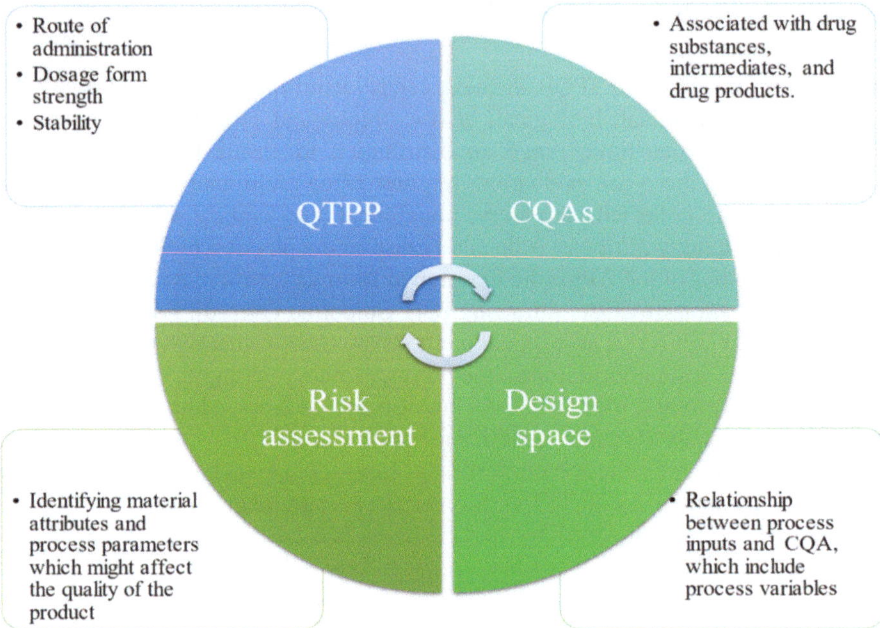

Fig. 14.3 The relationship between QTPP, CQA, Risk assessment, and Design of Space

consistency. Process capability is a statistical measurement of the inherent process variability for attributes. Six sigma is the most commonly used formula for process capability index is defined as the value of the tolerance specified for a particular characteristic divided by the process capability, which is defined as follows (Table 14.1):

Process capability index = Upper limit of specification – The lower limit of specification\6 standard deviation

14.4 Risk Assessment

The practice of connecting material features and process parameters to CQAs is known as risk assessment. According to ICH Q9 Quality Risk Management, a drug product's production, and usage entail a certain amount of risk. The quality risk management process should involve effort, formality, and paperwork appropriate to the amount of risk. It is relatively simple for the producer to choose which studies to do if a risk assessment is done before developing a pharmaceutical product [12]. Risk assessment studies can be done with the help of various simple methods like flow charts, process mapping, cause and effect diagrams, check sheets, etc (Fig. 14.4).

Table 14.1 CPPs and CQAs involved in tablet preparation

Operations during tableting	Critical process parameters	Critical quality attributes
Wet granulation	Mixing time Temperature	Granule size and distribution Moisture content
Drying	Drying time Inlet airflow	Bulk/tapped density Moisture content
Milling	Milling speed Feeding rate	Flow properties Particle size distribution
Mixing	Mixing time Order of addition	Blend uniformity
Compression	Compression force Dwell time Hopper design Ejection force	Hardness Weight variation Friability Content uniformity assay Dissolution Disintegration
Coating	Time Temperature Spray pattern and rate Inlet airflow	Hardness Appearance

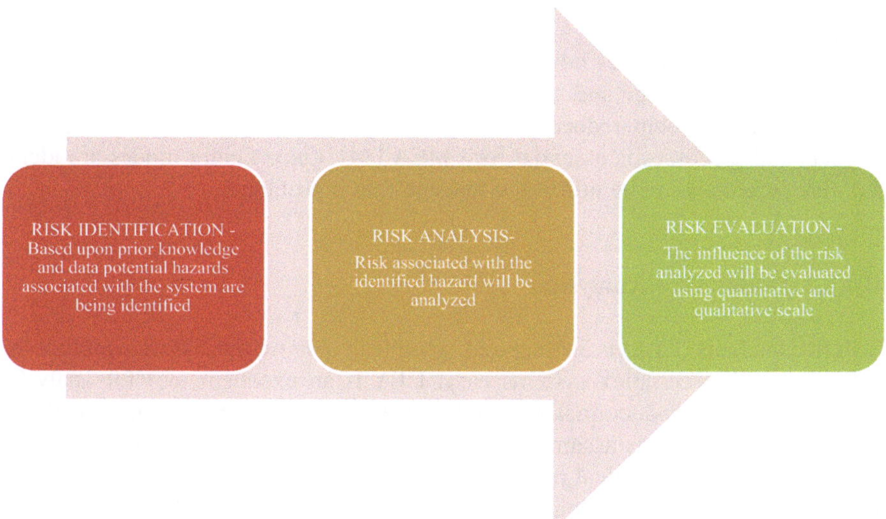

Fig. 14.4 Various risk assessment methods

14.4.1 Significant Elements of Risk Assessment

Cause-and-effect diagrams, also known as Ishikawa diagrams or fishbone diagrams, are used to represent all factors that have an impact on a critical product's quality. The diagram will have a horizontal line whose endpoints are in the direction of the

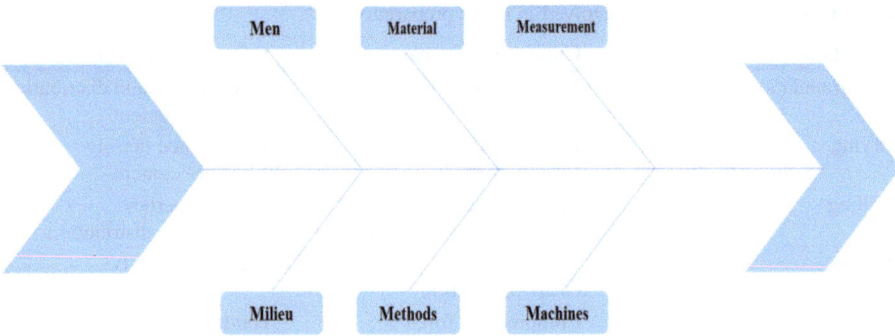

Fig. 14.5 The fishbone diagram for the risk assessment study

affected product quality, followed by diagonal lines for the major influencing factors and sublines for the diagonal lines for critical process parameters and critical material attributes [13] (Fig. 14.5).

14.4.1.1 Failure Mode Effect Analysis

This tool can find product development flaws. The technique predicts failure by identifying, prioritizing, and estimating the risk. Failure mode effects analysis (FMEA) can also help reduce the risk of modifying a process. Understanding products and processes is essential for FMEA [14]. Once failure modes are identified, risk reduction can be utilized to manage future problems.

14.4.1.2 Fault Tree Analysis

Fault tree analysis (FTA) is a structured, graphical, quantitative assessment tool that makes modelling complex systems easy. FTA is an excellent tool for analysing several aspects of product quality [15]. It is helpful for both monitoring and RA. A logic tree's procedure is assumed to be its root. There will only be one top event (origin) present. Each critical process parameter or critical material characteristic is added to the tree as a set of logic expressions. From fault trees, failure probabilities can be calculated using the software.

14.4.1.3 Design Space

The ICH Q8 (R2) States that the design space is a multidimensional combination and interaction of input variables (e.g., material attributes) and process parameters that have been demonstrated to assure quality [16]. Change is not regarded as occurring when working within the design space. A movement outside the design area would be considered a change and therefore be deemed the beginning of a regulatory post-approval change process. The applicant proposes the design space subject to regulatory review and approval. Because design space is potentially size and equipment dependent, the design space established at the laboratory scale may not be appropriate for the commercial-scale process. Therefore, commercial design-space verification becomes necessary until it is shown that the design space is scale-independent [17].

14.4.1.4 Control Strategy

A control strategy should be created and managed to guarantee the required product quality consistently. Specifications for raw materials, process checks, in-process testing, and end-product testing are all examples of control strategies [3]. The impact of process variables and material characteristics on product quality must be investigated during product development. The critical process parameters and critical material properties are determined from this data. To achieve the desired product quality, these essential inputs are then managed. The established control strategy needs to be supported and documented. It is necessary to determine where the variability is coming from, according to ICH Q8 (R2). The control strategy is "a planned set of controls, derived from current product and process understanding that assures process performance and product quality. The control strategy in the QbD paradigm is built with risk assessment and considers the importance of the CQA and process capacity. Procedural controls, in-process controls, lot release testing, process monitoring, characterization testing, comparability testing, and stability testing can all be included in the control strategy [18] (Fig. 14.6).

14.5 Product Lifecycle Management

Various efforts should be made to manage the quality of the product throughout its life cycle. Viewing the findings of exploratory and clinical development studies as beneficial contributions to product development is possible. Technology transfer activities produce knowledge valuable for manufacturing processes, control

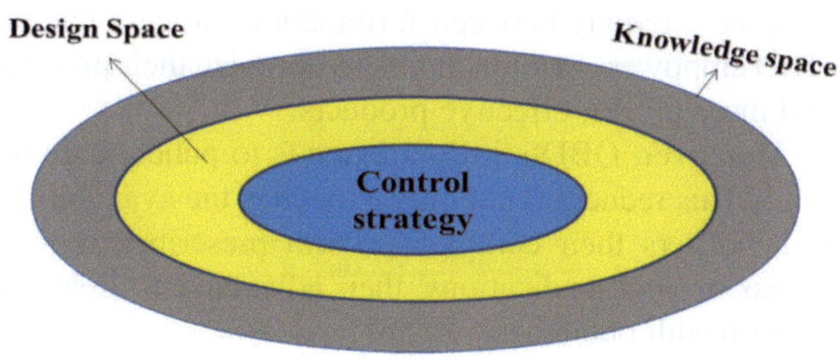

Fig. 14.6 The control strategy of analysing QbD

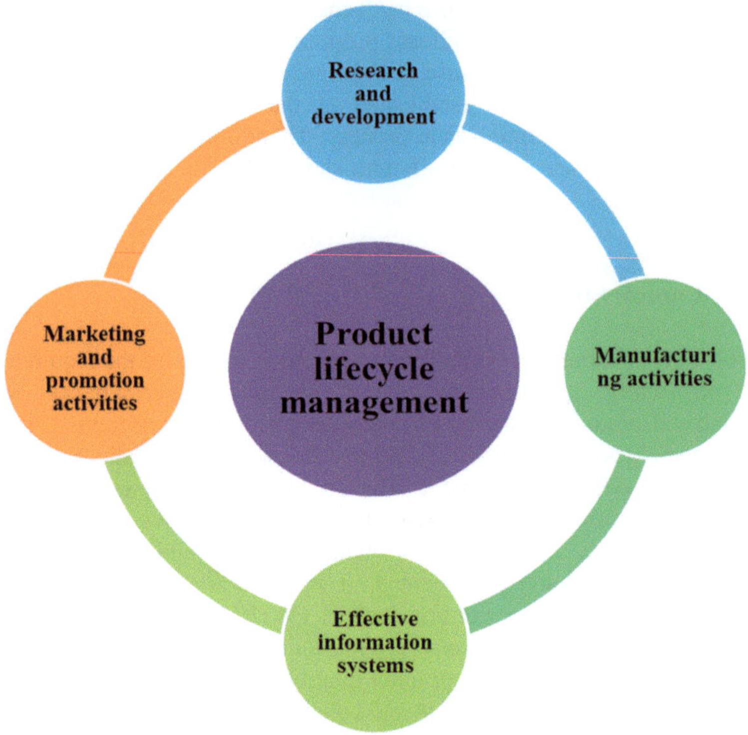

Fig. 14.7 Various stages of product lifecycle management

strategies, and process validation, the foundation for ongoing operation and product improvement [19]. The pharmaceutical quality system should discover and assess improvement opportunities during commercial manufacture [20]. The expertise gained from commercial manufacturing should also be used for continuous product enhancement. Continued product evaluation, including management of complaints and stability, should be done during product discontinuation [21] (Fig. 14.7).

Quality by Design (QBD) stands as a cornerstone in modern pharmaceutical development and manufacturing, offering a host of advantages that significantly enhance the industry's efficiency and effectiveness. One of its primary benefits is the tangible improvement in yield and product efficacy. By meticulously analyzing the intricate interplay between formulation, process parameters, and quality attributes, QBD empowers manufacturers to optimize their processes, resulting in higher yields and more potent, effective products.

Moreover, QBD's impact extends to patient care through reduced testing durations. This reduction not only expedites the availability of treatments to patients but also bolsters their compliance with prescribed regimens. As patients gain faster access to vital medications, their adherence to treatment plans improves, fostering better health outcomes.

The meticulous principles of QBD also give rise to remarkable batch-to-batch consistency. Through a comprehensive understanding and control of critical process parameters, manufacturers achieve unparalleled uniformity in product attributes such as potency, purity, and stability. This consistency is paramount for patient safety and ensures that each dose delivers the expected therapeutic effect.

Equally noteworthy is QBD‘s role in minimizing batch failures. By preemptively identifying and mitigating potential sources of variability, QBD curtails the likelihood of batch deviations and failures, reducing costly disruptions in production and maintaining a seamless supply chain.

The advantages of QBD transcend quality assurance, extending to the realm of economics and expediency. The meticulous optimization of processes leads to cost savings by streamlining resource utilization, while the profound understanding of the product and its attributes facilitates faster regulatory approvals and swifter product launches. This dual impact of cost-effectiveness and accelerated market entry underscores QBD‘s pivotal role in shaping a thriving pharmaceutical landscape, where both manufacturers and patients reap the rewards of enhanced products and improved access to treatments.

14.6 Quality by Design (QbD) in Clinical Trials

In a clinical trial for a personalized medicine approach targeting a specific genetic mutation, QbD principles were applied to optimize patient enrollment and stratification. The CQA in this case was the presence of the specific mutation. QbD techniques were utilized to identify appropriate inclusion and exclusion criteria, design a biomarker testing protocol, and implement a stratified randomization approach. By systematically considering these factors, the trial ensured that the patient population was properly characterized, minimizing variability and maximizing the chances of detecting a treatment effect in the target population.

Quality by Design (QbD) is a systematic approach that aims to ensure the quality and reliability of pharmaceutical products and processes by proactively designing and controlling them based on scientific understanding and risk management [2]. While QbD has been predominantly applied in drug development and manufacturing, its principles can also be extended to clinical trials to enhance the quality and validity of the generated data. This introduction will provide an overview of QbD in clinical trials, along with suitable examples and case reports that illustrate its application and benefits [22].

QbD principles in clinical trials encompass the identification and understanding of critical quality attributes (CQAs), design of experiments (DoE) techniques, risk management strategies, control strategies, and continuous improvement. By incorporating these principles, clinical trials can be designed and conducted to minimize bias, ensure patient safety, maximize statistical power, and generate reliable and meaningful results. Here are different roles of QbD in designing clinical trials for drug development [23].

14.6.1 Optimizing Clinical Trial Design

In a Phase III clinical trial for a new cardiovascular drug, QbD principles were applied to optimize the study design. The CQAs were identified as efficacy and safety endpoints, patient population characteristics, and data collection methods. By utilizing DoE techniques, the study team determined the appropriate sample size, randomization, and blinding strategies. This systematic approach allowed the trial design to minimize bias, enhance patient safety, and maximize statistical power, leading to robust and credible results [24].

14.6.2 Ensuring Data Quality and Reliability

In a multicenter clinical trial evaluating a novel cancer therapy, QbD principles were applied to ensure the quality and reliability of the collected data. A comprehensive control strategy was implemented, including standardized data collection procedures, rigorous monitoring for data integrity, and statistical methods for controlling confounding variables. This integration of QbD principles in data collection and analysis enhanced the quality and reliability of the trial's findings, providing valuable insights into the therapy's efficacy and safety [25].

14.6.3 Identification and Managing Potential Risks

In a Phase II clinical trial for a new investigational therapy, QbD principles were employed to identify and manage potential risks. A risk assessment was conducted to evaluate the impact of various factors on the trial's quality, safety, and integrity. Risk mitigation strategies were then implemented, such as training and monitoring of investigators, centralized data review, and independent data and safety monitoring committees. This QbD-based risk management approach minimized potential risks and improved the overall quality of the trial.

QbD begins with the identification of critical quality attributes, which are the characteristics that have a significant impact on the trial's quality, safety, and efficacy. CQAs can include endpoints, patient population characteristics, biomarkers, and data collection methods.

In a clinical trial for a new oncology drug, the CQAs may include tumor response rate, progression-free survival, and overall survival as primary endpoints. By identifying these critical quality attributes, the trial design can focus on evaluating the drug's effectiveness in improving patient outcomes.

In a vaccine trial, the CQAs may involve assessing the immunogenicity and safety of the vaccine in different patient populations, such as pediatric, adult, and elderly. By understanding the critical quality attributes, the trial design can address the specific needs and characteristics of each population.

DoE techniques are utilized to systematically study the impact of various factors on the trial's quality attributes. DoE allows for the optimization of trial design parameters, such as sample size, randomization, blinding, and stratification, to minimize bias and maximize statistical power.

A Phase III clinical trial for a novel diabetes drug employed DoE techniques to optimize the trial design. The study team utilized factorial designs to evaluate the impact of different dosage levels, treatment durations, and patient characteristics on glycemic control as a CQA. By systematically varying these factors, the trial design was optimized to identify the most effective dosage and treatment duration for different patient subgroups.

QbD emphasizes proactively identifying and managing risks throughout the trial's lifecycle. Risk management strategies are employed to assess and mitigate potential risks that may impact the trial's quality, safety, or integrity.

In a Phase II clinical trial for an investigational therapy, a risk assessment was conducted to identify potential risks related to patient safety, data quality, and trial execution. Risk mitigation strategies, such as investigator training, centralized data review, and independent data monitoring committees, were implemented to ensure patient safety and data integrity.

QbD promotes the development of a robust control strategy to ensure consistency and reliability in trial conduct, data collection, and analysis. Control strategies involve establishing procedures, specifications, and acceptance criteria to minimize variability and maintain trial quality.

A multicentre clinical trial evaluating a new medical device implemented a control strategy to ensure consistent data collection across study sites. Standardized protocols, training, and monitoring were employed to maintain uniformity in data collection procedures, reducing variability and enhancing the trial's reliability and validity.

A Phase IV post-marketing surveillance trial for a drug utilized continuous improvement principles to optimize patient safety monitoring. Adverse event reporting systems were regularly reviewed and updated, and real-time safety data analysis was performed to detect any emerging safety signals promptly. These continuous improvement efforts ensured patient safety and contributed to regulatory compliance.

14.7 Critical Quality Attributes (CQAs) in Clinical Trials

Critical Quality Attributes (CQAs) play a crucial role in clinical trials as they are the specific characteristics or parameters that significantly impact the quality, safety, and efficacy of the trial outcomes. By identifying and understanding CQAs, researchers can design trials that measure and evaluate these attributes to generate reliable and meaningful data [26]. This article will explain the concept of CQAs in clinical trials and provide suitable examples, and case reports to illustrate their importance and application.

CQAs are the measurable parameters that define the quality of a product or process. In the context of clinical trials, CQAs encompass various aspects, including endpoints, patient population characteristics, biomarkers, and data collection methods. Identifying CQAs helps in designing trials that effectively evaluate the impact of the intervention or treatment on these critical attributes [27].

In a vaccine clinical trial, safety endpoints such as the incidence of adverse events or the occurrence of specific adverse reactions are critical quality attributes. These attributes help evaluate the vaccine's safety profile and provide important data for assessing its risk-benefit profile. By identifying safety endpoints as CQAs, researchers can focus on monitoring and assessing the vaccine's safety throughout the trial [28].

14.7.1 Importance and Application of CQAs in Clinical Trials

14.7.1.1 Trial Design

Identifying CQAs during trial design allows researchers to focus on endpoints and attributes that are most relevant to the intervention being studied. This targeted approach enhances the trial's scientific validity and relevance [29].

14.7.1.2 Outcome Assessment

By defining CQAs, trial outcomes become more specific and meaningful. Evaluating the impact of an intervention on CQAs provides essential information to support regulatory decisions, clinical practice, and patient care [30].

14.7.1.3 Statistical Analysis

CQAs guide the selection of appropriate statistical methods and analyses to assess the treatment effect on the identified attributes. Statistical analyses can then be tailored to the specific CQAs, ensuring accurate and reliable interpretation of trial results [31].

14.7.1.4 Regulatory Considerations

Regulatory agencies often require the identification and measurement of CQAs to assess the quality, safety, and efficacy of a new drug or intervention. Demonstrating the impact of the intervention on CQAs strengthens the regulatory submission and increases the likelihood of approval [32].

14.7.1.5 Trial Optimization

Identifying CQAs allows researchers to focus resources, sample sizes, and endpoints on attributes that are most critical to the intervention's evaluation. This optimization improves trial efficiency, reduces costs, and enhances the chances of detecting treatment effects (Fig. 14.8).

Critical Quality Attributes (CQAs) are essential components of clinical trial design [33]. They define the specific parameters or characteristics that significantly impact the quality, safety, and efficacy of trial outcomes. By identifying and focusing on CQAs, researchers can design trials that generate reliable and meaningful data, supporting evidence-based decision-making in drug development and patient care. The examples and case reports provided in this article highlight the importance of CQAs in clinical trials, demonstrating their application in various therapeutic areas and emphasizing their role in optimizing trial design, outcome assessment, and statistical analysis.

14.8 Design of Experiments (DoE) in Clinical Trial Design

Design of Experiments (DoE) is a statistical methodology that plays a vital role in clinical trial design. It allows researchers to systematically study the impact of various factors or variables on critical trial outcomes. By utilizing DoE techniques, clinical trials can be optimized to enhance efficiency, minimize bias, and maximize the study's statistical power. This article will explain the concept of DoE in clinical trial design and provide suitable examples, and case reports to illustrate its application and benefits [34].

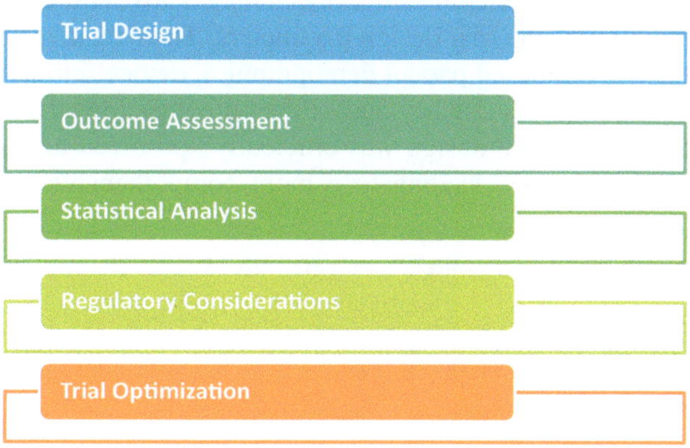

Fig. 14.8 CQA application in clinical trials

DoE is a systematic approach that involves manipulating and controlling independent variables or factors to observe their effect on dependent variables or outcomes. In the context of clinical trials, DoE techniques are employed to optimize trial design parameters and identify the most efficient and effective approaches to evaluate the intervention's impact on critical endpoints [34].

14.8.1 Factorial Design in Dose Optimization

In a clinical trial for a new drug, researchers can utilize a factorial design to study the impact of multiple factors simultaneously. For example, the trial may investigate the effect of different dosage levels and treatment durations on efficacy and safety outcomes. By varying these factors systematically, the trial design can identify the optimal dosage and treatment duration for maximum benefit with minimal adverse effects.

14.8.2 Response Surface Methodology in Optimization

In a clinical trial for an intervention targeting a specific disease, response surface methodology (RSM) can be applied to optimize the treatment parameters. RSM helps determine the appropriate combination of variables, such as dosage, frequency, and duration, to achieve the desired therapeutic response. By systematically varying these variables within a defined range, the trial can identify the optimal treatment regimen that maximizes efficacy while minimizing side effects [35].

14.8.3 Multifactorial Trial Design

In a multicenter clinical trial evaluating a novel medical device, DoE was employed to optimize the trial design parameters. The trial investigated the impact of multiple factors, including patient demographics, device settings, and procedural techniques, on the device's efficacy and safety outcomes [36]. By utilizing a fractional factorial design, the trial design was efficiently optimized to assess the combined effects of various factors, leading to a more comprehensive understanding of the device's performance.

14.8.4 *Optimization of Randomization and Blinding Strategies*

In a clinical trial comparing two interventions, DoE techniques were utilized to optimize the randomization and blinding strategies. The trial investigated the effect of different randomization ratios, allocation concealment techniques, and blinding approaches on minimizing selection bias and ensuring treatment allocation integrity. By systematically evaluating these factors, the trial design achieved an optimized randomization and blinding strategy that reduced bias and enhanced the validity of the study results [37].

DoE allows for an efficient allocation of resources by systematically varying factors within defined ranges. This optimization reduces the sample size requirements, time, and costs associated with the trial. By employing DoE techniques, clinical trials can identify and mitigate sources of bias by systematically varying and controlling factors that may influence the study outcomes. This enhances the internal validity of the trial. DoE helps maximize the trial's statistical power by optimizing trial design parameters. By efficiently allocating resources and considering various factors simultaneously, DoE ensures the trial is adequately powered to detect treatment effects. DoE allows for a comprehensive evaluation of the combined effects of multiple factors. It provides insights into interactions between variables, enabling a deeper understanding of the intervention's impact on critical trial outcomes. DoE helps optimize treatment parameters, such as dosage, frequency, and duration, to achieve the desired therapeutic response. This optimization improves patient outcomes and contributes to evidence-based treatment decisions.

Design of Experiments (DoE) is a powerful statistical methodology that enhances clinical trial design by systematically studying the impact of various factors on critical trial outcomes. By employing DoE techniques, trials can be optimized to efficiently allocate resources, minimize bias, maximize statistical power, and achieve comprehensive evaluations [38]. The examples and case reports provided in this article demonstrate the application of DoE in various clinical trial scenarios, highlighting its ability to optimize dose selection, treatment parameters, and randomization strategies. Incorporating DoE into clinical trial design enhances the scientific validity and efficiency of trials, leading to more reliable and meaningful results that support evidence-based decision-making in drug development and patient care [39].

QbD (Quality by Design) principles can be applied in patient enrollment and stratification in clinical trials to enhance the efficiency and effectiveness of recruitment, as well as to ensure the appropriate identification and characterization of patient subgroups. By incorporating QbD principles in patient enrollment, researchers can optimize trial design, increase patient engagement, and improve the overall quality of data generated [23]. This article will explain the applications of QbD in patient enrollment and stratification, along with suitable examples and case reports that illustrate their benefits [4]. QbD principles can be employed to optimize the inclusion and exclusion criteria in clinical trial enrollment. By identifying the critical quality attributes (CQAs) of the intervention, researchers can refine the criteria to focus on patient populations that are most likely to benefit from the treatment [40]. In a clinical trial for targeted therapy, QbD principles may be applied to refine the inclusion criteria based on specific genetic mutations or biomarker profiles. By focusing on patients who possess these CQAs, the trial can effectively evaluate the therapy's efficacy in the intended patient population [41]. QbD encourages patient-centric trial design, considering factors such as patient preferences, burden of participation, and patient-reported outcomes. By incorporating patient perspectives early in trial design, the trial can be designed to enhance patient engagement, retention, and adherence [42]. A QbD-driven trial for a chronic disease treatment involved patient input to optimize the trial design. Patient feedback on the frequency and mode of data collection, visit schedules, and treatment administration was collected. By integrating patient preferences, the trial design became more patient-centric, leading to improved patient recruitment and retention [43]. QbD principles can guide the stratification of patients based on biomarker profiles or genetic characteristics. By identifying CQAs related to specific biomarkers, researchers can stratify patients into subgroups to evaluate treatment response and improve the precision of the study results [44]. In a clinical trial for a cancer immunotherapy, QbD principles may be applied to stratify patients based on their tumor biomarker expression levels. This stratification allows for a more accurate evaluation of treatment response in different patient subgroups and enables personalized medicine approaches. QbD-driven risk-based stratification aims to identify patients who are at higher risk for adverse events or who are more likely to benefit from the intervention. By considering CQAs related to risk factors, patient characteristics, or comorbidities, researchers can optimize stratification strategies [45].

In a cardiovascular trial, QbD principles may be utilized to stratify patients based on their baseline risk profiles, such as age, hypertension, or cholesterol levels. By stratifying patients into risk categories, the trial can evaluate the intervention's efficacy and safety in specific patient subgroups.

14.8.5 QbD in Personalized Medicine Trials

In a clinical trial for a personalized medicine approach, QbD principles were employed to optimize patient enrollment and stratification. The trial focused on patients with a specific genetic mutation, identified as a CQA. QbD techniques were used to refine the inclusion criteria, biomarker testing protocol, and stratification methods. This QbD approach enabled the trial to effectively evaluate the treatment's efficacy in the target patient population [46].

14.8.6 Risk-Based Stratification in Oncology Trial

In an oncology trial evaluating a new treatment, QbD principles were applied to optimize patient stratification. The trial considered various risk factors and patient characteristics as CQAs. By implementing risk-based stratification, the trial aimed to identify patients who were at higher risk for disease progression and, thus more likely to benefit from the intervention. This QbD-driven stratification approach enhanced the trial's precision and treatment evaluation [47].

QbD-driven patient enrollment and stratification ensure that patients who are most likely to benefit from the intervention are included in the trial, leading to improved treatment evaluation and patient outcomes. By incorporating QbD principles, patient enrollment and stratification can be optimized to focus on specific patient subgroups, resulting in more precise and efficient evaluation of treatment effects. QbD encourages a patient-centric trial design, taking into account patient preferences and needs. This approach enhances patient engagement, retention, and adherence, ultimately improving the overall quality of trial data. QbD principles facilitate the implementation of personalized medicine approaches by stratifying patients based on biomarkers or risk factors. This allows for tailored treatment evaluation and supports the development of precision therapies. QbD principles in patient enrollment and stratification enhance the efficiency and effectiveness of clinical trials by optimizing trial design, focusing on patient-centric approaches, and improving the precision of treatment evaluation. Examples and case reports demonstrate the successful application of QbD in refining inclusion criteria, incorporating patient preferences, stratifying patients based on biomarkers or risk factors, and optimizing trial outcomes. By integrating QbD principles into patient enrollment and stratification, researchers can improve patient selection, enhance precision, and drive personalized medicine approaches in clinical trials.

QbD (Quality by Design) principles can be applied to data collection and analysis in clinical trials to ensure the quality, integrity, and reliability of the data generated. By incorporating QbD principles in these aspects, researchers can optimize data collection procedures, implement robust data management practices, and employ appropriate statistical methods for data analysis. This article will explain the application of QbD in data collection and analysis in clinical trials, along with suitable examples and case reports to illustrate their benefits [48].

14.8.7 Standardized Data Collection Procedures

QbD principles promote the use of standardized data collection procedures to ensure consistency and minimize variability. Standardization includes standardized data forms, protocols, and training for data collectors to enhance the reliability and integrity of the collected data. In a clinical trial for a novel drug, QbD principles may be applied to develop standardized electronic case report forms (eCRFs) and data collection protocols. This standardization ensures that data are collected consistently across study sites, minimizing data variability and facilitating data analysis [49].

14.8.8 Risk-Based Monitoring

QbD-driven risk-based monitoring identifies and prioritizes critical data elements and implements monitoring strategies accordingly. By focusing on critical data points, researchers can allocate monitoring resources effectively and ensure the quality and integrity of the data collected. In a clinical trial evaluating a medical device, QbD principles may be employed to conduct risk assessments and identify critical data elements. Based on this assessment, a risk-based monitoring plan can be implemented to ensure that critical data points are monitored more rigorously, minimizing the risk of data quality issues.

14.8.9 QbD in Data Analysis

QbD principles guide the planning and execution of statistical analysis in clinical trials. By considering critical quality attributes (CQAs) and trial design parameters, researchers can determine appropriate statistical methods and analyses to evaluate the treatment effect accurately and robustly. In a clinical trial for a new therapy, QbD principles may be applied to identify CQAs, such as primary endpoints, and design appropriate statistical analyses, such as superiority or non-inferiority testing, to assess treatment efficacy. Researchers can generate reliable and meaningful results by aligning statistical analysis plans with the CQAs. QbD principles help researchers develop strategies for handling missing data to minimize biases and maintain the integrity of the analysis. This includes implementing data monitoring, data quality checks, and appropriate statistical techniques for handling missing data. In a longitudinal clinical trial with missing data due to participant dropout, QbD principles may be applied to develop strategies for handling missing data [50]. Techniques such as multiple imputation or sensitivity analyses can be employed to account for missing data and minimize potential biases. QbD in Centralized Data Review In a multicenter clinical trial, QbD principles were applied to implement centralized data review processes. This involved establishing data quality checks, standardized data monitoring procedures, and a centralized data review committee. By employing QbD in data collection and analysis, the trial ensured consistency and reliability in data across study sites, enhancing the overall quality of the trial data.

14.8.10 QbD in Adaptive Design Analysis

In an adaptive design clinical trial, QbD principles were utilized in the statistical analysis planning. The trial employed an adaptive randomization scheme, and QbD principles guided the selection of appropriate statistical methods for analyzing the treatment effect under the adaptive design framework. By aligning the analysis plan with the trial design, the trial generated robust and reliable results.

QbD principles ensure the collection of high-quality data by implementing standardized procedures, risk-based monitoring, and robust data management practices. This enhances the reliability, integrity, and validity of the trial data. By aligning data collection and analysis with critical quality attributes and trial design parameters, QbD reduces biases and ensures that data analysis accurately evaluates treatment effects. QbD optimizes resource allocation by prioritizing critical data elements, conducting risk-based monitoring, and employing appropriate statistical techniques. This improves resource efficiency and allows for focused data collection and analysis [51]. By applying QbD principles, the quality and integrity of the trial data are enhanced, leading to reliable and meaningful results. This strengthens the evidence base for decision-making in drug development and patient care. QbD principles can significantly improve data collection and analysis in clinical trials by enhancing data quality, minimizing bias, optimizing resource allocation, and generating reliable and meaningful results. Examples and case reports demonstrate the successful application of QbD in standardizing data collection procedures, implementing risk-based monitoring, planning statistical analyses, and handling missing data. By incorporating QbD principles into data collection and analysis, researchers can ensure the quality and integrity of trial data, leading to more robust evidence for decision-making and ultimately improving patient care.

14.9 QbD Perspective: Risk Management in Clinical Trials

Risk management is a critical aspect of clinical trials, and when approached from a Quality by Design (QbD) perspective, it enhances the overall quality, safety, and integrity of the trial. QbD principles provide a systematic framework for identifying, assessing, mitigating, and managing risks throughout the trial's lifecycle. This article will explain risk management in clinical trials from a QbD perspective and provide suitable examples and case reports to illustrate its application and benefits [52]. QbD-driven risk management begins with the identification of potential risks that may affect the trial's quality, safety, or integrity. Risks can be categorized as clinical, operational, or statistical and should be evaluated in the context of critical quality attributes (CQAs) and trial design parameters.

In a clinical trial evaluating a new drug, potential risks may include adverse events, participant non-compliance, data integrity issues, or protocol deviations. By systematically identifying these risks, researchers can ensure they are appropriately managed throughout the trial. QbD principles guide assessing and prioritizing identified risks based on their potential impact and likelihood. A risk assessment matrix can be utilized to assign risk levels and determine the need for risk mitigation strategies. In a multicenter clinical trial, a risk assessment identified inadequate investigator training and inconsistent data collection as high-risk areas. By prioritizing these risks, researchers can allocate appropriate resources for training and implement standardized data collection procedures to mitigate the identified risks effectively. QbD-driven risk management focuses on developing robust risk mitigation and control strategies. These strategies aim to reduce or eliminate risks and ensure trial data quality, safety, and integrity. In a clinical trial for a medical device, risk mitigation strategies may involve implementing appropriate training programs for investigators, establishing data monitoring processes, conducting regular site visits, and utilizing independent data and safety monitoring committees [6]. These strategies help mitigate risks related to investigator performance, data integrity, and participant safety. QbD principles emphasize the ongoing monitoring and evaluation of identified risks throughout the trial. Regular monitoring activities help ensure that risk mitigation strategies are effective and that emerging risks are identified and addressed promptly. In Phase III clinical trial, risk monitoring processes may include regular site visits, monitoring of adverse events, data quality checks, and participant safety reviews. These activities provide continuous assessment of risks and enable proactive management to maintain trial quality and participant safety. QbD in Risk Management of a Complex Trial In a clinical trial evaluating a combination therapy for a rare disease, QbD principles were applied to risk management. The trial involved complex treatment regimens, potential drug-drug interactions, and a small patient population. A comprehensive risk assessment was conducted, and risk mitigation strategies, including participant education, careful monitoring of drug interactions, and close collaboration with treating physicians, were implemented. This QbD-driven risk management approach ensured the safe and effective conduct of the trial.

14.9.1 QbD in Risk-Based Monitoring

In a multinational clinical trial for a novel therapy, QbD principles guided the risk-based monitoring strategy. The trial involved multiple study sites, data collection from various sources, and potential challenges in data quality. A risk assessment identified critical data points and data collection processes as high-risk areas. Risk mitigation strategies, such as centralized data review, site training, and data quality checks, were implemented. This QbD-driven risk-based monitoring approach ensured the integrity and reliability of the trial data.

QbD-driven risk management ensures the quality, safety, and integrity of the trial, leading to reliable and meaningful results. By identifying and addressing potential risks early in the trial's lifecycle, QbD-driven risk management allows for proactive mitigation strategies, minimizing the impact of risks on trial outcomes. QbD-driven risk management prioritizes participant safety through appropriate risk assessment, mitigation strategies, and monitoring processes [53]. By focusing resources on high-risk areas, QbD-driven risk management optimizes resource allocation, improving the efficiency of trial conduct. Risk management in clinical trials, from a QbD perspective, is essential for ensuring trial data quality, safety, and integrity. Examples and case reports demonstrate the successful application of QbD principles in risk identification, assessment, mitigation, and monitoring. By adopting a systematic approach to risk management, researchers can proactively address potential risks, optimize resource allocation, and ensure the reliability and validity of trial outcomes. QbD-driven risk management enhances the overall quality of clinical trials and contributes to evidence-based decision-making in drug development and patient care.

14.10 Control Strategies in QbD-Based Clinical Trials

Control strategies in QbD (Quality by Design)-based clinical trials refer to the measures and procedures implemented to ensure the consistency, reliability, and quality of the trial processes, data collection, and analysis. These strategies are designed to control critical quality attributes (CQAs) and minimize variability, bias, and errors in order to generate robust and meaningful results. This article will explain control strategies in QbD-based clinical trials and provide suitable examples and case reports to illustrate their application and benefits [54].

14.10.1 Standard Operating Procedures (SOPs)

Standard operating procedures outline the step-by-step instructions and protocols for various trial activities, such as data collection, participant recruitment, and safety monitoring. Implementing standardized procedures helps ensure consistency and minimizes variability across study sites. In a multicenter clinical trial, QbD principles may be employed to develop SOPs for data collection, adverse event reporting, and laboratory procedures. These standardized procedures ensure uniformity in data collection and handling, minimizing variability and enhancing the reliability of trial results. Training programs are essential to ensure that all trial personnel, including investigators, site staff, and data collectors, have a consistent understanding of the trial protocols, procedures, and quality standards. Proper training helps ensure compliance with the established control strategies. In a clinical trial for a novel therapy, QbD principles may be applied to develop a comprehensive training program for investigators and site staff. The program would include training on trial procedures, data collection protocols, adverse event reporting, and ethical considerations. By providing thorough training, the trial can enhance adherence to control strategies and maintain the integrity of the trial data.

14.10.2 Quality Control Checks

Regular quality control checks are conducted to ensure the accuracy and reliability of trial data. These checks involve reviewing data for completeness, consistency, and adherence to predefined quality standards. Deviations from the established control strategies can be identified and corrected promptly. In a clinical trial involving electronic data capture, QbD principles may be utilized to implement quality control checks on a regular basis. These checks involve reviewing data entry accuracy, verifying the completeness and consistency of entered data, and identifying any data anomalies or errors. By conducting quality control checks, the trial ensures the reliability and integrity of the collected data [55].

14.10.3 Independent Data Monitoring Committees (IDMCs)

IDMCs are external expert committees responsible for monitoring the conduct of the trial and reviewing the trial data to ensure participant safety and data integrity. These committees provide an independent assessment of the trial processes, data quality, and adherence to control strategies. In a clinical trial for a high-risk intervention, QbD principles may be employed to establish an independent data monitoring committee. The committee would regularly review the trial data, assess participant safety, and evaluate adherence to control strategies. By having an independent oversight body, the trial can ensure the integrity and reliability of the trial data and maintain participant safety. QbD-based Data Quality Control In a Phase III clinical trial evaluating a new therapy, QbD principles were applied to implement robust data quality control measures. The trial utilized electronic data capture (EDC) systems, and comprehensive quality control checks were conducted at regular intervals. Data entry accuracy, completeness, and consistency were verified, and any identified errors or anomalies were corrected promptly [56]. This QbD-driven data quality control approach ensured the reliability and integrity of the trial data.

Independent Data Monitoring Committee in a Large-Scale Trial In a multinational, multicenter clinical trial involving a large patient population, QbD principles guided the establishment of an independent data monitoring committee. The committee was responsible for reviewing the trial data, monitoring participant safety, and ensuring adherence to control strategies. Their independent oversight ensured the reliability and validity of the trial outcomes, and their recommendations contributed to the overall quality and integrity of the trial.

14.10.4 Consistency and Uniformity

Control strategies ensure consistency and uniformity in trial processes, data collection, and analysis across study sites, reducing variability and enhancing the reliability of trial results.

14.10.5 Minimization of Bias and Errors

Control strategies help minimize biases and errors by establishing standardized procedures, conducting quality control checks, and implementing independent oversight.

14.10.6 Enhanced Data Quality

By implementing control strategies, trial data quality is improved, leading to reliable and meaningful results. This enhances the credibility of the trial findings and supports evidence-based decision-making.

14.10.7 Participant Safety and Ethical Conduct

Control strategies, such as independent oversight and training programs, contribute to participant safety and ensure ethical conduct throughout the trial.

Control strategies in QbD-based clinical trials play a vital role in ensuring the consistency, reliability, and quality of trial processes, data collection, and analysis. Examples and case reports illustrate the successful application of control strategies in various trial scenarios. By implementing standardized procedures, conducting quality control checks, providing thorough training, and establishing independent oversight, control strategies enhance the integrity of trial data and support evidence-based decision-making. QbD-driven control strategies contribute to the overall quality, reliability, and validity of clinical trial outcomes and ultimately improve patient care.

14.11 QbD Considerations in Randomization and Blinding

QbD (Quality by Design) considerations in randomization and blinding are crucial to ensure the integrity, reliability, and validity of clinical trial outcomes. Randomization and blinding are essential components of trial design that minimize bias and confounding factors, allowing for accurate treatment comparisons. By applying QbD principles to randomization and blinding, researchers can optimize these processes, enhance trial quality, and improve the strength of evidence generated. This article will explain QbD considerations in randomization and blinding and provide suitable examples and case reports to illustrate their application and benefits [57].

14.11.1 Randomization Method Selection

QbD principles guide the selection of appropriate randomization methods based on trial objectives, design, and potential sources of bias. The chosen method should ensure equal distribution of patient characteristics and minimize selection bias. In a clinical trial evaluating a new drug, QbD principles may be employed to select a randomization method that balances important prognostic factors, such as age, disease severity, or comorbidities. This helps ensure that patient characteristics are evenly distributed between treatment groups, reducing bias and increasing the validity of treatment comparisons.

14.11.2 Randomization Implementation

QbD-driven randomization implementation involves optimizing the randomization process to enhance its efficiency, transparency, and reliability. This includes developing robust randomization algorithms, utilizing central randomization systems, and implementing appropriate allocation concealment techniques. In a multicenter clinical trial, QbD principles may be applied to implement a centralized randomization system. This system ensures that treatment assignment is performed in a consistent and transparent manner across study sites, minimizing the potential for selection bias and enhancing trial integrity [58].

14.11.3 Blinding Method Selection

QbD principles guide the selection of blinding methods to ensure the integrity of trial outcomes and minimize bias. Blinding can be single-blind (participant or investigator), double-blind (both participant and investigator), or triple-blind (participant, investigator, and data analyst). In a placebo-controlled clinical trial, QbD principles may be applied to implement double-blinding, where both participants and investigators are unaware of the treatment assignment. This ensures unbiased assessment of treatment effects and minimizes the potential for subjective biases.

14.11.4 Blinding Maintenance and Assessment

QbD-driven blinding maintenance involves strategies to maintain blinding integrity throughout the trial. This includes monitoring for unblinding events, implementing measures to minimize unblinding, and assessing the success of blinding through blinding indices or questionnaires [59].

In a clinical trial evaluating a surgical intervention, QbD principles may be employed to implement measures to minimize unblinding, such as the use of sham procedures or placebo interventions. Additionally, blinding indices or questionnaires can be utilized to assess the success of blinding among participants and investigators. QbD in Randomization and Blinding in a Large-Scale Trial In a multinational, large-scale clinical trial, QbD principles guided the randomization and blinding processes. The trial employed a centralized randomization system that ensured equal distribution of patient characteristics across treatment groups. Additionally, a double-blind design was implemented, with participants and investigators unaware of the treatment assignment. Regular monitoring for unblinding events and blinding indices were utilized to assess blinding integrity. These QbD-driven randomization and blinding strategies contributed to the high-quality and reliable trial outcomes.

14.11.5 QbD in Adaptive Randomization and Blinding in an Adaptive Design Trial

In an adaptive design clinical trial, QbD principles were applied to optimize randomization and blinding strategies. The trial employed adaptive randomization methods to adjust treatment allocation based on interim analyses. The blinding process was carefully maintained throughout the trial, and measures were implemented to minimize unblinding events. These QbD-driven randomizations and blinding considerations ensured the integrity and validity of treatment comparisons in the adaptive design trial.

14.11.5.1 Benefits of QbD Considerations in Randomization and Blinding

14.11.5.1.1 Minimization of Bias

QbD-driven randomization and blinding considerations minimize selection bias and subjective biases, enhancing the internal validity of treatment comparisons.

14.11.5.1.2 Improved Trial Integrity

By optimizing randomization and blinding processes, QbD considerations enhance the overall trial integrity, reliability, and validity of trial outcomes.

14.11.5.1.3 Enhanced Treatment Comparisons

QbD-driven randomization and blinding strategies improve the accuracy of treatment comparisons by ensuring equal distribution of patient characteristics and unbiased assessment of treatment effects.

14.11.5.1.4 Consistency and Transparency

QbD considerations provide consistency and transparency in randomization and blinding processes, enabling efficient trial conduct and enhancing the credibility of trial findings. QbD considerations in randomization and blinding are critical to ensure the integrity, reliability, and validity of clinical trial outcomes. Examples and case reports highlight the successful application of QbD principles in optimizing randomization methods, implementing centralized randomization systems, selecting appropriate blinding methods, and maintaining blinding integrity throughout the trial. By incorporating QbD-driven randomization and blinding considerations, researchers can minimize bias, enhance trial integrity, and generate high-quality evidence for evidence-based decision-making in drug development and patient care.

14.12 Statistical Methods in QbD-Driven Clinical Trials

Statistical methods play a crucial role in QbD (Quality by Design) driven clinical trials. These methods are employed to analyze trial data, assess treatment effects, and draw meaningful conclusions. By applying statistical techniques within a QbD framework, researchers can optimize trial design, control variability, and enhance the reliability and validity of trial results [60].

14.12.1 Sample Size Determination

QbD principles guide the selection of appropriate statistical methods for sample size determination. By considering critical quality attributes (CQAs), effect sizes, variability, and significance levels, researchers can calculate the required sample size to achieve sufficient statistical power. In a clinical trial comparing two interventions, QbD principles may be applied to determine the sample size required to detect a clinically significant difference in a CQA, such as response rate or mean change in a biomarker. Statistical power calculations help ensure that the trial is adequately powered to detect treatment effects.

14.12.2 Design and Analysis of Experiments

QbD-driven clinical trials often employ experimental designs to study the impact of various factors on trial outcomes. Statistical methods, such as factorial designs, response surface methodology, and optimization techniques, are used to design experiments, analyze data, and optimize trial parameters. In a clinical trial evaluating a medical device, QbD principles may be applied to design experiments using fractional factorial designs. Statistical methods are then used to analyze the data and identify the optimal combination of device settings, patient characteristics, and procedural techniques that maximize device performance.

14.12.3 Control of Variability and Bias

Statistical methods are utilized to control variability and minimize bias in QbD-driven clinical trials. Techniques such as randomization, stratification, covariate adjustment, and propensity score analysis are employed to ensure comparability between treatment groups and reduce confounding factors. In a randomized controlled trial for a new drug, statistical methods, such as simple randomization or stratified randomization, are used to allocate participants to treatment groups. This helps control variability and minimizes selection bias, enhancing the internal validity of the trial [61].

14.12.4 Analysis of Treatment Effects

Statistical methods are employed to analyze treatment effects and assess the significance of observed differences between treatment groups. Techniques such as t-tests, analysis of variance (ANOVA), regression analysis, and survival analysis are commonly used in QbD-driven clinical trials. In a clinical trial comparing three treatment arms, statistical methods, such as ANOVA or regression analysis, may be employed to compare the mean response or hazard rates across treatment groups. These methods provide insights into the treatment effects and allow for appropriate comparisons. Statistical Analysis of Biomarker Response in a Personalized Medicine Trial In a QbD-driven clinical trial evaluating a targeted therapy for cancer, statistical methods were used to analyze biomarker response [61, 62]. The trial employed a stratified randomization approach based on biomarker status. Statistical analyses, including analysis of covariance (ANCOVA), were conducted to assess the treatment effect within different biomarker subgroups. These statistical methods allowed for personalized treatment evaluations based on biomarker response. Optimization of Dosage Regimen in a Dose-Finding Trial In a dose-finding clinical trial for a new drug, statistical methods were applied to optimize the dosage regimen. The trial utilized a dose-escalation design with predefined stopping rules. Bayesian adaptive methods were employed to model the dose-response relationship and estimate the optimal dose. These statistical methods allowed for efficient dose selection and optimization of treatment parameters.

14.12.4.1 Benefits of Statistical Methods in QbD-Driven Clinical Trials

14.12.4.1.1 Enhanced Trial Design

Statistical methods within a QbD framework optimize trial design parameters, such as sample size, randomization, and stratification, leading to improved trial efficiency and robustness.

14.12.4.1.2 Control of Variability and Bias

Statistical methods control variability, minimize bias, and address confounding factors, ensuring the validity of treatment comparisons, and reducing the risk of spurious findings.

14.12.4.1.3 Evidence-Based Decision-Making

Statistical methods provide a rigorous framework for analysing trial data and drawing meaningful conclusions, supporting evidence-based decision-making in drug development and patient care.

14.12.4.1.4 Optimization of Treatment Parameters

Statistical methods, such as experimental designs and optimization techniques, facilitate the optimization of treatment parameters, leading to more effective and efficient interventions.

Statistical methods are essential in QbD-driven clinical trials for analyzing trial data, assessing treatment effects, and drawing meaningful conclusions. Examples and case reports highlight the successful application of statistical methods in sample size determination, experimental designs, control of variability and bias, and analysis of treatment effects. By employing statistical techniques within a QbD framework, researchers can optimize trial design, control variability, and enhance the reliability and validity of trial results. QbD-driven statistical methods contribute to the generation of high-quality evidence for evidence-based decision-making in drug development and patient care.

14.13 QbD Approaches for Endpoint Selection in Clinical Trials

QbD (Quality by Design) approaches for endpoint selection in clinical trials involve systematic strategies to identify and select appropriate endpoints that reflect the desired clinical outcomes and align with the trial objectives. These approaches aim to ensure that the selected endpoints are relevant, reliable, sensitive, and clinically meaningful. By applying QbD principles to endpoint selection, researchers can optimize trial design, enhance the interpretability of trial results, and improve decision-making in drug development and patient care. This article will explain QbD approaches for endpoint selection in clinical trials and provide suitable examples and case reports to illustrate their application and benefits [63].

14.13.1 Identification of Critical Quality Attributes (CQAs)

QbD-driven endpoint selection begins with the identification of CQAs, which are the specific characteristics or attributes of the intervention that are critical to its therapeutic effect. CQAs can include clinical outcomes, biomarkers, patient-reported outcomes, or surrogate endpoints. In a clinical trial for a new cancer therapy, QbD principles may be employed to identify CQAs such as overall survival, progression-free survival, or tumor response rate. These CQAs represent the desired clinical outcomes that the intervention aims to achieve.

14.13.2 Clinical Relevance and Meaningfulness

QbD considers the clinical relevance and meaningfulness of the selected endpoints. Endpoints should capture outcomes that are important to patients, clinicians, regulatory authorities, and other stakeholders. They should also be directly associated with the intervention's intended therapeutic effect. In a clinical trial for a new cardiovascular drug, QbD principles may be applied to select clinically relevant endpoints such as cardiovascular events (e.g., heart attack, stroke, hospitalization for heart failure) or improvement in exercise capacity. These endpoints reflect the clinical impact of the intervention and are meaningful to patients and clinicians [64].

14.13.3 Endpoint Reliability and Sensitivity

QbD-driven endpoint selection emphasizes the reliability and sensitivity of the selected endpoints. Reliability refers to the consistency of measurement, while sensitivity refers to the ability of the endpoint to detect meaningful differences between treatment groups. In a clinical trial for an asthma medication, QbD principles may be employed to select a reliable and sensitive endpoint such as forced expiratory volume in 1 s (FEV1). FEV1 is a widely accepted measure of lung function that is reliable and sensitive to changes in respiratory health.

14.13.4 Validation and Regulatory Considerations

QbD approaches consider the validation and regulatory requirements for the selected endpoints. Validation ensures that the chosen endpoints are scientifically sound, have an established measurement methodology, and are accepted by regulatory authorities for evaluating treatment effects. In a clinical trial for a new diagnostic test, QbD principles may be applied to select an endpoint that has been validated against the gold standard diagnostic method. The selected endpoint should have a well-established measurement protocol and be accepted by regulatory authorities as a valid measure of diagnostic accuracy [65].

14.13.5 QbD-Driven Endpoint Selection in an Oncology Trial

In a clinical trial evaluating a new targeted therapy for lung cancer, QbD principles were applied to select relevant and meaningful endpoints. The trial considered CQAs such as overall survival, progression-free survival, and tumor response rate. Through stakeholder engagement and regulatory discussions, these endpoints were validated and accepted as appropriate measures of treatment efficacy. The QbD-driven endpoint selection facilitated the evaluation of the therapy's clinical benefit and supported evidence-based decision-making.

14.13.6 QbD-Driven Endpoint Selection in a Pain Management Trial

In a clinical trial for a new pain management intervention, QbD principles were employed to select clinically relevant and sensitive endpoints. The trial identified CQAs such as pain intensity scores, functional improvement, and patient-reported outcomes. Through literature review, patient input, and scientific discussions, these endpoints were determined to capture the clinical impact of the intervention accurately. The QbD-driven endpoint selection ensured that the trial outcomes were meaningful and relevant to patients and clinicians.

14.13.6.1 Benefits of QbD Approaches for Endpoint Selection in Clinical Trials

14.13.6.1.1 Improved Trial Design

QbD-driven endpoint selection optimizes trial design by selecting endpoints that align with trial objectives, capture clinically meaningful outcomes, and are sensitive to treatment effects.

14.13.6.1.2 Enhanced Interpretability

QbD approaches improve the interpretability of trial results by selecting endpoints that are directly associated with the therapeutic effect and are relevant to patients, clinicians, and regulatory authorities.

14.13.6.1.3 Stakeholder Engagement

QbD principles encourage stakeholder engagement in endpoint selection, ensuring that the selected endpoints align with the needs and perspectives of patients, clinicians, and regulatory agencies.

14.13.6.1.4 Regulatory Compliance

QbD-driven endpoint selection considers regulatory requirements, ensuring that the selected endpoints are scientifically valid, have a well-established measurement protocol, and are accepted by regulatory authorities.

QbD approaches for endpoint selection in clinical trials play a crucial role in optimizing trial design, enhancing the interpretability of trial results, and improving decision-making in drug development and patient care. Examples and case reports highlight the successful application of QbD principles in identifying CQAs, selecting clinically relevant and sensitive endpoints, considering reliability and regulatory requirements, and engaging stakeholders. By incorporating QbD-driven approaches in endpoint selection, researchers can ensure that trial outcomes are meaningful, reliable, and aligned with the therapeutic goals of the intervention. QbD-driven endpoint selection strengthens the evidence base for decision-making and supports the development of effective and patient-centered interventions.

14.14 QbD in Biomarker Testing and Personalized Medicine Trials

QbD (Quality by Design) principles play a crucial role in biomarker testing and personalized medicine trials. These trials aim to identify and validate biomarkers that can predict treatment response, guide therapy selection, and improve patient outcomes. By applying QbD principles to biomarker testing and personalized medicine trials, researchers can optimize assay development, ensure assay reliability and accuracy, and enhance the overall quality and validity of trial results. This article will explain the application of QbD in biomarker testing and personalized medicine trials and provide suitable examples and case reports to illustrate their benefits [66].

14.14.1 Assay Development and Optimization

QbD-driven biomarker testing begins with assay development and optimization. QbD principles guide the systematic approach to assay design, including assay methodology, reagents, controls, and instrumentation, to ensure reliable and accurate biomarker measurement. In a personalized medicine trial evaluating a biomarker for a targeted therapy, QbD principles may be applied to develop and optimize an assay for biomarker detection. This involves selecting appropriate antibodies, optimizing assay conditions, and validating the assay's performance to ensure accurate and reproducible measurement of the biomarker.

14.14.2 Analytical Performance Characterization

QbD principles emphasize the characterization of assay analytical performance, including precision, accuracy, sensitivity, and specificity. These parameters provide a quantitative assessment of the assay's reliability and ability to detect and quantify the biomarker accurately [49].

In a biomarker testing trial for cancer treatment, QbD principles may be employed to characterize the analytical performance of an assay that measures a specific genetic mutation. This involves evaluating the assay's precision (reproducibility), accuracy (closeness to the true value), sensitivity (limit of detection), and specificity (ability to detect only the intended biomarker).

14.14.3 Control Strategies and Quality Assurance

QbD-driven biomarker testing incorporates control strategies and quality assurance measures to ensure the reliability and accuracy of biomarker measurements. These strategies include the use of appropriate positive and negative controls, validation of assay performance, and implementation of quality control checks. In a personalized medicine trial utilizing a biomarker to guide therapy selection, QbD principles may be applied to establish control strategies. This involves the use of validated reference samples with known biomarker status as positive and negative controls. Regular quality control checks and validation of assay performance are implemented to ensure the reliability and accuracy of biomarker measurements.

14.14.4 Biomarker Validation and Regulatory Compliance

QbD principles guide the validation of biomarkers to ensure their clinical utility and regulatory compliance. Validation involves assessing the biomarker's association with clinical outcomes, evaluating its performance characteristics, and determining its predictive value. In a personalized medicine trial evaluating a biomarker as a predictive marker for drug response, QbD principles may be employed to validate the biomarker. This involves assessing the biomarker's association with treatment outcomes, evaluating its performance characteristics (e.g., sensitivity, specificity), and demonstrating its predictive value through statistical analyses. Validation ensures the clinical utility and regulatory compliance of the biomarker [62].

QbD in Assay Development for a Companion Diagnostic Test In a personalized medicine trial for a targeted therapy, QbD principles were applied to develop an assay for a companion diagnostic test. QbD-guided assay development involved optimizing assay conditions, selecting appropriate controls, and validating the assay's performance. The QbD-driven assay development ensured accurate and reliable measurement of the biomarker, allowing for precise patient stratification and therapy selection. QbD in Biomarker Validation for Treatment Response Prediction In a biomarker-driven trial for cancer treatment, QbD principles were employed to validate a biomarker as a predictive marker for drug response. QbD-guided validation involved assessing the biomarker's association with treatment outcomes, evaluating its performance characteristics, and demonstrating its predictive value. The QbD-driven biomarker validation facilitated personalized treatment decisions based on the biomarker status, improving patient outcomes.

14.14.4.1 Benefits of QbD in Biomarker Testing and Personalized Medicine Trials

14.14.4.1.1 Improved Assay Reliability

QbD-driven biomarker testing ensures the development and optimization of assays that are reliable, accurate, and reproducible, enabling precise biomarker measurements.

14.14.4.1.2 Enhanced Assay Performance Characterization

QbD principles facilitate the characterization of assay performance parameters, such as precision, accuracy, sensitivity, and specificity, ensuring the analytical validity of biomarker measurements.

14.14.4.1.3 Robust Control Strategies

QbD-driven biomarker testing incorporates control strategies and quality assurance measures, such as appropriate controls and regular quality control checks, ensuring the reliability and accuracy of biomarker measurements.

14.14.4.1.4 Validated Biomarkers for Personalized Medicine

QbD-guided biomarker validation ensures the clinical utility and regulatory compliance of biomarkers, supporting their integration into personalized medicine approaches for improved treatment selection and patient outcomes.

QbD principles have a significant impact on biomarker testing and personalized medicine trials by optimizing assay development, ensuring assay reliability and accuracy, and validating biomarkers for clinical utility. Examples and case reports demonstrate the successful application of QbD in assay development, analytical performance characterization, control strategies, and biomarker validation. By applying QbD-driven approaches, researchers can enhance the overall quality and validity of trial results, facilitate personalized treatment decisions, and improve patient outcomes in the era of personalized medicine. QbD-driven biomarker testing plays a crucial role in advancing precision medicine and improving patient care.

14.15 QbD and Data Quality Assurance in Clinical Trials

QbD (Quality by Design) principles and data quality assurance are essential components of clinical trials. QbD provides a systematic framework for designing and conducting trials that ensure data integrity, reliability, and validity. Data quality assurance involves implementing strategies and processes to control and monitor data quality throughout the trial lifecycle. By applying QbD and data quality assurance, researchers can optimize data collection, minimize errors and bias, and enhance the overall quality and credibility of trial results. This article will explain the application of QbD and data quality assurance in clinical trials and provide suitable examples and case reports to illustrate their benefits [48].

14.15.1 Designing Data Collection Processes

QbD principles guide the design of data collection processes, including the selection of appropriate data elements, data collection instruments, and data entry methods. By considering critical quality attributes (CQAs) and trial objectives, researchers can optimize data collection to ensure the capture of accurate and relevant information. In a clinical trial for a new therapy, QbD principles may be employed to design data collection processes, including electronic case report forms (eCRFs) with pre-defined data fields and data validation checks. These measures enhance data completeness, accuracy, and consistency [67].

14.15.2 Data Quality Control Measures

Data quality assurance involves implementing control measures to ensure data accuracy, consistency, and completeness. This includes implementing data validation checks, conducting regular data quality checks, and implementing standard operating procedures (SOPs) for data handling and management. In a multicenter clinical trial, data quality control measures may involve implementing range checks, logic checks, and consistency checks within the eCRFs. Regular data quality checks, including data cleaning and query resolution, are performed to identify and rectify any data anomalies or errors [68].

14.15.3 Training and Standardization

QbD-driven data quality assurance emphasizes training and standardization to ensure consistent data collection and entry. Training programs for trial personnel, including investigators and data collectors, promote adherence to standardized data collection protocols and data entry procedures [69].

In a clinical trial, QbD principles may be applied to develop a comprehensive training program for trial personnel. The program includes training on trial protocols, data collection procedures, and data entry guidelines. Standardized operating procedures for data handling, query resolution, and data management are also established.

14.15.4 Data Monitoring and Quality Checks

QbD-driven data quality assurance involves regular monitoring of data to identify and address any potential issues promptly. Data monitoring activities include data quality checks, source data verification, and data review for completeness, consistency, and accuracy.

In a clinical trial, QbD principles may be applied to conduct regular data monitoring activities, such as monitoring for missing data, outliers, or inconsistencies. Source data verification is performed to ensure data accuracy and reliability. These activities help identify and rectify data quality issues, ensuring the integrity of trial data.

QbD and Data Quality Assurance in a Large-Scale Trial In a multinational, large-scale clinical trial, QbD principles and data quality assurance measures were applied to ensure data integrity. The trial utilized electronic data capture (EDC) systems with built-in data validation checks. Training programs were conducted for trial personnel to ensure standardized data collection and entry. Regular data quality checks and source data verification were performed to monitor data quality. These QbD-driven measures enhanced data integrity, reliability, and validity in the trial.

QbD and Data Quality Assurance in a Real-Time Monitoring Trial In a trial utilizing real-time monitoring technology, QbD principles and data quality assurance were employed to ensure data accuracy and completeness. The trial implemented real-time data checks and alerts to identify and resolve data issues immediately. Training programs were conducted to ensure proper utilization of the monitoring technology and adherence to data quality standards. These QbD-driven measures facilitated real-time data quality assurance, ensuring the reliability and timeliness of trial data.

14.15.4.1 Benefits of QbD and Data Quality Assurance in Clinical Trials

14.15.4.1.1 Improved Data Integrity

QbD and data quality assurance measures enhance data integrity by minimizing errors, inconsistencies, and missing data, ensuring accurate and reliable trial results.

14.15.4.1.2 Enhanced Data Validity

By implementing QbD-driven data quality assurance, researchers can enhance the validity of trial data, making it more credible and suitable for evidence-based decision-making.

14.15.4.1.3 Reliable Results

QbD-driven data quality assurance measures ensure reliable and consistent data collection, minimizing variability and bias, and enabling the generation of robust and meaningful results.

14.15.4.1.4 Compliance and Regulatory Alignment

QbD and data quality assurance support compliance with regulatory requirements and guidelines, ensuring the credibility and acceptability of trial data by regulatory authorities.

QbD principles and data quality assurance play a vital role in clinical trials by optimizing data collection, ensuring data integrity, and enhancing the overall quality and credibility of trial results. Examples and case reports illustrate the successful application of QbD and data quality assurance in trial design, data collection processes, training, standardization, data monitoring, and quality checks. By implementing QbD-driven data quality assurance measures, researchers can minimize errors, improve data accuracy, and generate reliable and valid trial results. These approaches contribute to evidence-based decision-making in drug development and patient care, ultimately improving patient outcomes.

14.16 QbD in Clinical Trial Supply Chain Management

QbD (Quality by Design) principles are applicable not only to clinical trial design and execution but also to clinical trial supply chain management. Effective supply chain management is crucial for ensuring the availability, integrity, and quality of investigational products throughout the trial. By applying QbD principles to clinical trial supply chain management, researchers can optimize the supply chain processes, minimize risks, and enhance the reliability and quality of trial materials. This article will explain the application of QbD in clinical trial supply chain management and provide suitable examples and case reports to illustrate their benefits [70].

14.16.1 Risk Assessment and Mitigation

QbD-driven supply chain management involves a systematic assessment of risks associated with the trial materials, such as investigational drugs, devices, or biological samples. Potential risks, including supply shortages, temperature excursions, or product integrity issues, are identified, and appropriate mitigation strategies are implemented.

In a clinical trial involving the transportation of temperature-sensitive investigational drugs, QbD principles may be applied to assess the risks associated with temperature excursions during transportation. Mitigation strategies may include the use of validated packaging systems, temperature monitoring devices, and contingency plans in case of unforeseen events.

14.16.2 Supplier Qualification and Oversight

QbD-driven supply chain management emphasizes the qualification and oversight of suppliers and vendors involved in the production, storage, and distribution of trial materials. Supplier selection criteria, quality agreements, and ongoing performance monitoring are established to ensure the reliability, compliance, and quality of the supplied materials. In a clinical trial requiring specialized laboratory testing, QbD principles may be applied to qualify and oversee the performance of the laboratory supplier. Supplier qualification may include assessments of analytical method validation, equipment calibration, and adherence to good laboratory practices. Regular performance monitoring ensures ongoing compliance and quality [71].

14.16.3 Supply Chain Process Optimization

QbD principles guide the optimization of supply chain processes to ensure efficient and timely delivery of trial materials. This includes optimizing inventory management, implementing just-in-time manufacturing, and establishing appropriate storage and distribution processes.

In a clinical trial with multiple study sites, QbD principles may be applied to optimize inventory management and distribution processes. Centralized inventory control and demand forecasting help minimize inventory waste and ensure timely delivery of trial materials to study sites [72].

14.16.4 Documentation and Record-Keeping

QbD-driven supply chain management emphasizes documentation and record-keeping practices to ensure traceability, accountability, and compliance with regulatory requirements. Comprehensive records are maintained to document the chain of custody, transportation conditions, and storage conditions of trial materials [73].

In a clinical trial involving biological sample collection, QbD principles may be applied to establish comprehensive documentation practices. This includes maintaining accurate and complete records of sample collection, transportation, storage conditions, and chain of custody, ensuring traceability and compliance with regulatory requirements.

14.16.5 QbD in Risk Assessment and Mitigation for Clinical Trial Supply Chain

In a multinational clinical trial involving investigational drugs, QbD principles were applied to assess and mitigate risks associated with the supply chain. Risk assessments were conducted to identify potential risks, such as supply shortages or temperature excursions. Mitigation strategies were implemented, including establishing alternative suppliers, implementing temperature monitoring systems, and developing contingency plans. These QbD-driven risk assessment and mitigation measures ensured the reliability and availability of trial materials throughout the trial.

QbD in Supplier Qualification and Oversight for Clinical Trial Supply Chain In a clinical trial requiring specialized laboratory testing, QbD principles were employed to qualify and oversee the laboratory supplier. Supplier qualification assessments were conducted to evaluate the supplier's capabilities, compliance with regulatory requirements, and adherence to quality standards. Ongoing oversight and performance monitoring ensured that the supplier maintained the necessary quality and compliance throughout the trial. The QbD-driven supplier qualification and oversight measures contributed to the reliability and integrity of laboratory testing results.

14.16.5.1 Benefits of QbD in Clinical Trial Supply Chain Management

14.16.5.1.1 Risk Mitigation

QbD-driven supply chain management helps identify potential risks and implement mitigation strategies to ensure the availability and integrity of trial materials, minimizing supply disruptions and product quality issues.

14.16.5.1.2 Supplier Reliability

QbD principles guide the qualification and oversight of suppliers, ensuring their reliability, compliance, and adherence to quality standards, thereby minimizing the risk of substandard materials or services.

14.16.5.1.3 Process Efficiency

QbD-driven supply chain management optimizes supply chain processes, reducing inventory waste, improving delivery timelines, and ensuring efficient utilization of resources.

14.16.5.1.4 Compliance and Traceability

QbD-driven supply chain management emphasizes documentation and record-keeping practices, ensuring compliance with regulatory requirements and enabling traceability of trial materials throughout the supply chain.

QbD principles are valuable in optimizing clinical trial supply chain management processes, mitigating risks, and enhancing the reliability and quality of trial materials. Examples and case reports demonstrate the successful application of QbD in risk assessment and mitigation, supplier qualification and oversight, supply chain process optimization, and documentation practices. By incorporating QbD-driven approaches, researchers can minimize risks, ensure supplier reliability, optimize supply chain processes, and enhance compliance and traceability in clinical trial supply chain management. These efforts contribute to the overall quality and integrity of clinical trial conduct and facilitate the reliable and timely availability of trial materials, ultimately supporting the generation of high-quality evidence for evidence-based decision-making in drug development and patient care.

14.17 QbD Applications in Adaptive Clinical Trial Designs

QbD (Quality by Design) principles find valuable applications in adaptive clinical trial designs. Adaptive trials are flexible and dynamic, allowing modifications to key design elements based on accumulating data while preserving the trial's integrity. QbD approaches enhance the efficiency, reliability, and validity of adaptive trial designs by optimizing the decision-making process, ensuring control of risks, and facilitating evidence-based modifications. This article will explain the QbD applications in adaptive clinical trial designs and provide suitable examples and case reports to illustrate their benefits [74].

14.17.1 Decision-Making Framework

QbD principles guide the development of a decision-making framework for adaptive trials. This includes establishing clear decision criteria, defining decision rules, and ensuring transparency and accountability in decision-making.

In an adaptive trial for a new therapy, QbD principles may be applied to define decision criteria based on specific efficacy or safety measures. Decision rules are established in advance to determine whether and how to modify the trial design based on interim analyses. This QbD-driven decision-making framework ensures transparency and consistency in adaptive trial modifications.

14.17.2 Statistical Methodology

QbD-driven adaptive trial designs employ statistical methods to analyze interim data, make informed decisions, and control Type I error rates. Statistical techniques such as group sequential methods, Bayesian methods, or predictive modeling are utilized to estimate treatment effects and update trial parameters [75].

In an adaptive trial comparing two treatment arms, QbD principles may be applied to employ Bayesian statistical methods to update the treatment effect estimate based on accumulating data. These methods allow for adaptive modifications to sample size, randomization ratios, or treatment allocation probabilities based on interim analyses.

14.17.3 Risk Control and Monitoring

QbD-driven adaptive trial designs focus on risk control and monitoring mechanisms to maintain trial integrity and ensure patient safety. Risk assessments are conducted to identify potential risks associated with adaptive modifications, and risk mitigation strategies are implemented.

In an adaptive trial, QbD principles may be employed to conduct risk assessments of potential biases or operational challenges associated with adaptive modifications. Risk mitigation strategies, such as adjustment for multiplicity or implementation of adaptive randomization with appropriate control, are applied to maintain trial integrity and control risks.

14.17.4 Stakeholder Engagement

QbD-driven adaptive trial designs emphasize stakeholder engagement to ensure alignment of trial objectives, decision-making criteria, and modifications. Stakeholder involvement enhances transparency, builds consensus, and ensures that adaptive modifications are driven by scientific evidence.

In an adaptive trial for a rare disease, QbD principles may be applied to engage patient representatives, clinicians, statisticians, and regulatory authorities in the decision-making process. Stakeholder input is sought to define adaptive modifications, including changes to enrollment criteria or treatment arms, ensuring that the trial remains patient-centered and scientifically rigorous.

QbD in Decision-Making Framework in an Adaptive Trial In a multicenter adaptive trial for a new medical device, QbD principles were employed to develop a decision-making framework. Decision criteria were established based on efficacy and safety measures. Decision rules were defined to guide modifications to sample size, randomization ratios, or interim analyses based on accumulating data. The QbD-driven decision-making framework allowed for adaptive modifications while maintaining trial integrity and transparency.

QbD in Risk Control and Monitoring in an Adaptive Trial In an adaptive trial evaluating a novel therapy, QbD principles were applied to assess risks associated with adaptive modifications. Risk assessments identified potential biases or operational challenges related to interim analyses and adaptive modifications. Risk mitigation strategies, such as adjustment for multiplicity and careful monitoring for data quality and integrity, were implemented. The QbD-driven risk control and monitoring measures ensured the reliability and validity of trial results.

14.17.4.1 Benefits of QbD Applications in Adaptive Clinical Trial Designs

14.17.4.1.1 Efficiency and Resource Optimization

QbD-driven adaptive trial designs optimize resource utilization by enabling efficient decision-making based on accumulating data. This reduces the need for extensive sample sizes and increases trial efficiency.

14.17.4.1.2 Enhanced Trial Integrity

QbD applications in adaptive trials ensure control of risks, maintaining the integrity of the trial design and validity of the results.

14.17.4.1.3 Evidence-Based Modifications

QbD principles guide adaptive modifications based on scientific evidence, allowing for informed decisions that maximize the trial's scientific rigor and validity.

14.17.4.1.4 Stakeholder Engagement

QbD-driven adaptive trial designs promote stakeholder engagement, ensuring alignment of trial objectives, decision criteria, and modifications while incorporating diverse perspectives.

QbD principles provide valuable applications in adaptive clinical trial designs, enhancing decision-making, controlling risks, and facilitating evidence-based modifications. Examples and case reports illustrate the successful application of QbD in decision-making frameworks, statistical methodology, risk control, and stakeholder engagement. By incorporating QbD-driven approaches, researchers can optimize adaptive trial designs, improve trial efficiency, and ensure the integrity and validity of trial results. QbD applications in adaptive clinical trials contribute to evidence-based decision-making in drug development, ultimately benefiting patients and improving healthcare outcomes.

14.18 QbD and Regulatory Considerations in Clinical Trials

QbD (Quality by Design) principles play a vital role in addressing regulatory considerations in clinical trials. Regulatory authorities require the demonstration of a systematic and science-based approach to ensure the quality, safety, and efficacy of investigational products and trial conduct. By incorporating QbD principles into clinical trial design, execution, and data analysis, researchers can meet regulatory requirements, enhance regulatory interactions, and facilitate the approval process. This article will explain the application of QbD and regulatory considerations in clinical trials and provide suitable examples and case reports to illustrate their benefits [76].

14.18.1 Quality Risk Management

QbD-driven clinical trials employ quality risk management (QRM) principles to identify, assess, and mitigate risks throughout the trial lifecycle. QRM ensures that potential risks to trial quality, data integrity, and patient safety are proactively addressed.

In a clinical trial for a new drug, QbD principles may be applied to conduct a risk assessment of critical trial processes, such as randomization, blinding, or data collection. Mitigation strategies are implemented to control risks and ensure the integrity and reliability of trial data.

14.18.2 Documentation and Record-Keeping

QbD-driven clinical trials emphasize comprehensive and accurate documentation practices to meet regulatory requirements. Detailed records are maintained to document trial design, protocol deviations, adverse events, and data management processes.

In a clinical trial, QbD principles may be applied to establish documentation practices that ensure traceability and compliance with regulatory requirements. This includes maintaining accurate and complete records of trial protocol, amendments, adverse event reporting, and trial conduct.

14.18.3 Standard Operating Procedures (SOPs)

QbD-driven clinical trials establish SOPs to ensure consistent and standardized trial conduct. SOPs define processes, responsibilities, and procedures to meet regulatory requirements and ensure the reproducibility and reliability of trial outcomes [77].

In a clinical trial, QbD principles may be applied to develop SOPs for key trial processes, such as randomization, blinding, data management, and adverse event reporting. These SOPs ensure compliance with regulatory requirements and consistency in trial conduct across study sites.

14.18.4 Regulatory Interactions and Submissions

QbD principles guide interactions with regulatory authorities and support the preparation of regulatory submissions. A systematic and science-based approach, consistent with QbD principles, enhances the credibility and acceptability of trial data and facilitates the regulatory review and approval process.

In a clinical trial, QbD principles may be applied to compile regulatory submissions, such as investigational new drug (IND) applications or marketing authorization applications (MAAs). QbD-driven submissions include comprehensive data analysis, risk assessment, and justification of trial design, ensuring alignment with regulatory expectations.

QbD and Regulatory Considerations in a Phase III Trial In a phase III clinical trial for a new therapeutic intervention, QbD principles were applied to address regulatory considerations. Quality risk management was implemented, assessing potential risks to trial conduct and data integrity. Comprehensive documentation practices ensured traceability and compliance with regulatory requirements. SOPs were established for key trial processes, ensuring standardized and consistent trial conduct. The QbD-driven approach facilitated regulatory interactions, resulting in a successful regulatory submission and approval.

QbD and Regulatory Considerations in a Medical Device Trial In a clinical trial evaluating a new medical device, QbD principles were employed to address regulatory considerations. Quality risk management was applied, identifying potential risks associated with device performance or patient safety. Comprehensive documentation practices ensured compliance with regulatory requirements. SOPs were established for device handling, adverse event reporting, and data management. The QbD-driven approach facilitated regulatory interactions and submissions, leading to regulatory approval for the medical device.

14.18.4.1 Benefits of QbD and Regulatory Considerations in Clinical Trials

14.18.4.1.1 Compliance with Regulatory Requirements

QbD-driven clinical trials ensure compliance with regulatory requirements, addressing the expectations of regulatory authorities and increasing the likelihood of successful approval.

14.18.4.1.2 Enhanced Regulatory Interactions

QbD principles facilitate effective and meaningful interactions with regulatory authorities, as a systematic and science-based approach aligns with regulatory expectations, enhancing the credibility and acceptability of trial data.

14.18.4.1.3 Robust Trial Conduct and Data Integrity

QbD-driven clinical trials implement quality risk management, comprehensive documentation, and standardized processes, ensuring robust trial conduct and data integrity, which are essential for regulatory approval.

14.18.4.1.4 Streamlined Regulatory Approval Process

QbD-driven clinical trials support efficient and streamlined regulatory review and approval processes by providing clear and transparent documentation, addressing regulatory requirements, and facilitating regulatory interactions.

QbD principles and regulatory considerations are crucial in the successful design, execution, and regulatory approval of clinical trials. Examples and case reports demonstrate the application of QbD in quality risk management, documentation practices, SOPs, and regulatory interactions. By incorporating QbD-driven approaches, researchers can ensure compliance with regulatory requirements, enhance regulatory interactions, and facilitate the regulatory review and approval process. These efforts contribute to the overall quality, reliability, and credibility of clinical trial data, supporting evidence-based decision-making in drug development and patient care.

14.19 QbD Tools and Techniques for Clinical Trial Optimization

QbD (Quality by Design) principles provide a systematic approach to optimize clinical trials by ensuring the quality, reliability, and efficiency of trial design and execution. Various tools and techniques are available to implement QbD in clinical trial optimization, enabling researchers to make data-driven decisions, mitigate risks, and enhance trial outcomes. This article will explain the QbD tools and techniques for clinical trial optimization and provide suitable examples and case reports to illustrate their benefits [78].

14.19.1 Failure Mode and Effects Analysis (FMEA)

FMEA is a QbD tool used to systematically identify and analyze potential failure modes in trial design and execution. It helps identify risks, evaluate their potential impact, and prioritize mitigation strategies [79].

In a clinical trial, FMEA may be applied to identify potential failure modes, such as incomplete enrollment, data collection errors, or protocol deviations. Mitigation strategies, such as enhanced site training or data monitoring procedures, can be implemented to address these risks.

14.19.2 Design of Experiments (DoE)

DoE is a statistical tool that allows researchers to systematically evaluate the impact of multiple factors on trial outcomes. It helps optimize trial design, identify influential factors, and determine the optimal levels of these factors. In a clinical trial comparing two treatments, DoE may be used to identify the optimal dosage levels of the investigational drug, considering factors such as safety, efficacy, and patient response. This optimization process helps identify the most effective dosage regimen for the trial.

14.19.3 Process Mapping and Flowcharts

Process mapping and flowcharts visually represent the sequence of steps and interactions in trial processes. These tools help identify process inefficiencies, bottlenecks, and opportunities for improvement. In a clinical trial, process mapping and flowcharts can be used to depict the flow of activities from patient recruitment to data collection and analysis. By visually representing the trial processes, inefficiencies and potential areas for optimization can be identified and addressed.

14.19.4 Statistical Analysis and Modeling

Statistical analysis and modeling techniques play a crucial role in QbD-driven clinical trial optimization. They allow researchers to analyze trial data, identify trends, and make data-driven decisions to optimize trial design and execution. In a clinical trial, statistical analysis and modeling techniques, such as survival analysis or logistic regression, can be employed to analyze patient outcomes, identify prognostic factors, and optimize patient stratification strategies. This analysis helps tailor the trial design to the target patient population.

14.19.5 Risk Assessment and Management

Risk assessment and management techniques help identify potential risks in trial design and execution and develop strategies to mitigate those risks. QbD-driven risk assessment ensures proactive identification and control of risks. In a clinical trial, risk assessment techniques, such as risk matrices or risk registers, can be employed to identify and assess potential risks to patient safety, trial integrity, or data quality. Mitigation strategies, such as monitoring plans or contingency measures, can be developed to minimize these risks. QbD Tools and Techniques in Patient Enrollment

Optimization In a clinical trial aiming to optimize patient enrollment, QbD tools and techniques were employed. Process mapping and flowcharts were used to visualize the enrollment process and identify bottlenecks. Statistical analysis and modeling techniques were applied to analyze patient characteristics and identify factors influencing enrollment. Risk assessment techniques were used to identify potential barriers to enrollment and develop mitigation strategies. These QbD-driven tools and techniques enhanced patient enrollment efficiency, resulting in improved trial timelines.

QbD Tools and Techniques in Protocol Optimization In a clinical trial, QbD tools and techniques were applied to optimize the trial protocol. Failure Mode and Effects Analysis (FMEA) was employed to identify potential protocol deviations and risks. Design of Experiments (DoE) was utilized to optimize dosage levels and treatment regimens. Statistical analysis techniques were employed to analyze previous trial data and identify opportunities for protocol improvement. These QbD-driven tools and techniques led to an optimized protocol design, resulting in improved trial outcomes.

14.19.5.1 Benefits of QbD Tools and Techniques in Clinical Trial Optimization

14.19.5.1.1 Data-Driven Decision-Making

QbD tools and techniques enable data-driven decision-making, allowing researchers to optimize trial design, identify influential factors, and make informed choices based on statistical analysis and modeling.

14.19.5.1.2 Risk Mitigation

QbD-driven tools and techniques, such as FMEA and risk assessment, help identify potential risks and develop strategies to mitigate them, ensuring patient safety, trial integrity, and data quality.

14.19.5.1.3 Efficiency and Resource Optimization

QbD tools and techniques, such as process mapping and DoE, enable researchers to identify inefficiencies, streamline processes, and optimize resource utilization, improving trial efficiency and reducing costs.

14.19.5.1.4 Protocol Optimization

QbD tools and techniques facilitate the optimization of trial protocols by identifying factors that influence trial outcomes, tailoring treatments to patient characteristics, and optimizing dosage levels and treatment regimens.

QbD tools and techniques provide valuable approaches for optimizing clinical trials, enhancing trial design, decision-making, risk assessment, and protocol optimization. Examples and case reports demonstrate the successful application of tools such as FMEA, DoE, process mapping, statistical analysis, and risk assessment in clinical trial optimization. By incorporating QbD-driven approaches, researchers can make data-driven decisions, mitigate risks, and optimize trial outcomes. These efforts contribute to the overall quality, efficiency, and reliability of clinical trial design and execution, ultimately supporting evidence-based decision-making in drug development and patient care.

14.20 QbD in Safety and Pharmacovigilance in Clinical Trials

QbD (Quality by Design) principles are essential in ensuring safety and pharmacovigilance in clinical trials. Safety monitoring and pharmacovigilance are critical components of clinical research, aimed at detecting and assessing adverse events and ensuring patient safety throughout the trial. By incorporating QbD into safety and pharmacovigilance processes, researchers can proactively identify and mitigate risks, enhance data quality, and improve patient safety. This article will explain the application of QbD in safety and pharmacovigilance in clinical trials and provide suitable examples and case reports to illustrate their benefits [80].

14.20.1 Risk Assessment and Management

QbD-driven safety and pharmacovigilance processes involve risk assessment and management to proactively identify and mitigate potential risks to patient safety. This includes the identification of potential risks, evaluation of their severity and likelihood, and the implementation of risk mitigation strategies. In a clinical trial, QbD principles may be applied to conduct risk assessments for potential safety issues associated with the investigational product or trial procedures. Mitigation strategies, such as enhanced monitoring or dose adjustments, can be implemented to address identified risks.

14.20.2 Safety Data Collection and Reporting

QbD principles guide the design and implementation of safety data collection and reporting processes. This involves the identification of critical safety data elements, the development of standardized data collection instruments, and the establishment of reporting procedures and timelines.

In a clinical trial, QbD principles may be applied to design safety data collection forms that capture critical safety parameters and adverse events. Standardized data collection instruments facilitate data consistency and ensure the collection of accurate and relevant safety information.

14.20.3 Data Quality and Integrity

QbD-driven safety and pharmacovigilance processes focus on data quality and integrity to ensure reliable and credible safety assessments. This includes implementing data validation checks, conducting regular data quality checks, and ensuring proper documentation and traceability of safety data [48].

In a clinical trial, QbD principles may be applied to implement data validation checks within safety databases to identify inconsistent or incomplete safety data. Regular data quality checks and documentation practices ensure the accuracy, completeness, and reliability of safety data.

14.20.4 Signal Detection and Risk Assessment

QbD principles facilitate signal detection and risk assessment processes to identify emerging safety concerns and assess their potential impact. This includes statistical analysis techniques, data mining, and signal detection algorithms to identify potential safety signals. In a clinical trial, QbD-driven signal detection and risk assessment processes may employ statistical analysis techniques to identify potential safety signals by analyzing safety data in real time. Data mining and signal detection algorithms can be utilized to identify patterns or trends indicative of potential safety concerns.

QbD in Safety and Pharmacovigilance in a Phase III Trial In a phase III clinical trial for a new drug, QbD principles were applied to enhance safety and pharmacovigilance processes. Risk assessments were conducted to identify potential safety risks associated with the investigational product and trial procedures. Data collection forms were designed to capture critical safety parameters and adverse events. Data validation checks and regular data quality checks were implemented to ensure the accuracy and reliability of safety data. These QbD-driven processes facilitated the early detection and proper management of safety issues, ensuring patient safety throughout the trial.

QbD in Signal Detection and Risk Assessment in a Medical Device Trial In a clinical trial evaluating a medical device, QbD principles were employed to improve safety and pharmacovigilance processes. Signal detection and risk assessment techniques were utilized to identify potential safety signals associated with the device. Statistical analysis and data mining were employed to analyze safety data in real-time, identifying patterns indicative of potential safety concerns. These QbD-driven processes facilitated the early detection and appropriate management of device-related safety issues, ensuring patient safety and compliance with regulatory requirements.

14.20.4.1 Benefits of QbD in Safety and Pharmacovigilance in Clinical Trials

14.20.4.1.1 Proactive Risk Identification and Mitigation

QbD-driven safety and pharmacovigilance processes enable the proactive identification and mitigation of potential safety risks, ensuring patient safety throughout the trial.

14.20.4.1.2 Enhanced Data Quality and Reliability

QbD principles ensure the implementation of data validation checks, regular data quality checks, and documentation practices, enhancing the quality and reliability of safety data.

14.20.4.1.3 Early Signal Detection and Risk Assessment

QbD-driven signal detection and risk assessment processes enable the early identification of safety signals and potential safety concerns, facilitating timely risk assessment and appropriate risk management actions.

14.20.4.1.4 Compliance with Regulatory Requirements

QbD principles support compliance with regulatory requirements for safety monitoring and pharmacovigilance, ensuring the credibility and acceptability of safety data by regulatory authorities.

QbD principles are crucial in ensuring safety and pharmacovigilance in clinical trials. Examples and case reports illustrate the application of QbD in risk assessment, safety data collection and reporting, data quality and integrity, and signal detection.

By incorporating QbD-driven approaches, researchers can proactively identify and mitigate safety risks, ensure the quality and reliability of safety data, and facilitate early signal detection and appropriate risk assessment. These efforts contribute to the overall patient safety and support evidence-based decision-making in drug development and patient care.

14.21 QbD and Patient-Centricity in Clinical Trial Design

QbD (Quality by Design) principles can significantly contribute to patient-centricity in clinical trial design. Patient-centric clinical trials focus on the needs, preferences, and experiences of patients, ensuring their active involvement and maximizing the relevance and impact of trial outcomes. By incorporating QbD approaches into trial design, researchers can optimize patient-centered endpoints, streamline trial processes, enhance patient recruitment and retention, and improve overall patient satisfaction. This article will explain the application of QbD and patient-centricity in clinical trial design and provide suitable examples and case reports to illustrate their benefits [81].

14.21.1 Patient-Reported Outcomes (PROs)

QbD-driven clinical trial design emphasizes the selection and optimization of patient-reported outcomes (PROs), which capture patients' experiences, symptoms, and quality of life. PROs ensure that trial outcomes reflect patients' perspectives and provide meaningful insights [82].

In a clinical trial for a chronic condition, QbD principles may be applied to select PRO measures that capture relevant symptoms, functional limitations, and treatment-related side effects. PROs can be optimized by involving patient representatives in the selection process, ensuring their relevance and meaningfulness.

14.21.2 Patient Engagement and Input

QbD-driven clinical trial design involves active patient engagement and input throughout the trial lifecycle. Patients' perspectives and preferences are solicited and incorporated into trial design decisions, ensuring patient-centeredness. In a clinical trial, QbD principles may be applied to engage patient representatives in the trial design process. Patient input can be sought in the development of trial protocols, informed consent forms, and patient recruitment materials. This ensures that trial procedures and materials are patient-friendly, understandable, and aligned with patients' needs and preferences [83].

14.21.3 Trial Process Optimization

QbD-driven trial design focuses on optimizing trial processes to enhance patient experience and minimize burdens. Streamlining trial procedures, reducing unnecessary visits, and improving patient convenience are key considerations. In a clinical trial, QbD principles may be applied to optimize trial processes by reducing the frequency of study visits, implementing remote monitoring technologies, or simplifying data collection procedures. These optimizations enhance patient convenience, reduce the burden of participation, and improve overall patient experience [84].

14.21.4 Patient Recruitment and Retention

QbD principles guide patient recruitment and retention strategies, ensuring efficient and patient-centric approaches to maximize enrollment and minimize dropout rates. Patient-centered recruitment materials, tailored communication strategies, and personalized engagement contribute to improved recruitment and retention. In a clinical trial, QbD-driven patient recruitment and retention strategies may involve the development of patient-friendly recruitment materials, personalized communication approaches, and targeted support services. These strategies address patient needs, improve engagement, and enhance recruitment and retention rates. QbD and Patient-Centricity in a Rare Disease Trial In a clinical trial for a rare disease, QbD principles were applied to ensure patient-centricity. Patient representatives were actively engaged in the trial design process, providing input on trial procedures, informed consent forms, and patient recruitment materials. Patient-reported outcomes (PROs) were carefully selected and optimized to capture relevant symptoms and treatment effects. Trial processes were streamlined, minimizing the number of study visits and implementing remote monitoring technologies. These patient-centric approaches improved patient experience, recruitment, and retention in the trial. QbD and Patient-Centricity in an Oncology Trial In an oncology clinical trial, QbD principles were employed to enhance patient-centricity [85]. Patient input was sought in the development of trial protocols and patient materials, ensuring relevance and understandability. PROs were selected and optimized to capture meaningful endpoints related to symptoms, treatment-related toxicities, and quality of life. Trial processes were optimized to reduce the burden on patients, such as implementing home-based data collection and minimizing clinic visits. These QbD-driven approaches improved patient-centeredness, adherence, and overall trial experience.

14.21.4.1 Benefits of QbD and Patient-Centricity in Clinical Trial Design

14.21.4.1.1 Meaningful Trial Outcomes

QbD-driven patient-centric clinical trial design ensures that trial outcomes capture patients' perspectives and experiences, providing meaningful insights for decision-making.

14.21.4.1.2 Improved Patient Experience

QbD approaches optimize trial processes, reduce burden, and enhance convenience, leading to improved patient experience and satisfaction throughout the trial.

14.21.4.1.3 Enhanced Recruitment and Retention

Patient-centred trial design, recruitment materials, and personalized engagement strategies improve recruitment and retention rates by addressing patient needs and preferences.

14.21.4.1.4 Relevant and Patient-Friendly Endpoints

QbD principles guide the selection and optimization of patient-reported outcomes (PROs), ensuring that trial endpoints are relevant, patient-friendly, and aligned with patients' experiences. QbD principles play a crucial role in promoting patient-centricity in clinical trial design. Examples and case reports demonstrate the successful application of QbD in optimizing patient-reported outcomes, engaging patients in the trial design process, streamlining trial procedures, and improving patient recruitment and retention. By incorporating QbD-driven approaches, researchers can enhance the relevance, efficiency, and patient-centeredness of clinical trials. These efforts contribute to meaningful trial outcomes, improved patient experiences and ultimately support evidence-based decision-making in drug development and patient care.

14.22 Future Directions and Emerging Trends in QbD-Based Clinical Trials

Future directions and emerging trends in QbD (Quality by Design)-based clinical trials are driven by advancements in technology, regulatory frameworks, and evolving patient needs. These developments aim to enhance further the efficiency, reliability, and patient-centricity of clinical trial design and execution. This article will explore the future directions and emerging trends in QbD-based clinical trials, providing suitable examples and case reports to illustrate their potential impact [86].

14.22.1 Integration of Real-World Data (RWD) and Real-World Evidence (RWE)

The integration of RWD and RWE into QbD-based clinical trials allows for a more comprehensive understanding of treatment effects, patient outcomes, and safety profiles in real-world settings. RWD from electronic health records, wearable devices, and other sources can inform trial design and help identify patient subpopulations that may benefit most from a treatment. In a QbD-based clinical trial for a chronic disease, RWD from electronic health records can be integrated to understand the natural history of the disease, identify relevant endpoints, and optimize patient selection criteria. RWE can provide insights into treatment effectiveness and safety in diverse patient populations, helping to guide adaptive trial design [87].

14.22.2 Implementation of Digital Health Technologies

Digital health technologies, including mobile applications, wearable devices, and remote monitoring tools, are increasingly being incorporated into QbD-based clinical trials. These technologies facilitate remote data collection, real-time monitoring, and patient engagement, improving trial efficiency and patient convenience. In a QbD-based clinical trial evaluating the efficacy of a digital therapeutics intervention, a mobile application can be used to collect PROs, monitor treatment adherence, and capture real-time data. This implementation streamlines data collection, reduces participant burden, and enhances patient engagement.

14.22.3 Advanced Analytics and Artificial Intelligence (AI)

The application of advanced analytics and AI techniques in QbD-based clinical trials enables more sophisticated data analysis, predictive modeling, and decision support.

These tools help identify patterns, predict patient responses, and optimize trial design and outcomes. In a QbD-based clinical trial, advanced analytics and AI can be used to analyze large datasets, identify treatment-response patterns, and predict patient outcomes. This analysis can inform adaptive trial design, treatment stratification, and personalized medicine approaches.

14.22.4 Emphasis on Patient-Reported Outcomes (PROs)

QbD-based clinical trials increasingly prioritize the incorporation of patient-reported outcomes (PROs), ensuring that trial outcomes align with patients' perspectives and experiences. PROs capture patients' quality of life, symptoms, treatment satisfaction, and functional status. In a QbD-based clinical trial for a rare disease, PROs can be integrated into trial design to assess treatment impact on patients' daily functioning, psychological well-being, and social interactions. This patient-centric approach enhances trial outcomes' relevance and patient satisfaction.

14.22.5 Greater Collaboration and Data Sharing

Future QbD-based clinical trials are expected to emphasize collaboration and data sharing among researchers, industry partners, and regulatory authorities. Enhanced data sharing facilitates cross-study comparisons, meta-analyses, and robust evidence generation. In a QbD-based clinical trial, data sharing agreements can be established among multiple research centers, enabling the pooling of data for larger sample sizes and more robust analyses. This collaboration promotes evidence-based decision-making and fosters scientific advancements.

Integration of Wearable Devices in a QbD-based Clinical Trial In a QbD-based clinical trial evaluating an exercise intervention, wearable devices (such as activity trackers) were integrated into the trial design. These devices collected real-time data on participants' physical activity, heart rate, and sleep patterns. The integration of wearable devices facilitated remote data collection, enhanced trial efficiency, and provided objective measurements for outcome assessments.

AI-Assisted Adaptive Trial Design in a QbD-based Clinical Trial In a QbD-based clinical trial investigating a new drug, AI techniques were employed to analyze patient data, identify subpopulations with a higher likelihood of treatment response, and guide adaptive trial modifications. AI algorithms continuously analyzed incoming data, allowing for dynamic adjustments in treatment allocation and sample size to optimize trial outcomes.

14.22.5.1 Benefits of Future Directions and Emerging Trends in QbD-Based Clinical Trials

14.22.5.1.1 Enhanced Efficiency and Data Quality

Future QbD-based clinical trials leverage advanced technologies and analytics to streamline trial processes, optimize data collection, and improve data quality.

14.22.5.1.2 Improved Patient Experience and Engagement

Patient-centric technologies and approaches improve participant convenience, engagement, and adherence, ultimately enhancing the overall patient experience.

14.22.5.1.3 Personalized Medicine Approaches

Future QbD-based trials incorporate advanced analytics and AI techniques to enable personalized medicine approaches, tailoring treatments to individual patient characteristics and optimizing treatment outcomes.

14.22.5.1.4 More Robust Evidence Generation

Greater collaboration and data sharing among researchers and stakeholders facilitate larger sample sizes, meta-analyses, and more robust evidence generation, contributing to evidence-based decision-making.

14.23 Conclusion

Future directions and emerging trends in QbD-based clinical trials encompass the integration of RWD and RWE, adoption of digital health technologies, implementation of advanced analytics and AI, emphasis on PROs, and greater collaboration and data sharing. These advancements aim to optimize trial design, enhance patient-centricity, improve trial efficiency, and generate robust evidence. Examples and case reports demonstrate the potential impact of these future directions in driving innovation and improving patient outcomes. By embracing these trends, researchers can further advance QbD principles in clinical trial design and contribute to evidence-based decision-making in drug development and patient care.

References

1. Holm, P., Allesø, M., Bryder, M. C., & Holm, R. (2017). Q8(R2). In *ICH quality guidelines* (pp. 535–577). Wiley. https://doi.org/10.1002/9781118971147.ch20
2. Yu, L. X., Amidon, G., Khan, M. A., Hoag, S. W., Polli, J., Raju, G. K., et al. (2014). Understanding pharmaceutical quality by design. *The AAPS Journal, 16*(4), 771–783. https://doi.org/10.1208/s12248-014-9598-3
3. Pramod, K., Ma, T., Charoo, N., Ansari, S., & Ali, J. (2016). Pharmaceutical product development: A quality by design approach. *International Journal of Pharmaceutical Investigation, 6*(3), 129. Retrieved from http://www.jpionline.org/index.php/ijpi/article/view/234
4. ter Horst, J. P., Turimella, S. L., Metsers, F., & Zwiers, A. (2021). Implementation of Quality by Design (QbD) principles in regulatory Dossiers of medicinal products in the European Union (EU) between 2014 and 2019. *Therapeutic Innovation & Regulatory Science, 55*(3), 583–590. https://doi.org/10.1007/s43441-020-00254-9
5. Hasnain, M. S., Ahmed, S. A., Khatoon, A., Afzal, M., Ansari, M. T., Khatoon, S., et al. (2021). Pharmaceutical product development: A quality by design (QbD) approach. In *Advances and challenges in pharmaceutical technology* (pp. 131–146). Elsevier. Retrieved from https://linkinghub.elsevier.com/retrieve/pii/B9780128200438000177
6. Gandhi, A. R. C. (2016). Quality by Design (QbD) in pharmaceutical industry: tools, perspectives and challenges. *PharmaTutor, 4*(11), 12–20.
7. Zhang, L., & Mao, S. (2017). Application of quality by design in the current drug development. *Asian Journal of Pharmaceutical Sciences, 12*(1), 1–8. Retrieved from https://linkinghub.elsevier.com/retrieve/pii/S1818087616300575
8. Vishwasrao, S. S., & Singh, S. (2016). Current perspective on opportunities and adoption challenges of QbD implementation in pharmaceutical product development. *Pharmaceutical Process Development, 33*, 34–41.
9. Singh, G. (2018). Target product profile and clinical development plan. In *Pharmaceutical medicine and translational clinical research* (pp. 65–80). Elsevier. Retrieved from https://linkinghub.elsevier.com/retrieve/pii/B9780128021033000055
10. Ahmed, T., Kollipara, S., Boddu, R., & Bhattiprolu, A. K. (2023). Biopharmaceutics risk assessment—connecting critical bioavailability attributes with in vitro, in vivo properties and physiologically based biopharmaceutics modeling to enable generic regulatory submissions. *The AAPS Journal, 25*(5), 77. https://doi.org/10.1208/s12248-023-00837-y
11. Sun, M., Yang, J., Cao, W., Shao, J., Wang, G., Qu, H., et al. (2020). Critical process parameter identification of manufacturing processes of Astragali Radix extract with a weighted determination coefficient method. *Chinese Herbal Medicines, 12*(2), 125–132. Retrieved from https://linkinghub.elsevier.com/retrieve/pii/S1674638419300954
12. Rausand, M. (2013). *"Chapter 1: Introduction". Risk assessment: Theory, methods, and applications* (pp. 1–28). Wiley.
13. Bajwa, N., Mahal, S., Naryal, S., Singh, P. A., & Baldi, A. (2022). Development of novel solid nanostructured lipid carriers for bioavailability enhancement using a quality by design approach. *AAPS PharmSciTech, 23*(7), 253. https://doi.org/10.1208/s12249-022-02386-7
14. Liu, H.-C., You, J.-X., Fan, X.-J., & Lin, Q.-L. (2014). Failure mode and effects analysis using D numbers and grey relational projection method. *Expert Systems with Applications, 41*(10), 4670–4679. Retrieved from https://linkinghub.elsevier.com/retrieve/pii/S0957417414000554
15. Kabir, S. (2017). An overview of fault tree analysis and its application in model based dependability analysis. *Expert Systems with Applications, 77*, 114–135. Retrieved from https://linkinghub.elsevier.com/retrieve/pii/S0957417417300714
16. Peterson, J. J. (2008). A Bayesian approach to the ICH Q8 definition of design space. *Journal of Biopharmaceutical Statistics, 18*(5), 959–975. https://doi.org/10.1080/10543400802278197
17. Van Buskirk, G. A., Asotra, S., Balducci, C., Basu, P., DiDonato, G., Dorantes, A., et al. (2014). Best practices for the development, scale-up, and post-approval change control of IR and MR

dosage forms in the current quality-by-design paradigm. *AAPS PharmSciTech, 15*(3), 665–693. https://doi.org/10.1208/s12249-014-0087-x
18. Sister Janet Hunting, M.B.H.A. (1978). Sister at Christies Beach Community Health Centre. *The Australasian Nurses Journal, 8*(2), 16. Retrieved from http://www.fda.gov/downloads/AboutFDA/CentersOffices/CDER/ucm103526.pdf
19. Gao, J., & Nee, A. Y. (2018). An overview of manufacturing knowledge sharing in the product development process. *Proceedings of the Institution of Mechanical Engineers Part B Journal of Engineering Manufacture, 232*(13), 2253–2263. https://doi.org/10.1177/0954405418759375
20. Rathore, A. S., Garcia-Aponte, O. F., Golabgir, A., Vallejo-Diaz, B. M., & Herwig, C. (2017). Role of knowledge management in development and lifecycle management of biopharmaceuticals. *Pharmaceutical Research, 34*(2), 243–256. https://doi.org/10.1007/s11095-016-2043-9
21. Dziallas, M., & Blind, K. (2019). Innovation indicators throughout the innovation process: An extensive literature analysis. *Technovation, 80–81*, 3–29. Retrieved from https://linkinghub.elsevier.com/retrieve/pii/S0166497217301402
22. Sangshetti, J. N., Deshpande, M., Zaheer, Z., Shinde, D. B., & Arote, R. (2017). Quality by design approach: Regulatory need. *Arabian Journal of Chemistry, 10*, S3412–S3425. Retrieved from https://linkinghub.elsevier.com/retrieve/pii/S1878535214000288
23. Beg, S., Hasnain, M. S., Rahman, M., & Swain, S. (2019). Introduction to Quality by Design (QbD): fundamentals, principles, and applications. In *Pharmaceutical quality by design* (pp. 1–17). Elsevier. Retrieved from https://linkinghub.elsevier.com/retrieve/pii/B9780128157992000010
24. Evans, S. R. (2010). Fundamentals of clinical trial design. *Journal of Experimental Stroke & Translational Medicine, 3*(1), 19–27. https://doi.org/10.6030/1939-067X-3.1.19
25. Knepper, D., Fenske, C., Nadolny, P., Bedding, A., Gribkova, E., Polzer, J., et al. (2016). Detecting data quality issues in clinical trials: current practices and recommendations. *Therapeutic Innovation & Regulatory Science, 50*(1), 15–21. https://doi.org/10.1177/2168479015620248
26. Smiatek, J., Jung, A., & Bluhmki, E. (2020). Towards a digital bioprocess replica: computational approaches in biopharmaceutical development and manufacturing. *Trends in Biotechnology, 38*(10), 1141–1153. Retrieved from https://linkinghub.elsevier.com/retrieve/pii/S0167779920301426
27. Fiteni, F., Westeel, V., Pivot, X., Borg, C., Vernerey, D., & Bonnetain, F. (2014). Endpoints in cancer clinical trials. *Journal of Visceral Surgery, 151*(1), 17–22. Retrieved from https://linkinghub.elsevier.com/retrieve/pii/S1878788613001318
28. Kaur, R. J., Dutta, S., Bhardwaj, P., Charan, J., Dhingra, S., Mitra, P., et al. (2021). Adverse events reported from COVID-19 vaccine trials: A systematic review. *Indian Journal of Clinical Biochemistry, 36*(4), 427–439. https://doi.org/10.1007/s12291-021-00968-z
29. Nair, B. (2019). Clinical trial designs. *Indian Dermatology Online Journal, 10*(2), 193. https://doi.org/10.4103/idoj.IDOJ_475_18
30. Walton, M. K., Powers, J. H., Hobart, J., Patrick, D., Marquis, P., Vamvakas, S., et al. (2015). Clinical outcome assessments: Conceptual foundation—Report of the ISPOR clinical outcomes assessment—Emerging good practices for outcomes research task force. *Value in Health, 18*(6), 741–752. Retrieved from https://linkinghub.elsevier.com/retrieve/pii/S1098301515020343
31. Mishra, P., Pandey, C., Singh, U., Keshri, A., & Sabaretnam, M. (2019). Selection of appropriate statistical methods for data analysis. *Annals of Cardiac Anaesthesia, 22*(3), 297. https://doi.org/10.4103/aca.ACA_248_18
32. Wang, X., Piantadosi, S., Le-Rademacher, J., & Mandrekar, S. J. (2021). Statistical considerations for subgroup analyses. *Journal of Thoracic Oncology, 16*(3), 375–380. Retrieved from https://linkinghub.elsevier.com/retrieve/pii/S1556086420311345
33. Saitoh, S. (2018). The identification of critical quality attributes (cqas) for the development of antibody drugs. *Yakugaku Zasshi, 138*(12), 1475–1481. Retrieved from https://www.jstage.jst.go.jp/article/yakushi/138/12/138_18-00020-1/_article/-char/ja/

34. Politis, S. N., Colombo, P., Colombo, G., & Rekkas, D. M. (2017). Design of experiments (DoE) in pharmaceutical development. *Drug Development and Industrial Pharmacy, 43*(6), 889–901. https://doi.org/10.1080/03639045.2017.1291672
35. Díaz-García, J. A., & Ramos-Quiroga, R. (2001). An approach to optimization in response surface methodology. *Communications in Statistics - Theory and Methods, 30*(5), 827–835. https://doi.org/10.1081/STA-100002260
36. Chung, K. C., & Song, J. W. (2010). A guide to organizing a multicenter clinical trial. *Plastic and Reconstructive Surgery, 126*(2), 515–523. Retrieved from http://journals.lww.com/00006534-201008000-00021
37. Craig, P., Dieppe, P., Macintyre, S., Michie, S., Nazareth, I., & Petticrew, M. (2008). Developing and evaluating complex interventions: The New Medical Research Council guidance. *BMJ, 337*, a1655. https://doi.org/10.1136/bmj.a1655
38. Jankovic, A., Chaudhary, G., & Goia, F. (2021). Designing the design of experiments (DOE)—An investigation on the influence of different factorial designs on the characterization of complex systems. *Energy and Buildings, 250*, 111298. Retrieved from https://linkinghub.elsevier.com/retrieve/pii/S037877882100582X
39. Bowden, G. D., Pichler, B. J., & Maurer, A. (2019). A design of experiments (DoE) approach accelerates the optimization of copper-mediated 18F-fluorination reactions of Arylstannanes. *Scientific Reports, 9*(1), 11370. Retrieved from https://www.nature.com/articles/s41598-019-47846-6
40. Patino, C. M., & Ferreira, J. C. (2018). Inclusion and exclusion criteria in research studies: definitions and why they matter. *Jornal Brasileiro de Pneumologia, 44*(2), 84–84. Retrieved from http://www.scielo.br/scielo.php?script=sci_arttext&pid=S1806-37132018000200084&lng=en&tlng=en
41. Polley, M. Y. C., Korn, E. L., & Freidlin, B. (2019). Phase III precision medicine clinical trial designs that integrate treatment and biomarker evaluation. *JCO Precision Oncology, 3*, 1–9. https://doi.org/10.1200/PO.18.00416
42. Faulkner, S. D., Somers, F., Boudes, M., Nafria, B., & Robinson, P. (2023). Using patient perspectives to inform better clinical trial design and conduct: Current trends and future directions. *Pharmaceutical Medicine, 37*(2), 129–138. https://doi.org/10.1007/s40290-022-00458-4
43. Zurdo, J., Arnell, A., Obrezanova, O., Smith, N., Gómez de la Cuesta, R., Gallagher, T. R. A., et al. (2015). Early implementation of QbD in biopharmaceutical development: A practical example. *BioMed Research International, 2015*, 1–19. Retrieved from http://www.hindawi.com/journals/bmri/2015/605427/
44. Ou, F.-S., Michiels, S., Shyr, Y., Adjei, A. A., & Oberg, A. L. (2021). Biomarker discovery and validation: statistical considerations. *Journal of Thoracic Oncology, 16*(4), 537–545. Retrieved from https://linkinghub.elsevier.com/retrieve/pii/S1556086421016634
45. Girwar, S. M., Jabroer, R., Fiocco, M., Sutch, S. P., Numans, M. E., & Bruijnzeels, M. A. (2021). A systematic review of risk stratification tools internationally used in primary care settings. *Health Science Reports, 4*(3), e329. https://doi.org/10.1002/hsr2.329
46. Abrahams, E., & Silver, M. (2009). The case for personalized medicine. *Journal of Diabetes Science and Technology, 3*(4), 680–684. https://doi.org/10.1177/193229680900300411
47. Sorbye, H., Köhne, C.-H., Sargent, D. J., & Glimelius, B. (2007). Patient characteristics and stratification in medical treatment studies for metastatic colorectal cancer: A proposal for standardization of patient characteristic reporting and stratification. *Annals of Oncology, 18*(10), 1666–1672. Retrieved from https://linkinghub.elsevier.com/retrieve/pii/S0923753419418814
48. Dietrich, J., Forrest, A., Meeker-O'Connell, A. (2018). QbD in clinical trials: A focused approach to quality assurance and risk management. Guest Column, Clinical Leader
49. Sprenger, S., Nickerson, D., Meeker-O'Connell, A., & Morrison, B. (2012). Quality by Design in clinical trials: A collaborative pilot with the FDA. *Therapeutic Innovation & Regulatory Science, 47*(2), 161–166.

50. Bhutani, H., Kurmi, M., Singh, S., Beg, S., & Singh, B. (2004). Quality by design (QbD) in analytical sciences: an overview. *Quality Assurance, 3*, 39–45.
51. Stagner, W. C., & Haware, R. V. (2019). QbD innovation through advances in PAT, data analysis methodologies, and material characterization. *AAPS PharmSciTech, 20*, 1–2.
52. Singh, L., & Sharma, V. (2015). Quality by Design (QbD) approach in pharmaceuticals: Status, challenges and next steps. *Drug Delivery Letters, 5*(1), 2–8.
53. Kelley, B., Cromwell, M., & Jerkins, J. (2016). Integration of QbD risk assessment tools and overall risk management. *Biologicals, 44*(5), 341–351.
54. Cimarosti, Z., Bravo, F., Castoldi, D., Tinazzi, F., Provera, S., Perboni, A., Papini, D., & Westerduin, P. (2010). Application of the QbD principles in the development of the casopitant mesylate manufacturing process. Process research studies for the definition of the control strategy of some drug substance-CQAs for stages 2a, 2b, and 2c. *Organic Process Research and Development, 14*(4), 805–814.
55. Schweitzer, M., Pohl, M., Hanna-Brown, M., Nethercote, P., Borman, P., Hansen, G., Smith, K., & Larew, J. (2010). Implications and opportunities of applying QbD principles to analytical measurements. *Pharmaceutical Technology, 34*(2), 52–59.
56. Richert, L. Statistical approaches and methodological developments to optimize clinical vaccine research (Doctoral dissertation, Université de Bordeaux).
57. Choudhary, V. (2021). Design and conduct of clinical studies by integration of Quality by Design (QbD) into Quality Management System (QMS)—A pragmatic approach. *Clinical Research: Open Access, 7*(2).
58. Williams, J. *Improving generalizability of clinical trials by including underrepresented populations*.
59. Wechsler, J. (2012). Quality approaches key to transforming research. *Applied Clinical Trials, 21*(2), 18.
60. Beg, S., Saini, S., Bandopadhyay, S., Katare, O. P., & Singh, B. (2018). QbD-driven development and evaluation of nanostructured lipid carriers (NLCs) of Olmesartan medoxomil employing multivariate statistical techniques. *Drug Development and Industrial Pharmacy, 44*(3), 407–420.
61. Rizg, W. Y., Naveen, N. R., Kurakula, M., Safhi, A. Y., Murshid, S. S., Mushtaq, R. Y., Abualsunun, W. A., Alharbi, M., Bakhaidar, R. B., Almehmady, A. M., & Salawi, A. (2022). Augmentation of antidiabetic activity of glibenclamide microspheres using S-Protected Okra Powered by QbD: Scintigraphy and in vivo studies. *Pharmaceuticals, 15*(4), 491.
62. Duncan, R. (2017). Polymer therapeutics at a crossroads? Finding the path for improved translation in the twenty-first century. *Journal of Drug Targeting, 25*(9–10), 759–780.
63. Moradi, H., Schneider, M., Streja, E., & Cooper, D. (2021). Feasibility and acceptability of a structured quality by design approach to enhancing the rigor of clinical studies at an academic health center. *Journal of Clinical and Translational Science, 5*(1), e175.
64. Mishra, V., Thakur, S., Patil, A., & Shukla, A. (2018). Quality by design (QbD) approaches in current pharmaceutical set-up. *Expert Opinion on Drug Delivery, 15*(8), 737–758.
65. Wu, H., Tawakkul, M., White, M., & Khan, M. A. (2009). Quality-by-Design (QbD): An integrated multivariate approach for the component quantification in powder blends. *International Journal of Pharmaceutics, 372*(1–2), 39–48.
66. Jhawat, V., Gulia, M., Gupta, S., Maddiboyina, B., & Dutt, R. (2020). Integration of pharmacogenomics and theranostics with nanotechnology as quality by design (QbD) approach for formulation development of novel dosage forms for effective drug therapy. *Journal of Controlled Release, 327*, 500–511.
67. Nadpara, N. P., Thumar, R. V., Kalola, V. N., & Patel, P. B. (2012). Quality by design (QBD): A complete review. *International Journal of Pharmaceutical Sciences Review and Research, 17*(2), 20–28.
68. Chen, W., Tian, W., & Liang, Y. (2022). Research on quality control strategy of CAR-T cell therapy products based on QbD. *International Journal of Drug Regulatory Affairs, 10*(4), 62–74.

69. Aksu, B., De Beer, T., Folestad, S., Ketolainen, J., Lindén, H., Lopes, J. A., de Matas, M., Oostra, W., Rantanen, J., & Weimer, M. (2012). Strategic funding priorities in the pharmaceutical sciences allied to Quality by Design (QbD) and Process Analytical Technology (PAT). *European Journal of Pharmaceutical Sciences, 47*(2), 402–405.
70. Patel, K. T., & Chotai, N. P. (2010). Vendor qualification for pharmaceutical excipients–GMP requirements and approach. *Die Pharmazie-An international Journal of Pharmaceutical Sciences, 65*(11), 783–790.
71. Zurdo, J., Arnell, A., Obrezanova, O., Smith, N., Gómez de la Cuesta, R., Gallagher, T. R., Michael, R., Stallwood, Y., Ekblad, C., Abrahmsén, L., & Höidén-Guthenberg, I. (2015). Early implementation of QbD in biopharmaceutical development: a practical example. *BioMed Research International, 2015*, 605427.
72. Zheng, L., & Wang, L. Comprehensive information management system for a medical research cohort biobank based on quality by design. *BMC Medical Informatics and Decision Making, 23*(1), 222.
73. Lee, S. H., Kim, J. K., Jee, J. P., Jang, D. J., Park, Y. J., & Kim, J. E. (2022). Quality by Design (QbD) application for the pharmaceutical development process. *Journal of Pharmaceutical Investigation, 52*(6), 649–682.
74. Calhan, S. D., Eker, E. D., & Sahin, N. O. (2017). Quality by design (QbD) and process analytical technology (PAT) applications in pharmaceutical industry. *European Journal of Chemistry, 8*(4), 430–433.
75. Jain, S. (2014). Quality by design (QBD): A comprehensive understanding of implementation and challenges in pharmaceuticals development. *International Journal of Pharmacy and Pharmaceutical Sciences, 6*, 29–35.
76. Mitchel, J. T. *Regulatory considerations for digitally driven trials.*
77. Cunha, S., Costa, C. P., Loureiro, J. A., Alves, J., Peixoto, A. F., Forbes, B., Sousa Lobo, J. M., & Silva, A. C. (2020). Double optimization of rivastigmine-loaded nanostructured lipid carriers (NLC) for nose-to-brain delivery using the quality by design (QbD) approach: Formulation variables and instrumental parameters. *Pharmaceutics, 12*(7), 599.
78. Jayagopal, B., & Murugesh, S. (2020). QbD-mediated RP-UPLC method development invoking an FMEA-based risk assessment to estimate nintedanib degradation products and their pathways. *Arabian Journal of Chemistry, 13*(9), 7087–7103.
79. Hariry, R. E., Barenji, R. V., & Paradkar, A. (2022). Towards Pharma 4.0 in clinical trials: A future-orientated perspective. *Drug Discovery Today, 27*(1), 315–325.
80. Bhoop, B. S. (2014). Quality by Design (QbD) for holistic pharma excellence and regulatory compliance. *Pharmacy Times, 46*(8), 26–33.
81. Monda, J., Keipeer, J., & Were, M. C. (2012). Data integrity module for data quality assurance within an e-Health system in Sub-Saharan Africa. *Telemedicine and e-Health, 18*(1), 5–10. https://doi.org/10.1089/tmj.2010.0224
82. Bottomley, A., Jones, D., & Claassens, L. (2009). Patient-reported outcomes: Assessment and current perspectives of the guidelines of the Food and Drug Administration and the reflection paper of the European Medicines Agency. *European Journal of Cancer, 45*(3), 347–353. Retrieved from https://linkinghub.elsevier.com/retrieve/pii/S0959804908007922
83. Weldring, T., & Smith, S. M. S. (2013). Article commentary: Patient-reported outcomes (PROs) and patient-reported outcome measures (PROMs). *Health Services Insights, 6*, HSI.S11093. https://doi.org/10.4137/HSI.S11093
84. Rathert, C., Huddleston, N., & Pak, Y. (2011). Acute care patients discuss the patient role in patient safety. *Health Care Management Review, 36*(2), 134–144. Retrieved from https://journals.lww.com/00004010-201104000-00005
85. Chappidi, S. R., Bhargav, E., Marikunte, V., Chinthaginjala, H., Vijaya Jyothi, M., Pisay, M., et al. (2019). A cost effective (QbD) approach in the development and optimization of rosiglitazone maleate mucoadhesive extended release tablets—In vitro and ex vivo. *Advanced Pharmaceutical Bulletin, 9*(2), 281–288. Retrieved from https://apb.tbzmed.ac.ir/Abstract/apb-19473

86. Narbutas, Š., York, K., Stein, B. D., Magsanoc-Alikpala, K., Majima, Y., Kalo, Z., et al. (2017). Overview on patient centricity in cancer care. *Frontiers in Pharmacology, 8*, 698. https://doi.org/10.3389/fphar.2017.00698/full
87. Ziegler, A., Koch, A., Krockenberger, K., & Großhennig, A. (2012). Personalized medicine using DNA biomarkers: A review. *Human Genetics, 131*(10), 1627–1638. https://doi.org/10.1007/s00439-012-1188-9

Chapter 15
Applications of Quality by Design in Pharmaceutical Product Development Lifecycle

Anjali Kumari, Geeta Aggarwal, and Amanpreet Kaur

Abstract Quality by Design (QbD) is a contemporary methodology that guarantees the quality of pharmaceuticals by implementing a methodical and evidence-based approach throughout the pharmaceutical product development lifecycle (PPDL). The implementation of QbD throughout the PPDL has emerged as a key focus area for both regulatory bodies and the pharmaceutical sector. The foundation of QbD lies in the guidelines established by the International Council for Harmonization of Technical Requirements for Pharmaceuticals for Human Use (ICH) specifically, (ICH Q8, Q9, Q10 & Q11) for pharmaceutical development, quality risk management, pharmaceutical quality and development & manufacture of drug substances respectively. The utilization of QbD begins at the early stages of product development, where critical quality attributes (CQAs) are identified based on the desired therapeutic effect and patient needs. These CQAs serve as the foundation for subsequent formulation and process design. Design of experiments (DOE) and risk assessment are two QbD methods that assist in identifying and optimizing key process parameters influencing CQAs. Additionally, the implementation of process analytical technology (PAT) enables monitoring in real time and control of critical process parameters (CPPs) facilitating the manufacturing of high-quality products. Furthermore, the use of QbD in the PPDL brings numerous benefits, including enhanced product quality, reduced variability, improved process understanding, and increased regulatory compliance. In this chapter, some valuable insights on QbD applications viz.; API, development of formulation, quality assurance, packaging, clinical trial, biopharmaceutics and biosimilars etc. are described along with case studies.

A. Kumari
School of Pharmaceutical Sciences, Delhi Pharmaceutical Sciences and Research University, New Delhi, India

G. Aggarwal · A. Kaur (✉)
School of Pharmaceutical Sciences, Delhi Pharmaceutical Sciences and Research University, New Delhi, India

Centre for Advanced Formulation and Technology, Delhi Pharmaceutical Sciences and Research, New Delhi, India

N. K. Jain, N. Bajwa (eds.), *Introduction to Quality by Design (QbD)*,
https://doi.org/10.1007/978-981-99-8034-5_15

Keywords Quality by Design · Design of Experiments · Critical process parameters · Critical quality attributes · Pharmaceutical product development lifecycle

15.1 Introduction

The production of pharmaceuticals is a complicated process that involves interactions between the raw materials used and the process conditions. Despite continued progress, low quality and manufacturing standards have led to certain setbacks in the quality of finished product. Research and clearance for a newly developed drug can take many years and require a lot of expenditure. The considerable documentation needed for the review and approval of new product application has caused delays and making it a matter of great concern [1]. The quality of final product will depend upon careful optimization of product and process variables.

The conventional approach one factor at a time (OFAT)/one variable at a time (OVAT) for optimization of formulation has significant limitations. The OFAT/OVAT is used to manage modifications to input materials, processes, and equipment to account for anticipated variances. Adopting this development approach is not cost effective and may lead to a lack of identification of the product and the process. Additionally, this may result in rigid manufacturing processes that cannot account for normal variations in raw materials, manufacturing processes, production equipment, and instrumentation used in the laboratory. Furthermore, as per literature, the Quality by Testing (QbT) methodology demands thorough testing to adhere to strict Food and Drug administration (FDA) approved standards. Traditional pharmaceutical production typically relies on batch processing and quality control testing of finished product samples [2]. However, innovations in understanding of product and process development, manufacturing testing, and process control can offer substantial prospects for improvement in pharmaceutical product development, production, and quality assurance. Process modifications are anticipated to keep up with the changes as technology and improvements progress, and post-approval changes may be required due to various factors [3]. However, the increasing production of products and changes made after approval have resulted in a significant burden on regulatory bodies and industry folks. To assure product safety and efficacy, the way forward is to improve comprehension of product and process knowledge utilizing quality risk assessment method and pharmaceutical quality systems.

In recent years, the pharmaceutical sector has seen significant developments in risk management, quality management systems, and production information. To modernize and enhance the standards that control the production and ensuring the quality of pharmaceutical products, the FDA announced changes to the current Good Manufacturing Practice (cGMP) requirements in 2002 [4]. ICH, implemented the concept of Quality by Design (QbD) idea in the industry, offering recommendations for standardizing the technical specifications related to pharmaceutical products. The ICH guidelines released a new set of quality criteria (Q8, Q9, Q10, and Q11) in relation to the QbD idea. The pharmaceutical sector now has new dimensions because of these advancements.

15.1.1 ICH Q8

The ICH Q8 section is dedicated to the development of pharmaceutical drug products, as stated in module third of the standard technical specification [5]. Through scientific methods and quality risk management (QRM), the goal of this section is to offer a thorough comprehension understanding manufacturing process and final product. The guideline also specifies instances in which improved pharmaceutical and manufacturing sciences knowledge may result in more adaptable regulatory strategies. The guidelines are primarily written for the starting marketing original applications and further amendments as to reflect new information as obtained during the course of the PPDL.

15.1.2 ICH Q9

QbD is supported by ICH guidelines, Q9 providing guidelines for QRM, which can be implemented across various aspects of drug quality. Q9 offers guidance on risk evaluation and management tools that can be used throughout pharmaceutical operations, and its aim is to develop a common understanding of concept and techniques used in risk management for regulators and industry.

15.1.3 ICH Q10

Based on the ISO quality principle, ICH Q10 provides a versatile framework for establishing an efficient quality system for pharmaceuticals in accordance with ICH Q8 and ICH Q9 guidelines. It can be applied throughout the different phases of a product development life cycle and is optional to regional GMP requirements. ICH Q10 supports the enhancement of medicine quality and availability worldwide and enables innovation and continuous improvement in the pharmaceutical industry. The guideline should be applied appropriately to each product's life cycle phase [6].

15.1.4 ICH Q11

In 2012, ICH released ICH Q11, a publication on the development and manufacturing of the drug substances, encompassing both biotechnological and biological entities. ICH provides guidelines for the structure and content of the common technical document (CTD) and provides as a guidance for the kind of information that should be presented in Module third in this scenario. The third module of the CTD provides information on the pharmaceutical development and drug substance of a medicinal product [7].

15.2 Roadmap for QbD

Dr. Joseph M. Juran, a well-known name in the field of quality management, is credited with first introduction of QbD idea in the early 1990s. According to Dr. Juran, product quality should be considered from the design phase because many quality problems result from errors in the early product conception. In essence, he thought that by putting it into a product's design from the start, its quality might be improved, and quality must be ingrained or embedded within the product [8]. The idea of "Quality by Design"aims to develop a scientific knowledge of essential processes and product characteristics. It involves establishing controls and conducting testing within the scientifically defined range of knowledge. This knowledge is gained throughout the PPDL and utilized to continuously enhance the manufacturing environment.

ICH Q8 [5] defines QbD as "a systematic approach to development that begins with predefined objectives and emphasizes product and process understanding and process control, based on sound knowledge of science and quality risk management." QbD ensures product quality by identification CPPs and CQAs throughout the PPDL.

The component of QbD include CQAs, Critical Material Attributes (CMAs), and CPPs. The tools that are often used in QbD include Process Analytical Technology (PAT), experiment design, and risk assessment [2]. A complete QbD study involves the defining of quality target product profile (QTPP), development of product and processes to meet predefined goals, identifying high-risk attributes and variability sources through risk assessment, and developing a control strategy for consistent product quality [8]. Implementing QbD-based strategies in pharmaceutical development can result in faster product development, flexibility in regulatory compliance, and ensure patient safety by meeting CQAs [9].

15.2.1 Elements of QbD

15.2.1.1 Defining QTPP

Before starting the development process, it is crucial for an organization to engage in methodical planning to determine the product development objectives. The QbD method places a strong emphasis on the necessity of creating a QTPP, which lists the target product's high-quality attributes. Important product characteristics such as dosage form type, dose quantity, mode of administration, pharmacokinetics and stability are all included in the QTPP [10]. To guarantee that the finished product satisfies the required quality standards, it is essential to closely monitor these attributes throughout the entire product development process.

15.2.1.2 Identifying CQAs

A QbD-based strategy entails determining the major CQAs that are most similar to the QTPP in order to assure product quality. CQAs are measures of a product's performance in terms of product quality and they are selected by an organization depending on their relevance to the intended purpose of particular product and dosage form. These qualities might be chemical, physical, biological, or microbiological properties that need to be closely monitored within predetermined parameters to avoid variations in product quality. Design of Experiments (DOE) and QRM approaches are used in the QbD method to identify CMAs or CPPs in order to meet desired CQAs goals. Multivariate experimental designs are then used to optimize these parameters.

15.2.1.3 Identifying CMAs and CPPs

CMAs are the method developer that can focus on optimizing these parameters to make sure that the method aligns with the required quality standards and CPPs are the method parameters that impact the quality of the results significantly. These may include factors such as accuracy, precision, specificity, selectivity, and linearity. To define fixed standards for the quality of a target product, a design space is determined. The CQAs of the product are directly influenced by CMAs or CPPs, creating a multidimensional region known as the design space. Through risk assessment or factor screening investigations, CMAs/CPPs are distinguished from the "so many" factor. During the factor screening stage, certain experimental designs viz., fractional factorial (FFrD), Taguchi, Plackett-Burman design (PBD) are commonly employed. For risk detection purposes, techniques like the risk estimate matrix and failure mode effect analyses are utilized to identify potential risks. In an industrial context, these vital processes are essential for making sure the quality of the intended result.

15.2.1.4 Design Experiments to Evaluate Method Performance

DOE is a key tool in QbD, which allows method developers to assess the effect of several factors on the effectiveness of the approach. By conducting a systematic evaluation of method parameters, the developer can identify the optimal conditions for the method and ensure that it is robust and reliable.

15.2.1.5 Develop a Control Strategy

A control strategy is a set of procedures and specifications used to ensure that the method is operating within acceptable limits. The control strategy may include factors such as calibration, validation, and routine performance monitoring. The pharmaceutical industry must build a control strategy that incorporates several types of controls that ensure product quality in order to enable continuous improvement throughout the product development process. The performance of the product is monitored by these controls, which also include in-process and end-product testing. These stages of the product's life cycle include development, technology transfer, and commercial manufacture. A possible quality crisis may be avoided by developing a control plan at these crucial stages, which also helps with change management and reduces the impact of changes on the product's quality [11]. Documenting the method development process is essential to ensure that the method is reproducible and can be transferred to other laboratories. The documentation should include a detailed description of the method, as well as any validation and performance data.

15.2.2 Tools of QbD

For the successful development and manufacturing of top-quality products, it is important to understand the components of QbD and the stages for its execution.

Apart from the underlying science of design and manufacturing, QbD relies on the use of some tools such as QRM, DOE, and PAT, to ensure the quality of the product remains consistent throughout its lifecycle. Familiarity with these tools is key for effective implementation of QbD principles [5].

15.2.2.1 Process Analytical Technology

"PAT is a technique that uses real-time measurements during production to enable data-driven decision-making, enabling the production of a final product that consistently meets predetermined quality and performance standards." The three-step method for optimizing pharmaceutical manufacturing and production involves monitoring, research, and design. The monitoring stage involves real-time measurement of all CPPs and CQAs through a systematic procedure/ techniques/ analysis are employed to thoroughly investigate identified quality attributes. Experiments are carried out throughout the design stage to identify key factors and their effects on the finished product quality [2]. The data is utilized to determine the essential requirements for CQAs, CPPs, and QTPP necessary for the development of a process control system based on PAT.

The FDA defines PAT as "a system for designing, analyzing, and controlling manufacturing through timely measurements (i.e., during processing) of critical

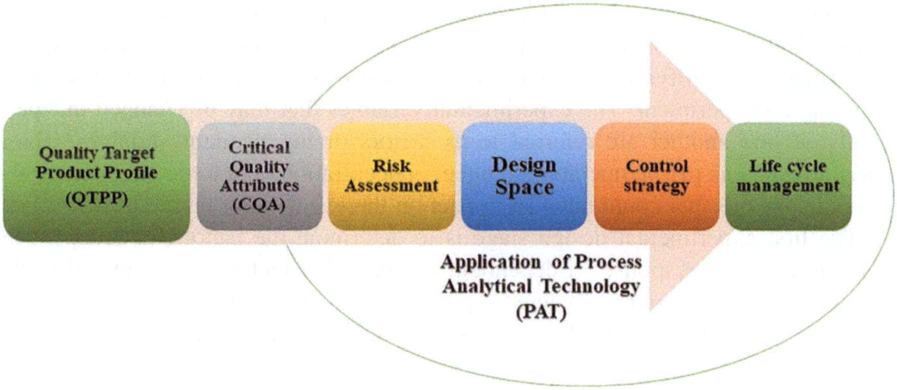

Fig. 15.1 Components of QbD

quality and performance attributes of raw and in-process materials and processes, with the goal of ensuring final product quality". The four primary drivers for the use of PAT are real-time evaluation and greater manufacturing process understanding; less variability and risk; enhanced efficiency and lower costs; and a shorter time for product release into the clinic or market. Figure 15.1 depicts the essential QbD components that form the foundation of the product development process. The measures are indispensable and represent the minimum requirements for obtaining regulatory approval for pharmaceutical products.

Applications of PAT are most of the time, spectroscopic methods involving UV-VIS, Raman and nuclear magnetic resonance (NMR) spectroscopy. In the pharmaceutical manufacturing industry, several additional PAT are employed for real-time monitoring, including Focused Beam Reflectance Measurements (FBRM), Near Infrared spectroscopy (NIR), Nanometric Temperature Measurement (MTM) and Tunable Diode Laser Absorption Spectroscopy (TDLAS) [12]. These technologies play critical roles in ensuring efficient and effective manufacturing processes.

15.2.2.2 Design of Experiments

Traditionally, optimization involved monitoring the impact of individual factors on an experimental response, leading to more experiments, reagents, materials, and time required for the study. However, in recent years, researchers have turned to multivariate approaches to enhance performance during method development. DOE is the multivariate and valuable tool in this context, it needs less effort, time and resources compared to univariate approaches [13]. In a predetermined number of trials, DOE permits the simultaneous evaluation of several elements at various levels, enabling the collection of a considerable quantity of data while reducing the required number of tests.

Experimental design plays an important role in optimizing methods by assisting in the identification of the experimental conditions that produce highest potential performance. Identifying the factors and response of interest is the first stage in planning an experiment. The experimental variables that may be changed independently of one another are referred to as factors or independent variables. These variables such as pH, temperature, reagent concentration, and ionic strength are the possible reasons that could have an impact on the event being researched.

The first experimental design stage is to determine the factors, or independent variables, and the response, or dependent variable. The factors are the experimental variables that are altered or changed separately from one another and are thought to influence the response variable, which is the study's primary outcome or the result being examined.

The process of optimizing a method using DOE is divided into two parts: screening designs and optimization designs. Full factorial designs (FFD), FFrD and PPBD are commonly used in screening experiments. These designs are used by researchers to effectively assess a large number of parameters in a limited number of tests. These designs are typically used to determine the most important factors, usually assessed on two levels that are significant for the development of product. On the other side, the design for optimization includes Box- Behnken Design (BBD), D-Optimal Design, Central composite design (CCD) and Doehlert design etc. [14]. The best levels of the most important factors previously chosen during the screening stage are varied as per the experiment runs suggested by the experimental design. The effects of factors on response variables (CQA) are studied using contour plots or response surface methodology (RSM). Various QbD-assisted software tools, such as Design-Experts (Stat Ease), Sartorius, Minitab, Fusion Product Development, and others, are widely used in analytical research and development [15]. Table 15.1 lists the various experiment design as per DOE [16, 17].

15.2.2.3 Risk Assessment

An outline of a typical QRM process is provided by ICH Q9 QRM guidelines. Risk assessment includes risk identification, risk analysis and risk evaluation. Other parts covered in this guideline are risk control, cover risk reduction and risk acceptance and at last checking the output and result of QRM process [18]. ICH Q9 offers the following 9 typical risk management tools, in the list as depicted in Fig. 15.2 [19]. These guidelines provide a structure for initiating and carrying out a risk management process, and they also provide the suitable QRM tools such as fault tree analysis (FTA), failure mode effects analysis (FMEA), hazard analysis and critical control points (HACCP), failure mode, effects and criticality analysis (FMECA) etc. [20]. DOE is utilized to prioritize risk assessment as a part of QbD implementation [21]. The formulation process factors were prioritized based on their level of risk using a risk estimation matrix (REM), which evaluated each factor's low, medium, and high-risk potential. This enabled the selection of CMAs and CPPs via the REM process. The factors were then screened for optimization studies based

Table 15.1 Various experimental design as per DOE

Sr. no.	Name of design	Type of DOE	Formula	Application
1	Full Factorial Design (FFD)	Screening design	$N = L^k$ (*k* is factors which can be set to *L* different level, *N* is depicts to experiments number)	It is observed that as the number of levels and factors increases, the number of feasible combinations increases significantly.
2	Fractional Factorial Design (FFrD)	Screening designs	$N = L^{k-p}$ (*p* is the number of independent variables, chosen to fractionate the design)	Full factorial studies may need several runs, particularly when there are many factors and levels. It can be extremely time-taking and expensive to run all potential combinations of the factor and levels, and in certain circumstances it might not even be viable or possible. For the sample size of the fractional factorial design, we pick a fraction such as 1/2 or 1/4 etc. of a full factorial design.
3	Placket-Burman Design (PBD)	Screening designs	$K = N - 1$ (*N* is a multiple of 4, such as 8, 12, 16 etc.)	The number of runs in fractional factorial designs is often a multiple of four. Plackett-Burman designs are frequently used for screening tests to assess the robustness of procedures while they are still being developed. If the primary effects are only calculated using a lesser number of trials then other interactions may be completely ignored.
4	Taguchi orthogonal array Design (TOA)	Screening designs	–	This type of design, which comes from Taguchi's textbook, is known as orthogonal array design. All key consequences are taken into account in their design, but no interactions are included.
5	Box-Behnken Design (BBD)	Optimization designs	$N = 2k(k - 1) + Cp$ (*k* is denote to number of factors and *Cp* denoted to number of center points)	The BBD is an instance of an independent, and quadratic experimental design. It does not incorporate a fractional or embedded factorial design, in contrast

(continued)

Table 15.1 (continued)

Sr. no.	Name of design	Type of DOE	Formula	Application
				to other designs. All of the experimental sites in the Box-Behnken design are situated inside a sphere with a specific radius. It is a rotating design. It has a three-level building design that necessitates changing the elements at levels −1, 0, and +1. The BBD must be repeated for each combination of the categorical fractional levels when categorical factors are added.
6	D-Optimal Design (OD)	Optimization designs	–	Based on a predetermined model, a design for categorical factors is produced, and it entails choosing a subset of all feasible factor combinations. This design's generation aims to reduce the inaccuracy connected to the model coefficients.
7	Central composite design (CCD)	Optimization designs	$N = 2^k + 2k + 1$ (k is factor)	Take the two-level factorial design and add: A center point (the middle point between all the factors), Axial points (the center point except with one variable changed to be at +/−, an extra value). Do this for all variables.
8	Doehlert design (DD)	Optimization designs (response surface methodology)	$N = k^2 + k + Cp$ (k denoted number of factors and Cp denoted to central points)	In comparison to other experimental designs, the Doehlert approach is more practical and affordable since it needs fewer experiments. Doehlert designs have different number of levels for each factor like 3 and 5 levels for the two factors, while 3, 5, and 7 are used for the three factors.

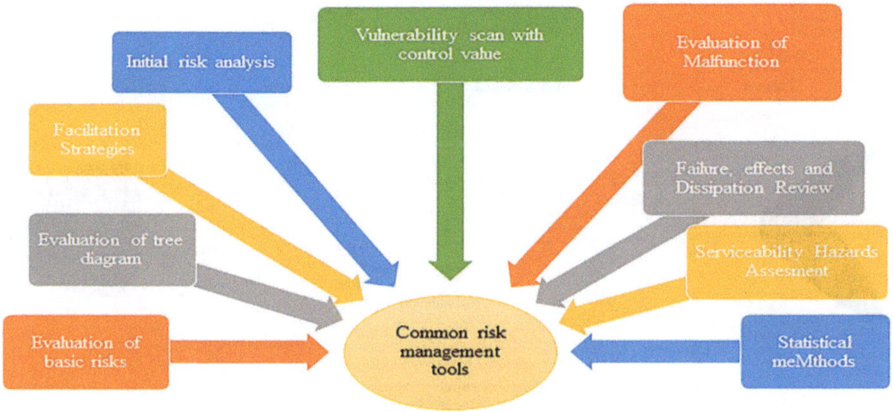

Fig. 15.2 QbD used as common risk management tools

on their risk priority number (RPN), which was calculated through a FMEA using a quantitative ranking system between 1 and 10 for severity, detectability, and occurrence. Factors with higher RPN scores were identified for optimization studies as CMAs/CPPs [22].

$$\mathrm{RPN} = \mathrm{Severity}\ \ (S) * \mathrm{Detectability}\ \ (D) * \mathrm{Occurrence}\ \ (O).$$

Overall, QbD and Q9 promote incorporation of quality into product-development phase rather than depending on testing to ensure quality. Moreover, they provide a systematic approach to evaluating and managing risk throughout the PPDL [18].

In the 1960s, famous Japanese quality control statistician Dr. Kaoru Ishikawa created a quality management diagram that later came to be known as the "Ishikawa diagram." Ishikawa diagram and risk control techniques that analyze outcomes and losses are often employed, and they are frequently used separately or in combination, in various fields. The Ishikawa diagram, sometimes known as the fishbone diagram due to its resemblance with a fish skeleton, has an easy-to-understand structure that makes it a useful tool for brainstorming. The main purpose of a fishbone diagram is to analyze an issue, outcome, or effect by examining, identifying, and illustrating the potential sources of the problem, outcome, or effect [23]. Figure 15.3 shows the Ishikawa diagram for the different issues, consequences, and results during formulation of pharmaceutical products also called 6Ms: men, material, method, machine, milieu, measurement.

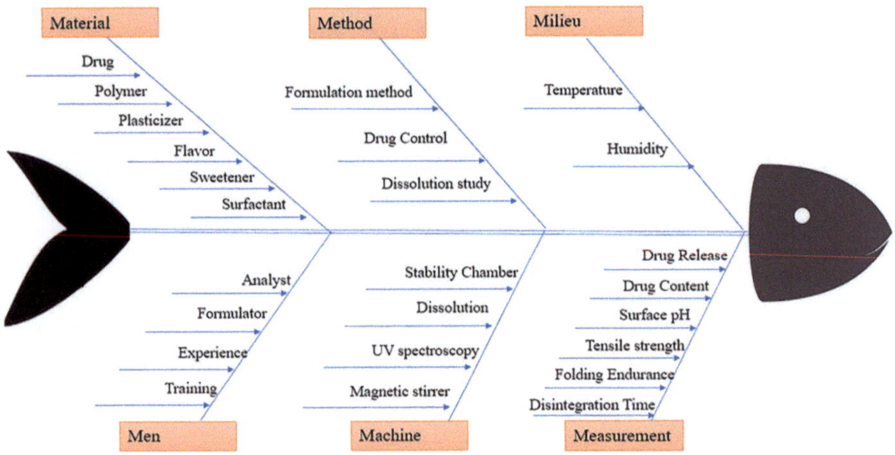

Fig. 15.3 Ishikawa Fishbone Diagram

15.3 Pharmaceutical Product Development Lifecycle (PPDL)

The steps of developing a new pharmaceutical product, from the initial concept through the finished product for sale, are referred to as the "**pharmaceutical product development lifecycle.**" The stages typically include discovery and research, preclinical development, clinical development, regulatory approval, manufacturing, commercialization.

Throughout the PPDL, various quality assurance measures are used to ensure that the product complies with the prescribed standards of quality. These measures may include QbD approaches, good manufacturing practices (GMP), and testing and validation of CQAs and CPPs.

15.3.1 Lifecycle Approach to Pharmaceutical QbD

The idea of QbD, in addition to formulation development, has been extended to other crucial aspects of pharmaceutical development. Such as development of analytical methods, the creation of pharmacological substances, stability and bioequivalence testing, clinical trials, and more. By implementing QbD in these areas, it is possible to identify CMAs, CPPs, and CQAs, as well as to create QTPPs to ensure the consistency and quality of the final product. The pharmaceutical sector may improve efficiency, lower risks, and improve patient safety by applying QbD concepts to four crucial areas of drug development**.** With QbD, formulation scientists can benefit from its advantages even after a product's commercial launch by implementing post-marketing surveillance. Therefore, QbD is an omnipresent concept that is vital to each stage of pharmaceutical product development as depicts in Fig. 15.4.

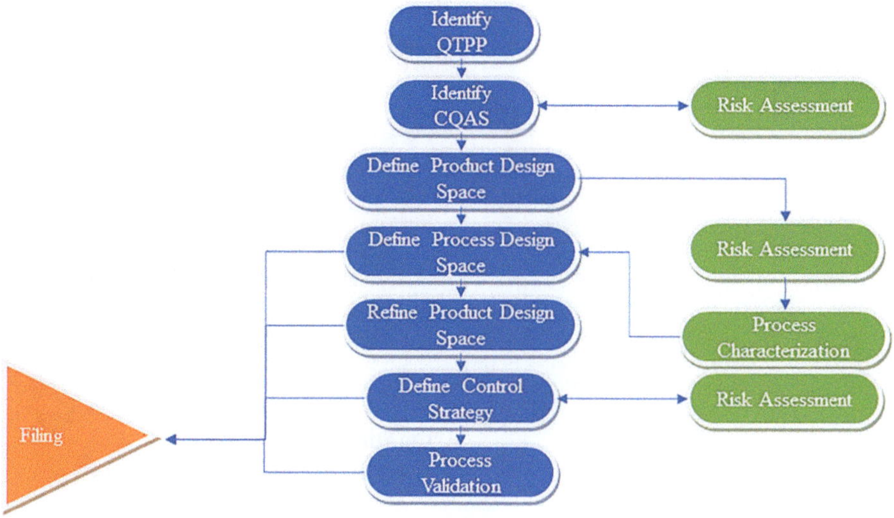

Fig. 15.4 QbD lifecycle

15.4 Applications of QbD

QbD principles may be used by different industries like pharma, food, cosmetics and hardware industries to develop quality in products. In pharmaceutical sector, QbD concepts can be applied to various areas including active pharmaceutical ingredient (API) manufacturing, formulation development, development of biotherapeutics, quality assurance, packaging, and clinical trial depicts in Fig. 15.5.

15.4.1 The Application of QbD to API

APIs are the key components of pharmaceutical products that provide therapeutic benefits. The quality of the API plays important role in terms of safety, effectiveness and stability of the final drug substance. By applying QbD principles to API manufacturing, manufacturers can ensure that the API meets the desired CQAs consistently. Every synthesis requires quality control of the produced impurities. This was done by evaluating the finished product and establishing acceptability standards for the purity and properties of the intermediates and API. The QbD approach involves identifying the CQAs of an API such as appearance, morphic form, particle size, water content, impurities, residual solvent, heavy metals, assay, residue on ignition and how they are impacted by various CPPs and understanding their relationship [20]. This understanding allows manufacturers to optimize the process and control the critical parameters within their design space to consistently

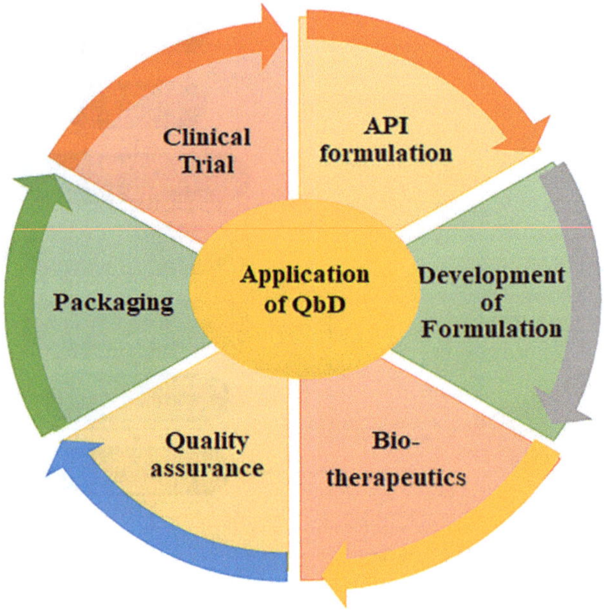

Fig. 15.5 Application of QbD in different stage of PPDL

produce APIs with the desired CQAs. Additionally, QbD principles are applicable to the development process of APIs consisting of raw materials selection, analytical technique development, and the establishment of specifications for the API. QbD can also help manufacturers reduce costs and minimize waste by optimizing the manufacturing process, reducing variability, and improving efficiency. Furthermore, the use of QbD in API manufacturing can minimize the risk of product failures and regulatory issues, reducing the need for costly and time-consuming post-approval changes. Overall, the application of QbD principles in API manufacturing can lead to improved product quality, consistency, and cost-effectiveness, while reducing the risk of regulatory issues and post-approval changes.

Case Study of Torcetrapib API

This case study described the establishment of the design space for the manufacturing process of drug (Torcetrapib) with the help of CQAs. The researchers highlight that impurity profile, assay, and impurity limits are the key CQAs that directly impact on the safety and efficacy of the drug. Other specifications like appearance, physical parameters and characteristics are the most important for the manufacturing of drug but it does not affect the safety and efficacy of drug.

The design space for the torcetrapib API manufacturing process is being developed with the goal of setting boundaries for controlling impurities. The researchers suggest that if the combination of synthesis conditions is favorable, impurity limits in the API specification may be achieved by expulsion during the synthesis, rendering specific testing limits unnecessary.

Torcetrapib API is manufactured synthetically using six chemical stages and four separate intermediates. Each with several focus areas (FAs) viz.; reaction, quench, work up, solvent displacement or replacements, crystallization, filtration and drying. Risk assessment is conducted within each FA by evaluating process parameter variability, functional interactions, and their impact on quality attributes (QAs). Understanding and controlling the QAs of synthetic intermediates throughout the manufacturing process help identify and control additional CQAs and CPPs. The risk assessment for the torcetrapib drug product involves fewer FAs but carries the possibility of CQAs being influenced by early FAs if certain QAs are not met. In following FAs or processing procedures, the effects of unfulfilled QAs might not be minimized or decreased. This reveals a key distinction between the production of drug substances and drug products.

To address the potential impact on CQAs of the torcetrapib API, a multivariate experimental design was conducted specifically focusing on the reaction and crystallization processes. The reaction and crystallization steps were identified as critical for controlling impurities, and their process parameters were investigated to establish robust control strategies.

The experimental investigations provided insights into impurity control, reaction rates, and yield optimization. The relationship between process parameters and impurity levels was analyzed through contour plots and statistical analysis. The experimental design and data analysis allowed for the characterization of process factors and the identification of optimal process conditions within the defined design space.

Overall, the study emphasizes the importance of identifying CQAs, establishing design spaces, and controlling impurities and process parameters to ensure the safety, efficacy, and quality of torcetrapib API and drug product. The experimental investigations contribute to procedure understanding and supply a basis for optimizing the manufacturing process to achieve the desired impurity control [24].

Figures 15.6 and 15.7 depict the contours of impurity levels in relation to process parameters. No new impurities were detected, and known impurities were not depicted at higher levels within the specified ranges. As compared to sodium phosphate, sodium carbonate exhibited a better control in forming a specific impurity. Extended process times and increasing quantities of a specific reagent resulted in effective control of the impurity below the specified limit. This control over extended times was also plotted, which showed an unexpected finding of stalling of the reactions with sodium carbonate, leaving unreacted starting material. This was due to a degradation pathway secondary when ratio of reactants approached 1, leading to a shortage of a specific reagent. However, this stalling did not impact the quality of the isolated intermediate, as a process acceptance range (PAR) ensured the purging of unreacted starting material during crystallization. The observation of this phenomenon is important for process understanding but does not affect the safety or quality of the final product. DOE results helped in establishing the relationship between process parameters and impurity quality attributes. Statistical analyses rank variables based on their impact on impurity purging. The evaluation of process parameters for crystallization determines design space boundaries and identifies the

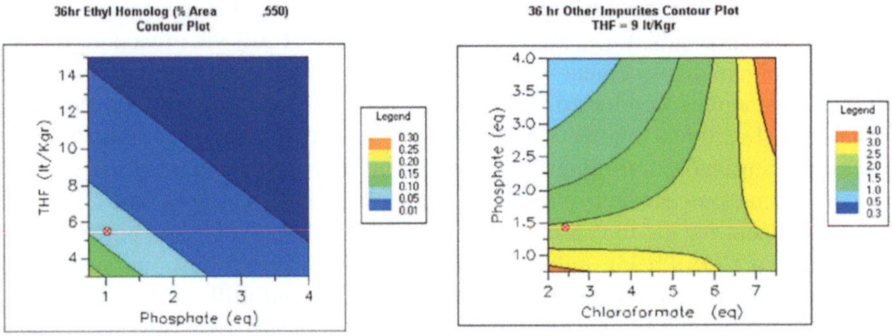

Fig. 15.6 The levels of the ethyl homolog and total impurities in sodium phosphate after an extended hold time of 36 h being assessed. The red dot represents the standard process conditions. The measurements are expressed as a percentage of the quality assurance (QA). (Reproduced with permission)

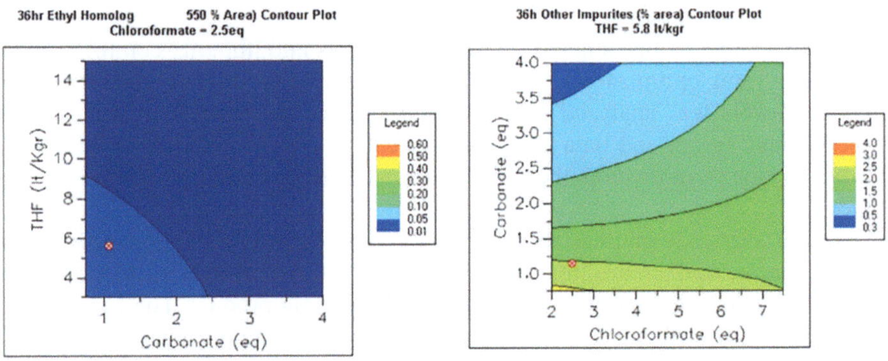

Fig. 15.7 The ethyl homolog and total impurities in sodium carbonate were evaluated after an extended hold time of 36 h. The left contour indicates that the generation of ethyl homolog is minimal, less than 0.05% across the entire range, under fixed ECF (Ethyl chloroformate) levels. On the other hand, the right contour suggests a slight sensitivity to low carbonate levels, which leads to slightly higher total impurities. The measurements are expressed as a percentage of the QA. (Reproduced with permission)

impact of specific parameters on impurity levels. Multiple variables are controlled within boundaries, and the interaction of variables influences impurity levels, leading to the creation of a multivariate design space. Results for process variables influencing impurity quality attributes, aiding in process optimization were also compiled. Furthermore, agitation and cooling rates for filtration rates, and alternate polymorphs were assessed but not found in the experimental design. Laboratory-scale experiments simulate manufacturing performance, considering scale-up sensitivity. For slow homogeneous chemistry, laboratory-scale predictions match scale results, while crystallizations may be affected by different flow environments.

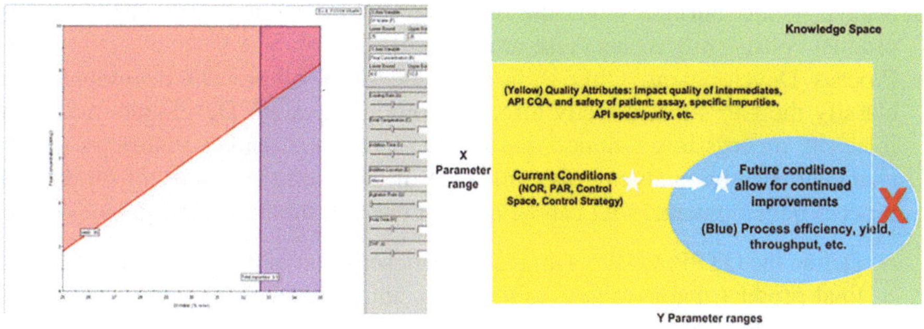

Fig. 15.8 (**a**) Process operating window is shown by the overlay plot of yield and quality for crystallization and (**b**) Design Space. (Reproduced with permission)

Filtration rate measurements are used to evaluate scale sensitivity, and further studies would be conducted if sensitivity is detected.

Figure 15.8a overlays yield and quality, guiding process improvement. The white area represents optimal conditions under current manufacturing conditions. The pink area indicates lower yield but excellent quality, possibly explored with future technology. The purple area approaches the process acceptance range limit for water content. The upper limit likely represents an end-of-range failure based on prior knowledge. In Fig. 15.8b the design space would be represented by areas in yellow or by regions where yellow and blue overlap and would not apply to any areas outside of the yellow zone. If the design space evolves, the process lifecycle should support continued improvements throughout the duration of the product, even after submission, to enable continuous improvement.

15.4.2 *The Application of QbD to Formulation Development*

Pharmaceutical quality assurance system relies on testing to ensure product quality. This system includes testing of the raw materials, production of the active pharmaceutical ingredient, a set process for producing the pharmaceutical drug product, testing of the in-process materials, and testing of the finished product [19]. These tests are conducted at various steps in order to make sure that the product possesses the necessary quality attributes. By conducting these tests, manufacturers can identify any deviations from the established control strategy and take corrective actions in order to maintain quality in final product [22].

QbD has several applications in the PPDL, including:

- Product Development: QbD can be used during the initial stages of product development to identify CQAs and develop a risk-based control strategy. By identifying the CQAs of the product, developers can design a product with the desired quality attributes, and the risk-based control strategy can ensure that the product consistently meets these attributes.

- Process Development: QbD is used at the stage of process development to classify the CPPs that directly affect the product quality. Developers may optimize the process to guarantee that the product consistently satisfies its quality characteristics by analyzing the influence of CPPs on the quality of product. In process development, QbD can be applied through the following steps:
  - Define the QTPP: QTPP is an inclusive description of the quality feature in the final product that is related to its safety and efficacy. It includes information such as dosage form, strength, purity, potency, stability, and other relevant characteristics.
  - Identify CQAs: CQAs are defined as the chemical, physical, biological and microbiological characteristics of product that have significant effect on its quality. The identification of CQAs is essential in defining the process parameters that will ensure that the product meets the QTPP. Typical CQAs of formulated drug product like tablet are organoleptic characteristic such as appearance, thickness, hardness, uniformity of dosage, content uniformity dissolution, microbiological limits and assay etc. [20].
  - Define CPPs: CPPs are the parameters of the manufacturing process which have direct effect on the CQAs. CPPs can be defined through a systematic risk assessment process.
  - Design experiments to understand the impact of CPPs on CQAs: In this step we design experiments to study the effect of CPPs on CQAs. This can be done using tools such as DOE.
  - Establish a design space: The CPPs'design space is a set of values that guarantees the CQAs fall within the required range. It is determined depending on the outcomes of the trials carried out in the previous step.
  - Define control strategy: This strategy is a series of instructions for the manufacturing process to make certain that the procedure adheres to the design space. The control strategy includes monitoring of CPPs, process validation, and change control procedures. By applying QbD principles in process development, a strong and reliable manufacturing process is developed that ensures consistent quality of the final product.

- Technology Transfer: QbD can be utilized for technology transfer to achieve desired quality within the product and ensure that the product maintains its CQAs during the transfer from one manufacturing site to another.
- Manufacturing: QbD can utilized during manufacturing to achieve desired quality within the product. By monitoring CPPs, manufacturers can identify any deviations from the established control strategy and take corrective actions to maintain the quality of product.
- Regulatory Submissions: QbD is useful in regulatory submissions to holistic knowledge of the product development process, which includes identifying key quality attributes, CPPs, and risk-based control strategies. This information can

assist regulators in measuring the quality of the product and make informed decisions about product approval. QbD, thus, provides a holistic approach to development of pharmaceutical products, ensuring the quality throughout the entire PPDL. Some case studies are mentioned in Table 15.2. Figure 15.9 depicts the combining QRM and DOE approaches to factor screening and optimization of product development.

Case Study of Erythropoietin-Loaded Solid Lipid Nanoparticles (EPO-Loaded SLNs)

In this case study, the EPO-loaded SLN formulation was optimized using Box-Behnken design. After successfully incorporation of EPO, a relatively large protein molecule, into Solid Lipid Nanoparticles (SLNs), the duration of homogenization was identified as the significantly influential variable in determining the size and distribution rate of the nanoparticles. Additionally, the concentration of surfactant and EPO were found to significantly influence the encapsulation efficiency (EE) and several tests were performed to study the stability and compatibility of the methods with the protein. An optimized formulation (130 mg Span®80, 152.5 μL EPO) was obtained with 1.9 min homogenization (high shear) which resulted in 280 nm SLNs with a narrow size distribution, high entrapment efficiency, and no cytotoxicity. The *in vivo* studies showed that 2500 U EPO-loaded SLN compared to the 5000 U EPO solution, SLNs exhibited comparable or even superior effects on raising the amounts of red blood cells (RBC), hemoglobin, and hematocrit. Overall, the study proposed a suitable method for preparing EPO-loaded SLNs as a protein-based medication delivery method. The polydispersity index (PDI) was inversely proportional the homogenization time, while the amount of drug and emulsifier did not show any significant impact on PDI. Higher concentrations of drug and emulsifier resulted in increased entrapment efficiency (EE). Figure 15.10 depicts response 3D graph of effect of dependent variables (surfactant content, EPO quantity and homogenization time) on response variables (particle size and % entrapment efficiency).

The optimization formulation achieved particle size (280 nm), PDI (0.282), and EE (43.4%), values that were in accordance with the predicted values [33].

15.4.3 The Application of QbD to Development of Analytical Method

Defining the analytical target profile (ATP): The ATP specifies intended use, performance criteria, and specifications for the analytical method. This step ensures that the method is fit for purpose and aligns with the overall quality goals.

Identifying CMAs: CMAs are the method parameters that have a substantial effect on the standard of the outcomes. These may include factors such as accuracy, precision, specificity, selectivity, and linearity. By identifying CMAs, the method

Table 15.2 QbD for optimization of pharmaceutical product development

Unit operations	Drug formulation	Active pharmaceutical ingredient	DOE	CMA	CPP	CQA
	Liposome	Cefoperazone sodium	BBD	Pulmonary route of administration, dry powder	Sonication time, hydration time	Particle size and % entrapment efficiency [25]
	Liposome	Lamotrigine	Response screening design	Liquid formulation, Nasal route, one dose volume, time, vesicle size, pH, dissolution profile/ absorption	Vacuum evaporation, hydration (temperature), ultrasonication (temperature)	Vesicle/particle size, constant vesicle size, no aggregation [26]
Hot- melt extrusion	SLN	Fenofibrate	PBD	Surfactant, lipid content	Speed of screw, temperature of barrel, liquid addition zone	Particle size, PDI, Zeta potential, EE [27]
	SLN	Rivastigmine	FFD	Ratio of drug: Lipid, concentration of surfactant	Homogenization time	Particle size, PDI, EE [28]
Homogenate membrane method	ORO-dispersible film	Theophylline	CCD	% of glycerol, % of HPMC	Drying temperature	Tensile strength, elongation at break, DT [29]
Enhanced bio-availability and biodistribution	Nano-micelles	Galantamine	TOA	Concentration of conjugation phospholipid-polymer, emulsifier concentration, formulation medium (aqueous/alcoholic)	Type of, order of mixing, stirrer type, stirring time, evaporator stirring speed, cooling temperature and probe stirring speed	Particle size, Zeta potential, EE and in vitro drug release [30]
	Cubosomes	Luliconazole	CCD (3-factor, 5- level)	Lipid content, surfactant content, co-surfactant content		Particle size, entrapment efficiency [31]
Enteric coated	Nanoparticles	Metronidazole	FFrD	Concentrations of chitosan, tripolyphosphate, and acetic acid,	Stirring speed and stirring time	Particle size, entrapment efficiency [32]

	Hydrogel	Ketoconazole loaded cubosomes	3^2 factorial designs	Surfactant concentration, stabilizer concentration	Stirring time, stirring speed, dilution of pre- emulsion, temp. Of stabilizer solution, lipophilic component dissolve in organic solvent	Particle size, EE, *in vitro* drug release [22]
	SLN	Erythropoietin	BBD 3^3 factorial designs	Emulsifier concentration, drug concentration	Homogenization time	Particle size, PDI, EE [33]
	SLN	Carvedilol	PBD	Emulsifier concentration, charge-inducing agent concentration, lecithin concentration lipid concentration	Homogenization time, sonication time	Particle size, PDI, EE, Zeta potential, in vitro drug release [34]
	Nano lipid carrier	Tenofovir disoproxil fumarate	CCD	Ratio of drug: Lipid	Sonication time, pH,	Particle size, PDI, EE [35]

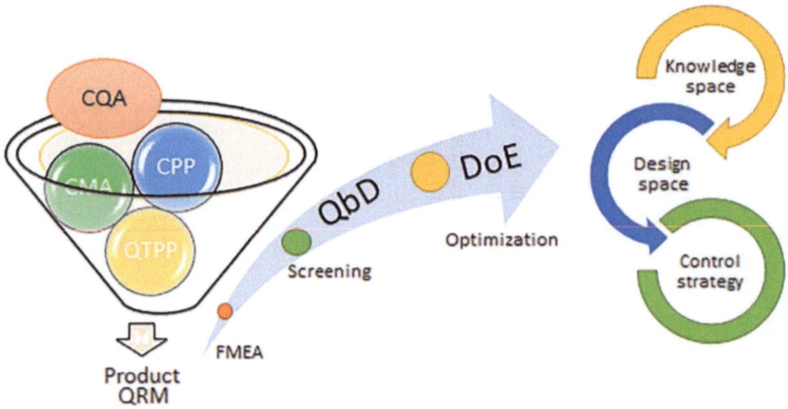

Fig. 15.9 Combining the QRM and DOE approaches to factor screening and optimization

developer can focus on optimizing these parameters to make sure the technique fulfils the essential requirements for quality.

Design experiments to evaluate method performance: DOE is a key tool in QbD, allowing method developers to assess how several factors affect the effectiveness of a procedure. By conducting a systematic evaluation of method parameters, the developer can identify the optimal conditions for the method and ensure that it is robust and reliable [36].

Develop a control strategy: A control strategy is a series of procedures and specifications used to make sure that the method is operating within a reasonable range. The control strategy includes factors such as calibration, validation, and routine performance monitoring.

Documenting the method development process: Documenting the method development process is necessary to be sure that the method is reproducible and can be transferred to other laboratories. The documentation should include a detailed description of the method, as well as any validation and performance data. Figure 15.11 shows the flow diagram of analytical method development by QbD.

Case Study for Analytical Method Development for Raloxifene Hydrochloride This study focussed on creation of an analytical method using reversed phase HPLC which shall be economical, sensitive, and reliable for accurately and precisely quantifying Raloxifene hydrochloride (RLX). The aim was to reduce elution time following the directives of analytical quality by design (AQbD) and ICH guidelines Q2(R1), which would allow authentication of the method effectively. Before the implementation of the AQbD approach, the preliminary step is defining the Quality Target Method Profile (QTMP), which outlines the desired quality attributes of the intended analytical method. The critical analytical attributes (CAAs) to be used during QMTP are tailing factor (TF), % assay and theoretical plate count (TPC). Various experimental design risk assessment studies were performed to acquire insight into the factors impacting the QTMP. These evaluations sought to

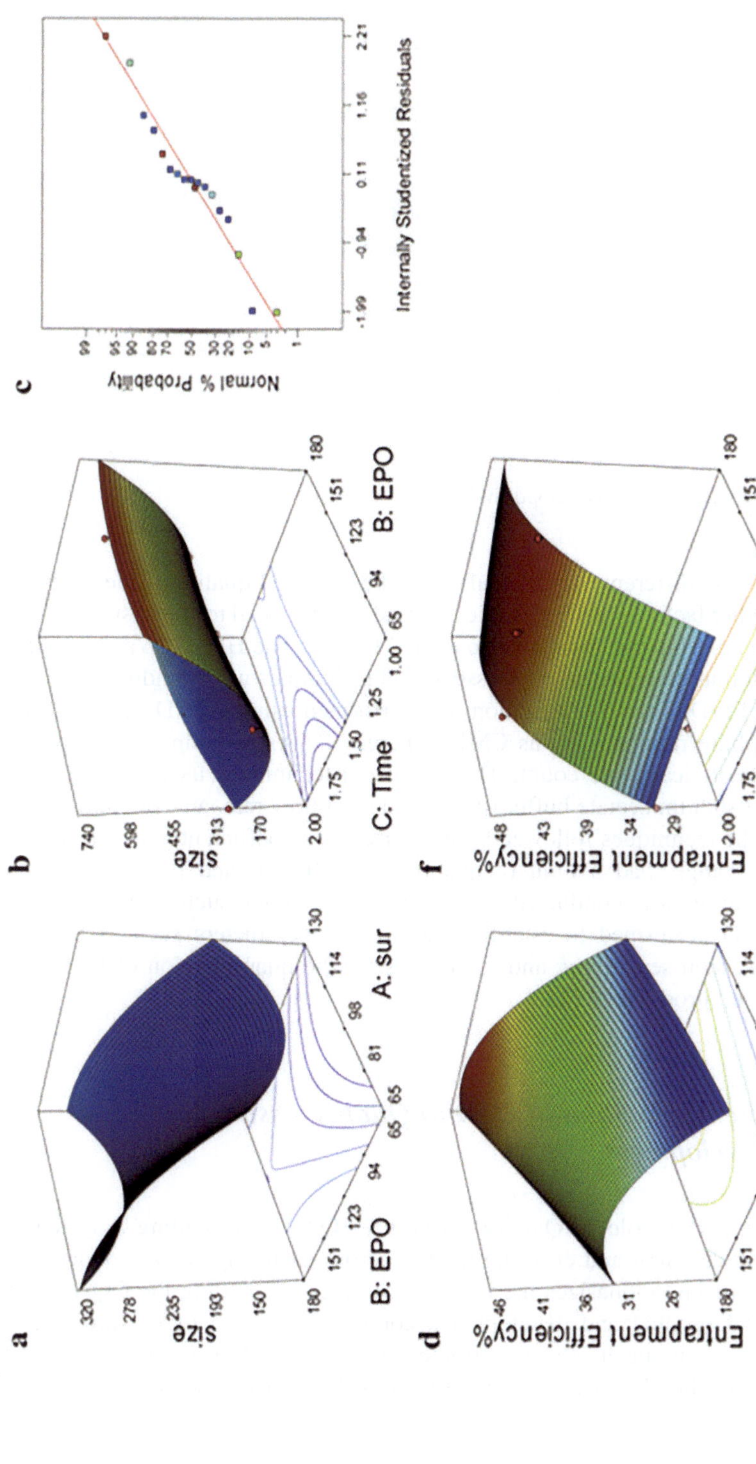

Fig. 15.10 **(a, b, d, f)** Response 3D graphs demonstrating the interaction impact of **(A)** surfactant concentration, **(B)** EPO quantity, and **(C)** homogenization duration on particle size and entrapment efficiency% (c) Normal probability plot of residuals. (Reproduced with permission)

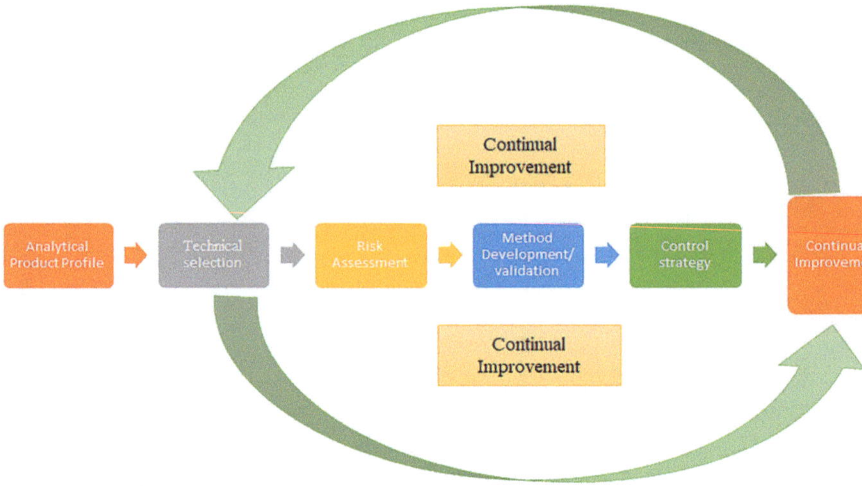

Fig. 15.11 Analytical method Development in QbD

comprehend how different variables affected the intended quality of the analytical procedure. Other factor screening studies were also performed to shortlist "vital few" input parameters from the "plausible so many" factors. The HPLC (Waters acquity) used C18 column screening and risk assessment techniques were conducted with the help of QbD FFrD. Then, optimization was performed using a CCD, with flow rate and mobile phase ratio serving as CMPs whereas CAAs as tailing factor, percent assay, and theoretical plate count. The optimized techniques used 50:50 ratio of methanol and sodium acetate buffer (pH 4) and set a flow rate, an oven temperature and a λ_{max}. The techniques followed by the correlation coefficient with linear in the concentration range, had a limit of quantification (LOQ) and limit of detection (LOD). Validation was conducted according to guidelines such as ICH Q2(R1), and the AQbD performed to establish the various parameters such as linearity, accuracy, precision sensitivity, and robustness for the quantification of bulk drugs and commercial products [37].

15.4.4 The Application of QbD to Quality Assurance Development

QbD plays a crucial role in Quality Assurance (QA) by providing a systematic approach to developing and ensuring quality across each stage of a product's life. The QbD approach emphasizes the importance of understanding the CQAs of the pharmaceutical product and developing a control strategy to make sure that the product consistently meets those attributes. In QA, QbD is used to define the specifications of the pharmaceutical product, develop a risk-based control strategy, and establish a continuous improvement program. By using QbD, the pharmaceutical product‘s compliance with strict standards for quality, safety, and efficacy may be confirmed through QA.

Case Study of QbD in Quality Assurance in the Development of a Novel Powder-Based Inhalation Therapy
The QbD approach was employed to optimize the formulation’s quality attributes and ensure product consistency. The investigation revealed the CQAs of the formulation, which included particle size, drug content uniformity, and aerodynamic behavior. The CPPs affecting these CQAs were defined and analyzed by DOE technique. The DOE study allowed for the identification of the key CPPs, such as the concentration of the drug, the milling time, and the spray-drying parameters. The data collected from the establishing a design space with the help of a DOE study ensured the formulation’s quality attributes remained within the desired range [38]. This design space allowed for the determination of the most effective process parameters and their acceptable ranges. The QbD approach helped to achieve a robust and consistent formulation that met the desired quality attributes of the safe and efficacious.

15.4.5 The Application of QbD to Packaging Development

The application of QbD in packaging is critical to ensure that after packaging the drug product is protected during storage by microbiological and physiochemical changes caused by environmental factors and physical damage during distribution and shipping. QbD can help to identify the CQAs of the packaging material, such as the moisture barrier, light protection, and oxygen permeability, and develop a control strategy to ensure that the packaging material consistently meets those attributes.

QbD can also help to optimize the packaging process by identifying CPPs that can affect the quality of the packaging material. It is essential to carefully choose containers before starting formal stability tests in order to ensure a science based and QbD approach to product development. This underlines how crucial it is to choose containers carefully when a product is still in the early phases of development. However, by monitoring and controlling these CPPs, packaging suppliers can make sure that the packaging material is of high quality and preserves the drug product during storage. Here a DOE case study is discussed which chooses appropriate packaging techniques for drug products that are susceptible to light and moisture.

Moisture permeability is a CQA of packaging systems, and it is often evaluated by measuring the moisture vapor transmission rate (MVTR). The MVTR is an essential parameter used to evaluate the effectiveness of packaging systems during the process of conducting selection studies. It is worth noting that MVTR/unit data is a reliable indicator of packaging system moisture protection equivalency and can be used into regulatory applications. Depending on the pharmaceutical product and packaging method, additional characteristics such as oxygen permeability rate and light transmission rate can also be taken into account. The chapter continues by mentioning a case study pertaining to the choice of appropriate packaging techniques for pharmaceuticals that are sensitive to moisture and light [39].

Case Study on Selection of Packaging for Moisture Sensitive Drug Product Badawy et al. (2001) [40], studied the effect of different packing arrangements on the stability of Roxifiban tablets, (an ester prodrug). This drug is moisture-sensitive and susceptible to ester hydrolysis in the solid form. The authors used FFrD to study the effect of different packaging materials on quality of drugs. In terms of QbD, interaction effects, another method for locating crucial factors and optimizing process parameters, may be contrasted with the nested design utilized in this study. In terms of statistics, when each level of the first factor is solely related with one level of the second factor, the first factor is termed nested within the second factor. In this experiment, two different sizes of HDPE bottles (30 and 60 cm^3) were used. Additionally, other factors such as desiccant quantity (silica gel) with varying amounts (1.0, 1.5, 2.0, and 3.0 g), strength of tablet (0.5, 1, 2 mg) and tablet counts per bottle (50 and 100) were modified for both bottle sizes. This resulted in eight different packaging configurations. FFrD of this study indicates 8 experimental runs and 4 packaging variables including bottle size (A), strength of tablets (B), tablet counts in each bottle (C) and amount of desiccant in bottle (D). However, there is no interaction among these variables. The findings of this investigation revealed that the size of the HDPE bottle had no major influence on tablet stability and while the larger bottle had slightly higher permeability, it was not associated with increased degradation. The study also found that an increase in desiccant quantity per bottle resulted in a significant decrease in degradation rate, as the predicted RH values decreased. These findings offer support for the recently planned USP (United States Pharmacopeia) criteria of MVTR (moisture vapor transmission rate) assessment and offer insights into head space comparison for rationalized interpretation of stability results.

15.4.6 The Application of QbD to Clinical Development

Identifying the needs of QbD in clinical trials encompasses identifying and promoting methods that can enhance the effectiveness and quality of clinical trials. One of the main challenges in clinical trials is to make sure of the safety and efficacy of novel drugs or treatments. Clinical development is a complicated and expensive process that involves several steps, including preclinical testing, clinical trials, and regulatory approval [41]. The key challenge in clinical development is trial design which includes complexity, feasibility as well as delays from avoidable amendment and trial oversight includes focus on the accuracy of single data points, over 3 reliance on retrospective oversight methods as well as one size-fits- all approach [42].

The QbD approach can help to identify the CQAs of the drug product and develop a control strategy to make sure that the product consistently meets those attributes. QbD can also help to optimize the clinical trial process by identifying CPPs that have the potential to impact the quality of the drug product. By monitoring and controlling these CPPs, clinical trial sponsors can ensure that the drug product is both safe and effective and that the clinical trial is conducted in a reproducible manner [43].

Case Study of Implemented QbD Principles in Clinical Research
This study focused on the use of QbD principles in clinical study conducted at an academic health center [44]. The researchers utilized multidisciplinary design studios to identify and prioritize critical to quality (CTQ) factors, which are essential elements for patient safety and data integrity in clinical trials. They reviewed eight different clinical research protocols using the CTQ design studio (CTQ-DS), conducted post-meeting interviews and surveys to assess the efficacy and acceptability of this intervention [45]. The protocols examined in the study comprised of both nonrandomized, investigator-initiated investigations and multicenter, randomized clinical trials. Through a developmental evaluation, the CTQ-DS process was refined iteratively [46]. The CTQ-DS technique was well regarded by principal investigators (PIs), according to post-meeting questionnaires, and it resulted in several protocol revisions [45, 47].

This study demonstrated how the use of QbD principles may successfully direct the design and execution of clinical research inside an academic health center. This was achieved through the utilization of multidisciplinary design studios, which facilitated the identification and prioritization of critical elements [48, 49]. The participants, including study PIs, responded positively to this approach. However, additional research is required to assess the efficiency of this strategy in raising the standard of clinical research [46].

15.4.7 Application of QbD in Biotechnology Product Development

The field of biotechnological drug development is continuously growing and becoming more important with each passing day. Due to the irreversible and costly manufacturing of biotechnological products, as well as the numerous critical parameters involved in the process, relying solely on quality control tests for the final product is not an effective strategy. As a result, maintaining predetermined quality becomes crucial. In the biotechnology sector, the concepts of QbD have been used at every level of process development [50, 51]. A case study is discussed below.

Case Study of QbD Implementation for a Biosimilar Product's Downstream Process Development
This case study demonstrates the implementation of QbD in the downstream process development of a biosimilar product like Granulocyte Colony-Stimulating Factor

(GCSF). The first phase of this study involves to determines the CQAs by performing extensive literature review, studying the monograph GCSF. GCSF is a form of glycoprotein which helps in stimulation of bone marrow leading to the production of granulocyte which are thus released into the circulation system. As we discussed earlier, different CQAs of GCSF are product concentration, potency, host cell protein, DNA, endotoxin, purity, reduced GCSF, oxidized GCSF etc. CQAs effects the manufacturing of product design. They also have an impact on characterization of manufacturing drug substances [52].

After identification of CQAs of GCSF, the product design space was identified with the help of information available in the monograph of GCSF. Further the process design space was defined with multilevel approaches such as refolding step optimization, develop a high throughput process with the use of chromatography, evaluation and at last process integration. Information gained by this data was applied in commercial scale production without any verification. QbD based DOE was used to optimize the refolding process. Following that, risk analysis was conducted to determine the process parameter necessary for the experimental analysis and then various factors like temperature, refolding, pH, and urea, cysteine and dithiothreitol (DTT) and concentration of inclusion bodies were analyzed. It is pertinent to mention that FFrD was used for examination. Based on the results, pH, concentration of urea and DTT were further analyzed with the help of FFD to explain its impact on product quality and step yield interaction between them. Finally design space was established to obtain permissible range for CQAs. The CPPs illustrates the advantages of using the QbD methodology with its high step recovery and reasonably large design area. Process effectiveness is mandatory before implementation of design space at commercial level. After this purification of post refolding was started. According to an industry study, cation exchange chromatography and anion exchange chromatography were found to be the preferred modes for the commercially available GCSF technologies [53].

Identification of the control strategy was the last phase of this study. It was fully dependent on CQAs of the product such as oxidized, reduced and aggregated species which ultimately effect the downstream process. The quality of the product formed was analyzed and monitored by using analytical technique viz. refolding and chromatography process. The significance of consistent product quality under the QbD paradigm for the development of biosimilars was discussed in this case study. In order to ensure product comparability, a thorough analytical comparability exercise was carried out to compare the biosimilar GCSF product to the innovator product and other biosimilars on the market. The test showed that the GCSF biosimilar product and its original counterpart were very comparable in terms of product quality, supporting the decision to approve it for therapeutic use. The next phase was to compare the product with the reference product by the parameters like identification of product by using mass spectrophotometer, characterizing the host cell protein by using combination of 2D gel electrophoresis and UPLC. A similar examination of the quantity of high- and low-molecular-weight species was performed using SDS-PAGE (non-reducing) gel electrophoresis. The findings were discovered to be in concordance with those from size exclusion HPLC in that neither the commercial GCSF nor the biosimilar contained many high-molecular-weight species. The biosimilar GCSF was tested in vitro to determine its biological activity, and the results showed that it had an 89.3% higher biological potency than commercial GCSF. This closely fits the product's specification, which is between 80 and 20%, demonstrating that the biosimilar GCSF maintains functional integrity in addition to structural integrity [54].

15.5 Advantage of QbD to the Pharmaceutical Industry

- QbD encourages a greater knowledge of both the process and the product, which is one of its key advantages. The CPPs and CQAs of the pharmaceutical product must be recognized and understood, according to QbD. The development team acquires a greater understanding of how modifications to the procedure or product may impact its quality in this way. This knowledge reduces the possibility of product failures and unpredictability, and it may also lead to the creation of production procedures that are more effective, affordable with meagre batch failures [55].
- Another advantage of QbD is that it encourages a greater comprehension of the risk related to product development and production. The use of risk management tools and technology is emphasized by QbD to identify, assess, and mitigate potential risks throughout the PPDL. By identifying and addressing probable risks early on throughout the development process, this can minimize the risk of product failure or recall and ensure that pharmaceutical product is safely and effectively functioning for the purpose for which it is intended. This approach also helps to reduce expensive delays and increase the effectiveness and efficiency of the development process and accommodate the possibility of handling unexpected issues or problems.
- QbD helps in minimizing the variations in the production process to achieve consistency in quality of the product. The CPPs that most significantly affect the final pharmaceutical product quality should be determined and controlled as per QbD methodology. The development team may ensure that the product is produced in a consistent manner and adheres to the necessary quality requirements by constantly monitoring and managing these factors. This method aids in lowering manufacturing process variability and enhances the end product quality and consistency, leading to higher market share and improved customer satisfaction.
- QbD has the potential for reducing the number of post-approval submissions that regulatory bodies would appreciate and expect from the manufacturers. QbD places a strong emphasis on the necessity of incorporating quality in a product from the outset of the development procedure and of employing a risk-based strategy to detect and resolve any quality problems. By following these actions, the development team may lessen the chance that unanticipated quality problems will surface after the product has been certified and placed on the market. As a

result of this strategy, regulatory agencies may receive fewer post-approval filings, require less time and resources for both the development team and the regulatory bodies. This provides greater regulator confidence of robust products.

- The implementation of novel process validation techniques that are more effective and efficient than conventional validation techniques might be encouraged by QbD. The CPPs and CQAs of the product must be understood to design the manufacturing process such that these parameters may be controlled within acceptable limits. By keeping the process under control, manufacture can improve the process by continuous monitoring and uphold the desired quality criteria. This strategy may lead to a validation process that is more effective and efficient, saving time and money for both the development team and the regulatory bodies.
- The benefit of QbD is that it can result in more drug accessibility and fewer product recalls. QbD places a strong emphasis on adopting a risk-based approach to identify and manage possible quality concerns as well as the importance of incorporating quality into a product from the initial stages of its development. The development team may lessen the possibility that unanticipated quality problems would surface after the product has been authorized and placed on the market by performing these procedures. This strategy can lead to fewer drug recalls and supply disruptions, which helps ensure that the drug is more broadly accessible to people who need it. A more reliable and consistent medicine supply can also increase patient confidence in the brand and the product with timely launch of products.
- The concept of "right the first time and every time" is accomplished through several steps, including developing a design space and control techniques, recognizing CQAs and CPPs, and the use of PAT to monitor and real-time management of the process.
- One of the advantages of QbD involves a process of ongoing enhancement across all stages of the PPDL. Continuous improvement includes ongoing product and process examination, monitoring, and optimization with the aim of identifying and implementing changes that improve product quality, efficacy, and safety. This is accomplished through utilizing the concept of PAT, SPC, and FMEA as well as other risk management techniques.
- Real-time release (RTR) is a method for producing pharmaceuticals that permits the release of therapeutic products in real-time based on the constant observation and management of CQAs utilizing PAT. A benefit of implementing RTR through PAT is that cycle durations are shortened, processes are more effective, and product quality and consistency are raised. Additionally, it makes it possible for pharmaceutical firms to react swiftly to shifts in demand or interruptions in the supply chain, lowering the risk of drug shortages.
- Implementing QbD concepts and real-time release (RTR) into practice using PAT can result in considerable cost reductions and improved return on investment (ROI). QbD decreases the likelihood of product rejection and recall, boosts productivity and process effectiveness, and eliminates the need for expensive testing. Real-time monitoring and control made possible by RTR through PAT

shorten cycle durations and eliminate the requirement for end-product testing. Overall, QbD and RTR through PAT can enhance client happiness and brand reputation, resulting in longer-term increases in revenue and sales.

- Technology transfer is a critical process in the pharmaceutical industry that involves transferring knowledge and technology from one organization or site to another for the purpose of scaling up production or transferring a product from development to commercialization. The effectiveness of technology transfers may be increased in many of ways with the use of QbD principles. PAT makes it possible to monitor and regulate crucial process variables and quality characteristics in real-time, which eliminates the need for further testing and validation [56]. Overall, QbD and PAT implementation can cut down on the time and expense involved in technology transfer.

15.6 Conclusion

The application of QbD approaches, which can tackle a range of quality concerns and challenges, continues to rise in prevalence throughout the many stages of PPDL. The chemistry, manufacturing, and regulatory control processes may be transformed into a scientific and risk-based evaluation through the implementation of QbD principles. In this chapter, the applications and case studies of using QbD approach in various development fields such as API, product formulation, analytical method development, quality assurance, clinical trials, and packaging have been discussed in detail. QbD offers the collection of information in a systematic manner allowing the definition of a robust procedure that will continuously produce the specified quality of the product, reducing manufacturing costs as well as shorten development timelines. By implementing QbD throughout the entire PPDL, from early development phase to post-approval phase, manufacturers can better understand the CQAs of their products, optimize the manufacturing process, CPPs, developed a control strategy and make sure that the finished pharmaceutical product continuously satisfies the intended quality criteria. Advanced development of QbD promises to provide a high degree of quality assurance with a strong risk-based science as the basis of its foundation. Industry must work with the FDA to continue bridging the gap between conventional and advanced QbD based formulation research. Continued development of QbD frameworks and concepts, which can be used to design relevant QbD-centric laws and successful agency policy. However, the successful case studies of QbD in pharmaceutical development have demonstrated the efficiency of this strategy and its potential to revolutionize the industry. Therefore, incorporating QbD into all stages of PPDL should be a mandate for all pharmaceutical industry to stay and maintain competitive edge in the market and provide high-quality, safe, and effective product to the patients.

References

1. Rahman, M., Suman, S., & Beg, S. (2017). Pharmaceutical QbD: Omnipresence in the product development lifecycle. *European Pharmceutical Review* (1).
2. Sharma, N., Singh, S., Behl, T., Gupta, N., Gulia, R., & Kanojia, N. (2022). Explicating the applications of quality by design tools in optimization of microparticles and nanotechnology-based drug delivery systems. *Biointerface Research in Applied Chemistry, 12*(4), 4317–4336. https://doi.org/10.33263/BRIAC124.43174336
3. Ranjan, D., & Bodla, R. B. (2018). QbD: Key to quality pharmaceuticals. *Scholars Academic Journal of Pharmacy*. https://doi.org/10.21276/sajp.2018.7.6.8
4. *Pharmaceutical CGMPS for the 21 ST century-a risk-based approach final report*. 2004.
5. *International conference on harmonisation of technical requirements for registration of pharmaceuticals for human use pharmaceutical development Q8(R2)*. 2009.
6. *International conference on harmonisation of technical requirements for registration of pharmaceuticals for human use pharmaceutical quality system Q10*. 2009.
7. Aksu, B., Aydogan, M., Kanik, B., & Aksoy, E. (2013). A flexible regulatory approach for different raw materials suppliers using QbD principles. *Research Journal of Pharmaceutical, Biological and Chemical Sciences, 4*, 358–372.
8. Yu, L. X., et al. (2014). Understanding pharmaceutical quality by design. *AAPS Journal, 16*(4), 771–783. https://doi.org/10.1208/s12248-014-9598-3
9. Aksu, B., Sezer, A. D., Yegen, G., & Kuşçu, L. (2016). QbD implementation in biotechnological product development studies. In *Special topics in drug discovery*. INTECH. https://doi.org/10.5772/66296
10. Namjoshi, S., Dabbaghi, M., Roberts, M. S., Grice, J. E., & Mohammed, Y. (2020). Quality by design: Development of the quality target product profile (QTPP) for semisolid topical products. *Pharmaceutics, 12*(3), 287. https://doi.org/10.3390/pharmaceutics12030287
11. Csóka, I., Pallagi, E., & Paál, T. L. (2018). Extension of quality-by-design concept to the early development phase of pharmaceutical R&D processes. *Drug Discovery Today, 23*(7), 1340–1343. https://doi.org/10.1016/j.drudis.2018.03.012
12. Reich, G. (2005). Near-infrared spectroscopy and imaging: Basic principles and pharmaceutical applications. *Advanced Drug Delivery Reviews, 57*(8), 1109–1143. https://doi.org/10.1016/j.addr.2005.01.020
13. Patwardhan, D. M., Amrutkar, S. S., Kotwal, T. S., & Wagh, M. P. (2017). Application of quality by design to different aspects of pharmaceutical technologies. *International Journal of Pharmaceutical Sciences and Research, 8*(9), 3649–3662. https://doi.org/10.13040/IJPSR.0975-8232.8(9).3649-62
14. Gujral, G., Kapoor, D., & Jaimini, M. (2018). An updated review on design of experiment (DOE) in pharmaceuticals. *Journal of Drug Delivery and Therapeutics, 8*(3), 147–152. https://doi.org/10.22270/jddt.v8i3.1713
15. Jena, B. R., Panda, S. P., Umasankar, K., Swain, S., et al. (2020). Applications of QbD-based Software's in analytical research and development. *Current Pharmaceutical Analysis, 17*(4), 461–473.
16. Müller, A. L. H., De Oliveira, J. A., Prestes, O. D., Adaime, M. B., & Zanella, R. (2019). Design of experiments and method development. In *Solid-phase extraction* (pp. 589–608). https://doi.org/10.1016/B978-0-12-816906-3.00022-4
17. Butreddy, A., Bandari, S., & Repka, M. A. (2021). Quality-by-design in hot melt extrusion based amorphous solid dispersions: An industrial perspective on product development. *European Journal of Pharmaceutical Sciences, 158*, 105655. https://doi.org/10.1016/j.ejps.2020.105655
18. Step, "*Committee for Human Medicinal Products ICH guideline Q9 on quality risk management*," 2015. [Online]. Retrieved from www.ema.europa.eu/contact.

19. Zhang, L., & Mao, S. (2017). Application of quality by design in the current drug development. *Asian Journal of Pharmaceutical Sciences, 12*(1), 1–8. https://doi.org/10.1016/j.ajps.2016.07.006
20. Abhinandana, P., Abhinandana, P., & Nadendla, R. Application of quality by design and its parameters for pharmaceutical products drug-excipient compatability study view project implementing of analytical quality by design for high quality pharmaceutical products. *International Journal of Pharma and Chemical Research I, 3*(3) Retrieved from www.ijpacr.com.
21. Boussès, C., Ferey, L., Vedrines, E., & Gaudin, K. (2015). Using an innovative combination of quality-by-design and green analytical chemistry approaches for the development of a stability indicating UHPLC method in pharmaceutical products. *Journal of Pharmaceutical and Biomedical Analysis, 115*, 114–122. https://doi.org/10.1016/j.jpba.2015.07.003
22. Rapalli, V. K., et al. (2021). QbD-driven formulation development and evaluation of topical hydrogel containing ketoconazole loaded cubosomes. *Materials Science and Engineering C, 119*, 111548. https://doi.org/10.1016/j.msec.2020.111548
23. Liliana, L. (2016). A new model of Ishikawa diagram for quality assessment. In *IOP Conference Series: Materials Science and Engineering*. Institute of Physics Publishing. https://doi.org/10.1088/1757-899X/161/1/012099
24. Ende, D., et al. (2007). API quality by design example from the torcetrapib manufacturing process. *Journal of Pharmaceutical Innovation, 2*(3–4), 71–86. https://doi.org/10.1007/s12247-007-9015-x
25. Ghodake, V., Vishwakarma, J., Vavilala, S. L., & Patravale, V. (2020). Cefoperazone sodium liposomal formulation to mitigate *P. Aeruginosa* biofilm in cystic fibrosis infection: A QbD approach. *International Journal of Pharmaceutics, 587*, 119696. https://doi.org/10.1016/j.ijpharm.2020.119696
26. Pallagi, E., Jójárt-Laczkovich, O., Németh, Z., Szabó-Révész, P., & Csóka, I. (2019). Application of the QbD-based approach in the early development of liposomes for nasal administration. *International Journal of Pharmaceutics, 562*, 11–22. https://doi.org/10.1016/j.ijpharm.2019.03.021
27. Patil, H., Feng, X., Ye, X., Majumdar, S., & Repka, M. A. (2015). Continuous production of Fenofibrate solid lipid nanoparticles by hot-melt extrusion technology: A systematic study based on a quality by design approach. *AAPS Journal, 17*(1), 194–205. https://doi.org/10.1208/s12248-014-9674-8
28. Shah, B., Khunt, D., Bhatt, H., Misra, M., & Padh, H. (2015). Application of quality by design approach for intranasal delivery of rivastigmine loaded solid lipid nanoparticles: Effect on formulation and characterization parameters. *European Journal of Pharmaceutical Sciences, 78*, 54–66. https://doi.org/10.1016/j.ejps.2015.07.002
29. Visser, J. C., Dohmen, W. M. C., Hinrichs, W. L. J., Breitkreutz, J., Frijlink, H. W., & Woerdenbag, H. J. (2015). Quality by design approach for optimizing the formulation and physical properties of extemporaneously prepared orodispersible films. *International Journal of Pharmaceutics, 485*(1–2), 70–76. https://doi.org/10.1016/j.ijpharm.2015.03.005
30. Lohan, S., et al. (2021). QbD-steered development of mixed nanomicelles of galantamine: Demonstration of enhanced brain uptake, prolonged systemic retention and improved biopharmaceutical attributes. *International Journal of Pharmaceutics, 600*, 120482. https://doi.org/10.1016/j.ijpharm.2021.120482
31. Mahmood, A., Rapalli, V. K., Waghule, T., Gorantla, S., & Singhvi, G. (2021). Luliconazole loaded lyotropic liquid crystalline nanoparticles for topical delivery: QbD driven optimization, in-vitro characterization and dermatokinetic assessment. *Chemistry and Physics of Lipids, 234*, 105028. https://doi.org/10.1016/j.chemphyslip.2020.105028
32. Sreeharsha, N., et al. (2020). Development of metronidazole loaded chitosan nanoparticles using QBD approach—A novel and potential antibacterial formulation. *Pharmaceutics, 12*(10), 1–22. https://doi.org/10.3390/pharmaceutics12100920

33. Dara, T., et al. (2019). Erythropoietin-loaded solid lipid nanoparticles: Preparation, optimization, and in vivo evaluation. *Colloids and Surfaces. B, Biointerfaces, 178*, 307–316. https://doi.org/10.1016/j.colsurfb.2019.01.027
34. El-Say, K. M., & Hosny, K. M. (2018). Optimization of carvedilol solid lipid nanoparticles: An approach to control the release and enhance the oral bioavailability on rabbits. *PLoS One, 13*(8), e0203405. https://doi.org/10.1371/journal.pone.0203405
35. Sarma, A., & Das, M. K. (2019). Formulation by design (FbD) approach to develop Tenofovir Disoproxil fumarate loaded nanostructured lipid carriers (NLCs) for the aptness of nose to brain delivery. *Journal of Drug Delivery and Therapeutics, 9*(2), 148–159. https://doi.org/10.22270/jddt.v9i2.2391
36. Ranga, S., Jaimini, M., Sharma, S. K., Chauhan, B. S., & Kumar, A. (2014). A review on Design OF Experiments (DOE). *International Journal of Pharmaceutical and Chemical Sciences, 3*, 216–224. Retrieved from www.ijpcsonline.com
37. Jain, A., Sharma, T., Sharma, G., Khurana, R. K., Katare, O. P., & Singh, B. (2019). QbD-driven analytical method development and validation for Raloxifene hydrochloride in pure drug and solid oral dosage form. *Analytical Chemistry Letters, 9*(4), 463–477. https://doi.org/10.1080/22297928.2019.1624193
38. *Catalent Pharma Solutions*. 2014. Retrieved from www.ondrugdelivery.com.
39. Singh, B. N. Product development, manufacturing, and packaging of solid dosage forms under QbD and PAT paradigm: DOE case studies for industrial applications. *AAPS PharmSciTech, 20*(8), 313. https://doi.org/10.1208/s12249-019-1515-8
40. Farag Badawy, S. I., Gawronski, A. J., & Alvarez, F. J. (2001). Application of sorption-desorption moisture transfer modeling to the study of chemical stability of a moisture sensitive drug product in different packaging configurations. *International Journal of Pharmaceutics, 223*, 1–13. Retrieved from www.elsevier.com/locate/ijpharm
41. *ICH-E6 Good Clinical Practice (GCP)*. [Online]. Retrieved from https://ich.org/page/formal-ich-procedure.
42. *E8(R1) General considerations for clinical studies guidance for industry*. 2022. Retrieved from https://www.fda.gov/vaccines-blood-biologics/guidance-compliance-regulatory-information-biologics/biologics-guidances.
43. Ioannidis, J. P. A. (2016). Why Most clinical research is not useful. *PLoS Medicine, 13*(6), e1002049. https://doi.org/10.1371/journal.pmed.1002049
44. Bernard, G. R., et al. (2018). A collaborative, academic approach to optimizing the national clinical research infrastructure: The first year of the trial innovation network. *Journal of Clinical and Translational Science, 2*(4), 187–192. https://doi.org/10.1017/cts.2018.319
45. Walters, S. J., et al. (2017). Recruitment and retention of participants in randomised controlled trials: A review of trials funded and published by the United Kingdom Health Technology Assessment Programme. *BMJ Open, 7*, 15276. https://doi.org/10.1136/bmjopen-2016
46. Moradi, H., Schneider, M., Streja, E., & Cooper, D. (2021). Feasibility and acceptability of a structured quality by design approach to enhancing the rigor of clinical studies at an academic health center. *Journal of Clinical and Translational Science, 5*(1), 535–541. https://doi.org/10.1017/cts.2021.837
47. Selker, H. P. (2017). The continued evolution of team science. *Journal of Clinical and Translational Science, 1*(1), 7–7. https://doi.org/10.1017/cts.2017.4
48. Patrick-Lake, B. (2018). Patient engagement in clinical trials: The clinical trials transformation initiative's leadership from theory to practical implementation. *Clinical Trials, 15*(1_suppl), 19–22. https://doi.org/10.1177/1740774518755055
49. Daudelin, D. H., et al. (2020). Implementing common metrics across the NIH clinical and translational science awards (CTSA) consortium. *Journal of Clinical and Translational Science, 4*(1), 16–21. https://doi.org/10.1017/cts.2019.425
50. Böhl, O. J., Schellenberg, J., Bahnemann, J., Hitzmann, B., Scheper, T., & Solle, D. (2021). Implementation of QbD strategies in the inoculum expansion of a mAb production process. *Engineering in Life Sciences, 21*(3–4), 196–207. https://doi.org/10.1002/elsc.202000056

51. Finkler, C., & Krummen, L. (2016). Introduction to the application of QbD principles for the development of monoclonal antibodies. *Biologicals, 44*(5), 282–290. https://doi.org/10.1016/j.biologicals.2016.07.004
52. Brot, N. (1991). Biochemistry of methionine sulfoxide residues in proteins.3(2):91-6. PMID: 1910456. *BioFactors, 3*(2), 91–96.
53. Rapalli, V. K., Khosa, A., Singhvi, G., Girdhar, V., Jain, R., & Dubey, S. K. (2019). Application of QbD principles in nanocarrier-based drug delivery systems. In *Pharmaceutical quality by design: principles and applications* (pp. 255–296). Elsevier. https://doi.org/10.1016/B978-0-12-815799-2.00014-9
54. Rathore, A. S. (2016). Quality by Design (QbD)-based process development for purification of a biotherapeutic. *Trends in Biotechnology, 34*(5), 358–370. https://doi.org/10.1016/j.tibtech.2016.01.003
55. Kadam, V. R., Patil, M. P., Pawar, V. V., & Kshirsagar, S. (2017). A review on: Quality by Design (QbD). *Asian Journal of Research in Pharmaceutical Sciences, 7*(4), 197–204.
56. Waghule, T., Dabholkar, N., Gorantla, S., Rapalli, V. K., Saha, R. N., & Singhvi, G. (2021). Quality by design (QbD) in the formulation and optimization of liquid crystalline nanoparticles (LCNPs): A risk based industrial approach. *Biomedicine and Pharmacotherapy, 141*, 111940. https://doi.org/10.1016/j.biopha.2021.111940

Chapter 16
Quality by Design and Marketing

Virender Kumar, Suchitra Nishal, Parmita Phougat, Vandana Garg, and Harish Dureja

Abstract Critical scientific, technological, and regulatory constraints in pharmaceuticals' study, development, and marketing must be addressed using a more comprehensive, systematic, integrated science and risk-based approach, such as Quality by Design (QbD). Consistent, dependable, and high-quality pharmaceutical goods are ensured by a greater understanding of the product and process of those results from applying the QbD approach. Additionally, this strategy encourages innovation and ongoing product improvement across the whole product lifecycle. Regulatory agencies strongly advise QbD-based submission for the effective transition of pharmaceuticals from laboratory-scale research to the pharmaceutical market with the necessary quality, safety, and efficacy standards. Two crucial steps in generating customer value are identifying quality standards through relationship quality management and transitioning from marketing innovation. Customer value creation is the goal of quality management, according to QbD tools. The current chapter deals with a complete understanding of QbD tools and their role in the management and advancement of the marketing of pharmaceuticals in each category.

V. Kumar (✉)
College of Pharmacy, Pandit Bhagwat Dayal Sharma University of Health Sciences, Rohtak, Haryana, India

Department of Pharmaceutical Sciences, Maharishi Dayanand University, Rohtak, Haryana, India
e-mail: virenderkumar.copharmacy@uhsr.ac.in

S. Nishal
College of Pharmacy, Pandit Bhagwat Dayal Sharma University of Health Sciences, Rohtak, Haryana, India

P. Phougat
School of Medical and Sciences, G D Goenka University, Gurugram, Haryana, India

V. Garg · H. Dureja
Department of Pharmaceutical Sciences, Maharishi Dayanand University, Rohtak, Haryana, India

N. K. Jain, N. Bajwa (eds.), *Introduction to Quality by Design (QbD)*,
https://doi.org/10.1007/978-981-99-8034-5_16

Keywords Quality by design · Quality product · Regulatory · Marketing · Management

16.1 Introduction

QbD refers to the development of products and processes that meet predefined quality goals by designing and developing them accordingly. Increasing product quality and reducing production costs have made it a significant topic of interest in the pharmaceutical industry. Additionally, QbD has been successful in improving the quality of products and services in other industries, including food, cosmetics, and services. We discuss the benefits, challenges, and future directions of QbD in marketing in this article. QbD is increasingly being used by companies to develop products. With QbD, quality is designed into products prior to production rather than testing and inspecting them afterward. It is essential to understand customer preferences and needs to develop high-quality products. Recalls and other quality problems can be reduced through Quality by Design [1–3]. When products are developed using a QbD approach, marketing plays an important role in their success. It is possible for companies to build a strong reputation for providing high-quality and reliable products by promoting the quality and reliability of those products. Customers can also be educated about the benefits of QbD through marketing and why it is important when choosing products. To design successful products, companies benefit from QbD by making them more likely to succeed. It is possible for companies to reduce costs and delays associated with non-compliant products by developing products that meet regulatory standards from the start. The product development process can be improved by reducing risk and increasing efficiency. QbD benefits can be promoted to customers through marketing. A company can differentiate itself from competitors and build a reputation for delivering high-quality products by highlighting their quality and reliability. Customers can also be educated about QbD benefits and why it is important to consider when choosing a product through marketing [4–8].

16.2 Principles of QbD

To guide product development and marketing, companies can use several QbD principles Following are some of the principles of QbD (Fig. 16.1).

- Designing products with the end user in mind: A company's products are more likely to succeed in the marketplace if they are designed with the end user in mind. In addition to increasing customer satisfaction, this can also increase sales and profitability [9, 10].
- Incorporating customer feedback into the product development process: A company can ensure that its products are meeting the needs and expectations of its

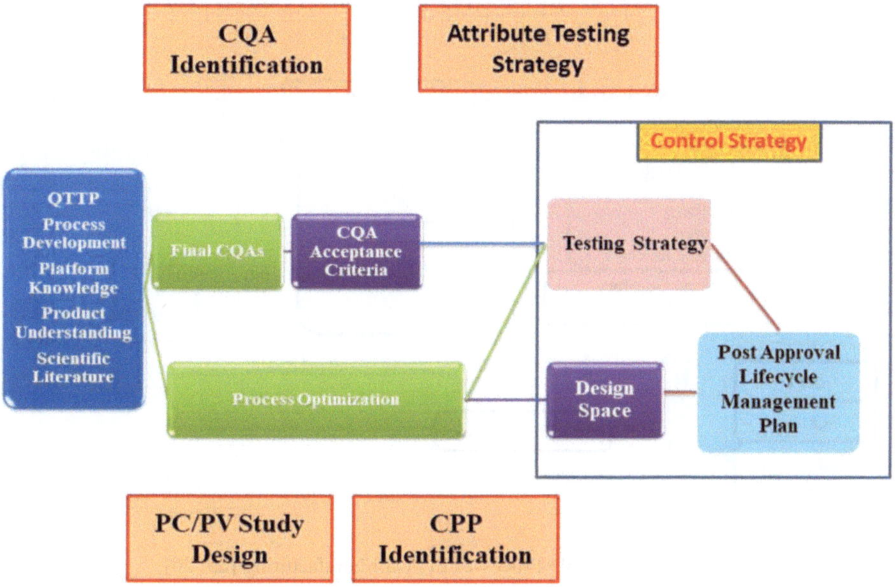

Fig. 16.1 Principles of Quality by Design [19–22]

target markets by soliciting feedback from customers throughout the product development process. Eventually, this can make the product more reliable and of higher quality [11, 12].

- Using data and analytics to inform decision-making: Using data analysis tools and analytics tools can help companies develop and market products more effectively. Over time, this can result in significant cost savings due to the reduction of waste and improvement of efficiency [13–15].
- Collaborating across functional areas: To achieve QbD, researchers and engineers must coordinate with manufacturing, marketing, and quality assurance teams. Identifying and resolving quality issues is easier when teams work together. Quality by Design (QbD) is an approach to product development that emphasizes the importance of designing quality into a product from the outset. In addition to complying with regulatory requirements, this approach helps companies avoid costly recalls and other quality problems. The marketing department plays a crucial role in the success of products developed with a QbD approach by promoting their quality and reliability and educating consumers about their advantages [14, 16–18].

Marketing and design are the cornerstones of successful businesses. Designing and developing products is made more efficient by understanding the relationship between key processes and critical characteristics. A product, process, and system solution must be designed to achieve optimal quality. Understanding the needs of the customer, the product requirements, as well as the constraints of the design process are essential. Marketing plays a crucial role in the success of any product. Customer

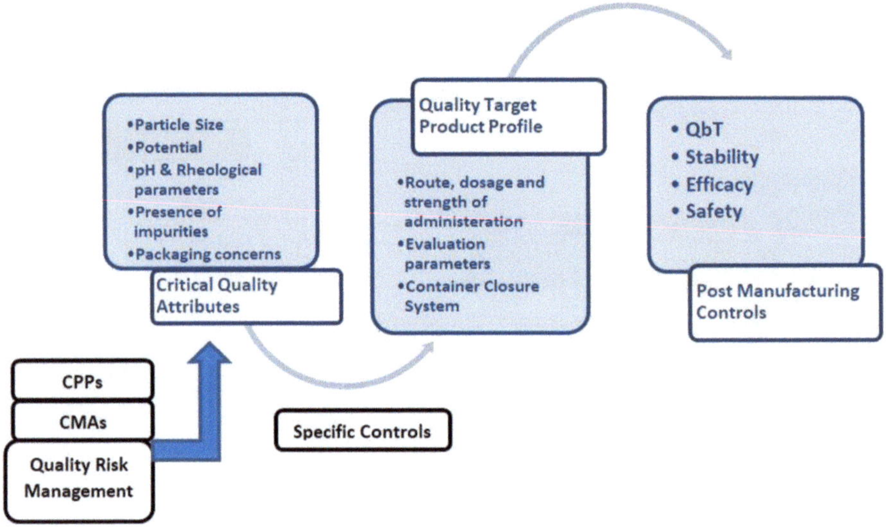

Fig. 16.2 Depicting the role of QbD Processes in Post Manufacturing [25, 26]

needs and wants must be met to make your product successful, followed by effective marketing. To create an effective marketing plan, you must know your target audience, plan how to connect with them, and know how to reach them. It is important to analyze the market, your customers, and your competitors before developing a marketing plan. It is also important to choose an appropriate pricing and promotion strategy. Quality by design is the result of collaboration between marketers and designers. The purpose of marketing is to promote a product in an effective manner. Our Quality by Design process ensures that our products meet the needs and wishes of our customers. The implementation of Quality by Design and Marketing should go hand in hand to ensure that the product is successful [10, 23, 24]. QbD processes exhibits an essential role in post manufacturing (Fig. 16.2).

16.2.1 Using QbD in Marketing Can Look Like This

- Developing unique selling propositions: Developing unique selling propositions can be facilitated by Quality by Design. By designing products from the outset with quality in mind, a company can differentiate itself from its competitors and communicate its value to customers.
- Reducing costs and increasing efficiency: QbD can increase marketing efficiency and reduce costs as well. Moreover, companies can maintain their brand reputation when they design products with quality in mind from the start. Recalls and product failures can damage a company's reputation, impacting its revenue and profits.

- Improve customer satisfaction: By designing products that meet quality and performance specifications, companies can meet or exceed customer expectations. Quality should be built into products from the start if companies want to increase customer loyalty [27–30].

16.3 What Is Quality by Design?

QbD is a systematic approach to product and process development that starts with a clear understanding of the desired quality attributes. It involves identifying critical quality attributes (CQAs) and critical process parameters (CPPs) that can affect the quality of the product or process (Fig. 16.3). QbD also emphasizes the use of statistical tools and risk management strategies to ensure that the product or process consistently meets the predefined quality goals [11, 26, 31, 32].

Key Elements of QbD in Marketing (Fig. 16.4)

- *Customer Focus*: Understanding customer preferences and needs is central to QbD in marketing. Researching the market, analyzing customer data, and incorporating feedback from customers is essential to developing marketing strategies. It is possible for companies to ensure that their products and services meet the needs and expectations of their customers by incorporating their input into the marketing process.

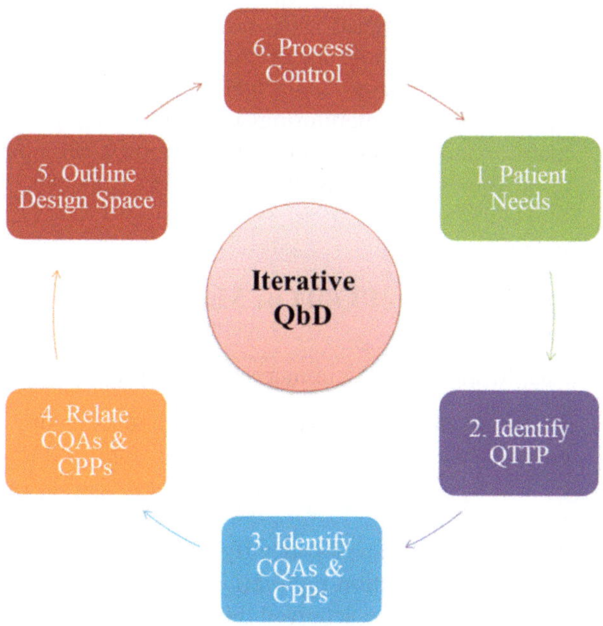

Fig. 16.3 Describing Cycle for Developing QbD Paradigm [33, 34]

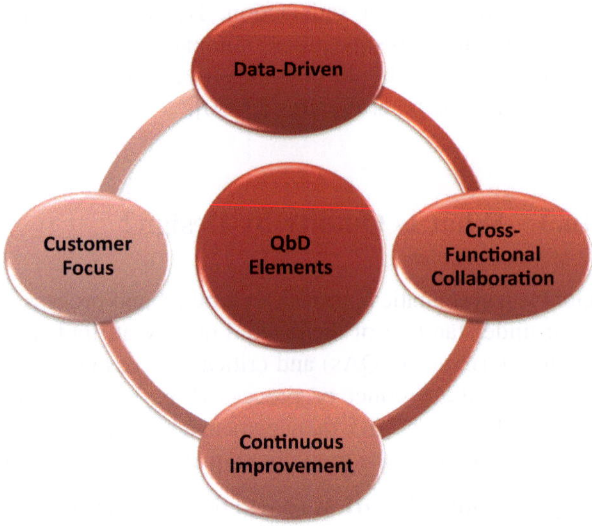

Fig. 16.4 Overview of QbD Elements

- *Data-Driven*: Marketing decisions are informed by data and analytics in QbD. Analyzing market trends, tracking customer behavior, and using predictive analytics are all parts of this process. Companies can improve marketing effectiveness by leveraging data to make more informed decisions.
- *Cross-Functional Collaboration*: QbD in marketing requires collaboration across different functional areas, including marketing, sales, product development, and customer service. By working together, these teams can ensure that marketing strategies are aligned with business objectives and that customer needs are addressed throughout the product lifecycle.
- *Continuous Improvement*: QbD in marketing is a continuous process that involves ongoing monitoring and evaluation. By tracking marketing performance metrics and analyzing customer feedback, companies can identify areas for improvement and adjust their marketing strategies [35, 36]

16.4 Benefits of QbD in Marketing

The application of QbD in marketing can provide several benefits, including:

- *Improved customer satisfaction*: By understanding customer needs and preferences, QbD can help companies develop products and services that meet or exceed customer expectations. Customer loyalty and satisfaction are likely to increase as a result.
- *Increased efficiency*: By identifying and optimizing key process parameters, QbD can make production more efficient. Reducing production costs and increasing profitability can be achieved this way.

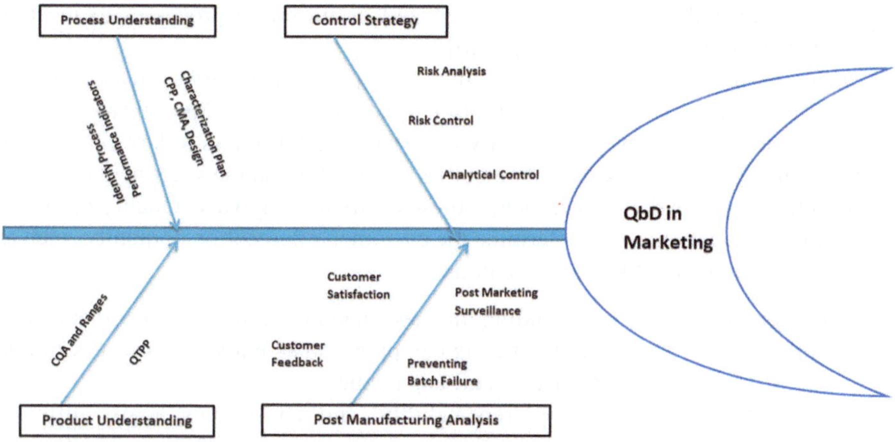

Fig. 16.5 Ishikawa-Fish Bone Diagram Depicting Relation among QbD and Marketing [30, 39]

- *Faster time-to-market*: By identifying potential issues early in the development process, QbD can help companies reduce development time and launch costs.
- *Better risk management*: Identifying potential risks and developing mitigation plans are important aspects of QbD, such as failure mode and effects analysis (FMEA). In this way, companies can protect their reputations and avoid costly product recalls.

In marketing, products, and services must meet customers' needs, the value proposition must be communicated to customers, and the needs of the customer must be understood. A key component of the marketing process is quality, which affects customer perception and decision-making. The relation among QbD and marketing can best be explained with the help of Ishikawa-Fish Bone diagram (Fig. 16.5) [37, 38].

Marketers can benefit from QbD by designing and implementing a quality marketing process that focuses on customer value and continuous improvement. Following are the steps involved in the QbD approach to marketing:

- *Define the target customer*: Identifying the target customer and their needs is the first step in QbD in marketing. In marketing, Quality by Design (QbD) targets organizations that want to provide their customers with high-quality products or services. Marketers include organizations that engage in marketing activities, such as advertising, branding, market research, and engagement with customers, as their target customers. Consumer goods, healthcare, technology, finance, and retail are among the industries where these organizations operate. Typically, QbD in marketing targets organizations that value customer satisfaction and have a continuous improvement focus. Ensure quality is a critical component of compliance and patient safety for organizations operating in highly regulated industries, such as pharmaceuticals and medical devices. Customers of QbD in marketing are generally organizations that are conscious of the importance of quality and willing to invest in design and optimization processes to meet or exceed customer expectations [27, 40].

- *Define the quality attributes*: Customer perception of quality is determined by quality attributes, such as reliability, functionality, safety, and usability. For high-quality products or services, Quality by Design (QbD) refers to key characteristics of marketing processes and outcomes. Depending on the specific context and objectives of a marketing campaign, these attributes may vary: Quality attributes are the characteristics that determine the customer's perception of quality, such as reliability, functionality, safety, and usability.
  - *Customer Focus*: Understanding the needs and preferences of customers while designing and executing marketing campaigns. The marketing process should focus on customer satisfaction and engagement.
  - *Consistency*: Predictability, repeatability, and consistency across different contexts and channels in marketing processes and outcomes. Continuous improvement, quality control, and standardization are all addressed here.
  - *Clarity*: A marketing message's clarity, conciseness, and ease of comprehension. Ensure that the messaging and branding are relevant, consistent, and engaging for the target audience.
  - *Creativity*: A marketing effort that is innovative, original, and differentiates itself from those of its competitors. Developing new and effective marketing strategies involves creative thinking, problem-solving, and experimentation.
  - *Compliance*: Compliance with legal, regulatory, and ethical standards regarding marketing efforts. Marketing practices should focus on transparency, accuracy, and accountability.
  - *Effectiveness*: Measures the success of a marketing campaign through the achievement of its intended goals and the production of a positive return on investment. A key performance indicator and business objective should be used to measure and evaluate marketing performance. To consistently deliver high-quality products or services, organizations can focus on these quality attributes and optimize marketing processes accordingly. A better customer experience, increased engagement, and improved business results can be achieved as a result [13, 36].

- *Design the process*: The marketing process must be designed to ensure that the product or service meets the target values once the quality attributes are defined. It is important that the process is designed to meet the critical quality attributes consistently and to control any variability. By using Quality by Design (QbD) in marketing, marketing processes are designed, optimized, and validated in such a way that they consistently satisfy or exceed customer expectations. QbD in marketing involves the following steps:
  - *Identify Critical Quality Attributes (CQAs)*: A high-quality product or service is dependent on identifying the critical quality attributes that pertain to the marketing process and outcomes. The key to identifying your customers' key

engagement and satisfaction drivers is to understand their needs and preferences.

  - *Design and Optimize* Marketing *Processes*: Marketing processes must be designed and optimized to consistently deliver the identified CQAs. Customer insights, data, and analytics are used to create customer-focused marketing processes. Through testing and experimentation, you can find out whether a marketing strategy is effective.
  - *Validate Marketing Processes*: To validate marketing processes, they need to be designed and optimized, followed by testing and verification. Marketing processes are tested under different conditions and statistical methods are used to ensure they consistently deliver the identified CQAs.
  - *Monitor and Control*: It is crucial that the marketing processes are monitored and controlled over time to maintain the identified CQAs. The use of data, analytics, and key performance indicators is essential to maintaining quality and customer satisfaction.
  - *Continuous Improvement*: A final component of the QbD process is continuous improvement to maintain the effectiveness and efficiency of the marketing processes over time. The marketing process must be iterated to optimize performance using feedback, data, and analytics. Using QbD in marketing from a systematic perspective allows organizations to design and optimize marketing processes that consistently deliver products and services of the highest quality. Customers are likely to be more satisfied, engagement is likely to be higher, and business outcomes will be better as a result [41].

- *Implement the process*: By applying the Quality by Design process to marketing, products and services will consistently meet or exceed customer expectations, based on the designed and validated marketing processes. To implement the QbD process in marketing, the following steps need to be taken:

  - *Training and Education*: The first step to implementing QbD in marketing is to provide training and education on the designed and optimized marketing processes. To achieve this, you will need to educate them about the marketing campaign's goals, quality attributes, and expected results. During this training program, employees learn how to deliver high-quality products and services to customers.
  - *Standardization*: To ensure consistency and quality across different teams, departments, and channels, the next step is to standardize the marketing processes. Creating standard operating procedures (SOPs) is one way to ensure the marketing processes are carried out correctly, quality standards are met, and the expected results are achieved. As a result of this step, marketing processes are executed consistently across the organization.
  - *Implementation*: The next step is to implement QbD in marketing once the training and standardization steps have been completed. In this case, marketing processes must be executed according to the designed and validated processes and monitored and controlled to ensure that high-quality products are being delivered.

  - *Performance Monitoring*: To ensure that the marketing processes continue to meet the identified quality attributes, the next step is to monitor their performance. To maintain quality and customer satisfaction, key performance indicators (KPIs) need to be used to track and analyze marketing processes.
  - *Continuous Improvement*: Finally, QbD in marketing involves continuous improvement to maintain the efficiency and effectiveness of marketing processes. By using feedback, data, and analytics, you can identify areas for improvement and make iterative changes to your marketing processes to improve results. This process enables organizations to deliver high-quality products or services consistently, exceeding customer expectations by implementing the QbD process in marketing. Customers will be more satisfied, engagement will be higher, and business results will be better as a result [5, 30].

- *Continuously improve*: A key component of Quality by Design (QbD) in marketing is continuous improvement. In this process, ongoing efforts are made to ensure that marketing processes are continuously optimized so that high-quality products and services meet or exceed customer expectations. To continuously improve the Quality by Design process in marketing, the following steps must be taken:

  - *Review and Analysis*: Reviewing and analyzing the marketing processes is the first step in continuous improvement. Data, analytics, and customer and employee feedback are used to identify areas for improvement. Marketing processes may be evaluated according to critical quality attributes and benchmarked against industry standards during the review process.
  - *Root Cause Analysis*: Once areas for improvement are identified, the next step is to conduct root cause analysis to identify the underlying causes of the issues. This involves using problem-solving techniques to identify the root causes of the issues and developing solutions to address them.
  - *Implement Changes*: To address these issues, changes need to be made to the marketing processes once solutions have been identified. Standard operating procedures may need to be adjusted, marketing strategies may need to be updated, or new tools or technologies may need to be utilized to improve performance.
  - *Test and Verify*: The next step after implementing changes is to test and verify that they address the issues. To assure that the changes are not detrimental to the critical quality attributes and that they produce the expected results, testing and experimentation may be necessary.

- *Monitor and Control*: Monitor and control marketing processes are the final steps in continuous improvement. To maintain high-quality products and services, the changes must be effective. Tracking performance and adjusting as needed requires data, analytics, and key performance indicators.

A continuous improvement of the QbD process in marketing can help organizations continue to deliver high-quality products or services that meet or exceed customer expectations. This can lead to improved customer satisfaction, increased engagement, and better business outcomes [41, 42].

16.5 Challenges of QbD in Marketing

In addition to its potential benefits, QbD can also present several challenges, including:

16.5.1 Investment in Time and Resources

QbD principles can be time-consuming and resource-intensive to implement in marketing. When implemented correctly, QbD can provide enhanced customer satisfaction, reduce risk, and increase profitability. Implementing QbD in marketing requires the following resources and time:

1. Training and Education: As part of QbD implementation, staff members must be trained and educated in QbD philosophy and tools, as well as in marketing theory and practice. Various types of training are available, such as learning how to analyze customer needs, develop products, optimize processes, maintain quality control, manage risks, and improve continuously.
2. Customer Research: Customer needs and preferences must be deeply understood for QbD to be successful. Surveys, focus groups, and market research can be conducted to gather information about customer needs and preferences.
3. Product Design: Quality by Design is about designing products that are optimally performing and meet customer needs. To achieve this, it is necessary to invest in research and development, prototyping, and testing.
4. Process Optimization: By optimizing marketing processes, QbD reduces waste and improves efficiency. To accomplish this, tools, and technologies for mapping processes, improving processes, and monitoring processes will be needed.
5. Quality Control: From product design to customer satisfaction, QbD requires a focus on quality. In addition to investing in quality control tools and technologies, employees can be trained and educated about quality.
6. Continuous Improvement: It is essential that QbD is committed to continuous improvement and learning from customer feedback, market research, and other sources of data. Investments can be made in tools that collect and analyze data, such as data analytics tools and customer feedback systems.

A company's level of commitment to QbD practices and principles will impact the investment needed for QbD implementation in marketing. Many marketing organizations can benefit from QbD implementation in the long run through improved customer satisfaction, reduced risks, and increased profitability [43, 44].

16.5.2 Data Availability

For a quality-oriented marketing approach based on Quality by Design (QbD), data availability is crucial to supporting decision-making and continuous improvement. The implementation of QbD in marketing poses several challenges in terms of data availability. The following are a few of the challenges we face:

1. Data Fragmentation: It is difficult to access and integrate marketing data when it is fragmented across a wide range of sources. The data may reside in a variety of systems, including customer relationship management platforms (CRM), sales platforms, and marketing automation platforms, as well as in external sources, such as social media, web analytics, and third-party sources. Customer and marketing processes can become fragmented because of this fragmentation.
2. Data Quality: High-quality data is essential for QbD to support continuous improvement and decision-making. The analysis and decision-making process can be undermined by incomplete, inaccurate, or inconsistent marketing data. Cleaning, validating, and integrating data can take a lot of time and resources.
3. Data Privacy: Buying behavior data, personal identification data, and product usage data can be found in marketing data. It can be challenging to access and analyze marketing data while ensuring compliance with privacy regulations like GDPR and CCPA.
4. Data Access: Depending on organizational policies, data silos, and individual job roles, access to marketing data may be restricted. Customer data, for instance, may only be accessible to customer service representatives, while sales data may only be accessible to the sales team. Obtaining a complete picture of the customer journey and marketing process can be challenging due to this.
5. Data Analysis: Analyzing data is required to identify patterns, trends, and opportunities for continual improvement during QbD implementation. It can be time consuming and difficult to analyze marketing data, as it requires specialized skills and tools. Investment in tools for data visualization, data modeling, and statistical analysis can be particularly challenging when analyzing data from multiple sources and formats. QbD implementation in marketing can pose significant challenges related to data availability. A commitment to data-driven decision making and continuous improvement efforts is necessary to address these challenges.

16.5.3 Organizational Culture

Quality by Design (QbD) principles in marketing are successfully implemented when organizational culture is strong. Creating and maintaining an organizational culture focused on QbD can, however, be challenging. There are several challenges to overcome, including:

1. Resistance to Change: As a result of QbD implementation, traditional marketing practices need to be changed, resulting in resistance from employees. People may resist change because of a fear of the unknown, lack of understanding of the benefits of QbD, or because they believe that current methods are effective.
2. Siloed Departments: Teams that work independently in marketing departments without collaborating with colleagues from other departments can cause silos. As a result, there may not be a shared vision and communication breakdowns, which may hinder the implementation of QbD.
3. Lack of Training: Analyzing data, optimizing processes, and pursuing continuous improvement are all necessary for understanding customer needs. QbD implementation may be hindered by employees' lack of training. Training in these areas can be time consuming and costly, making it challenging to build a skilled workforce.
4. Lack of Leadership Support: Leadership commitment and support are essential for successful QbD implementation in marketing. The opposition to change, lack of understanding of QbD principles, and conflicting priorities may hinder leaders' adoption of QbD. As a result, QbD implementation efforts may not receive the support they need.
5. Reward Systems: Incentives for certain behaviors and outcomes often influence organizational culture. It is highly likely that employees will not prioritize QbD implementation efforts if reward systems are not aligned with QbD principles and practices. This could result in low levels of commitment and enthusiasm for QbD on the part of employees.

A QbD implementation in marketing can face significant challenges related to organizational culture. A QbD-oriented incentive and reward system, along with leadership support and commitment, will help to tackle these challenges. The culture of collaboration and communication needs to be created, silos need to be broken down, and a skilled and empowered workforce must be developed.

16.5.4 Lack of Regulatory Requirements

Regulatory requirements and guidelines that apply specifically to marketing activities make it challenging to implement Quality by Design (QbD) principles in marketing. The QbD method is not mandated by regulatory bodies or guidelines like it is in other industries, including pharmaceuticals and medical devices. Regulatory requirements are lacking, which can lead to several challenges, such as:

1. Lack of Standards: Due to the lack of regulatory requirements and guidelines, QbD implementation in marketing has no established standards. Because of this lack of consistency, organizations may not implement QbD principles as quickly as they should.
2. Resistance to Change: The lack of clear regulatory requirements or guidelines may discourage organizations from investing time and resources in QbD

implementation. A lack of commitment from employees and leadership can result from this.
3. Limited Resources: It may be difficult for organizations to allocate enough resources to QbD implementation initiatives if there is no regulatory requirement, resulting in a lack of focus and a limited amount of progress toward implementing QbD principles.
4. Limited Benefits: QbD implementation in marketing may not benefit organizations without regulatory requirements, resulting in improved customer satisfaction, increased efficiency, and reduced costs. As a result, QbD principles may not be implemented with motivation.

Despite these challenges, organizations can still benefit from QbD principles in marketing. QbD can improve marketing campaigns, engage customers, and improve results through more efficient and effective campaigns. As a result, organizations are able to market themselves as quality-driven organizations that are committed to continuous improvement.

By establishing best practices, standards, and guidelines, organizations can overcome the lack of regulatory requirements. Training and education can also help build a skilled and knowledgeable workforce. Organizations can benefit from QbD implementation even if there are no specific regulatory requirements [28, 45].

16.6 Future Research Directions

Companies can benefit substantially from QbD in marketing, despite the challenges it faces. To improve QbD implementation in marketing, further research is needed in a variety of areas, including:

16.6.1 Importance of Customer-Centric Marketing

In marketing, Quality by Design (QbD) emphasizes customer-centricity. Quality by Design emphasizes meeting customer needs and preferences right from the start in marketing efforts. The goal of customer-centric marketing is to design marketing strategies that are focused on the customer in order to deliver optimal value to the customer. A customer-centric approach to marketing is integral to QbD. Focusing on the needs and preferences of customers can help a company develop more effective marketing strategies. By doing so, you will gain satisfied customers and increase the likelihood of repeat business. By examining customer feedback, market research, and social media monitoring, companies can gain an understanding of their target customers' needs and preferences. Marketing strategies can be tailored based on the needs and preferences of the target market by analyzing this data. This can include developing targeted messaging, personalized offers, and customized marketing campaigns. Another key benefit of customer-centric marketing in QbD is that it can help companies differentiate themselves from their competitors. A targeted message can be developed, and marketing campaigns can be customized. In addition to helping companies differentiate themselves from their competitors, customer-centric marketing can help them grow their businesses. Superior customer service can establish a strong brand reputation and enable companies to stand out from competitors who are focused more on internal processes or products. The implementation of customer-centric marketing in QbD requires cross-functional teams that include marketing, sales, customer service, and product development. In this way, it is possible to ensure that customer needs are addressed throughout the entire product life cycle, from the design phase to the marketing phase and finally to the sales phase. Marketing QbD is characterized by customer-centricity. Marketing strategies can be more effective, differentiated, and customer-focused when companies focus on the needs and preferences of customers. As a result, customers will be happier, loyal, and repeat customers will return, as well as their reputation will grow.

16.6.2 Growth of Data-Driven Decision-Making

In marketing, quality by design is heavily reliant on data-driven decisions. Technology advances and increased availability of data have contributed to an increase in data-driven decision-making in recent years. In recent years, data-driven decision-making has gained significant traction despite not being a new concept. When data-driven decision-making is used in QbD, it can provide better insight and enable better decisions instead of gut feelings and intuition. Analysis of customer data, market trends, and other relevant information can be used to develop more effective marketing strategies. Marketing strategies will be better targeted to the needs and preferences of target customers as a result. By analyzing data, QbD also assists in measuring and tracking the effectiveness of marketing strategies. The performance of marketing campaigns can be monitored and adjusted using clear metrics and KPIs. Marketing campaigns can be monitored and adjusted as needed using a clear set of metrics and KPIs. Also, companies can identify new opportunities and potential growth areas with the help of data-driven decision-making. Utilizing data on customer behavior and preferences, companies can identify new product opportunities and market segments that may otherwise have gone unnoticed. Data availability and advanced analytics tools have facilitated the growth of data-driven decision-making in QbD. Social media, web analytics, and customer feedback can now be used to collect and analyze large amounts of data. It is, however, also challenging to implement data-driven decision-making in a QbD environment. Data accuracy and reliability are major challenges when using this type of information. A robust data collection process and the use of advanced analytics tools are required to ensure accurate and clean data collection. It is also difficult to collaborate across departments within the organization. A range of stakeholders must be involved in data-driven decision-making, including marketing, sales, product development, and

IT. For these departments to work effectively together, they must ensure effective communication. It is expected that data-driven decision-making will continue to grow in QbD in the coming years, despite these challenges. Increasingly competitive markets will require companies to make data-driven decisions to differentiate themselves. The use of data and analytics tools helps companies develop more effective marketing strategies tailored to their target customers' preferences and needs.

16.6.3 Role of QbD in Developing Effective Marketing Strategies

As a key component of developing effective marketing strategies, Quality by Design (QbD) can help companies understand their target customers, identify their needs and preferences, and design marketing campaigns tailored specifically to meet those needs and preferences. A key principle of QbD is to understand the customer's needs from the beginning of product development. To develop effective marketing strategies, it is essential to understand the needs and preferences of the customer. To inform marketing strategies, companies can collect and analyze information about customer behavior, preferences, and feedback. QbD can also assist companies in identifying and prioritizing the most effective marketing channels and tactics. The most effective channels for driving customer engagement and conversion can be determined by analyzing data on customer behavior and engagement. In addition to social media, email marketing, influencer marketing, and search engine optimization, these channels can be used. Developing effective marketing strategies also requires continuous improvement, a key component of QbD. This same approach can be applied to marketing as QbD emphasizes ongoing testing, monitoring, and refinement of products and processes. To improve performance, companies need to constantly monitor and test marketing campaigns. The use of QbD can also streamline marketing efforts to align them with business goals. Companies can better align marketing strategies to their business objectives by understanding how marketing drives revenue growth, customer acquisition, and brand awareness. Lastly, QbD can assist companies in ensuring compliance with relevant regulations and guidelines. Compliance considerations can help companies design marketing campaigns that are effective and ethical, thereby avoiding costly compliance issues. QbD is responsible for ensuring that organizations develop effective marketing strategies based on data-driven insights and a thorough understanding of their target customers. Businesses can develop more effective marketing campaigns that are tailored to their target customers' needs and preferences, and drive business results, if they apply QbD principles.

16.7 Comparative Evaluation of QbD Components in Marketing Appliances

In April 2017, the FDA released the results of its mutual trial initiative involving European Medicine Agency (EMA) meant for parallel assessment of applications' units pertaining to QbD. The pilot program's objectives were to:

- Sharing information
- Encourage consistent application of QbD principles in global guidelines
- Encourage accessibility of high-quality medications across U.S. and European Union.

The purpose of this programme was to make it easier for the principles of QbD established through ICH Q8, Q9, & Q10 publications to be executed uniformly through both areas, as well as to achieve highest degree of regulatory decision-making harmonization. This was made possible by the exchange of opinions between US and EU assessors/ reviewers regarding the application of ICH ideas besides pertinent regulatory requests exercising real applications that requested admission to programme. Initial duration of the programme was 3 years. Both agencies decided to continue it beyond its initial phase for an additional 2 years to provide additional harmonization of relevant QbD- allied domains. Officially, this programme was finished in April of 2016. Organizations got 16 applications for participation during this time. One submission was denied because it did not just apply to QbD applications, while another was never submitted by the applicant and was therefore never reviewed. Nine scientific recommendations, three variations/ supplements, two MAA, and two NDA were all examined through this programme. One MAA/ NDA was evaluated using concurrent appraisal process, and the others used consultative guidance. FDA along with EMA created as well as released 3 groups of Q&A materials based on the lessons learned throughout the pilot. These documents also addressed suggestions for furthering harmonization made by Japanese, Pharmaceuticals and Medical Devices Agency (PMDA), that took part as viewer. These Q&A documents had the dual purposes of creating review guidelines for the assessors/ reviewers and informing academia and business about the results of the pilot [46, 47]. These records also recorded any variation in regulatory brought on by regional needs, such as the information of process validation details in the dossier. In each of the three Q&A documents, the following subjects were covered:

- On August 20, 2013, following subjects were covered in Q&A: QTTP & CQA, Criticality, detail in descriptions of industrial processes, QbD for analytical techniques.
- On Nov. 1, 2013, a Q&A on Design Space Verification was published. It covered definition, presentation, rationale as well as substantiation of design spaces for active ingredients besides finished goods.

- On December 19, 2014, Q&A covered: Risk assessment, DOE, and design space documentation detail levels.

First EMA-FDA parallel appraisal of QbD components of first MAA together with some consultation advice procedures have been completed successfully. The pilot programme was deemed to be of the utmost value by both agencies for information exchange, facilitating uniform application of ICH principles, and achieving the greatest degree of regulatory decision-making harmonization [47–50]. To make future parallel assessments easier, procedural procedure enhancements have also been introduced. EMA & FDA clutched settlements as a result of this pilot programme on a variety of QbD-related issues, as seen in the Q&A shown in Table 16.1. The Q&A on parallel appraisal of QbD elements of marketing appliances are described in Tables 16.1, 16.2 and 16.3 on first parallel assessment, design space verification, risk assessments, respectively. Questions along with responses on depth in the regulatory filings are described in Table 16.3.

Design space verification shows that the suggested arrangement of input variables for process together with material descriptions is proficient in producing eminent goods at industrial level. In first section, there is general Q&A that coordinated FDA and EMA responses. Expectations particular to the EMA are in Appendix 1, and those related to FDA are in Appendix 2.

In "Generic Drug User Fee Act (GDUFA) II Commitment Letter," US FDA specifies complex products in a broader sense, including:

1. Products that contain complex ingredients i.e. peptides, complex mixture of APIs, polymeric compounds, natural components, composite formulations, multifaceted means of delivery.
2. Products with complicated drug-device combinations like auto0injectors and MDI.
3. Other products where early scientific collaboration might be advantageous due to complexity or uncertainty regarding regulatory pathway or potential alternative approach. Although FDA categories as well as defines complex products of drug, it has not yet given Non-Biological Complex Drugs *i.e.* NBCDs a formal description. Moreover, neither complex products nor non-biologic complex drugs are defined by EMA.

NBCDs cover assortment of pharmaceutical goods, including dendrimers, nanoparticles, glatiramoids, nanocrystals, liposomes, transferosomes, iron-carbohydrate complexes, micelles, nanoemulsions, etc. Despite its success and the rise in market approvals over the past few decades, complex drug product development in pharmaceutical industry believed to be hampered due to numerous technical, industrial in addition to regulatory obstacles in areas of innovation, manufacture, clinical assessment regulatory endorsement, together with growth management. Their responses frequently referenced "follow-on products" and "complex generics" pose additional difficulties due to ineffective efforts and high failure rates at every stage of manufacturing process, from initial to clinical research stage.

Table 16.1 Description of lessons learned and questions that came up during first parallel evaluation [51, 52]

S. no.	Questions	Responses
1	What do the Agencies anticipate from a regulatory QTTP?	QTTP, which explains prospectively the attributes of drug product, which should be reached to assure intended quality while taking into account the drug product's safety and efficacy, is what the Agencies anticipate applicants to give. Only finished product is covered under QTTP specification. QTTP information can be provided in any format but table format is preferable
2	Which CQAs are anticipated during regulatory proposal?	When applicable, applicants must submit outline of CQAs for the drug substance, the final product together with excipients. A justification for classifying these features as CQAs should be included in this list, together with acceptance thresholds for each CQAs. Additionally, based on past information, risk analysis, or experimental data, description of related CQAs of drug substance along with excipient accomplishing product CQAs should be included. Goal of the control approach should be to continuously hold onto CQAs of drug material and finished good within predetermined ranges
3	Would the Agencies agree to a 3-tired criticality scale for process parameters?	The applicant proposed a method for determining criticality and risk assessment that comprises a three level grouping for quality traits and process parameters: key, non-critical, and, critical. Submission would be examined by both EMA an FDA. This defined a critical element as something that caused an experiment to fail. A key element as something that caused an experiment to fail. A crucial component was one that, while not having caused failure within the range under study, may still have an effect on caliber of the final output. Key Process Parameters is not an ICH word, Agencies oppose its use
4	What requirements for descriptions of the production process are the Agencies looking for in regulatory submissions?	For 505(b)(1) products, a comparable detailed process summary can be presented to US FDA in place of a Master Production File. For generic and 505(b)(2) products, however, master production records on a projected or actual commercial scale are necessary. In all circumstances, the EU requires a thorough explanation of manufacturing process

(continued)

Table 16.1 (continued)

S. no.	Questions	Responses
5	What are the Agencies' opinions on using analytical target profile for analytical methods?	By drawing an analogy between an ATP and QTTP as specified in ICH Q8(R2), ATP can generally be accepted as qualification of anticipated scheme implementation. Agencies would not, however, equate analytical techniques with differing underlying principles (HPLC/ NIR) based just on ATP compliance. A candidate should not switch between these two sorts of procedures without first receiving the necessary regulatory application and permission
6	What do the Agencies anticipate from Method Operational Design Rages (MODR) regulation submissions?	MODRs may be subjected to similar guidelines and information needs as outlined in ICH Quality Working Group Points to count (R2), Section 6 "Design Space". The appraisal of authentication requisites stated in ICHQ2(R1) all through MODR along with verification of scheme applicability throughout the MODR are issues that deserve more emphasis

Quality assurance is one of the key obstacles to development of NBCDs. The NBCD class includes a variety of products based on assorted architecture, numerous modules, and specific layouts unique along with key characteristics. NBCDs are particularly prone to quality issues due to their complicated and inconsistent frameworks, dearth of biological and physical excipient compatibility, dynamic and metastable formulations, and issues in physical-chemical and fundamental assessment, extensive processing cycles, or meticulous controls over quality. In order to prove the therapeutic equivalent of complicated generic formulations, there exists significant regulatory need for qualitative (Q1) along with quantitative (Q2) constraints as well as bioequivalence investigations [30, 55–58].

16.8 Discussion

In marketing, QbD means creating and implementing marketing strategies in a systematic way so that quality is built into the process from the beginning. In order to maximize marketing effectiveness, reduce risk, and improve customer value, companies need to focus on their customers' needs and preferences, use data analytics to inform their decisions, and promote cross-functional collaboration. It does, however, require a considerable investment of time, resources, and organizational culture changes to implement QbD in marketing. In today's dynamic marketplace, QbD marketing will become an increasingly important tool for companies seeking to remain competitive as customer-centric marketing and data-driven

Table 16.2 of Q&A reflects the opinions of the EMA and FDA on the subject of design space verification [53]

S. no.	Questions	Responses
1	Why is a design space confirmed at different stages of a product's growth?	According to experience of both agencies, design space authentication at marketable level is not always finished at the time the application is submitted, but it should happen over the course of products along with process' lifetime. It is acknowledged that additional risk mitigation measures are not required for design space verification when an applicant can show that a design space is scale independent
2	What does commercial-scale design space verification serve to accomplish?	Design space verification indicates that scaling up impacts are controllable and are not detrimental to anticipated product quality at commercial size within design space boundaries
3	How is the initial development and commercial scale verification of a design space done?	Commercial design space restrictions may be based on scale-up correlations found in development studies and/ or testing. Additionally, computational simulations can be used to push boundaries of design space through conduction in Normal Operating Range
4	How can a design space be validated at an industrial level?	In theory, a design space may include many areas that are checked at the time of submission, but the design space may also, as necessary, undergo additional verification throughout the product lifecycle. Results of risk assessment on the possible impact of changes to scale-related parameters on product quality can be used to drive the design space verification strategy during the course of a product's lifespan
5	What needs to be covered in submission regarding the design space authentication practice?	A design space authentication methodology may theoretically contain the following: list of scale-specific variables that have an impact on CQAs has not been confirmed at a commercial scale, definitions of the potential scale-up risks to the CQAs, evaluation of the effectiveness of the control plan in mitigating these risks, and a description of any necessary additional controls
6	What happens if unanticipated outcomes or events emerge when doing design-space authentication experiments?	This could point to a problem with the design space or a problem with the evaluation or verification plan. According to local regulations, changes to design space's limits or description as well as any necessary adjustments to the design space

(continued)

Table 16.2 (continued)

S. no.	Questions	Responses
		verification methodology should be communicated to the Agencies using proper notification categories. Appendix 1 and 2 covers regional expectations and laws, and expectations from EU, respectively
7	How can a commercial design space for biological products be verified?	The same rules that apply to chemical products also apply to biological products
8	What distinguishes design space verification from process validation?	Confusion among process validation and design-space verification should be avoided. Both studies are carried out at a commercial size, taking into account past knowledge, and draw conclusions, but their scopes differ. Verification studies only apply to those operations where a design space has been proposed, as opposed to validation which refers to entire production process
9	How should the pharmaceutical quality system handle the design space verification approach?	FDA advises commercials to have a defined strategy for when and how to assess the requirement for design space verification as part of their pharmaceutical quality system. Such proposals for design-space authentication are expected to be available at the manufacturing facility, according to FDA

decision-making become increasingly important. Generic medications, which are the patent-expired copies of branded medications, have the same API, and used for the same indications as branded medications but cost is substantially less. Drugs must adhere to the same high standards of quality as their branded counterparts in order to be marketed. Pharmaceutical businesses should have had enough time to use QbD for the development of their medicinal products since it was adopted into ICH guidelines (Q8, 9, & 10). With just a tiny number of MAA submitted in Europe that include supporting QbD data, the EMA noted in 2014 that applications dossiers with QbD information are far from becoming a prevalent method. The use of QbD principles to industrial operations has benefited international pharmaceutical businesses in particular. As a result, these businesses have experienced higher production yields and efficiency, fewer outputs that fall outside of specifications and fewer production recalls. Leading pharmaceutical companies have set the bar high by adding QbD data in the dossiers needed to approve new drugs. However, the majority of pharmaceutical manufacturers have been hesitant to link their MAA to this greater level of quality control. The FDA's goal is to make sure that there are both safe and effective medications available. Both prescription and non-prescription medications are governed by CDER. This obligation goes beyond prescribing medications. For instance, dandruff shampoos, sunscreen,

Table 16.3 Description of Q&A on Extent of Depth in Regulatory submissions [54]

S. no.	Questions	Responses
1	Which risk assessments can be submitted to a regulatory body?	The applicant is responsible for choosing which RAs to include in the submission. The assessor or reviewer may find this information helpful in understanding how the applicant chose the particular formulation, production method, and controls. The Agencies have, for instance, seen risk assessments relating to choice of formulation variables, the outline of the effects of different process factors, and the choice of in-process controls
2	How much information ought to be included in a regulatory filing for a RA pertaining to process design?	The level of detail should correspond to how important RA's findings are to the control plan and the commercial manufacturing process. In these situations, it would be appropriate to add following details: • A declaration of RA tool that was used, or if a novel RA tool was employed, a definition of the tool and examples of its output • A thorough qualitative or quantitative summary of results of RA supporting and outlining the applicant's justification for the selection of variables that called for further investigation and those that were not
3	How much information has to be included in a regulatory submission for a RA pertaining to product design?	As it is designed to be used by the patient, the amount of detail should correspond to level of danger given by formulation and dose form. The details can include how formulation was chosen and optimized using risk assessment method, as well as formulation factors were ranked according to their possible impact on CQA of final product. The criteria mentioned may additionally take into account physical and chemical stability impacting shelf life
4	To what extent should a design of experiments (DOE) in a regulatory submission be considered?	The details that should be presented in these situations could include: • Kind of experimental design and parameter ranges examined. Justification for design selection may be helpful, especially if the design is not completely balanced • Tables that summarize batch size and inputs / outputs • An overview of parameters that were held constant throughout DOE • Justification for classifying elements

(continued)

Table 16.3 (continued)

S. no.	Questions	Responses
		as scale dependent or independent, such as experimental findings, logic from science, or prior knowledge • An explanation of the primary influences and interactions on response variables, together with p-value • Discussion of regression model validation parameters, including coefficient table, ANOVA regression analysis, residual plots, R^2, and projected responses with confidence ranges
5	In a regulatory submission, what level of specificity should be taken into account while defending and characterizing a design space?	The capability of control strategy to identify faults and level of commercial scale expertise in various design space areas can be additional considerations. When a surface is included in design space, regression model for that surface should be given

antiperspirants, and fluoride toothpastes are all categorized as drugs. To guarantee that patients obtain safe, efficient, and high-quality generic medications, the CDER's Office of Generic Drugs analyses generic drug applications. The CDER's Office of New drugs is in charge of making decisions on marketing approval for new drugs, including decisions relating to changes to existing marketed goods, it also provides regulatory oversight for investigational studies conducted during drug development. The use of the QbD technique in marketing authorization procedures is both favored and strongly advised. A more expedient and cost effective transition from research to market approval and scale up was made by the adoption of an upgraded QbD strategy to cover the early stages of pharmaceutical research operations [19, 30, 37, 40, 59, 60].

References

1. Bhise, J. J., Bhusnure, O. G., Mule, S. T., Mujewar, I. N., Gholve, S. B., & Jadhav, P. U. (2019). A review on quality by design approach (QBD) for pharmaceuticals. *Journal of Drug Delivery and Therapeutics, 9*, 1137–1146.
2. Grangeia, H. B., Silva, C., Simões, S. P., & Reis, M. S. (2020). Quality by design in pharmaceutical manufacturing: A systematic review of current status, challenges and future perspectives. *European Journal of Pharmaceutics and Biopharmaceutics, 147*, 19–37.
3. Kumar, V. P., & Gupta, N. V. (2015). A review on quality by design approach (QBD) for pharmaceuticals. *International Journal of Drug Development and Research, 7*, 52–60.
4. Chirino, A. J., & Mire-Sluis, A. (2004). Characterizing biological products and assessing comparability following manufacturing changes. *Nature Biotechnology, 22*, 1383–1391.
5. Juran, J. M., & Juran, J. (1992). *Juran on quality by design: The new steps for planning quality into goods and services*. Simon and Schuster.

6. Kehoe, D. F. (2012). *The fundamentals of quality management*. Springer Science & Business Media.
7. Mauch, P. D. (2009). *Quality management: Theory and application*. CRC.
8. Rathore, A. S., & Winkle, H. (2009). Quality by design for biopharmaceuticals. *Nature Biotechnology, 27*, 26–34.
9. Beg, S., Akhter, S., Rahman, M., & Rahman, Z. (2017). Perspectives of quality by design approach in nanomedicines development. *Current Nanomedicine (Formerly: Recent Patents on Nanomedicine), 7*, 191–197.
10. Kumar, D. D., Ancheria, R., Shrivastava, S., Soni, S. L., & Sharma, M. (2019). Review on pharmaceutical quality by design (QbD). *Asian Journal of Pharmaceutical Research and Development, 7*, 78–82.
11. Kadam, V. R., Patil, M., Pawar, V. V., & Kshirsagar, S. (2017). A review on: Quality by design (QbD). *Asian Journal of Pharmaceutical Sciences, 7*, 197–204.
12. Pramod, K., Tahir, M. A., Charoo, N. A., Ansari, S. H., & Ali, J. (2016). Pharmaceutical product development: A quality by design approach. *International Journal of Pharmaceutical Investigation, 6*, 129.
13. Nadpara, N. P., Thumar, R. V., Kalola, V. N., & Patel, P. B. (2012). Quality by design (QBD): A complete review. *International Journal of Pharmaceutical Sciences Review and Research, 17*, 20–28.
14. Rathore, A. S. (2014). QbD/PAT for bioprocessing: Moving from theory to implementation. *Current Opinion in Chemical Engineering, 6*, 1–8.
15. Su, Q., Bommireddy, Y., Shah, Y., Ganesh, S., Moreno, M., Liu, J., Gonzalez, M., Yazdanpanah, N., O'Connor, T., & Reklaitis, G. V. (2019). Data reconciliation in the quality-by-design (QbD) implementation of pharmaceutical continuous tablet manufacturing. *International Journal of Pharmaceutics, 563*, 259–272.
16. Aksu, B., De Beer, T., Folestad, S., Ketolainen, J., Lindén, H., Lopes, J. A., de Matas, M., Oostra, W., Rantanen, J., & Weimer, M. (2012). Strategic funding priorities in the pharmaceutical sciences allied to quality by design (QbD) and process analytical technology (PAT). *European Journal of Pharmaceutical Sciences, 47*, 402–405.
17. Hein, T., Winkler, M., Gulati, H., & Balakrishnan, A. (2012). Product lifecycle management for the pharmaceutical industry. *Oracle Life Sciences*, 1–2.
18. Patil, A. S., & Pethe, A. M. (2013). Quality by design (QbD): A new concept for development of quality pharmaceuticals. *International Journal of Pharmaceutical Quality Assurance, 4*, 13–19.
19. Finkler, C., & Krummen, L. (2016). Introduction to the application of QbD principles for the development of monoclonal antibodies. *Biologicals, 44*, 282–290.
20. Johnston, E. (2021). Optimizing tech transfer with advanced analytics. *Pharmaceutical Technology, 45*, 31–33.
21. Kelley, B., Cromwell, M., & Jerkins, J. (2016). Integration of QbD risk assessment tools and overall risk management. *Biologicals, 44*, 341–351.
22. Rapalli, V. K., Khosa, A., Singhvi, G., Girdhar, V., Jain, R., & Dubey, S. K. (2019). *Application of QbD principles in nanocarrier-based drug delivery systems*. Elsevier.
23. Chobisa, D. (2019). QbD-based development of pharmaceutical parenteral drug products: An overview. In *Pharmaceutical Quality by Design* (pp. 151–172).
24. Zurdo, J., Arnell, A., Obrezanova, O., Smith, N., Gómez de la Cuesta, R., Gallagher, T. R., Michael, R., Stallwood, Y., Ekblad, C., & Abrahmsén, L. (2015). Early implementation of QbD in biopharmaceutical development: a practical example. *BioMed Research International, 2015*, 605427.
25. Crowley, M. E., & Crean, A. M. (2015). Quality by design in an evolving manufacturing sector. *European Journal of Parenteral & Pharmaceutical Sciences, 20*, 63.
26. Pharm, B., & Pharm, M. (2013). Quality by design (QbD): Manufacturing and product quality of generics drugs perspective. *Journal of Global Trends in Pharmaceutical Sciences, 4*, 1257–1262.

27. Beg, S., Hasnain, M. S., Rahman, M., & Swain, S. (2019). Introduction to quality by design (QbD): Fundamentals, principles, and applications. In *Pharmaceutical quality by design*. Elsevier.
28. Sangshetti, J. N., Deshpande, M., Zaheer, Z., Shinde, D. B., & Arote, R. (2017). Quality by design approach: Regulatory need. *Arabian Journal of Chemistry, 10*, S3412–S3425.
29. Uhlenbrock, L., Sixt, M., & Strube, J. (2017). Quality-by-design (QbD) process evaluation for phytopharmaceuticals on the example of 10-deacetylbaccatin III from yew. *Resource-Efficient Technologies, 3*, 137–143.
30. Zagalo, D. M., Sousa, J., & Simões, S. (2022). Quality by design (QbD) approach in marketing authorization procedures of non-biological complex drugs: A critical evaluation. *European Journal of Pharmaceutics and Biopharmaceutics, 178*, 1–24.
31. Aksu, B., & Mesut, B. (2015). Quality by design (QbD) for pharmaceutical area. *Journal of Faculty of Pharmacy of Istanbul University, 45*, 233–251.
32. Purohit, P., & Shah, K. (2013). Quality by design (QBD): New parameter for quality improvement & pharmaceutical drug development. *Pharma Science Monitor, 4*, 12–23.
33. Dewi, M. K., Pratama, R., Arifka, M., & Chaerunisaa, A. Y. (2022). Quality by design: Approach to analytical method validation. *Sciences of Pharmacy, 1*, 33–40.
34. van de Berg, D., Kis, Z., Behmer, C. F., Samnuan, K., Blakney, A. K., Kontoravdi, C., Shattock, R., & Shah, N. (2021). Quality by design modelling to support rapid RNA vaccine production against emerging infectious diseases. *npj Vaccines, 6*, 65.
35. Boer, M. D. (2022). *Quality by design: Principles to develop successful products*. Retrieved from May 23, 2023, from https://www.scilife.io/blog/quality-by-design-principles.
36. Chen, C.-W. (2006). *Implementation of ICH Q8 and QbD—An FDA perspective*. PharmaForum.
37. Bhoop, B. S. (2014). Quality by design (QbD) for holistic pharma excellence and regulatory compliance. *Pharmacy Times, 46*, 26–33.
38. Raj, N., Fernandes, S., Charyulu, N. R., Dubey, A., & GS, R. & Hebbar, S. (2019). Postmarket surveillance: A review on key aspects and measures on the effective functioning in the context of the United Kingdom and Canada. *Therapeutic Advances in Drug Safety, 10*, 2042098619865413.
39. Tapkir, N., Soni, F., Sahu, A. K., Jadav, T., Tekade, R. K., & Sengupta, P. (2021). A comprehensive review on assessment and key control strategies for impurities in drug development with a special emphasis on post-marketing surveillance. *Journal of Pharmaceutical Innovation, 4*, 1–20.
40. Csóka, I., Pallagi, E., & Paál, T. L. (2018). Extension of quality-by-design concept to the early development phase of pharmaceutical R&D processes. *Drug Discovery Today, 23*, 1340–1343.
41. Yu, L. X., Amidon, G., Khan, M. A., Hoag, S. W., Polli, J., Raju, G., & Woodcock, J. (2014). Understanding pharmaceutical quality by design. *The AAPS Journal, 16*, 771–783.
42. Testas, M., Da Cunha Sais, T., Medinilha, L. P., Niwa, K. N. I., De Carvalho, L. S., Maia, S. D., Flores, A., Braz, L. P., Menezes, J. C., & Yamakawa, C. Y. (2021). An industrial case study: QbD to accelerate time-to-market of a drug product. *AAPS Open, 7*, 1–13.
43. Gandhi, A., & Roy, C. (2016). Quality by design (QbD) in pharmaceutical industry: Tools, perspectives and challenges. *PharmaTutor, 4*, 12–20.
44. Varu, R. K., & Khanna, A. (2010). Opportunities and challenges to implementing quality by design approach in generic drug development. *Journal of Generic Medicines, 7*, 60–73.
45. Winkle, H. N., & Nasr, M. M. (2011). *FDA perspectives: understanding challenges to quality by design*. Retrieved Jun 3, 2023, from https://www.pharmtech.com/view/fda-perspectives-understanding-challenges-quality-design.
46. FDA. (2017). *Final Report from the FDA-EMA pilot program for the parallel assessment of quality-by-design elements of marketing applications*. Retrieved May 27, 2023, from https://www.fda.gov/drugs/pharmaceutical-quality-resources/final-report-fda-ema-pilot-program-parallel-assessment-quality-design-elements-marketing.

47. Mohurle, M. S. M., Mda, J. A., DD, R. C., Kurian, M. J., & Bais, G. (2019). Quality by design (QbD): An emerging trend in improving quality and development of pharmaceuticals. *Saudi Journal of Medical and Pharmaceutical Sciences, 5*, 1132–1138.
48. Das, V., Bhairav, B., & Saudagar, R. (2017). Quality by design approaches to analytical method development. *Research Journal of Pharmacy and Technology, 10*, 3188–3194.
49. Hewitt, H., Rignall, A., Trone, M., Jackson, P., Chatfield, M., Argentine, M., Wang, Q., Karmarkar, S., Pless, A., & Williams, Z. (2017). Evaluating progress in analytical quality by design. *Pharmaceutical Technology, 41*, 52–59.
50. Schmitt, S., & Zaman, M. A. (2018). Regulatory guidance. In *Pharmaceutical Quality by Design: A practical approach* (pp. 321–333).
51. Bhutani, H., Kurmi, M., Singh, S., Beg, S., & Singh, B. (2004). Quality by design (QbD) in analytical sciences: An overview. *Quality Assurance, 3*, 39–45.
52. US Food and Drug Administration. (2013). *EMA-FDA pilot program for parallel assessment of quality-by-design applications: Lessons learnt and Q&a resulting from the first parallel assessment*. European Medicines Agency and US Food and Drug Administration.
53. FDA. (2013). *FDA-EMA Questions and Answers on Design Space Verification*. Retrieved May 29, 2023 from http://www.ema.europa.eu/docs/en_GB/document_library/Other/2013/11/WC500153784.pdf.
54. FDA. (2014). *Questions and answers on level of detail in the regulatory submissions EMA/59240/2014*. Retrieved May 29, 2023, from from:https://www.ema.europa.eu/en/documents/other/questions-answers-level-detail-regulatory-submissions_en.pdf.
55. Crommelin, D. J., Shah, V. P., Klebovich, I., McNeil, S. E., Weinstein, V., Flühmann, B., Mühlebach, S., & de Vlieger, J. S. (2015). The similarity question for biologicals and non-biological complex drugs. *European Journal of Pharmaceutical Sciences, 76*, 10–17.
56. Davit, B. M., Kanfer, I., Tsang, Y. C., & Cardot, J.-M. (2016). BCS biowaivers: Similarities and differences among EMA, FDA, and WHO requirements. *The AAPS Journal, 18*, 612–618.
57. von Stosch, M., Schenkendorf, R., Geldhof, G., Varsakelis, C., Mariti, M., Dessoy, S., Vandercammen, A., Pysik, A., & Sanders, M. (2020). Working within the design space: Do our static process characterization methods suffice? *Pharmaceutics, 12*, 562.
58. Wang, Y., & Grainger, D. W. (2022). Regulatory considerations specific to liposome drug development as complex drug products. *Frontiers in Drug Delivery, 2*, 10.
59. Kesselheim, A. S., & Gagne, J. J. (2015). Product-specific regulatory pathways to approve generic drugs: The need for follow-up studies to ensure safety and effectiveness. *Drug Safety, 38*, 849–853.
60. Oner, Z. G., Michel, S. L., & Polli, J. E. (2017). Equivalence and regulatory approaches of nonbiological complex drug products across the United States, the European Union, and Turkey. *Annals of the New York Academy of Sciences, 1407*, 26–38.

Conclusions and Future Perspective

Neha Bajwa and N. K. Jain
University Centre of Excellence in Research, Baba Farid University of Health Sciences, Faridkot, Punjab, India
Department of Pharmaceutical Sciences, Dr. H. S. Gour University, Sagar, Madhya Pradesh, India

In this book we highlighted significance of QbD in the pharmaceutical sector, emphasizing its role in ensuring high-quality products, managing product quality throughout its life cycle, and promoting continuous improvement in the industry.

QbD is a systematic approach that ensures quality is built into a product or process from its inception. It is a concept widely used in various industries, including pharmaceuticals, manufacturing, and software development. QbD focuses on understanding the product or process, identifying critical quality attributes, and designing appropriate controls to achieve the desired quality outcomes. The foundation of QbD lies in the principles of risk management, scientific understanding, and quality risk assessment. It encourages a proactive approach rather, where potential risks and opportunities addressed early in the development or manufacturing stages. By considering quality throughout, from design to implementation, QbD aims to minimize errors, reduce variability, and enhance efficiency and effectiveness. QbD emphasizes the collaboration and communication between different including scientists, engineers, quality assurance personnel, and regulatory authorities. It promotes the use of advanced analytical tools, statistical techniques, and scientific knowledge to gain a deep understanding process, ensuring that quality is not dependent solely on inspection and testing. Key elements of QbD include defining the target product profile (TPP) and critical quality attributes (CQAs), establishing a design space, setting appropriate specifications and controls, and employing continuous monitoring and improvement. By employing QbD principles, organizations can streamline their processes, reduce costs, enhance customer satisfaction, and ensure compliance with regulatory requirements. In summary, Quality by Design

N. K. Jain, N. Bajwa (eds.), *Introduction to Quality by Design (QbD)*,
https://doi.org/10.1007/978-981-99-8034-5

is a systematic approach that aims to ensure quality is built into a product or process from the beginning. By focusing on risk management, scientific understanding, and collaboration, QbD helps organizations achieve their desired quality outcomes while fostering innovation and efficiency.

Regulatory Aspects of Quality by Design (QbD) are essential for ensuring the safety, efficacy, and quality of pharmaceutical products. Various regulatory authorities, including the US Food and Drug Administration (USFDA), the Pharmaceuticals and Medical DevicesPMDA) in Japan, the International Council for Harmonisation (ICH), and the European Medicines Agency (EMA), have developed guidelines and frameworks that outline the requirements and expectations for implementing QbD principles. Here is a brief overview of the regulatory aspects of QbD as outlined by these authorities: (1) US Food and Drug Administration (USFDA): The USFDA encourages the use of QbD principles in pharmaceutical development and manufacturing. The FDA's QbD approach focuses on understanding the critical quality attributes of a product, identifying and controlling critical process parameters, and establishing appropriate specifications and controls. The FDA has released various guidance documents, such by Design for ANDAs: An Example for Release Dosage Forms and Pharmaceutical Quality for the twenty-first Century: A Risk-Based Approach," which provide detailed information on implementing QbD in the regulatory context. (2) Pharmaceuticals and Medical DevicesPMDA): The PMDA in Japan has also adopted QbD principles to enhance the quality and safety of pharmaceutical products. The PMDA's QbD framework emphasizes the importance of risk assessment and control strategy throughout the product lifecycle. They have issued guidelines such as "Quality by Design (QbD) for Drug and Drug-Device Combination Products" and "QbD Manufacturing of Drug Substances and Drug Products" to provide guidance on implementing QbD in the Japanese regulatory context. (3) International Council for Harmonisation (ICH): The ICH, a global organization that brings together regulatory authorities and the pharmaceutical industry, has developed guidelines that harmonize the requirements for QbD implementation. The ICH Q8, Q9, Q10, and Q11 guidelines are particularly relevant to QbD. These guidelines cover topics such as pharmaceutical development, risk management, quality management systems, and the development and manufacture of drug substances and drug products. The ICH guidelines provide a globally recognized framework for implementing QbD principles across different regions. Ines Agency (MA has also emphasized the importance of QbD in pharmaceutical development and manufacturing. The EMA's guideline on QbD, titled "Guideline on the Investigation of armaceutical Products in the Context of QbD,\" provides guidance on the application of QbD principles for the development and approval of pharmaceutical. The guideline emphasizes the need for a systematic approach to product development, identification of critical quality attributes, the use of risk assessments, and the establishment of appropriate control strategies. The EMA encourages the submission of QbD-related information in marketing authorization applications to demonstrate a thorough understanding of the product and its manufacturing process. In conclusion, the regulatory aspects of Quality by Design (QbD) play a vital role in ensuring the quality, safety, and efficacy of pharmaceutical products. Regulatory authorities such as the USFDA, PMDA, ICH, and EMA have developed guidelines and frameworks to support the implementation of QbD principles. Compliance with these regulatory requirements helps pharmaceutical companies achieve regulatory approvals and maintain the highest standards of quality in.

When selecting a design for a process or product, several criteria can be considered to ensure its effectiveness and success. Some common criteria include:Process Analytical Technology (PAT): This criterion focuses on the use of advanced analytical tools and techniques to monitor, control, and optimize the process. PAT enables real-time measurements and analysis, providing valuable insights into the process performance and facilitating timely adjustments if needed.

Critical Material Attributes (CMA) refers to the key characteristics of raw materials or ingredients that have a significant impact on the product's quality and performance. When selecting a design, considering the CMA ensures that the chosen materials meet the required specifications and contribute to the desired outcomes. Critical Process Parameters (CPP) represents the key operational variables that must be controlled within predefined limits to ensure product quality and safety. Selecting a design that incorporates appropriate CPPs and control strategies helps minimize process variability and ensures consistent and reproducible results. Critical Quality Attributes (CQA) are the measurable characteristics of a product that directly impact its quality, safety, and efficacy. The chosen design should consider the desired CQAs and ensure that they can be consistently achieved and controlled throughout the manufacturing process.

Risk Management and Error Analysis (RMEA) involves identifying and assessing potential risks associated with the design, process, or product and implementing appropriate mitigation strategies. A robust design should consider potential failure modes, their severity, and their likelihood, enabling proactive measures to prevent errors and reduce risks.

Continued Process Verification and Improvement (CPV/CPA) focuses on continuously monitoring and evaluating the process performance and identifying areas for improvement. The design should incorporate mechanisms for ongoing data collection, analysis, and feedback to facilitate continuous process verification and optimization.Compliance with Quality Assurance (CQA) Standards involves adhering to quality assurance standards and regulations specific to the industry or product. The chosen design should ensure compliance with relevant quality standards, guidelines, and regulatory requirements to ensure product safety, efficacy, and quality. By considering these criteria during the selection of a design, organizations can increase the likelihood of achieving desired quality outcomes, optimizing process performance, and ensuring regulatory compliance. It is important to evaluate these criteria based on the specific needs, objectives, and constraints of the organization and the product or process in question. Analytical Quality by Design (AQbD) is a systematic approach that applies the principles of Quality by Design (QbD) specifically to analytical methods used in the pharmaceutical and biopharmaceutical industries. AQbD focuses on understanding and controlling the sources of variability in analytical methods to ensure the reliability, accuracy, and reproducibility of the analytical results. The goal of AQbD is to design and develop robust analytical methods that meet predefined quality objectives and regulatory requirements. Similar to QbD, AQbD emphasizes a proactive approach, where critical method parameters and their impact on the analytical performance are identified and controlled during the method development process. By applying AQbD principles, organizations can develop analytical methods that are more reliable, accurate, and robust. This leads to improved quality control, enhanced regulatory compliance, reduced variation, and increased confidence in the analytical results generated. Ultimately, AQbD helps to optimize the analytical processes, leading to more efficient and effective decision-making in drug development and manufacturing.

API by Design (QbD) is a systematic approach used in the pharmaceutical industry to ensure the quality of Active Pharmaceutical Ingredients (APIs) from the early stages of development. QbD principles can be applied to the design, development, and manufacturing processes of APIs to enhance their quality, safety, and efficacy.

The main goal of API QbD is to understand the impact of various factors on the quality of the API and to design appropriate controls to achieve the desired quality attributes. This approach shifts the focus from quality testing and inspection at the end of the manufacturing process to proactive quality assurance throughout the entire API lifecycle.

By adopting API QbD principles, pharmaceutical companies can enhance the quality, efficiency, and safety of their APIs. This approach not only improves patient safety and product efficacy but also helps streamline manufacturing processes and reduce costs. Additionally, it facilitates regulatory compliance by demonstrating a thorough understanding of the critical quality attributes and control strategies for the API.

Quality by Design (QbD) is an essential approach in formulation development that ensures the quality of pharmaceutical products. It involves a systematic and science-based approach, focusing on understanding the product and its critical quality attributes (CQAs) to design a robust and reliable formulation. By following the QbD approach, formulation development can be more efficient, reliable, and cost-effective. It enables a deeper understanding of the formulation and its critical parameters, reduces the need for extensive testing, and facilitates regulatory compliance. Ultimately, QbD enhances the overall quality of the formulation and contributes to patient safety and satisfaction.

Quality by Design (QbD) is a systematic approach used in the pharmaceutical industry to ensure that the final product meets the desired quality requirements. When applied to oral formulations such as tablets and spheroids, QbD involves a comprehensive understanding of the formulation and manufacturing process, identifying critical quality attributes (CQAs) that are important for the product's safety and efficacy. By following the QbD approach, pharmaceutical companies can enhance product quality, reduce batch failures, and ensure consistent product performance. It allows for a more efficient and cost-effective development and manufacturing process, leading to safer and more effective oral formulations.

Quality by Design (QbD) is an approach used in the pharmaceutical industry to ensure that the final product meets the desired quality requirements. When applied to topical formulations, QbD involves a systematic understanding of the formulation and manufacturing process, identifying critical quality attributes (CQAs) that are important for the product's safety, efficacy, and overall performance. By applying the QbD approach to topical formulations, pharmaceutical companies can enhance product quality, reduce batch failures, and ensure consistent performance and safety. It allows for a more efficient development and manufacturing process, leading to safer and more effective topical products.

Quality by Design (QbD) is an approach used in the pharmaceutical industry to ensure that the final product meets the desired quality requirements. When applied to parenteral formulations (administered through injection or infusion), QbD involves a systematic understanding of the formulation and manufacturing process, identifying critical quality attributes (CQAs) that are important for the product's safety, efficacy, and overall performance. By applying the QbD approach to parenteral formulations, pharmaceutical companies can enhance product quality, reduce batch failures, and ensure consistent performance and safety.

The Quality by Design (QbD) approach can also be applied to novel drug delivery systems to ensure that the final product meets the desired quality requirements. QbD for novel drug delivery systems involves a systematic understanding of the formulation, manufacturing process, and the unique characteristics of the delivery system itself. By applying the QbD approach to novel drug delivery systems, pharmaceutical companies can enhance product quality, reduce batch failures, and ensure consistent performance and safety of the delivery system. It allows for a more efficient development and manufacturing process, leading to improved patient outcomes and better therapeutic outcomes.

Quality by Design (QbD) is an approach used in the pharmaceutical industry to ensure that the final product meets the desired quality requirements. When applied to herbal products/formulations, QbD involves a systematic understanding of the formulation and manufacturing process, identifying critical quality attributes (CQAs) that are important for the product's safety, efficacy, and overall performance.

Quality by Design (QbD) is a systematic approach to product development and quality assurance that focuses on designing quality into a product rather than testing it at the end of the process. It is a proactive and science-based methodology that aims to ensure that products meet predefined quality attributes and are consistently produced with a high level of quality. By incorporating QbD principles into Quality Assurance practices, companies can reduce the risk of product failures, minimize the need for costly and time-consuming investigations, and ensure a higher level of product quality, consistency, and patient safety. Additionally, QbD helps foster a culture of quality and continuous improvement within an organization.

Quality by Design (QbD) is an approach widely adopted in the pharmaceutical industry to ensure the quality of products throughout their lifecycle. It involves systematically designing and controlling the manufacturing processes to ensure the final product consistently meets the desired quality attributes. QbD principles are also applied to pharmaceutical packaging to ensure the integrity, safety, and effectiveness of the packaged drugs. By adopting QbD principles in pharmaceutical packaging, companies can enhance product quality, reduce the risk of recalls, improve patient safety, and maintain compliance with regulatory requirements.

Quality by Design (QbD) is also relevant and beneficial in the context of clinical trials. QbD principles can be applied to improve the overall quality, efficiency, and reliability of clinical trials, leading to better outcomes and more accurate data. By integrating QbD principles into clinical trials, pharmaceutical companies and researchers can enhance the quality and reliability of the data generated, leading to more informed decisions about drug efficacy, safety, and potential benefits for patients. Ultimately, this approach can help accelerate the drug development process and bring new treatments to patients more efficiently and effectively.

Quality by Design (QbD) principles are primarily focused on ensuring the quality and reliability of products and processes. While QbD is not directly related to marketing activities, it can indirectly influence marketing efforts and customer satisfaction. It's essential to recognize that while QbD can indirectly influence marketing, marketing efforts involve various other elements such as market research, advertising, pricing, and distribution. QbD mainly contributes to the foundation of a successful marketing strategy by ensuring that the product itself is of high quality, safe, and meets customer expectations. The overall marketing success still depends on how well these products are promoted and positioned in the market.

The future prospects of Quality by Design (QbD) in the pharmaceutical industry are promising, as this approach continues to gain traction and play a significant role in drug development and manufacturing. Several trends and developments suggest a continued and expanded application of QbD principles in the pharmaceutical sector:

Regulatory agencies worldwide, such as the U.S. FDA and the EMA, have been actively promoting QbD as an effective way to ensure pharmaceutical product quality and safety. As harmonization efforts progress, more countries are likely to adopt QbD guidelines, making it a global standard.

The pharmaceutical industry is witnessing significant technological advancements in areas such as process analytical technology (PAT), data analytics, artificial intelligence (AI), and machine learning. These technologies enable better process monitoring, real-time control, and data-driven decision-making, all of which align well with QbD principles.

Continuous manufacturing, a concept closely related to QbD, is gaining popularity in the pharmaceutical industry. It offers improved process control, reduced waste, and enhanced product quality. As continuous manufacturing becomes more mainstream, QbD will play a crucial role in its successful implementation.

The shift towards personalized medicine and targeted therapies requires a deep understanding of critical quality attributes to meet patient-specific needs. QbD principles are essential in developing and manufacturing these specialized medicines with consistency and precision.

Biopharmaceuticals and complex formulations, including gene therapies and cell-based therapies, are becoming more prevalent. The pharmaceutical industry is increasingly globalized, with companies outsourcing manufacturing to different regions. QbD facilitates a standardized and robust approach to ensure consistent product quality across various manufacturing sites.

QbD's emphasis on process optimization and risk reduction aligns well with the pharmaceutical industry's efforts to improve efficiency and reduce costs. By implementing QbD, companies can streamline processes, reduce waste, and minimize resource utilization.

The lifecycle approach, a core principle of QbD, focuses on continuous improvement throughout a product's lifecycle. This enables manufacturers to adapt to changing requirements and incorporate new knowledge, leading to better products and processes over time.

QbD promotes a culture of quality within organizations, with a strong focus on meeting patient needs and expectations. As patient-centricity becomes increasingly important, QbD will remain a critical component in delivering safe and effective medications to patients.

Overall, the future of QbD in the pharmaceutical industry looks promising, with an increasing emphasis on quality, safety, efficiency, and patient outcomes. The continued support from regulatory agencies, technological advancements, and the industry's evolving needs will drive the widespread adoption and integration of QbD principles in drug development and manufacturing processes.

Index

N. K. Jain, N. Bajwa (eds.), *Introduction to Quality by Design (QbD)*,
https://doi.org/10.1007/978-981-99-8034-5

MIX
Papier aus verantwortungsvollen Quellen
Paper from responsible sources
FSC® C105338

FSC
www.fsc.org

If you have any concerns about our products,
you can contact us on
ProductSafety@springernature.com

In case Publisher is established outside the EU,
the EU authorized representative is:
Springer Nature Customer Service Center GmbH
Europaplatz 3, 69115 Heidelberg, Germany

Printed by Libri Plureos GmbH
in Hamburg, Germany